VLSI Analog Circuits

VLSI Analog Circuits

Algorithms, Architecture, Modeling, and Circuit Implementation

Hongjiang Song, Ph.D.

Second Edition

McGraw Hill Education

New York Chicago San Francisco
Athens London Madrid
Mexico City Milan New Delhi
Singapore Sydney Toronto

Library of Congress Control Number: 2016945579

**VLSI Analog Circuits:
Algorithms, Architecture, Modeling,
and Circuit Implementation, Second Edition**

1 2 3 4 5 6 7 8 9 QVS 21 20 19 18 17 16

ISBN 978-1-25-964493-1
MHID 1-25-964493-6

Sponsoring Editor
Michael McCabe

Editorial Supervisor
Stephen M. Smith

Production Supervisor
Lynn M. Messina

Acquisitions Coordinator
Lauren Rogers

Project Manager
Apoorva Goel,
Cenveo® Publisher Services

Copy Editor
Yumnam Ojen Singh

Proofreader
Manish Tiwari,
Cenveo Publisher Services

Art Director, Cover
Jeff Weeks

Composition
Cenveo Publisher Services

About the Author

Hongjiang Song, Ph.D., is an adjunct professor in the School of Electrical, Computer, and Energy Engineering at Arizona State University, where he has taught a series of VLSI circuit design courses in analog signal processing, high-speed I/O, structural analog circuit design, and modulation circuits since 2002. Dr. Song earned his Ph.D. from ASU and has been working in the VLSI circuit industry as a senior design engineer and circuit architect with Intel Corporation since 1994. He holds 30 U.S. patents related to VLSI circuits. He is also author of a few VLSI circuit design textbooks.

Contents at a Glance

Contents

Preface

It has been many years since my first analog filter circuit design implementation based on commercial discrete electronic components. During these years, there have been significant changes in this field related to the design and fabrication processes of analog filter and signal processing circuits. These changes are featured by the so-called mixed-signal SOC applications, which integrated tens of millions of devices on a single chip, including both analog and digital circuit elements. These changes are related to both design methodologies and design applications and have forced designers to develop new design approaches and circuit technologies that are more suitable for high levels of VLSI integrations.

In this text, we will first attempt to bring together these new VLSI filter circuit design techniques with design examples in an understandable manner. Furthermore, we will discuss the most recent developments in VLSI analog filter circuit technologies, which are compatible with the VLSI fabrication process technologies targeted at SOC applications. Compared with past years when filter circuit developments were more focused on external design specifications that required little knowledge of the design and fabrication processes of the circuit components, today a complete understanding of circuit design techniques, circuit fabrication processes, and system application specifications is essential for the successful design and implementation of VLSI analog filter and analog signal processing circuits.

The material in this text is organized into three main areas with 24 chapters covering various topics related to the analysis, design, and implementation of various types of VLSI analog filter circuits, including passive, active-RC, MOS-C, Gm-C, CTI, SC, and SI filter circuit techniques.

The basic VLSI circuit techniques for VLSI signal processing systems are covered in the first part. The VLSI filter circuit implementations are covered in the second part. And advanced developments in VLSI analog filter circuits are covered in the third part.

This text was first developed in 2001 as the class notes for the graduate-level VLSI circuit design course (EEE598: VLSI Switched-Capacitor Filter and Analog Signal Processing Circuits) offered in the electrical engineering department at Arizona State University. There have been many modifications and enhancements made during the yearly course offerings since 2002 in both the covered material and the presentation methods, with much valuable feedback from the students and many other readers.

I would like to thank many of my colleagues and friends—especially Prof. Sayfe Kiaei (ASU), Prof. David R. Allee (ASU), Bobby Nikjou (Intel), Bart McDaniel (Intel),

Hedayat Daie (Intel), and Yan Song (Intel)—for their encouragement and their informative discussions during this text's development. I am also indebted to my many students on whom much of the material was tested. Finally, I express my gratitude to my family for their support during this text's development and the course offerings in the past years.

Hongjiang Song, Ph.D.

VLSI Analog Circuits

CHAPTER 1
Introduction

Analog signal processing circuits have been widely used in many VLSI circuit applications such as communication circuits, high-speed I/O circuits, and instrumentation circuits. The increasing demands for system mobility with higher packing density, smaller form factors, lower power dissipation, and the trend of system-on-a-chip (SOC) integrations in recent years are the main driving forces for VLSI-integrated analog signal processing circuits implemented in the standard CMOS process technologies. Such implementation trends are featured by the recent developments of various CMOS-specific analog processing circuit techniques.

The focuses of this book are the algorithms, the architecture, the modeling, and the VLSI circuit-implementation techniques of various CMOS analog signal processing circuits. The material is organized with two goals in mind: first, to bridge the existing gaps between the existing theoretical work and the implementations of the VLSI analog signal processing circuits and, second, to present the systematic design implementations and analysis methodologies for these VLSI signal processing circuits based on the state-of-the-art CMOS process technologies.

1.1 Organization of This Book

The book is organized into three major parts with 24 chapters covering various topics of the VLSI signal processing circuit design and analysis techniques.

Part 1 of the book has 8 chapters covering VLSI circuit design techniques for various CMOS analog circuit-implementation techniques, including the active-RC, MOS-C, Gm-C, continuous-time current-mode (CTI), switched-capacitor (SC), and switched-current (SI) circuit techniques. These circuit techniques provide the fundamental building components for various VLSI analog signal processing circuit and system implementations.

In Chap. 2, the theoretical basis of the continuous-time (CT) signal processing circuits and systems are introduced. CT linear analog signal processing circuits (or filters) and systems are linear circuit networks with selective responses to signals of different frequencies. Their responses can typically be derived from characteristic functions or circuit diagrams. Thus, it is important for the designers to be familiar with concepts relevant to the analog signal processing circuit characterizations. Analog signal processing circuits or systems can usually be represented in various forms, such as the schematic representation, the transfer-function representation, the Bode plots representation, and the signal flow graph (SFG), or a type of block diagram representation. The schematic representation, in general, provides a detailed structure representation of the system. However, the technology-dependent nature of such a representation makes it less effective in certain system descriptions. The transfer-function representation, on the other hand, provides a mathematic-based description of the circuits and systems; and it is useful for system analysis. However, such a presentation does not provide enough information for system implementation. Consequently, the SFG representation can be used to bridge the gaps between the schematic and mathematic representations of the systems. In this chapter, the relationship and conversion techniques among these representations for various VLSI analog signal processing systems are discussed in detail.

In Chap. 3, the VLSI active-RC circuit techniques, which are among the most basic and widely used VLSI circuits for the CT analog signal processing systems, are introduced. These circuits are usually constructed based on CMOS operational amplifiers (OPAMPs) and the VLSI passive resistors and capacitors. The advantages of such circuit techniques are their straightforward relation with respect to the traditional discrete active signal processing circuit's topologies and their low distortions. Major drawbacks of active-RC circuit techniques are their limited bandwidths mainly due to the OPAMP bandwidths' limitation. In addition, the frequency accuracy of the VLSI active-RC circuits, as determined by the RC constant of the resistors and capacitors, cannot be directly achieved in VLSI technologies. Consequently, frequency accuracies of these circuits are typically limited to less than 50 percent unless discrete tuning techniques with capacitor or resistor banks are used which can improve the circuit frequency accuracies to around 1–5 percent with some penalties on chip area and design complexity.

In Chap. 4, the VLSI MOS-C circuit techniques, based on MOS voltage-controlled resistors (VCRs), are introduced. These circuits offer resistive tuning capabilities, which can be used to achieve precision RC time constant in the CT analog signal processing circuit and system implementations. The second attractive feature of MOS-C filters is that they are closely related to the well-understood VLSI active-RC circuits. The main drawback of VLSI MOS-C circuit is its degraded linearity compared with the active-RC circuit counterpart, which mainly resulted from the nonlinearity of MOS VCRs. Consequently, a design focus of MOS-C circuit techniques is to develop circuit topologies such as the adaptive gate control and the two-transistor or four-transistor differential circuits, which provide effective compensation of the nonlinear effects in the MOS VCRs.

In Chap. 5, the VLSI Gm-C circuit technique is introduced. Major attractions of Gm-C circuits are their achievable wide bandwidths, their low allowable supply voltages, and their tuning capabilities. The wide bandwidth operations of the Gm cells are partly due to the fact that all internal nodes in Gm circuits have low impedances. The main drawback of Gm-C circuits is their limited linearity. Consequently, the focus of practical VLSI Gm-C circuit developments is the MOS Gm circuit structures, which offer improved linearity, flexible tuning capability, and enhanced output impedance.

In Chap. 6, the VLSI CTI circuit techniques are introduced. In these circuit techniques, the current-mode signals are used in the signal processing operations. Among the various VLSI CTI circuit techniques, current-mirror–based CTI circuits inherently have very few internal nodes, and they are subjected to minimum secondary parasitic pole effects. Therefore, such circuits offer potential advantages over the Gm-C circuit techniques in terms of lower power consumption, smaller operating voltage, and potentially higher bandwidth. In addition, the second-generation current conveyor (CCII) circuit–based CTI circuit techniques, or CCII circuits, present the alternative solution for analog signal processing system implementations, which traditionally have been based on VLSI voltage OPAMPs.

In Chap. 7, the fundamentals of the discrete-time (DT) or sampled-data analog signal processing systems are discussed. The DT signal systems serve the intermediate positions between the analog and digital systems. In this chapter, the important characteristics of the DT signals and systems, as well as the similarities and differences of the DT and CT systems, are discussed. The representation of the DT systems by using Laplace transform, z-transform, transfer function, block diagram, and SFG are discussed. The numeric approximation methods of the CT system using DT system and the s-to-z transformations are also addressed.

In Chap. 8, the VLSI SC circuit techniques are introduced. SC circuits, as the DT signal processing circuits, were first developed to be able to design accurate analog signal processing circuits at voice-band frequencies economically in fully VLSI-integrated form. VLSI SC circuits can be used to realize the RC constant accurately without using the resistors, which avoids the complicated tuning circuits as typically required in VLSI CT signal processing circuit techniques. In addition, very large-value resistors can easily be implemented in VLSI SC circuit forms, which solves the problem of VLSI analog filter in most low-frequency (such as audio) signal processing applications.

In Chap. 9, the VLSI switched-current (SI) circuit techniques, which provide fully digital VLSI CMOS process technology compatibility, are introduced. SI circuit techniques are based on current-mode DT signal processing operations employing the nonlinear MOS gate capacitors to store charge so as to avoid the high-quality poly-to-poly capacitors or MIM capacitors and the OPAMPs in the circuit implementations. In the past decades, there has been a tremendous interest in this technology because of the cost benefits of realizing signal processing functions in digital technologies.

Part 2 of the book has 10 chapters covering the design and implementation techniques of various CT and DT VLSI filter circuits employing the CMOS circuit techniques introduced in earlier chapters. The frequency manipulation techniques, the systematic design approaches based on the SFG of the signal processing system, are introduced.

In Chap. 10, various useful frequency response manipulation techniques for analog signal processing systems are introduced. Frequency-scaling methods are used to unify the analog signal processing system structures based on the same circuit topology having different characteristic frequency parameters. Frequency transformations, on the other hand, are used to create desired frequency responses such as the high-pass, the bandpass, and the bandstop responses from the low-pass signal processing circuit prototypes. These methods provide effective ways of reusing existing filter structures. Various methods of numeric approximation of the CT system based on s-to-z transformation techniques are also addressed in this chapter, which is to be used as the theoretical basis of numeric-based DT circuit implementations of the CT signal processing applications.

In Chaps. 11, 12, and 13, the VLSI CT signal processing system implementations based on VLSI active-RC, MOS-C, and Gm-C circuit techniques are discussed. The VLSI circuit implementations of the basic analog signal processing building blocks, such as integrators, adders, scalers, and the first- and second-order filter sections, are introduced. Several VLSI active-RC, MOS-C, and Gm-C filter design approaches, including the SFG-based and the cascade circuit–based techniques, are provided. These methods provide systematic paths for VLSI circuit implementations of a given CT signal processing system.

In Chap. 14, the VLSI CTI circuit implementations for the CT analog signal processing systems are introduced. The CTI integration, the addition, the scaling, and the first- and second-order section circuits are introduced. The SFG- and cascade-based VLSI CTI signal processing circuit design methods are studied.

In Chap. 15, the VLSI SC analog signal processing circuit-implementation techniques are discussed. The VLSI SC analog signal processing filter approaches, including the SFG- and cascade circuit–based approaches, the s-to-z transform, the integration mapping, and the exact Z-domain transfer-function implementation approaches are introduced. These techniques provide wide ranges of methods for the implementation of the VLSI analog signal processing circuit based on the SC circuit techniques.

In Chap. 16, the VLSI analog signal processing circuit implementation based on the VLSI SI circuit techniques employing the SFG, cascade, or exact design synthesis methods are introduced.

In Chap. 17, design techniques for mixed-mode analog signal processing applications are introduced. Based on the novel concept of fundamental I/V and V/I circuit elements partitioning, a VLSI analog signal processing circuit can be flexibly implemented in the voltage, the current, or the mixed domains. Design examples for the mixed-mode signal processing circuit based on various VLSI analog circuit techniques—including the active-RC, MOS-C, Gm-C, CTI, SC, and SI circuits—are provided.

In Chap. 18, the VLSI filter design based on the component simulation method is discussed. These methods are based on the equivalent passive circuit component simulations, which offer the advantages of lower sensitivity VLSI circuit implementation for high-order signal processing systems with lower spread of component values and higher reusability of well-developed passive circuit structures.

In Chap. 19, the VLSI active-RC, MOS-C, Gm-C, CTI, SC, and SI circuit implementations of the ladder networks are discussed. These techniques are featured by their very low sensitivities to variations in circuit parameters and their wide uses for the VLSI implementation of high-order analog signal processing circuits and systems.

Part 3 of the book has 5 chapters covering some recent developments in the VLSI analog signal processing circuits, such as VLSI polyphase filter circuits and RF signal processing circuits. In addition, the VLSI CT filter-tuning techniques and the practical design aspect of the VLSI analog filters are also addressed.

In Chap. 20, various tuning circuit architectures and techniques for CT VLSI analog signal processing systems are discussed. These tuning techniques are associated with the key limitations of VLSI technologies, which are difficult to achieve at the absolute accuracy of better than ±20~±50 percent for certain circuit parameters such as the on-chip resistance and capacitance. On the other hand, these techniques are based on the fact that it is fairly easy under VLSI technology to achieve relative (or ratio) accuracy (better than 1–0.1 percent) through matching of the same type of devices. The commonly used F-tuning, Q-tuning, resistive, capacitive, and Gm-tuning circuit techniques, which can yield improved signal processing circuit accuracy, are discussed.

In Chap. 21, the design sensitivity of VLSI analog signal processing circuit implementations under various variations, including the manufacture process, supply voltage, and application temperature, are provided. These techniques are commonly used for predicting the manufacture yields of the signal processing circuits that provide effective methods for selecting the optimal specifications, circuit architectures, and VLSI manufacture processes. In addition, signal dynamic range of the VLSI signal processing systems, including the system noise floor and signal processing circuit linearity, are also discussed.

In Chap. 22, the VLSI analog polyphase filters are introduced. Polyphase filters provide nonsymmetric gain response with respect to the positive and negative frequency, which gained increasing interest in recent years for the VLSI-integrated low-IF receiver RF architectures. In this chapter, the differences of the real and complex filters are discussed. Novel circuit designs and circuit analysis techniques for various VLSI polyphase filters are provided.

In Chap. 23, various basic VLSI RF signal process circuit techniques, including the VLSI inductor and varactor realizations, the impedance-matching methods, the frequency translation, the voltage-controlled oscillation, and the frequency-synthesis circuits are discussed.

Chapter 24 introduces a novel VLSI analog filter design concept employing the digital-based analog (DBA) circuit design method developed for the VLSI SOC analog filter applications. These techniques combine the available VLSI analog and digital signal processing circuit capabilities and are targeted at higher signal processing efficiency and robustness for lower cost and power dissipation.

References

1. O. Casas, R. Pallas-Areny, "Basics of Analog Differential Filters," *IEEE Trans. on Instrumentation and Measurement*, vol. 45, no. 1, pp. 275–279, February 1996.
2. R. Castello, F. Montecchi, F. Rezzi, A. Baschirotto, "Low-Voltage Analog Filters," *IEEE Trans. on Circuits and Systems I: Fundamental Theory and Applications*, vol. 42, no. 11, pp. 827–840, November 1995.

CHAPTER 2

VLSI Continuous-Time Signal Processing System Fundamentals

Linear VLSI CT analog signal processing (ASP) circuits are linear networks that offer frequency selection. These circuits can be characterized by various representations. Schematic representation is one of the most important representations that provide the detailed circuit structure, device parameters, and circuit connectivity essential for the VLSI implementations. At the behavior level, the frequency-domain (S-domain) transfer-function representation provides the mathematic representation of the circuit. Based on such representations, the behaviors of the CT ASP circuits, such as frequency responses and stabilities, can be characterized by the pole/zero locations of their transfer functions (TFs). Graphically, CT ASP circuits can be represented using their Bode plots. A Bode plot representation provides the gain (magnitude) and phase responses of circuits with respect to signals of different frequencies that are widely used in the system specifications. At the abstract level, CT ASP circuits can be represented by their signal flow graphs (SFGs) or block diagrams. As can be seen throughout this text, the technology-independent connectivity information offered by the SFG representation provides a very useful method to bridge the gaps between the behaviors and the VLSI implementations of the CT ASP circuits, and it offers systematic ways for VLSI circuit implementation. Mathematically, linear CT ASP circuits can also be represented by linear differential equations. This representation, when incorporated with the Fourier and Laplace transform techniques, can be used for effectively representing and designing signal processing systems with mixed-form signals (electrical, mechanical, optical, etc.).

2.1 Characterization of Linear Continuous-Time Systems

For the simple passive VLSI LRC circuit with input voltage $V_i(t)$ shown in Fig. 2.1, the output $V_o(t)$ of such circuit can, in general, be described in mathematical form using the following differential equations:

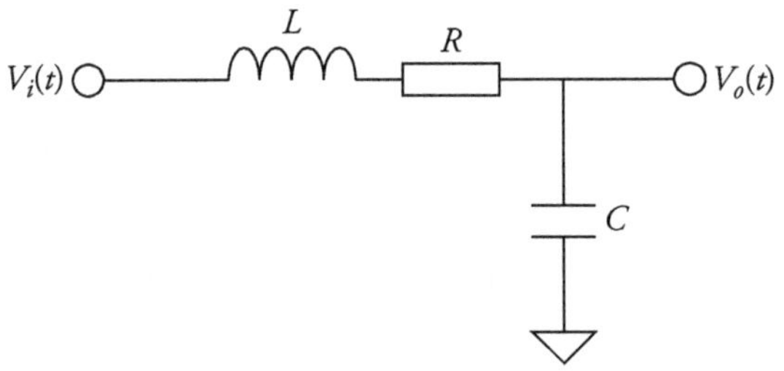

FIGURE 2.1 Passive LRC circuit network.

$$\begin{cases} LC\dfrac{d^2V_o(t)}{dt^2}+RC\dfrac{dV_o(t)}{dt}+V_o(t)=V_i(t) \\ \dfrac{d}{dt}V_o(t)\big|_{t=0}=\dfrac{d}{dt}V_o(0) \\ V_o(t)\big|_{t=0}=V_o(0) \end{cases} \tag{2.1}$$

Although it is not trivial to obtain the explicit form of the output signal using the above differential equations for the general signal input, it is easy to solve for the steady-state response of the circuit based on the following sinusoidal signal input:

$$\begin{cases} V_i(t)=\mathrm{Re}(Ae^{j\omega t})=A\cos(\omega t) \\ V_o(t)=\mathrm{Re}(Be^{j\omega t})=|B|\cos(\omega t+\theta) \end{cases} \tag{2.2}$$

Under such steady-state conditions, Eq. (2.1) can easily be solved as

$$\Rightarrow LC\frac{d^2}{dt^2}(Be^{j\omega t})+RC\frac{d}{dt}(Be^{j\omega t})+(Be^{j\omega t})=(Ae^{j\omega t}) \tag{2.3}$$

It is also easy to obtain the amplification response of the network to the signal of frequency ω as follows:

$$\Rightarrow H(j\omega)\equiv\frac{B(j\omega)}{A}=\frac{1}{\sqrt{(1-LC\omega^2)^2+(RC\omega)^2}}e^{-j\tan^{-1}\left(\frac{RC\omega}{(1-LC\omega^2)}\right)} \tag{2.4}$$

The steady-state output signal for the given input can be expressed as

$$\Rightarrow V_o(t)=\frac{A}{\sqrt{(1-LC\omega^2)^2+(RC\omega)^2}}\cos\left(\omega t-\tan^{-1}\left(\frac{RC\omega}{(1-LC\omega^2)}\right)\right) \tag{2.5}$$

It can be seen that the gain and the phase of the output signal, in general, depend on the frequency of the sinusoidal input signal.

2.1.1 Gain Response

The gain response of the network to the input signal can be expressed based on Eq. (2.4) as follows:

$$\Rightarrow |H(j\omega)| = \frac{1}{\sqrt{(1-LC\omega^2)^2+(RC\omega)^2}} \tag{2.6}$$

It is interesting to see that such a response is frequency dependent. From the above equation, it is obvious to obtain the gain responses of the network at several special frequency points as

$$\Rightarrow \begin{cases} |H(j\omega)|_{\omega\to 0} = 1 \\ |H(j\omega)|_{\omega\to\infty} = \dfrac{1}{LC\omega^2}\Big|_{\omega\to\infty} \to 0 \\ |H(j\omega)|_{\omega\to\omega_0 = \frac{1}{\sqrt{LC}}} = \dfrac{1}{R}\sqrt{\dfrac{L}{C}} \equiv Q \end{cases} \tag{2.7}$$

The gain response versus frequency of the network can also be expressed in graphic form as shown in Fig. 2.2. It can be seen that this network will pass low-frequency signals and attenuate high-frequency signals. It can also be seen that at the resonant

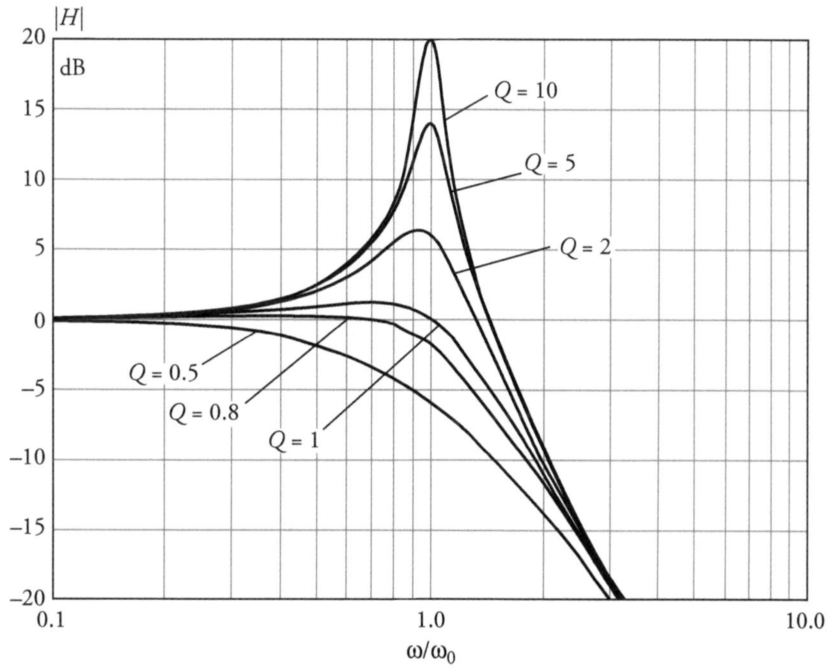

FIGURE 2.2 Gain response of the simple passive LRC network.

frequency [$\omega_0 = 1/(LC)^{0.5}$] of the circuit, the gain of the network only depends on the quality factor Q of the circuit.

2.1.2 Phase Response

The phase response or the phase shift of the output signal with respect to the input electrical signal can be obtained as follows:

$$\Rightarrow \theta = \angle H(j\omega) = -\tan^{-1}\left(\frac{RC\omega}{(1-LC\omega^2)}\right) \tag{2.8}$$

Similarly, at these special frequency points, we have the following:

$$\Rightarrow \begin{cases} \angle H(j\omega)_{\omega\to 0} = 0 \\ \angle H(j\omega)_{\omega\to\infty} = -180^\circ \\ \angle H(j\omega)_{\omega\to\omega_0 = \frac{1}{\sqrt{LC}}} = -90^\circ \end{cases} \tag{2.9}$$

The phase response of the network can also be plotted versus the frequency as shown in Fig. 2.3.

2.1.3 Fourier Transform

From the above discussion, it can be seen that it is easy to obtain the steady-state response of a linear CT system to the sinusoidal signal input of specific frequency. Such a method suggests that the response of a given linear CT network to a general signal input can be obtained based on the superposition technique using sinusoidal signal

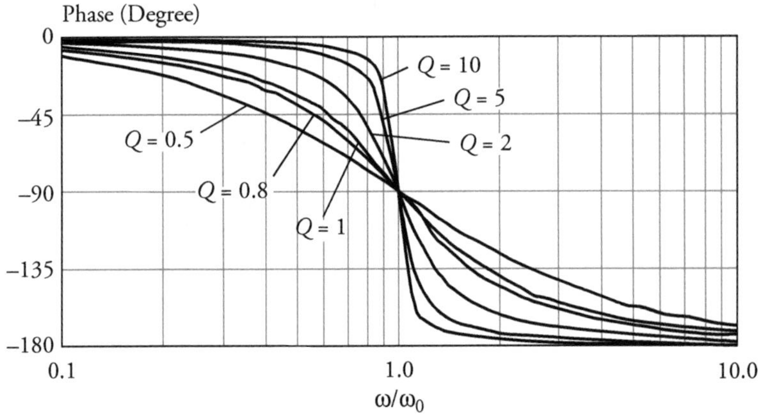

FIGURE 2.3 Phase response of the simple passive LRC circuit.

inputs of different frequency components. To implement the above idea, we may introduce the Fourier transform of a signal $f(t)$ as

$$\begin{cases} F(j\omega) \equiv F(f(t)) = \int_{-\infty}^{+\infty} f(t)e^{-j\omega t}\,dt \\ f(t) \equiv F^{-1}(F(j\omega)) = \dfrac{1}{2\pi}\int_{-\infty}^{+\infty} F(j\omega)e^{j\omega t}\,d\omega \end{cases} \tag{2.10}$$

where $F(j\omega)$ is the Fourier transform of the signal, and it is also known as the spectrum of $f(t)$. Based on the Fourier transform, a given signal can be expressed by a linear combination of sinusoidal signals of various frequencies.

We may now solve the response of the simple passive LRC circuit shown in Fig. 2.1 by applying the Fourier transformation on Eq. (2.1) as follows:

$$\Rightarrow -\omega^2 LCV_o(j\omega) + j\omega RCV_o(j\omega) + V_o(j\omega) = V_i(j\omega) \tag{2.11}$$

The frequency-domain transfer function of the circuit can easily be derived as follows:

$$\Rightarrow H(j\omega) \equiv \frac{V_o(j\omega)}{V_i(j\omega)} = \frac{1}{1-\omega^2 LC + j\omega RC} \tag{2.12}$$

Here, $H(j\omega)$ is the frequency transfer function of the given passive LRC circuit. Consequently, the output of the circuit can be obtained by recombining the output of all sinusoidal components based on inverse Fourier transform as

$$v_o(t) = \frac{1}{2\pi}\int_{-\infty}^{+\infty} H(j\omega)V_i(j\omega)e^{j\omega t}\,d\omega \tag{2.13}$$

2.1.4 Laplace Transform

Fourier transform has been very useful in analyzing the steady-state response of the circuit. Alternatively, we may introduce the Laplace transform, which can be used to analyze both the steady-state and transient responses of the CT system. The (one-side) Laplace transform of a signal $f(t)$ is defined as

$$\begin{cases} F(s) \equiv L(f(t)) = \int_{0}^{+\infty} f(t)e^{-st}\,dt \\ f(t) \equiv L^{-1}(F(s)) = \dfrac{1}{2\pi j}\int_{\sigma-\infty}^{\sigma+j\infty} F(s)e^{st}\,ds \end{cases} \tag{2.14}$$

where $F(s)$ is the Laplace transform of signal $f(t)$.

We can obtain the general response of the given passive LRC circuit shown in Fig. 2.1 in terms of parameters based on the Laplace transform as follows:

$$\Rightarrow LC(s^2V_o(s) - sv_o(0) - \frac{d}{dt}v_o(0)) - RC(sV_o(s) - v_o(0)) + V_o(s) = V_i(s) \tag{2.15}$$

or

$$\Rightarrow V_o(s) = \frac{1}{LCs^2 + RCs + 1} V_i(s) + \frac{1}{LCs^2 + RCs + 1}\left(LC(sv_o(0) + \frac{d}{dt}v_o(0)) + RCv_o(0)\right) \tag{2.16}$$

It is obvious to see that the first term of the above equation corresponds to the steady-state response of the circuit, and the second term, on the other hand, corresponds to the initial (or transient) conditions of the circuit.

We may now define the S-domain transfer function of the circuit based on the steady-state term of Eq. (2.16) as

$$H(s) \equiv \frac{V_o(s)}{V_i(s)} = \frac{1}{LCs^2 + RCs + 1} \tag{2.17}$$

It can be seen that Eqs. (2.12) and (2.17) are equivalent with respect to the steady-state response of the filter since

$$H(s)|_{s=j\omega} \equiv H(j\omega) \tag{2.18}$$

Such equivalence will be used without further explanation throughout this text.

2.2 Analysis of Continuous-Time Filters

In general, a linear CT system (electrical or nonelectrical) can be expressed using the following linear differential equation (or a group of linear equations):

$$\sum_{k=0}^{n} a_k \frac{d^k}{dt^k} v_o(t) = \sum_{k=0}^{m} b_k \frac{d^k}{dt^k} v_i(t) \tag{2.19}$$

By applying the Laplace transform to Eq. (2.19) and ignoring the initial conditions, we may obtain the S-domain transfer function of the CT system as

$$H(s) = \frac{V_o(s)}{V_i(s)} = \frac{\sum_{k=0}^{m} b_k s^k}{\sum_{k=0}^{n} a_k s^k} \tag{2.20}$$

The frequency response of the filter to a sinusoidal signal component can be derived as follows:

$$\frac{V_o(s)}{V_i(s)}|_{s=j\omega} = H(s)|_{s=j\omega} = |H(j\omega)|\, e^{j\angle H(j\omega)} \tag{2.21}$$

It can be seen that the steady-state response of the CT system is totally defined by its S-domain transfer function.

• Integrator	$\int dt \Leftrightarrow \frac{1}{s}$
• Differentiator	$\frac{d}{dt} \Leftrightarrow s$
• Resistor	$R \Leftrightarrow R$
• Capacitor	$C \Leftrightarrow \frac{1}{Cs}$
• Inductor	$L \Leftrightarrow Ls$

TABLE 2.1 Basic Continuous-Time Network Building Blocks and Their S-Domain Representations

2.2.1 S-Domain Representations of Functional Blocks

Although the transfer function of a given linear CT system can, in principle, be derived from its differential equation representation using the Laplace transform, in many cases, it is more efficient to obtain the transfer function of the system using the well-developed S-domain representations of low-level system building components. Table 2.1 shows a list of commonly used linear CT network building components and their S-domain representations.

The above Laplace transform pairs can be used to convert the differential equations into algebraic equations or to convert the LRC networks into pure resistive networks, which can be used to significantly simplify our system design and analysis processes.

For the passive linear LRC circuit shown in Fig. 2.4, we may now directly convert it into the S-domain form using Table 2.1 through the mapping method.

The transfer function of the circuit can now be easily derived as follows:

$$\Rightarrow H(s) = \frac{V_o(s)}{V_i(s)} = \frac{\frac{1}{Cs}}{Ls + R + \frac{1}{Cs}} = \frac{1}{LCs^2 + RCs + 1} \tag{2.22}$$

It can be seen that the result is the same as of Eq. (2.17), using the previously introduced method.

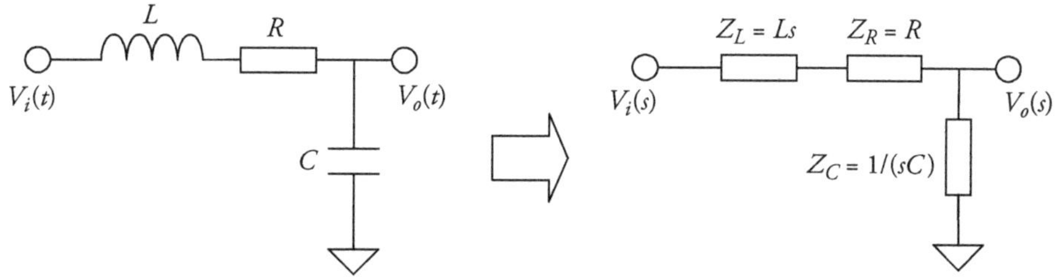

FIGURE 2.4 Mapping of real circuit to S-domain circuit.

2.2.2 Characteristics of the S-Domain Transfer Function

The behavior of a given linear CT system can also be specified by the locations of the poles and zeros of its s-domain transfer function. The locations of the poles and zeros not only determine the frequency responses (gain and phase) of the system but also determine the stability of the system. In practical design applications, poles and zeros of the transfer function of the system are commonly used to specify the system and are directly used for the system implementation, such as in the cascaded filter design approaches that will be discussed in later chapters.

2.2.3 Poles and Zeros

The s-domain transfer function can be expressed in the pole-zero form as

$$\Rightarrow H(s) = \frac{\sum_{k=0}^{m} b_k s^k}{\sum_{k=0}^{n} a_k s^k} = \frac{b_m \prod_{k=1}^{m} (s - z_k)}{a_n \prod_{k=1}^{n} (s - p_k)} \tag{2.23}$$

where $\{z_k, k = 1, 2, \ldots, m\}$ and $\{p_k, k = 1, 2, \ldots, n\}$ are the zeros and poles of the s-domain transfer function of the linear CT system.

For the passive LRC circuit shown in Fig. 2.1, the two poles are located at

$$p_{1,2} = -\frac{R}{2L} \pm \sqrt{\left(\frac{R}{2L}\right)^2 - \frac{1}{LC}} \tag{2.24}$$

2.2.4 Time-Domain Response and Stability Criteria

The s-domain output signal can be expressed based on the transfer function as

$$\Rightarrow V_o(s) = H(s)V_i(s) = \frac{a_m \prod_{k=1}^{m} (s - z_k)}{b_n \prod_{k=1}^{n} (s - p_k)} V_i(s) = \sum_{k=1}^{n} \frac{C_k}{(s - p_k)} + \sum_{k=n+1}^{L} \frac{C_k}{(s - p_k)} \tag{2.25}$$

where the second term is introduced by the input signal, and by definition, it is bounded (or stable). Therefore, we only need to investigate the behavior of the first term in the above equation. By taking the inverse Laplace transform of the above equation, we have that

$$\begin{aligned} \Rightarrow V_o(t) &= L^{-1}(V_o(s)) = L^{-1}\left(\sum_{k=1}^{n} \frac{C_k}{(s - p_k)} + \sum_{k=n+1}^{L} \frac{C_k}{(s - p_k)}\right) \\ &= \sum_{k=1}^{n} C_k L^{-1}\left(\frac{1}{(s - p_k)}\right) + \sum_{k=n+1}^{L} C_k L^{-1}\left(\frac{1}{(s - p_k)}\right) \\ &= \sum_{k=1}^{n} C_k e^{p_k t} + \sum_{k=n+1}^{L} C_k e^{p_k t} \end{aligned} \tag{2.26}$$

It can be seen that if and only if the first term is bounded in time, the CT system is stable. This suggests that for a given linear CT system to be stable, all poles of the s-domain transfer function of the system must be in the left half of the s-plane.

$$\mathrm{Re}(p_k) < 0 \tag{2.27}$$

For the linear passive LRC circuit example, we have the following:

$$\mathrm{Re}(p_{1,2}) = \begin{cases} -\dfrac{R}{2L} < 0 & \dfrac{1}{R}\sqrt{\dfrac{L}{C}} \geq 0.5 \\ -\dfrac{R}{2L} \pm \sqrt{\left(\dfrac{R}{2L}\right)^2 - \dfrac{1}{LC}} < 0 & \dfrac{1}{R}\sqrt{\dfrac{L}{C}} < 0.5 \end{cases} \tag{2.28}$$

It can be seen that such a circuit is always stable. In fact, we can also directly get the same answer from the basic physics theory that there are no active devices or energy generators in this circuit.

2.2.5 Bode Plots

Bode plots are commonly used in practical CT system specifications to provide graphical representation of the system gain and phase responses for different frequency input signal components. For the transfer function given in the following equation, we can generate the Bode plots according to the above rule as shown in Fig. 2.5.

$$H(s) = \frac{(s+1)(s+1000)}{(s+10)(s+100)} \tag{2.29}$$

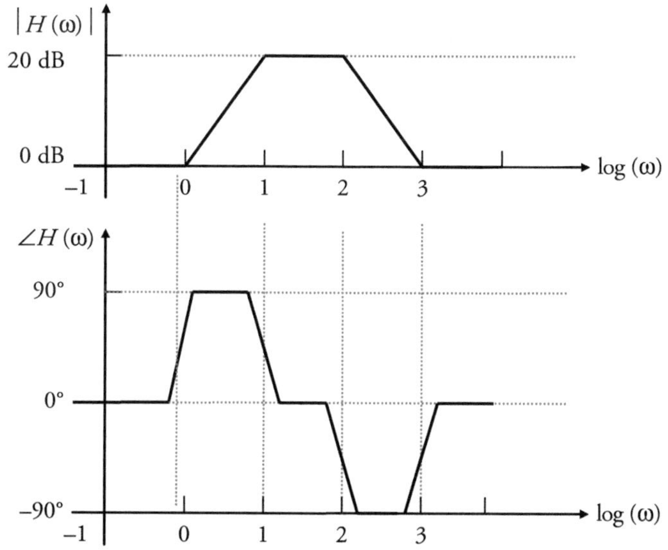

FIGURE 2.5 Bode plots of the specified transfer function.

2.3 Block Diagram and Signal Flow Graph

When dealing with a complex CT system, it is more effective to break down the system into its basic building blocks through the divide-and-conquer approach. Under such an approach, the functionality of each block is described with the details of the internal network of the blocks ignored. The detailed system properties, such as the impedance of a circuit, are not presented. This results in a high-level, simpler structure description of the system, which can be more easily understood.

Note that the block diagram representation of a CT system is not unique. This means that the same system can be implemented differently based on the requirement and the convenience of the implementation for a specified VLSI technology.

Figure 2.6 shows two equivalent system representations of the same second-order sigma-delta modulator circuit.

The transfer function of a CT system provides the behavioral representation of that system. However, transfer-function representation does not provide much topological or implementation information of that system.

There are many ways to select the basic building elements for a targeted CT system realization; a popular method is to use the integrator-based approach. In this method, the three basic building elements of choice are the integrator, the adder, and the scaler (or gain element). In principle, all linear CT systems or subfilter blocks can be implemented by this method.

The block diagrams and signal flow graph (SFG) representations of the three basic building elements are shown in Figs. 2.7, 2.8, and 2.9.

- The integrator
- The adder
- The scaler

2.3.1 Construction of Integrator-Based SFG

Based on the selection of the VLSI basic building elements, a given s-domain transfer function can be implemented in VLSI circuits with the help of the SFG representation.

From the s-domain transfer function of a linear CT system given as

$$H(s) \equiv \frac{Y}{X} = \frac{b_n s^{-n} + b_{n-1} s^{-(n-1)} + \cdots + b_1 s^{-1} + b_0}{a_n s^{-n} + a_{n-1} s^{-(n-1)} + \cdots + a_1 s^{-1} + 1} \tag{2.30}$$

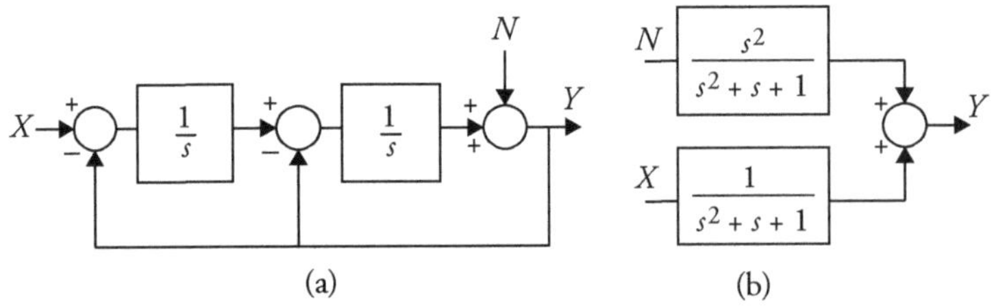

Figure 2.6 Equivalent block diagrams of a second-order ΣΔ A/D converter.

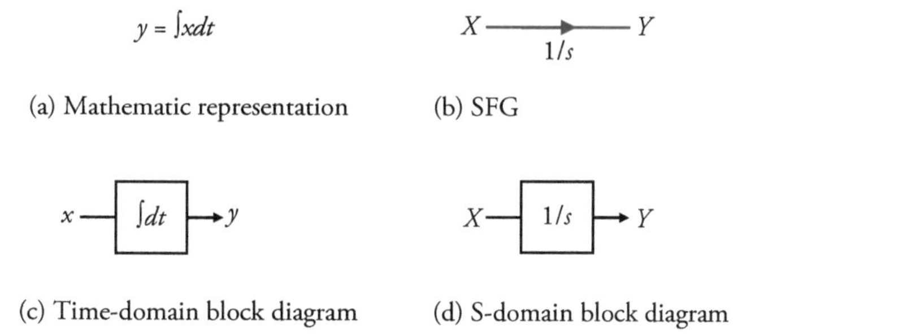

FIGURE 2.7 Integration representations.

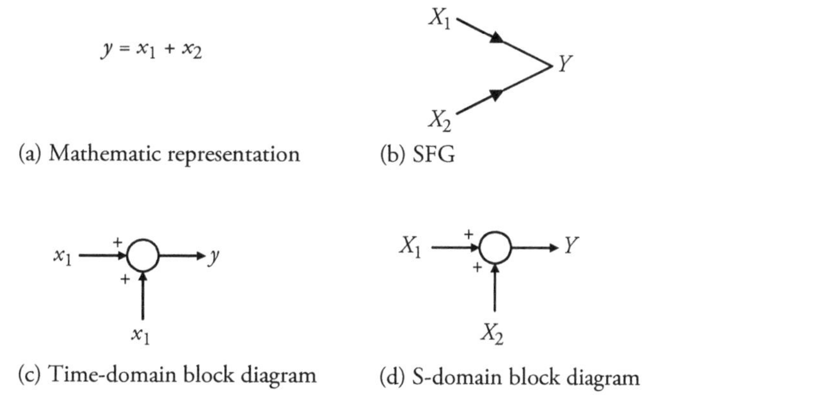

FIGURE 2.8 Addition representations.

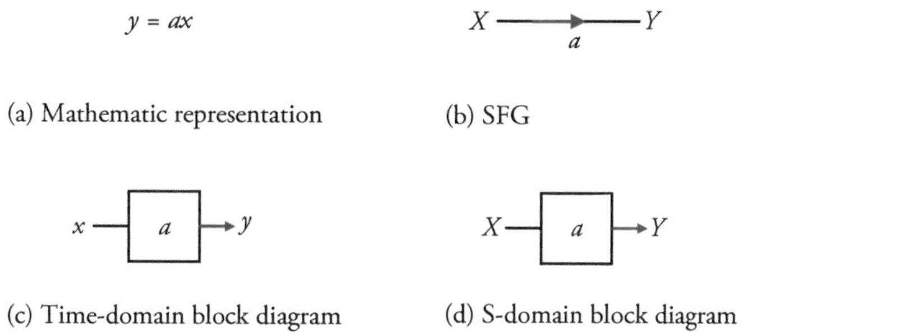

FIGURE 2.9 Scaler representations.

the input and output relation of such a transfer function can be rewritten in terms of the integrator, the adder, and the scaler elements as

$$\begin{aligned} &\Rightarrow (a_n s^{-n} + a_{n-1} s^{-(n-1)} + \cdots + a_1 s^{-1} + 1)Y \\ &= (b_n s^{-n} + b_{n-1} s^{-(n-1)} + \cdots + b_1 s^{-1} + b_0)X \end{aligned} \tag{2.31}$$

or

$$\begin{aligned} \Rightarrow Y = &(b_n X - a_n Y)s^{-n} + (b_{n-1} X - a_{n-1} Y)s^{-(n-1)} \\ &+ \cdots + (b_1 X - a_1 Y)s^{-1} + b_0 X \end{aligned} \tag{2.32}$$

Such an equation can be directly translated into the SFG representation as shown in Fig. 2.10. Alternatively, we may also express the transfer function by introducing a variable U as

$$\frac{Y}{X} = \frac{b_n s^{-n} + b_{n-1} s^{-(n-1)} + \cdots + b_1 s^{-1} + b_0}{a_n s^{-n} + a_{n-1} s^{-(n-1)} + \cdots + a_1 s^{-1} + 1} \cdot \frac{U}{U} \tag{2.33}$$

or

$$\begin{cases} U = X - (a_n s^{-n} + a_{n-1} s^{-(n-1)} + \cdots + a_1 s^{-1})U \\ Y = (b_n s^{-n} + b_{n-1} s^{-(n-1)} + \cdots + b_1 s^{-1} + b_0)U \end{cases} \tag{2.34}$$

Such an equation leads to an alternative integrator-based SFG representation as shown in Fig. 2.11.

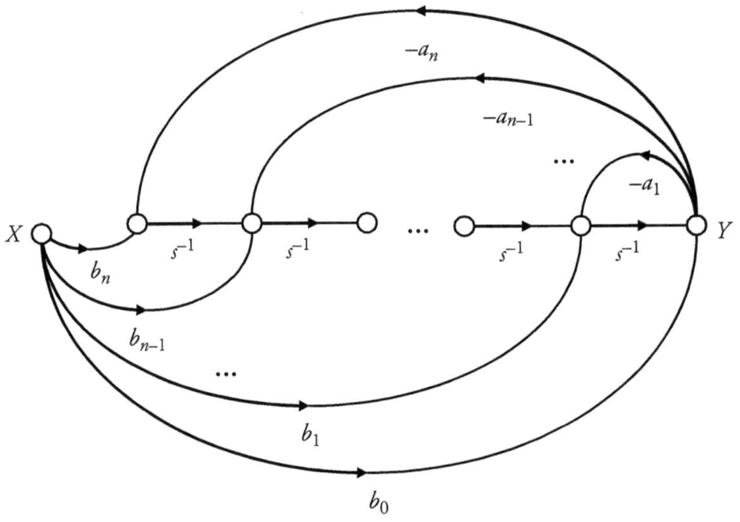

FIGURE 2.10 Integrator-based SFG representation of the general CT system transfer function.

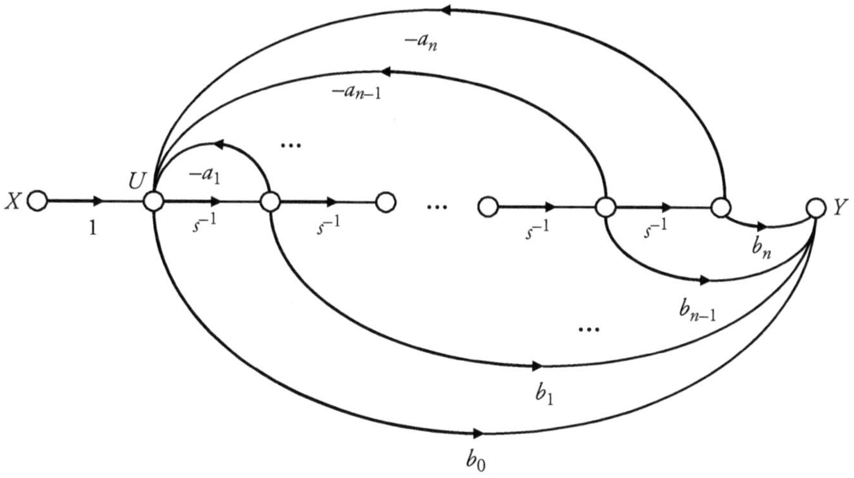

FIGURE 2.11 Alternative SFG representation of the general filter transfer function.

It can be seen that the SFG representation for the same transfer function is not unique even with the same set of basic building blocks. Therefore, different SFG representations may be constructed based on the way the basic building blocks are selected or the way the transfer function is expressed.

Among various SFG representations of the same CT system, there is one useful way of the SFG construction based on the highly reusable first- and second-system sections. In this approach, the transfer function can be expressed by the product of first- (for real pole) and second-order (for complex pole) transfer-function sections.

$$H(s)=\frac{\sum_{k=0}^{m} b_k s^k}{\sum_{k=0}^{n} a_k s^k}=\prod_k\left(\frac{A_k s^2+B_k s+C_k}{s^2+\dfrac{s\omega_k}{Q_k}+\omega_k^2}\right)\prod_k\left(\frac{D_k s+E_k}{s+p_k}\right) \tag{2.35}$$

In this case, the SFG of the system can be constructed by cascading the SFGs of first- and second-order sections.

2.3.2 Construction of General SFG

In many practical design applications, VLSI basic building blocks other than the integrator may actually be preferred for the optimal implementation of the CT system based on specific design constraints and targeted technologies. Such an effort can be easily realized using an innovative g-based transfer-function technique developed in this text. The procedure of this technique can be described as follows:

- A building block G, which represents the desired building block for implementing the targeted CT system, is introduced.

- The parameter s or $1/s$ in the s-domain transfer function of the targeted CT system is substituted using the following introduced G block:

$$g = f(s) \tag{2.36}$$

or

$$s = f^{-1}(g) \tag{2.37}$$

The targeted s-domain transfer-function $H(s)$ can now be expressed in terms of parameter g through variable substitution using g.

$$\frac{Y}{X} \equiv H(s)\big|_{s=f^{-1}(g)} \equiv \frac{d_n g^n + d_{n-1} g^{(n-1)} + \cdots + d_1 g + d_0}{c_n g^n + c_{n-1} g^{(n-1)} + \cdots + c_1 g + 1} \tag{2.38}$$

- Similarly, output signal Y can be reexpressed using g as the parameters.

$$\Rightarrow Y = (d_n X - c_n Y) g^n + (d_{n-1} X - c_{n-1} Y) g^{(n-1)} + \cdots + (d_1 X - c_1 Y) g + d_0 X \tag{2.39}$$

Similar to Fig. 2.12, a g-based SFG can be constructed based on the g-cell, adder, and scaler as the basic building components.

We may also obtain an alternative g-based SFG by introducing a variable U as shown in Fig. 2.13.

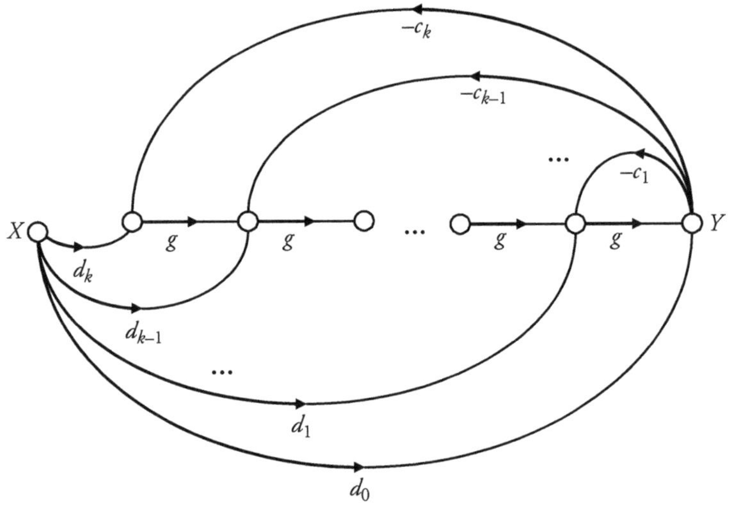

FIGURE 2.12 g-based SFG representation of the given transfer function.

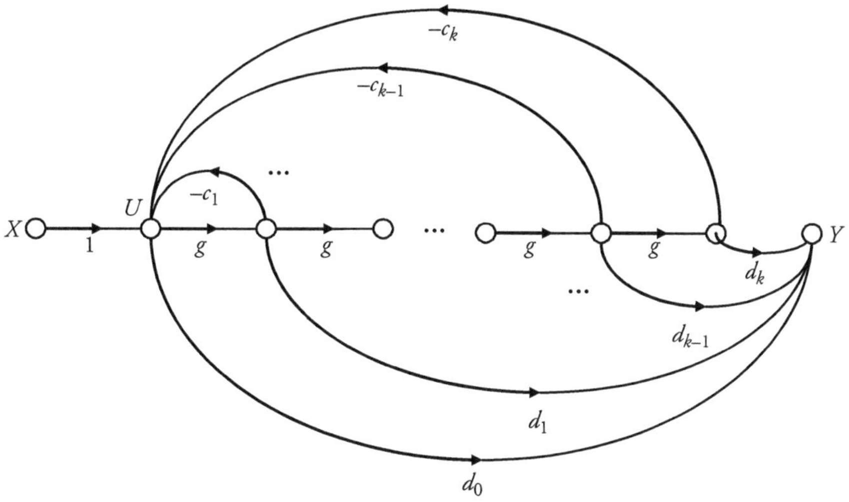

Figure 2.13 Alternative g-based SFG of the given transfer function.

References

1. G. A. Nenov, "On Some Equivalent Transformations of Electrical Network Signal-Flow Graphs and Related First-Order Transfer Function Sensitivities," *ICECS'96*, vol. 2, pp. 1235–1238, October 1996.
2. Y. Sun, J. K. Fidler, "Some Design Methods of OTA-C and CCII-RC Filters," *IEE Colloquium on Digital and Analogue Filters and Filtering Systems*, pp. 7/1–7/8, November 1993.
3. C. Acar, "Nth-Order Voltage Transfer Function Synthesis Using a Commercially Available Active Component: Signal-Flow Graph Approach," *Electronics Letters*, vol. 32, pp. 1933–1934, October 1996.
4. M. Helfenstein, A. Muralt, G. S. Moschytz, "Direct Analysis of Multiphase Switched-Current Networks Using Signal-Flow Graphs," *ISCAS'95*, vol. 2, pp. 1476–1479, May 1995.
5. P. Lin, "Topological Realization of Transfer Functions in Canonical Forms," *IEEE Trans. on Automatic Control*, vol. 30, no. 11, pp. 1104–1106, November 1985.
6. M. C. M. Teixeira, H. F. Marchesi, E. Assuncao, "Signal-Flow Graphs: Direct Method of Reduction and MATLAB Implementation," *IEEE Trans. on Education*, vol. 44, no. 2, pp. 185–190, May 2001.
7. M. H. Fino, J. E. Franca, A. Steiger-Garcao, "Automatic Symbolic Analysis of Switched-Capacitor Filtering Networks Using Signal Flow Graphs," *IEEE Trans. on Computer-Aided Design of Integrated Circuits and Systems*, vol. 14, no. 7, pp. 858–867, July 1995.
8. Wai-Kai Chen, "The Use of Matrix Signal-Flow Graph in State-Space Formulation of Feedback Theory," *ISCAS'91*, vol. 2, pp. 1005–1008, June 1991.

9. J. A. Hegt, "Signal Flow Graph Based Synthesis of Strays-insensitive Switched-Capacitor Filters," *ISCAS'90*, vol. 3, pp. 2177–2180, May 1990.
10. Wing-Hung Ki, "Signal Flow Graph Analysis of Feedback Amplifiers," *IEEE Trans. on Circuits and Systems I: Fundamental Theory and Applications*, vol. 47, no. 6, pp. 926–933, June 2000.
11. S. Ozoguz, E. O. Gunes, H. O. Elwan, T. B. Tarum, "CCII-Based Balanced Fully Integrated Continuous-Time Current-Mode Filter Synthesis: Signal-Flow Graph Approach," *1998 Midwest Symposium on Circuits and Systems*, pp. 412–414, August 1998.
12. R. S. Guindi, M. I. Elmasry, "High-level Analog Synthesis Using Signal Flow Graph Transformations," *The Eighth Annual IEEE ASIC Conference and Exhibit*, pp. 366–369, September 1995.
13. G. W. Roberts, A. S. Sedra, "Synthesizing Switched-Current Filters by Transposing the SFG of Switched-Capacitor Filter Circuits," *IEEE Trans. on Circuits and Systems*, vol. 38, no. 3, pp. 337–340, March 1991.
14. M. A. Tan, R. Scaumann, "Generation of Transconductance Grounded Capacitor Filters by Signal-Flow Graph Simulation of LC-ladders," *ISCAS'88*, vol. 3, pp. 2407–2410, June 1988.
15. C. Simone, "Equivalent Ladder Networks by the Use of Signal Flow Graphs," *IEE Trans. on Circuit Theory*, vol. 6, no. 1, pp. 75–81, March 1959.

CHAPTER 3

Introduction to VLSI Active-RC Circuit Techniques

Passive RLC circuits can, in principle, be integrated using CMOS technologies to provide low-power, low-noise, wide-wideband signal processing applications (RF, high-speed I/O, etc.). However, passive VLSI signal processing circuits suffer from fundamental limitations such as the low availability of on-chip inductance, the lack of power gain and effective isolation between cascaded circuit stages, and the lack of frequency-tuning capability for resolving poor absolute frequency accuracy. These limitations are main motivations of using VLSI-integrated active circuit structures.

Active-RC circuits based on the operational amplifiers (OPAMPs) used in negative feedback associated with the resistor and capacitor networks are the most basic and most widely used circuits among various available VLSI active signal processing circuit structures. VLSI OPAMPs in active-RC circuits provide the gain and necessary isolation for various signal processing circuit implementations. A key advantage of active-RC signal processing circuits is the achievable high linearity mainly due to the high-gain OPAMP and high-linearity passive components. However, the requirements for OPAMPs to drive resistive load in the active-RC circuits also lead to the drawback of additional power consumption for active-RC circuits compared with their passive circuit counterparts. In addition, the frequency accuracy of the signal processing circuit determined by the RC constants of the circuit depends on the absolute values of VLSI resistors and capacitors, which cannot be accurately controlled in VLSI technologies.

3.1 VLSI Operational Amplifier Circuit Structures

OPAMPs with the passive component (resistors and capacitors) negative feedbacks are used to realize basic circuit functions of integration, addition, and scaling as shown in Fig. 3.1 that support complex VLSI analog signal processing functions.

3.1.1 VLSI OPAMP Circuit Models

An ideal OPAMP circuit can be modeled as a voltage-controlled voltage source as shown in Fig. 3.2 that has infinite voltage gain, infinite input resistance, and zero output

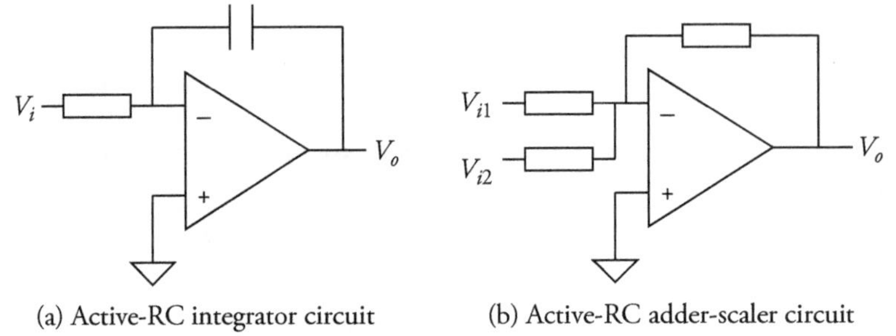

FIGURE 3.1 Basic VLSI active-RC building block circuits.

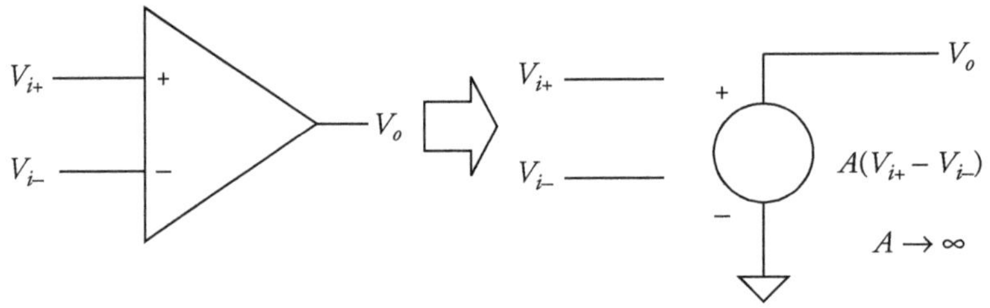

FIGURE 3.2 Ideal OPAMP circuit model.

resistance. Such a device is free of frequency and temperature dependence, distortion, and noise. Note that practical OPAMPs can only approximate such an ideal device.

For high-performance signal processing circuit applications, fully differential OPAMPs are commonly used to improve signal dynamic range and noise tolerance. Such an OPAMP circuit structure has two input and two output ports as shown in Fig. 3.3. Fully differential OPAMPs usually require common-mode feedback (CMF) to maintain the common-mode output signal level for proper circuit operation.

A typical VLSI OPAMP can offer close rail-to-rail output signal range between the negative and the positive supplies. However, the input signal range is usually limited, mainly due to the requirement of keeping input P- or NMOS devices in saturation operation as shown in Fig. 3.4. Typically, an OPAMP circuit with NMOS input devices suffers

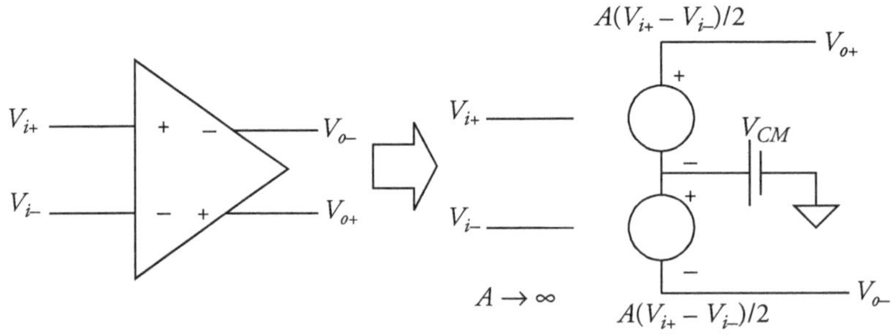

FIGURE 3.3 Fully differential OPAMP circuit model.

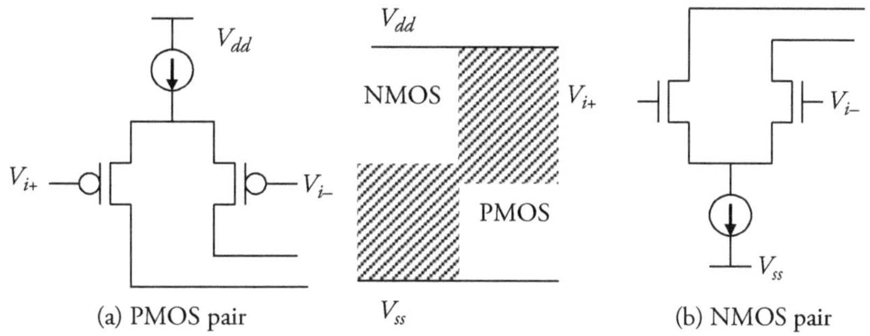

FIGURE 3.4 PMOS or NMOS input pair operation range.

from the minimum input signal voltage level limitation, and an OPAMP with PMOS input devices suffers from maximum input signal voltage level limitation.

In applications that require rail-to-rail OPAMP input range, both PMOS and NMOS devices are required in the OPAMP input stage.

Some important properties of the practical VLSI OPAMP circuits can be summarized as follows:

- Typical VLSI OPAMPs have limited dc gain of around 60–100 dB.
- The linear relation between the differential input and output is valid only for a limited range of outputs, usually determined by the power supply voltage.
- Most CMOS OPAMPs typically have an input offset voltage of a few millivolts caused by the device mismatch and systematic offset in the OPAMP circuit.
- The common-mode rejection ratio (CMRR) is defined as the ratio of gains of differential signal to common-mode signal. Typical values for practical CMOS OPAMPs are 60–80 dB.
- The gain bandwidth (GBW) is defined as the frequency at which the gain is 0 dB. Typical values for CMOS OPAMP circuits are 1–100 MHz.
- The phase margin (PM) is defined as the difference between the 180° and the output phase shift, and the frequency point the OPAMP circuit has unit gain. Typical PMs for CMOS OPAMPs are 45–90°, and >60° is usually desired.
- The slew rate (SR) is defined as the maximum rate of change of the output voltage for step input. Typical values for CMOS OPAMP circuits are 1–20 μV/μs.
- A practical OPAMP circuit has a finite output resistance, which is mainly limited by the output buffer device of the OPAMP.
- CMOS OPAMP circuits usually have higher noise than BJT OPAMPs. Typical values from the 10-Hz to 1-MHz frequency band are 10–50 μV rms.
- Signal dynamic range (DR) is defined as the difference between the highest input set by nonlinearity distortion and the lowest input set by the noise floor. In CMOS OPAMPs, a DR of 80–90 dB is typically achievable.
- Typical CMOS OPAMP circuits have the power supply rejection ratio (PSRR) of 60–80 dB.

In practical VLSI circuit design, the above performance parameters are strong functions of OPAMP structure, device sizing, circuit bias, and feedback conditions and should be optimized using proper techniques.

3.1.2 CMOS Two-Stage OPAMP Circuit

A widely used CMOS OPAMP circuit structure for active-RC circuit is the two-stage OPAMP circuit shown in Fig. 3.5, where the input MOS differential pair with active loads serves as the first gain stage, and the output device with the load serves as the second gain stage. The Miller capacitor compensation circuit in the OPAMP that employs R_c and C_c is usually required to ensure enough PM for stability. This compensation method makes the pole associated with the input stage the dominant pole. The compensation also pushes the pole associated with the output stage located beyond the unity gain frequency.

Because of the class A operation of the output stage, the two-stage OPAMP circuit structure offers good linearity in the normal operation condition. In addition, the class B operation offers a low-noise operation, resulting from the constant current drawn from power supply. However, such a circuit structure has the major drawback of higher operation power dissipation. A second drawback of such a VLSI OPAMP circuit structure is that it typically offers limited dc gain of 40–60 dB contributed by the limited output impedance of the first gain stage.

3.1.3 CMOS Folded Cascode OPAMP Circuit

The second widely used CMOS OPAMP circuit structure is the folded cascode OPAMP circuit employing the class AB output buffer as shown in Fig. 3.6. This OPAMP circuit structure consists of a folded cascode input stage and a class AB output stage. This OPAMP circuit structure offers an improved dc gain compared with the two-stage OPAMP structure. In addition, the class AB output stage provides significant reduction in the quiescent current of the output stage. The Miller compensation similar to the two-stage OPAMP is also usually required to ensure stability of the OPAMP circuit.

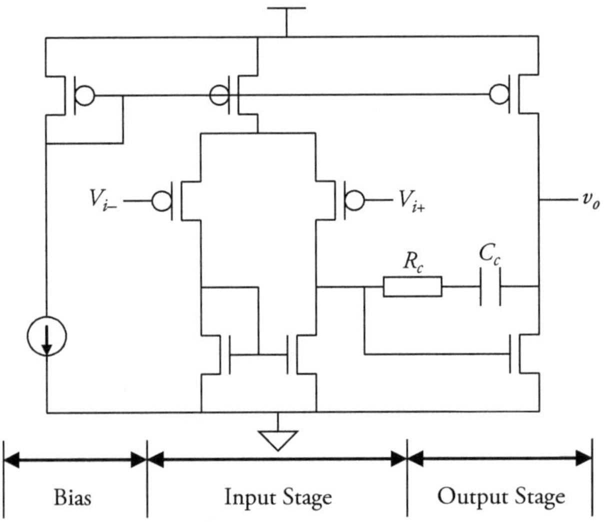

FIGURE 3.5 CMOS two-stage OPAMP circuit structure.

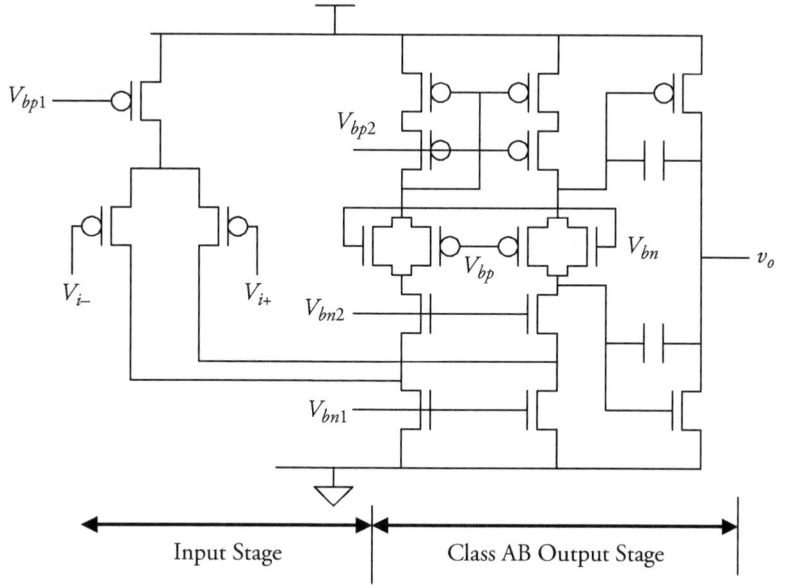

FIGURE 3.6 CMOS folded cascode OPAMP with class AB buffer.

3.1.4 CMOS Rail-to-Rail OPAMP Circuit

Figure 3.7 shows a CMOS OPAMP circuit structure that offers rail-to-rail input signal handling capability. In this OPAMP circuit structure, both the PMOS and NMOS input pairs are used to extend the input signal operation range.

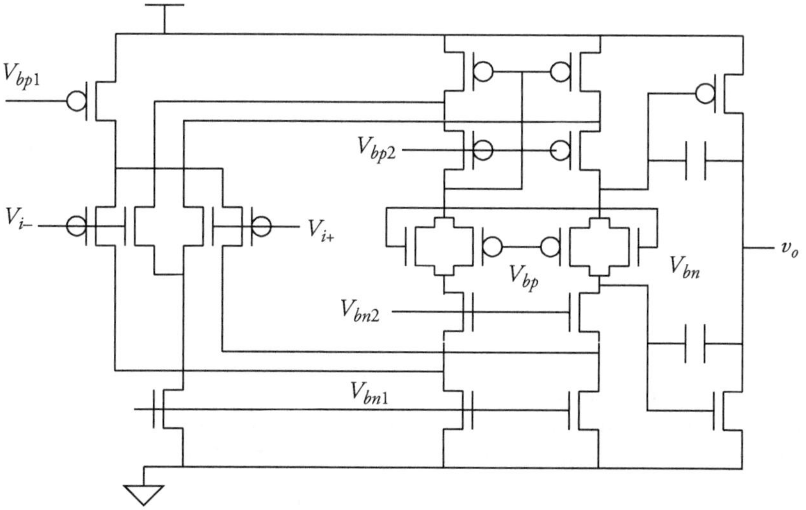

FIGURE 3.7 CMOS rail-to-rail input OPAMP.

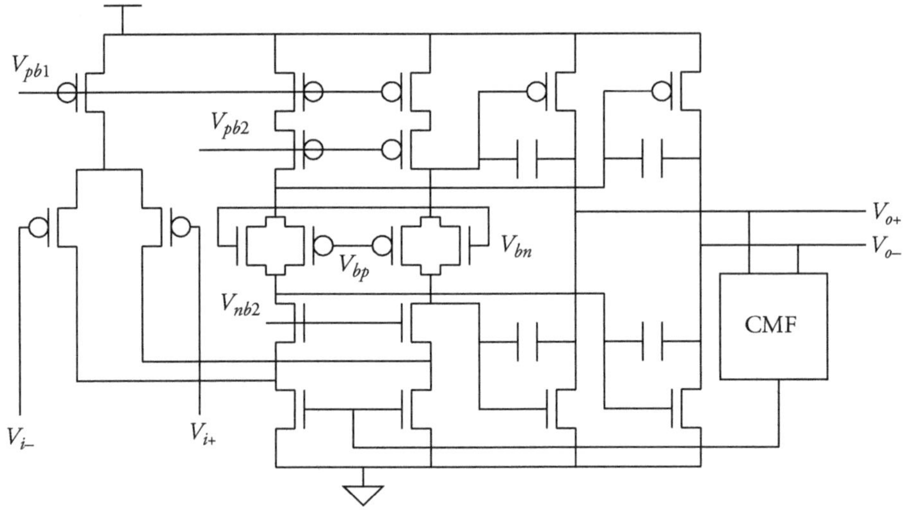

FIGURE 3.8 CMOS fully differential OPAMP circuit structure.

3.1.5 CMOS Fully Differential OPAMP Circuit

Many high-performance VLSI circuit applications require the OPAMP circuit to provide fully differential inputs and outputs as shown in Fig. 3.8 for better signal dynamic range and noise performance. Fully differential OPAMP circuit structures require internal CMF circuits to set the output common-mode voltages of the OPAMP circuits.

3.2 VLSI Resistor Structures

There are several available resistor types offered in the CMOS process technologies, including poly resistors, diffusion resistors, well resistors, and metal resistors. In general, poly resistors offer better accuracy than other resistor structures. Well resistors can be used to provide large resistance values if the high-sheet-resistance poly resistors are unavailable. Metal resistors, on the other hand, can be used to provide very low value resistors. Thin-film resistors can offer superior temperature stability. However, they are usually not available in the standard VLSI CMOS processes. The absolute tolerances of VLSI-integrated resistors are usually very poor with typical absolute variation in the range of ±30–50 percent. However, the matching of the same type of resistors can be better than ±0.1 percent. Laser-trimming techniques can be used for fine adjustment of integrated resistors. However, such capability is usually not available in standard VLSI process technologies.

The value of VLSI-integrated resistors can be calculated from their physical dimension (L/W ratio that is usually known as the number of "square"), and the sheet resistance R_s, which is usually given in the process technology files (see Fig. 3.9).

$$R = R_s\left(\frac{L}{W}\right) \tag{3.1}$$

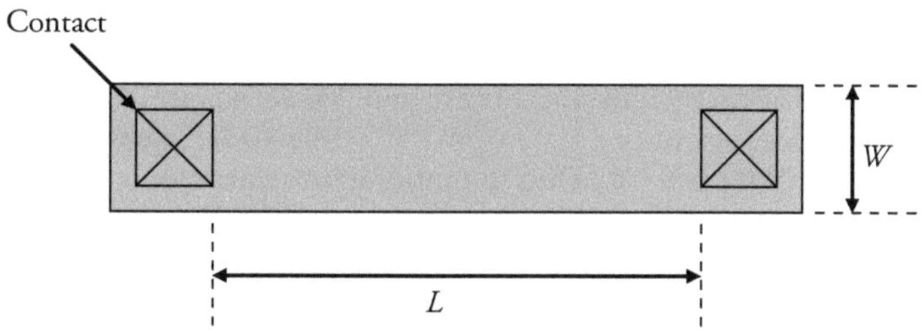

Figure 3.9 Physical dimension of VLSI resistor structures.

Corrections to the resistance calculation equation are sometimes required for better absolute design accuracy to account for the second-order effects such as oxide opening variation due to photolithography and etching, out-diffusion, nonuniform current flow, and contact resistance. In addition, the resistance can vary due to the resistor temperature coefficient and nonlinearity caused by self-heating, high-field velocity saturation, and depletion region encroachment. These variations can usually contribute to as high as ±25 percent sheet resistance variation and ±20 percent size variation for minimum geometry resistors.

Some general design guidelines of practical VLSI resistors are as follows:

- Minimum width resistors can be used with expected variations of about ±50 percent.
- Resistors of two to three times of the minimum feature size can be used to reduce the variations to about ±35 percent.
- Resistors of about five times of the minimum feature size can be used to reduce the variations to about ±30 percent.
- Replica resistors with same geometry and environments are required for matching.

Since no practical VLSI resistor is even isolated from its environment, capacitive and inductive coupling inevitably occurs at higher frequencies, and some types of resistors may also experience junction leakage effects.

3.2.1 Poly Resistors

A VLSI poly silicon layer usually provides the best on-chip resistor in the standard VLSI CMOS process technologies. There are several commonly available types of poly resistors on VLSI CMOS process technologies. Silicided poly that is commonly used for gates of digital MOS devices usually offers low-sheet resistance of a few ohms per square. Silicided poly may not provide enough resistance for a reasonable layout area, and therefore, silicide block mask is usually required on such type of processes to achieve high resistance. Heavily doped N- or P-type MOS gate poly usually has sheet resistance of a few tens of ohms per square. Lightly doped poly, on the other hand, may offer sheet resistance of a few hundreds to thousands of ohms per square. The doping level of a poly resistor may affect the poly resistor's temperature coefficient. For example, a lightly doped poly can usually be used to fabricate negative-temperature coefficient poly resistors while heavily doped poly can be used to fabricate positive-temperature coefficient poly resistors. As a result, a properly doped poly can be used to fabricate poly resistors with close-to-zero temperature coefficient. For example, for a 400-nm-thick poly, the zero-temperature-coefficient point is around 200 Ω/square. The 500 Ω/square poly has a temperature coefficient of –1000 ppm/°C while 70 Ω/square has a temperature coefficient of about 500 ppm/°C. Due to manufacture variations, the practically achievable temperature coefficient for poly resistor is around ±250 ppm/°C. VLSI poly resistors usually have good linearity since poly resistors do not experience tank modulation and conductivity modulation as long as the sheet resistance is low. In order to reduce parasitic capacitance between the resistor and the substrate, a poly resistor should be placed on top of the field oxide. This also ensures that the steps do not cause unexpected resistance variations.

3.2.2 Diffusion Resistors

Diffusion resistors can be fabricated using the N or P implant in all types of CMOS process technologies (see Fig. 3.10). These resistors usually exhibit a sheet resistance of a few tens of ohms per square, and they are almost immune to voltage modulation. N and P implants are relatively shallow diffusions that avalanche at relatively low voltage due to sidewall curvature. Diffusion resistors may not be constructed in the silicide process unless a silicide-blocking mask is used to prevent the diffusion to be silicided. Diffused resistors are rarely used because most CMOS processes offer poly resistors with equal- or greater-sheet resistances. Diffusion resistors can be used as ESD devices because their parasitic junction diodes can serve as voltage clamps. Diffusion resistors can also withstand larger power transients than poly resistors because thermal conductivity of silicon far exceeds that of the field oxide. However, high-sheet-resistance poly resistors are usually preferable to diffusion resistors in most other applications. The basic drawbacks of diffusion resistors include limited value range, high-temperature coefficient, and high sensitivity to operating voltage and current. For example, typically P-diffusion resistors have temperature coefficients around +1300 ppm/°C and voltage coefficients of about –350 ppm/V.

3.2.3 Well Resistors

Well (N-well or P-well) resistors are available in all types of CMOS process technologies. Well resistors can be fabricated using either N- or P-well with N^+- or P^+-well contacted at either ends where the N^+ (for N-well resistor; see Fig. 3.11) or P^+ (for P-well resistor) contacts serve as the ohmic contacts of the well resistor. Well resistors usually exhibit the sheet resistance of a few kilohms per square. Even higher resistance can be obtained by pinching the well with diffusions. The N-well resistor usually has a very large temperature coefficient of typically a few thousands of ppm/°C and high-voltage

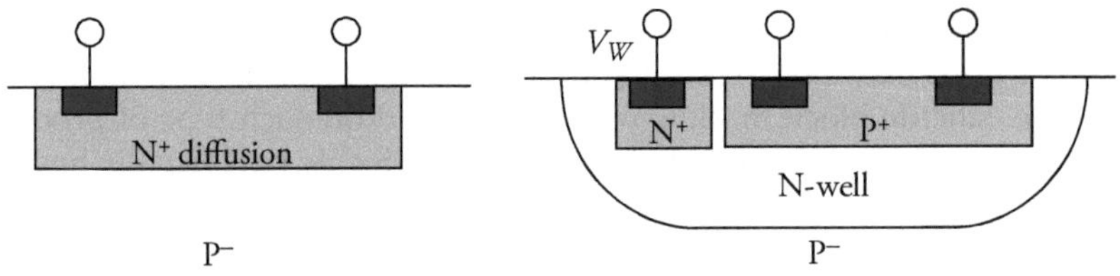

(a) N^+ diffusion resistor
(b) P^+ diffusion resistor inside N-well

Figure 3.10 VLSI diffusion resistor structure.

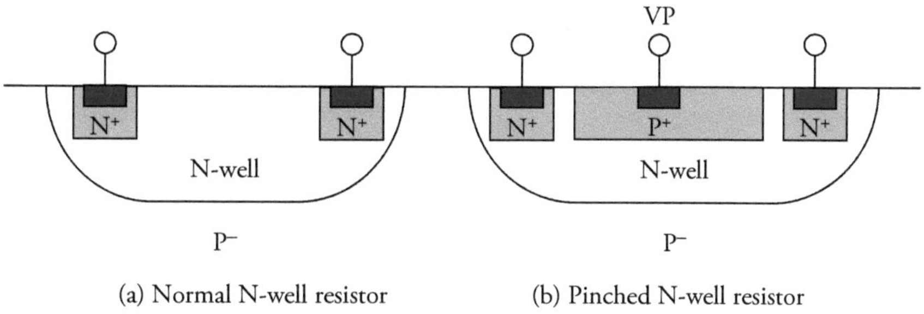

FIGURE 3.11 VLSI N-well resistor structure.

coefficient. One of the major limitations in well resistors is the side diffusion of the well, which usually limits the spacing of a well resistor. As a result, it will require a larger area for the layout of the well resistors since the well does not achieve its full junction depth unless the drawn geometry is at least twice as wide as the well is deep. N-well resistors with widths less than this will exhibit higher sheet resistance.

3.2.4 Metal Resistors

The sheet resistance of aluminum metallization is usually very low and typically in the order of a few tenth of ohms per square. Resistors in this range are typically used for current-sensing circuit applications. Accurate voltage sensing across metal resistors requires special techniques such as the Kelvin connection to minimize the resistance effects of the contact and routing. The values of metal resistors depend primarily upon the thickness of the metal layer. The sheet control of the resistor approximately equals the variation of its metal thickness. Most processes experience some ±20 percent variation in metal thickness over the field oxide, so a similar variation in metal sheet resistance can be expected.

3.2.5 Thin-Film Resistors

In some specialized VLSI analog process technologies, specialized materials are deposited as thin films on the surface of an integrated circuit to construct resistors. The thin-film resistors offer even better performance than poly resistors and diffusion resistors. These thin-film resistors are made from a wide variety of materials such as nichrome and sichrome. Thin-film resistors can achieve temperature coefficients of less than 100 ppm/°C with nearly perfect linearity, and they can be fabricated to tolerances better than ±0.1 percent by laser trimming. These resistors are desirable components for high-performance analog signal processing applications, but these specialized process technologies are costly and not usually available in standard CMOS process technologies.

3.3 VLSI Capacitor Structures

Most analog VLSI circuits contain at least one capacitor. Almost all of the capacitors used in the VLSI circuits are parallel-plate capacitors, which consist of two conductive plates. In the simplest parallel-plate capacitor, the two electrodes are assumed to have the same dimensions and reside directly opposite each other. VLSI capacitors can be constructed using thin insulating films or reverse-biased PN junctions as shown in

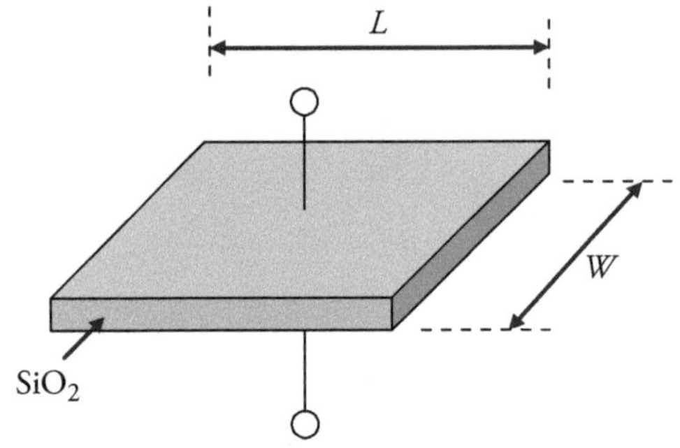

FIGURE 3.12 VLSI capacitor physical parameters.

Fig. 3.12. There are several typical types of integrated capacitances in CMOS technologies, including MOS capacitors, poly-poly capacitors, poly-metal capacitors, metal-metal capacitors, metal-insulator-metal (MIM) capacitors, and junction capacitors. MOS capacitors are available in all MOS process technologies. Many CMOS process technologies also offer highly linear poly-poly capacitor or MIM capacitors, which use thin oxide layers as dielectrics. Although the fabrications of these capacitors increase the process complexity and cost, the superior performance of these capacitors can often justify their inclusion. Properly constructed thin-film capacitors are less parasitic than junction capacitors, but they usually require additional processing steps. The absolute accuracy of the VLSI capacitor is also poor. Variations in doping and junction depth can cause junction capacitors to vary by up to ±30 percent. Physical dimension variation causes thin-film capacitors typically to vary by at least ±10 percent. Capacitors are difficult to trim because the usual trim structures add an excessive amount of parasitic capacitance. The microscopic dimensions of the VLSI circuit preclude the fabrication of more than a few hundred picofarads of capacitance. Even this tiny amount suffices for certain crucial applications, particularly for compensating feedback loops.

The value of the simple parallel-plate capacitor can be calculated using the following approximate equation as

$$C = C_S(WL) \tag{3.2}$$

where C_s is the capacitance per unit square that has typical values of 0.1–4 fF/μm^2 depending on the type of VLSI capacitor structures.

For more accurate design applications, the above capacitance equation usually requires certain corrections. Corrections to the capacitance shown in Eq. (3.2) are sometimes due to the nonideal effects such oxide thickness variation due to photolithography and etching, parasitic capacitance to nearby elements, and lead capacitance. Some commonly used design guidelines of VLSI capacitors are as follows:

- Arbitrary capacitor shape can be used to build capacitors with expected variations of about ±50 percent.
- A square capacitor with a large size can be used to reduce the variation to about ±30 percent.
- Replica capacitors with the same geometry and environments are required for matching.

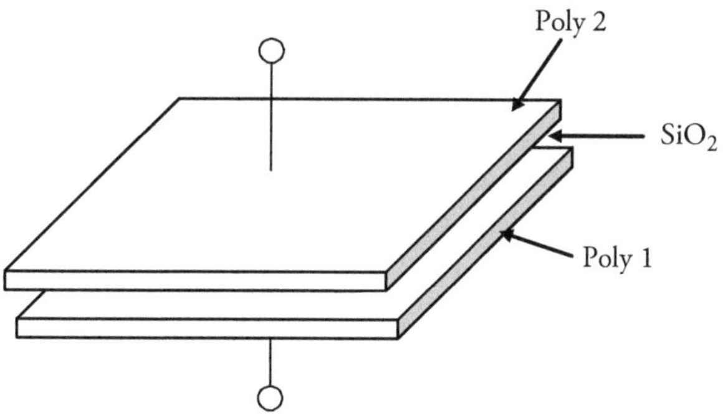

FIGURE 3.13 VLSI poly-poly capacitor structure.

3.3.1 Poly-Poly Capacitors

Some CMOS process technologies incorporate multiple poly layers, which allow the construction of poly-poly capacitors using two poly electrodes in combination with an oxide layer (see Fig. 3.13). Poly-poly capacitors always require at least one additional process step. Even if both the electrodes consist of existing depositions, the capacitor dielectric is unique to this structure and consequently requires a process extension. The voltage modulation of poly-poly capacitors is relatively small as long as both electrodes are heavily doped. An unsilicided poly-poly capacitor typically exhibits a voltage modulation of 150 ppm/V. The temperature coefficient of a poly-poly capacitor also depends on voltage modulation effects, and it is typically less than 250 ppm/°C. Both of these values will increase if either or both electrodes are lightly doped. The two plates of a poly-poly capacitor are not interchangeable. The top plate usually has less parasitic capacitance than the bottom plate.

3.3.2 Metal-Insulator-Metal Capacitors

Capacitors can also be constructed using two metal plates. One way to construct such type of capacitors is to use two normal interconnected metal layers, which are always available in all CMOS technologies. A limitation of such metal capacitor structures is the low capacitance density (typically 0.1–0.2 fF/μm^2), mainly due to the large space between the normal interconnected metal layers. Most modern VLSI technologies offer the metal-insulator-metal (MIM) capacitor structure as shown in Fig. 3.14, where an additional metal layer (top plate) is added between two normal, routine metal layers (metal N and metal $N + 1$). The added layer and the lower metal layer form the top and bottom plates of the MIM capacitor structure. The added metal layer is connected to upper metal layer through via. Since both top and bottom plates of the capacitor are made of metal, the resistance of the capacitor is significantly reduced. Furthermore, MIM capacitors are usually constructed at the very top of the routine metal layers, which is far away from the substrate; and consequently, the parasitic capacitor is also highly minimized. As a result, a MIM capacitor allows to construct high-density (>1 fF/μm^2) and high-quality factor (Q = 100–1000) capacitors in CMOS process technologies at RF, which has attracted great attention in analog VLSI and RF signal processing applications recently.

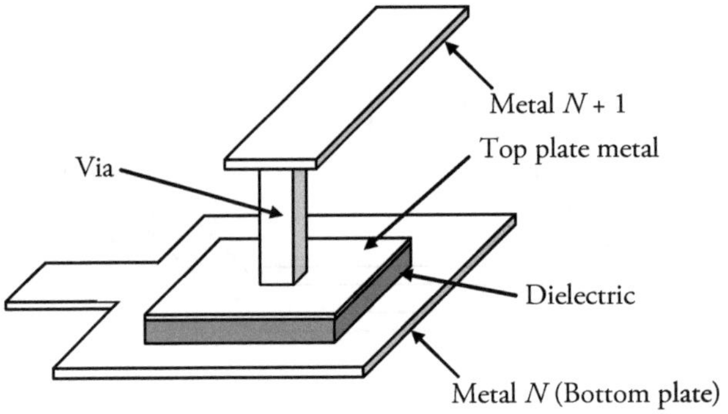

FIGURE 3.14 VLSI MIM capacitor structure.

3.3.3 MOS Gate Capacitors

MOS capacitor consists of a thin layer of oxide formed on silicon diffusion that serves as one of the electrodes. The other electrode consists of either metal or doped poly. If gate oxide is used to form an MOS capacitor, the resulting structure is often called a gate-oxide capacitor. MOS capacitors are particularly vulnerable to depletion modulation because their lower electrode is lightly doped and therefore easily depleted. Figure 3.15 shows typical capacitance curves of an NMOS transistor configured as an MOS capacitor. Majority carriers drawn up from the bulk silicon

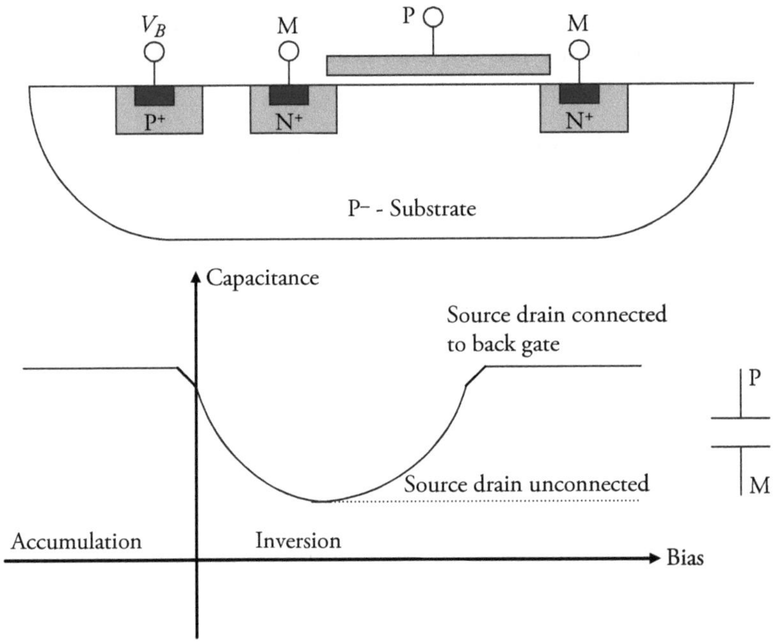

FIGURE 3.15 MOS capacitor structure.

accumulate beneath the gate oxide when the gate is biased negatively with respect to the back gate. The capacitor of the device in accumulation is determined solely by the gate dielectric. When the gate is biased positively, majority carriers are repelled from the surface, and a depletion region begins to form. As the bias increases, this depletion widens, and its capacitance decreases. Once the gate bias equals to the threshold voltage, sufficient minority carrier occurs. Larger forward biases only increase the concentration of minority carriers and do not affect the width of the depletion region. Therefore, the capacitance levels off at the new lower value. This minimum capacitance C_{min} may equal to less than 20 percent of the gate-oxide capacitance. The above analysis applies as long as the source and drain diffusions are absent or unconnected. If these diffusion terminals exist and are connected to the back gate, the conducting channel shorts the source and drains the terminals. This channel becomes the lower plate of the capacitor, and the capacitance rises again to approach the gate-oxide capacitance.

3.3.4 Junction Capacitors

Junction capacitors use the depletion region surrounding a reverse-biased junction as a dielectric. Due to a higher dielectric constant of the silicon than the oxide, a junction capacitor can obtain high capacitance per unit area, usually in the order of 0.8 fF/μm². However, this type of capacitor is highly nonlinear, caused by variations in the depletion region width with applied bias. The zero-bias capacitance serves as a measure of the value of the capacitor. As the reverse bias across the junction increases, its depletion region widens, and its capacitance decreases. The area of a diffused junction capacitor consists of three components: the area of the bottom surface, which approximately equals to the area of the oxide window shown in Fig. 3.16, the area of the sidewalls proportional to the perimeters of the drawn geometry, and the area of the corner fillet.

The following empirical equation can be used to determine the junction capacitance as

$$C_{\text{total}} = C_a A_d + C_p P_d \tag{3.3}$$

where the area capacitance C_a represents the capacitance per unit area and the peripheral capacitance C_p represents the capacitance per unit periphery.

Junction capacitors are sometimes represented as PN junction diodes with values given in terms of junction areas. This practice is understandable from simulation point of view since the junction capacitor is modeled as a PN diode.

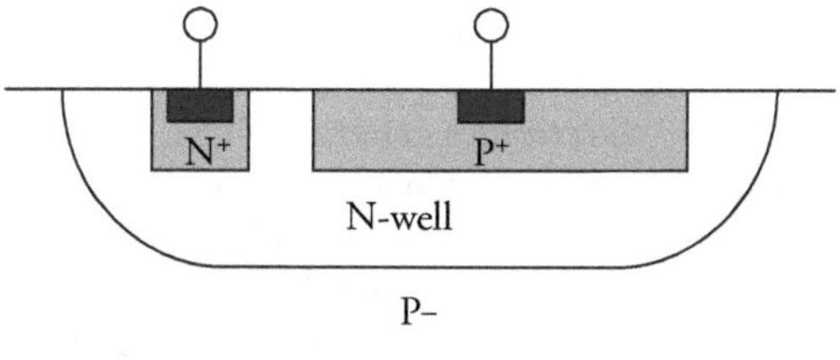

FIGURE 3.16 VLSI junction capacitor structure.

3.3.5 Metal-Metal Capacitors

For VLSI process technologies without MIM capacitor structures, multiple-layer metal-metal capacitor structures are commonly used to achieve high-density metal capacitors using normal interconnected layers. These VLSI metal capacitor structures achieve higher capacitance density by taking advantage of the lateral fringing field of the same layer and by cross-coupling with different layers. Modeling shows about six times of unity capacitance increase with metal-metal capacitors using fractal capacitor structures on 0.5-μm process compared with parallel capacitor structures. And it is expected that even greater unity capacitance enhancement comes along with the process of scaling down. Silicon data from a report show about 2.3× unity capacitance enhancement. Metal-metal capacitor may achieve a Q of 50–100 at RF (1 GHz+); therefore, it is suitable for RF applications.

References

1. R. Schaumann, "The Design of Continuous-Time Fully Integrated Filters: A Review," *IEE Proceedings, Part G*, vol. 136, pp. 184–190, August 1989.
2. A. M. Durham, W. Redman-White, J. B. Hughes, "Low Distortion VLSI Compatible Self-Tuned Continuous-Time Monolithic Filters," *IEEE ISCAS'91*, pp. 1448–1451, 1991.
3. T. Hollman, S. Lindfors, M. Lansirinne, K. Jussila, A. I. Halonen, "A 2.7-V CMOS Dual-Mode Baseband Filter for PDC and WCDMA," *IEEE J. Solid-State Circuits*, vol. 36, no. 7, pp. 1148–1153, July 2001.
4. R. Hogervorst, J. P. Tero, R. G. H. Eschauzier, J. H. Huijsing, "A Compact Power Efficient Rail-to-Rail Input/Output Operational Amplifier for VLSI Cell Libraries," *IEEE J. Solid-State Circuits*, vol. SC-29, pp. 1505–1512, December 1994.
5. R. G. H. Eschauzier, J. H. Huijsing, "A Programmable 1.5V CMOS Class-AB Operational Amplifier with Hybrid Nested Miller Compensation for 120dB Gain and 6 Mhz UGF," *IEEE J. Solid-State Circuits*, vol. SC-29, no. 12, pp. 1497–1504, December 1994.
6. P. V. A. Mohan, "Noise and Sensitivity Optimization of Actively Compensated Active RC Filters Based on Friend-Deliyannis SAB," *ISCAS'88*, vol. 3, pp. 2673–2676, June 1988.
7. M. J. Paschal, L. P. Huelsman, "Compensation of the Infinite-Gain Multiple-Feedback Low-Pass Active RC Filter," *Proceedings, the 37th Midwest Symposium on Circuits and Systems*, vol. 2, pp. 772–775, August 1994.
8. M. S. Zarnik, F. Novak, S. Macek, "Oscillation Test Structures of Active RC Filters," *IEE Proceedings, Circuits, Devices and Systems*, vol. 147, no. 5, October 2000.
9. H. Huang, E. K. F. Lee, "Low Voltage Technique for Active RC Filter," *Electronics Letters*, vol. 34, no. 15, July 1998.
10. C. Acar, H. Kuntman, "Limitations on Input Signal Level in Current-Mode Active-RC Filters using CCIIs," *Electronics Letters*, vol. 32, no. 16, pp. 1461–1462, August 1996.
11. M. Santo Zarnik, F. Novak, S. Macek, "Design of Oscillation-Based Test Structures for Active RC filters," *Proceedings, ED&TC*, p. 618, March 1997.
12. S. Signell, K. Mossberg, "Design of High-Frequency Active RC-Filters Using Passive Compensation Methods," *IEEE ISCAS'88*, vol. 3, pp. 2669–2672, June 1988.

13. A. Borys, N. Fliege, "Complementary Transformation in Mildly Nonlinear Feedback Systems," *European Conference on Circuit Theory and Design*, pp. 445–447, September 1989.
14. M. Soderstrand, S. Mitra, "Design of Active RC Filters with Zero Gain-Sensitivity Product," *IEEE Trans. on Circuits and Systems*, vol. 20, no. 4, pp. 441–445, July 1973.
15. K.-P. Pun, C.-S. Choy, C.-F. Chan, J. E. da Franca, "Digital Frequency Tuning Technique Based on Current Division for Integrated Active RC Filters," *Electronics Letters*, vol. 39, no. 19, pp. 1366–1367, September 2003.
16. M. Banu, Y. Tsividis, "Fully Integrated Active RC Filters in MOS Technology," *IEEE ISSCC*, vol. XXVI, pp. 244–245, February 1983.
17. D. Jurisic, G. S. Moschytz, N. Mijat, "Low-Noise Low-Power All pole Active-RC Filters Minimizing Resistor level," *Proceedings, ISCAS'03*, vol. 1, pp. I-469–I-472, May 2003.
18. J. Jussila, K. Halonen, "A 1.5V Active-RC Filter for WCDMA Applications," *Proceedings, ICECS*, vol. 1, pp. 489–492, September 1999.
19. N. G. Maratos, G. T. Ioannidis, "A Method for Dynamic Range Maximization for a Class of Integrated Active-RC Filters," *ISCAS'94*, vol. 5, pp. 763–766, May 1994.
20. S. Hirano, E. Hayahara, "A Compensation Technique for High Frequency Responses of Active RC Filter using Functional Trimming," *Proceedings, the 36th Midwest Symposium on Circuits and Systems*, vol. 2, pp. 1229–1232, August 1993.
21. H. Tang, R. J. Mack, "A Dictionary Approach to Fault Location in Linear Circuits," *IEEE ISCAS'91*, vol. 4, pp. 2064–2067, June 1991.
22. G. T. Volpe, L. M. Freeman, "Design of a Hybrid Integrated Circuit Elliptic Active RC Filter," *IEEE J. Solid-State Circuits*, vol. 9, no. 2, pp. 76–79, April 1974.
23. W. Khalil, T.-Y. Chang, X. Jiang, S. R. Naqvi, B. Nikjou, J. Tseng, "A Highly Integrated Analog Front-End for 3G," *IEEE J. Solid-State Circuits*, vol. 38, no. 5, pp. 774–781, May 2003.
24. A. M. Durham, W. Redman-White, J. B. Hughes, "High-Linearity Continuous-Time Filter in 5V VLSI CMOS," *IEEE J. Solid-State Circuits*, vol. 27, no. 9, pp. 1270–1276, September 1992.
25. Y. Ishibashi, "Active RC Filter Based on Simulation of Dissipative LC Filters," *IEE Proceedings, Circuits, Devices and Systems*, vol. 141, no. 2, pp. 101–104, April 1994.
26. D. Jurisic, N. Mijat, "Impedance Tapering Effects on 'Low-Q' SAB Band-Pass Active-RC Filter," *Proceedings, the First International Workshop on Image and Signal Processing and Analysis*, pp. 229–234, June 2000.
27. C. Frost, G. Levy, B. Allison, "A CMOS 2 MHz Self-Calibrating Bandpass Filter for Personal Area Networks," *ISCAS'03*, vol. 1, pp. I-485–I-488, May 2003.
28. J. K. Pyykonen, "A Low Distortion Wideband Active-RC Filter for a Multicarrier Base Station Transmitter," *ISCAS'01*, vol. 1, pp. 244–247, May 2001.
29. C. S. Wong, "A 3V GSM Baseband Transmitter," *Digest of Technical Papers, 1998 Symposium on VLSI Circuits*, pp. 120–123, June 1998.
30. O. Vainio, S. J. Ovaska, "A Class of Active-RC Filters with Predictive Characteristics," *Proceedings, IEEE, IMTC*, p. 358, April 1995.
31. J. Padalino, A. Tuszynski, "Note on the Analysis and Synthesis of Active RC Filters," *IEEE Trans. on Circuits and Systems*, vol. 15, no. 4, pp. 503–505, December 1968.
32. M. Soderstrand, S. Mitra, "Gain and Sensitivity limitations of Active RC Filters," *IEEE Trans. on Circuits and Systems*, vol. 18, no. 6, pp. 600–609, November 1971.

33. W. Holmes, "On Canonical Second-Order Active RC Filters," *IEEE Trans. on Circuits and Systems*, vol. 18, no. 6, pp. 721–723, November 1971.
34. D. Akerberg, K. Mossberg, "A Versatile Active RC Building Block with Inherent Compensation for the Finite Bandwidth of the Amplifier," *IEEE Trans. on Circuits and Systems*, vol. 21, no. 1, pp. 75–78, January 1974.
35. R. Srinivasagopalan, G. Martens, "Comparison of a Class of Active Filters with Respect to the Operational-Amplifier Gain-Bandwidth Product," *IEEE Trans. on Circuits and Systems*, vol. 21, no. 3, pp. 377–381, May 1974.
36. K. Laker, M. Ghausi, "Synthesis of a Low-Sensitivity Multiloop Feedback Active RC Filter," *IEEE Trans. on Circuits and Systems*, vol. 21, no. 2, pp. 252–259, March 1974.
37. P. A. BrackettSedra, "Active Compensation for High-frequency Effects in OPAMP Circuits with Applications to Active RC Filters," *IEEE Trans. on Circuits and Systems*, vol. 23, no. 2, pp. 68–72, February 1976.
38. M. Reddy, "An Insensitive Active-RC Filter for High Q and High Frequencies," *IEEE Trans. on Circuits and Systems*, vol. 23, no. 7, pp. 429–433, July 1976.
39. M. Reddy, "Operational-Amplifier Circuits with Variable Phase Shift and Their Application to High-Q Active RC-filters and RC-Oscillators," *IEEE Trans. on Circuits and Systems*, vol. 23, no. 6, pp. 384–389, June 1976.
40. S. Mitra, P. Vaidyanathan, B. Anderson, "A General Theory and Synthesis Procedure for Low-Sensitivity Active RC Filters," *IEEE Trans. on Circuits and Systems*, vol. 32, no. 7, pp. 687–699, July 1985.
41. M. Ishikawa, T. Kimura, N. Tamaki, "A CMOS Automatic Line Equalizer LSI Chip Using Active-RC Filtering," *IEEE J. Solid-State Circuits*, vol. 19, no. 4, pp. 506–509, August 1984.
42. C. S. Wong, "A 3-V GSM Baseband Transmitter," *IEEE J. Solid-State Circuits*, vol. 34, no. 5, pp. 725–730, May 1999.
43. D. Jurisic, G. S. Moschytz, N. Mijat, "Low-sensitivity Active-RC High- and Band-Pass Second-Order Sallen & Key Allpole Filters," *ISCAS'02*, vol. 4, pp. IV-241–IV-244, May 2002.
44. S. Natarajan, "A Comparison of Two Actively Compensated Active-RC Filters," *IEEE Trans. on Circuits and Systems*, vol. 37, no. 2, pp. 296–299, February 1990.
45. D. A. Johns, W. M. Snelgrove, A. S. Sedra, "Orthonormal Ladder Filters," *IEEE Trans. on Circuits and Systems*, vol. 36, no. 3, pp. 337–343, March 1989.
46. D. Jurisic, G. S. Moschytz, N. Mijat, "Low Active-Sensitivity All Pole Active-RC Filters using Impedance Tapering," *ISCAS'04*, vol. 1, pp. 89–92, May 2004.
47. C. Durdodt, A. Hanke, S. Heinen, U. Langmann, "Comparison of an Inductorless Low-IF and Zero-IF Receiver for Bluetooth," *The 2002 45th Midwest Symposium on Circuits and Systems*, vol. 1, pp. I-563–566, August 2002.
48. P. Aronhime, M. Desai, "Offset Independent Active RC Filters," *Proceedings of the 35th Midwest Symposium on Circuits and Systems*, vol. 2, pp. 1460–1463, August 1992.
49. D. Jurisic, G. S. Moschytz, "Low-Noise Active-RC Low-, High- and Band-pass. All Pole Filters using Impedance Tapering," *MELECON*, vol. 2, pp. 591–594, May 2000.

CHAPTER 4

Introduction to VLSI MOS-C Circuit Techniques

High-frequency precision VLSI active-RC circuits cannot be implemented without capacitor or resistor tuning, since the RC time constant of VLSI active-RC circuits varies significantly under the process, supply voltage, and application temperature (PVT) conditions. VLSI MOS-C circuit techniques are introduced to resolve the frequency-tuning limitation of active-RC circuits where MOS transistors serve as voltage-controlled resistors (VCRs), which can be adaptively adjusted based on a certain tuning-control circuit loops to provide precision RC time constants. Attractive features of MOS-C circuits are their low sensitivity to parasitic effects, their modularity and amenability to standard VLSI design, and their strong relation to well-understood active-RC–based signal processing circuit structures. The major drawback of MOS-C circuits employing MOS VCRs is the limitation in circuit linearity due to the fundamental nonlinear characteristic of the MOS transistors. Due to this reason, the major design and development focus of VLSI MOS-C circuit structures is to develop linear MOS VCR circuit structures.

4.1 Basic MOS Voltage-Controlled Resistors

Figure 4.1 shows the basic MOS VCR circuit structure. The device is biased in the linear operation mode through the gate voltage V_c. This control voltage can be used for tuning of the VCR resistance. The terminal voltages on the source and drain are given as V_s and V_d. The substrate is connected to a fixed dc voltage V_B, which is usually the ground voltage for circuit fabricated using the N-well process technologies.

The current flowing through the MOS VCR can be expressed based on the characteristic equation of the NMOS transistor in linear mode as

$$I = \beta\left(V_{gs} - V_T - \frac{V_{ds}}{2}\right)V_{ds} = \beta\left(V_g - V_T - \frac{V_s + V_d}{2}\right)(V_d - V_s) \tag{4.1}$$

where

$$\beta \equiv k_p\left(\frac{W}{L}\right) \tag{4.2}$$

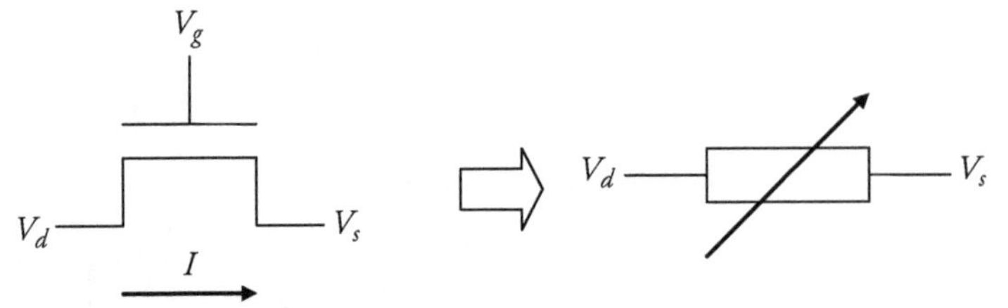

FIGURE 4.1 The basic NMOS voltage-controlled resistor circuit structure.

The effective conductance of the VCR can be derived from Eq. (4.1) as follows:

$$\Rightarrow G \equiv \frac{1}{R} = \beta\left(V_g - V_T - \frac{V_s + V_d}{2}\right) \tag{4.3}$$

Note that the linearity of such basic VCRs, in general, is limited since the conductance contains the terminal voltages. Equation (4.3) also suggests that if the source and drain voltages can be controlled symmetrical to a constant voltage V_{CM}, then the MOS transistor can serve as a linear-controlled resistor.

$$\begin{cases} V_s = V_{\mathrm{CM}} - \frac{1}{2}V \\ V_d = V_{\mathrm{CM}} + \frac{1}{2}V \end{cases} \tag{4.4}$$

$$\Rightarrow G \equiv \frac{1}{R} = \beta(V_g - V_T - V_{CM}) \tag{4.5}$$

In general cases, we may introduce a symmetry factor α as

$$\begin{cases} V_s = V_{\mathrm{CM}} - \alpha V \\ V_d = V_{\mathrm{CM}} + (1-\alpha)V \end{cases} \tag{4.6}$$

where $0 < \alpha < 1$ and $\alpha = 0$ or 1 indicates that the source or drain of the MOS device is fixed to constant voltage, and $\alpha = 0.5$ indicates the device is operated on fully symmetrical (differential) mode. We can now express the VCR current in terms of V across the VCR as follows:

$$\Rightarrow I = \beta\left[V_g - V_T - V_{\mathrm{CM}} + \left(\alpha - \frac{1}{2}\right)V\right]V = a_1 V + a_2 V^2 \tag{4.7}$$

The linear term (i.e., the conductance) and the second-order coefficients for the circuit can be derived from Eq. (4.7) as follows:

$$\begin{cases} a_1 \equiv G = \beta(V_g - V_T - V_{\mathrm{CM}}) \\ \dfrac{a_2}{a_1} = \dfrac{\left(\alpha - \frac{1}{2}\right)}{(V_g - V_T - V_{\mathrm{CM}})} \end{cases} \tag{4.8}$$

As can be seen, two general methods, the high gate driving voltage ($V_g - V_T - V_{CM} > \alpha - 0.5$) and the symmetrical terminal operation ($\alpha = 0.5$), can usually be used to improve the linearity of the basic MOS VCR circuit.

4.1.1 Single-Transistor Linear MOS VCR Circuit

With fixed gate control voltage, it is difficult to achieve high linearity since the two conditions suggested in Eq. (4.8) are generally difficult to achieve. However, Eq. (4.3) implies that the linearity of the basic MOS VCR can also be improved if the gate control voltage can be generated adaptively based on two terminal voltages as

$$V_g = \left(V_c + \frac{V_d + V_s}{2}\right) \tag{4.9}$$

Substituting Eq. (4.9) into Eq. (4.3) yields the following:

$$\Rightarrow G \equiv \frac{1}{R} = \beta(V_c - V_T) \tag{4.10}$$

Equation (4.10) suggests that a linear MOS VCR can be constructed by such a method. Shown in Fig. 4.2 is a conceptual implementation of the linear MOS VCR suggested by Eq. (4.9), where the MOS VCR gate control voltage is generated adaptively by summation of the tuning-control voltage V_c and the two scaled terminal voltages.

The linear MOS VCR based on the above technique can be realized in CMOS circuits as shown in Fig. 4.3. In the circuit shown in Fig. 4.3, M_1 and M_2, M_3 and M_4, and M_5 and M_6 are matched MOS transistor pairs, which are used to generate the adaptive gate control as stated in Eq. (4.9). Tuning of the VCR can be realized by adjusting the bias current I_B.

Keeping the current flow through M_3 and M_4 significantly lower than the current flow through M_1 and M_2, we have the following:

$$I_{M_3} = \frac{\beta_5}{2}(V_S - V_{Y_1} - V_T)^2 \tag{4.11}$$

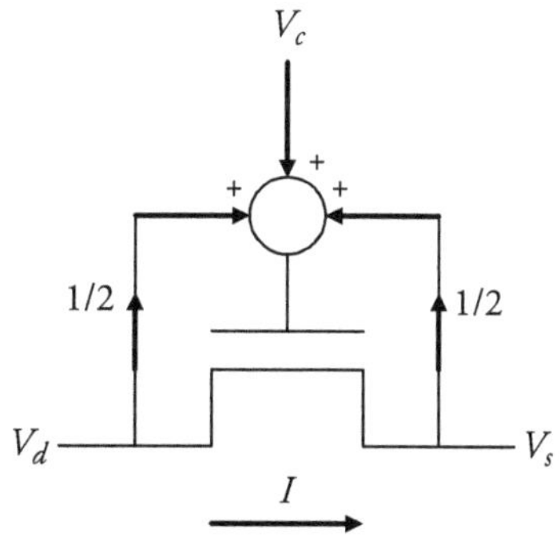

FIGURE 4.2 VLSI linear MOS VCR circuit structure.

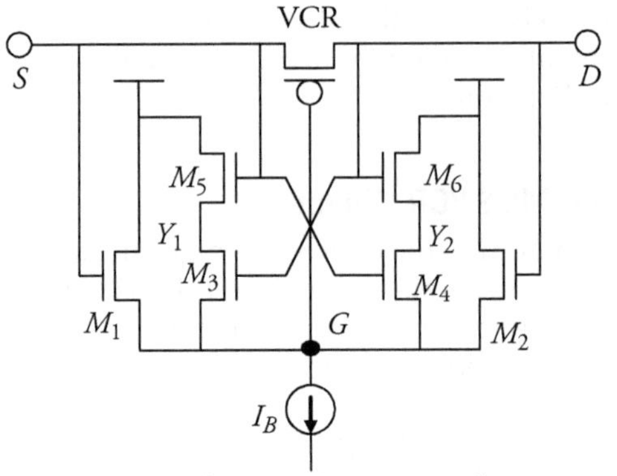

(a) VLSI PMOS linear VCR circuit implementation

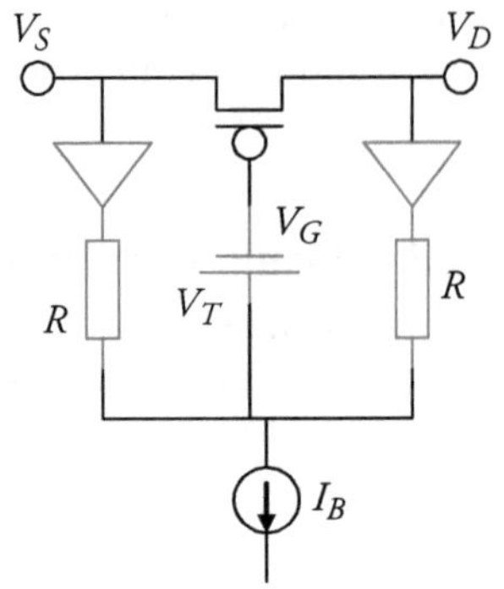

(b) Equivalent circuit

Figure 4.3 PMOS implementation of linear MOS VCR circuit.

By solving for V_{Y1}, we have

$$\Rightarrow V_{Y1} = V_S - V_T + \sqrt{\frac{2I_{M3}}{\beta_5}} \approx V_S - V_T \tag{4.12}$$

and similarly

$$\Rightarrow V_{Y2} = V_D - V_T + \sqrt{\frac{2I_{M3}}{\beta_6}} \approx V_D - V_T \tag{4.13}$$

By summing all currents of all four paths, we have the following:

$$\begin{aligned}
\Rightarrow I_B &= \frac{\beta_1}{2}(V_s - V_g - V_{\text{Tn}})^2 + \frac{\beta_1}{2}(V_d - V_g - V_{\text{Tn}})^2 \\
&\quad + \beta_3\left(V_d - V_g - V_{\text{Tn}} - \frac{V_{Y1} - V_g}{2}\right)(V_{Y1} - V_g) \\
&\quad + \beta_3\left(V_S - V_g - V_{\text{Tn}} - \frac{V_{Y2} - V_X}{2}\right)(V_{Y2} - V_g) \\
&\approx \frac{\beta_1}{2}[(V_d - V_g - V_{\text{Tn}})^2 + (V_s - V_g - V_{\text{Tn}})^2] \\
&\quad + \beta_3\left[\left(V_s - V_g - V_{\text{Tn}} - \frac{V_d - V_g - V_{\text{Tn}}}{2}\right)(V_d - V_g - V_{\text{Tn}})\right. \\
&\quad \left. + \left(V_s - V_g - V_{\text{Tn}} - \frac{V_s - V_X - V_{\text{Tn}}}{2}\right)(V_s - V_g - V_{\text{Tn}})\right] \\
&= \frac{\beta_1 - \beta_3}{2}[(V_d - V_g - V_{\text{Tn}})^2 + (V_s - V_g - V_{\text{Tn}})^2] \\
&\quad + 2\beta_3(V_d - V_g - V_{\text{Tn}})(V_s - V_g - V_{\text{Tn}})
\end{aligned} \tag{4.14}$$

If we select the size of M_1, M_2, M_3, and M_4 such that

$$\beta_1 = 3\beta_3 \tag{4.15}$$

we have

$$\Rightarrow I_B \approx \frac{\beta_1}{3}[(V_s - V_g - V_{\mathrm{Tn}}) + (V_d - V_g - V_{\mathrm{Tn}})]^2 \tag{4.16}$$

The desired gate control voltage for linear PMOS VCR can be generated as follows:

$$\Rightarrow V_g \approx \frac{V_d + V_s}{2} - V_{\mathrm{Tn}} - \sqrt{\frac{3I_B}{\beta_1}} \tag{4.17}$$

By substituting Eq. (4.16) into Eq. (4.3), a linear conductance of the VCR can be achieved as follows:

$$\Rightarrow G \equiv \frac{1}{R} = \beta_{\mathrm{VCR}}\left(V_{\mathrm{Tn}} + V_{\mathrm{Tp}} + \sqrt{\frac{3I_B}{\beta_1}}\right) \tag{4.18}$$

For the NMOS linear VCR implementation shown in Fig. 4.4, M_1, M_2, M_3, and M_4 are identical NMOS transistors in saturation modes, which are used to generate the desired gate control voltage of the linear VCR. M_5 serves as a voltage level shifter. The currents flowing through all three current paths are given as

$$\begin{cases} I_B = \frac{\beta}{2}(V_s - V_A - V_T)^2 + \frac{\beta}{2}(V_x - V_A - V_T)^2 \\ I_B = \frac{\beta}{2}(V_d - V_B - V_T)^2 + \frac{\beta}{2}(V_x - V_B - V_T)^2 \\ I_B = \frac{\beta}{2}(V_x - V_A - V_T)^2 + \frac{\beta}{2}(V_x - V_B - V_T)^2 \end{cases} \tag{4.19}$$

By subtracting the third equation from the first and the second equations in Eq. (4.19) respectively, we have the following:

$$\Rightarrow \begin{cases} \frac{\beta}{2}(V_s - V_A - V_T)^2 - \frac{\beta}{2}(V_x - V_B - V_T)^2 = 0 \\ \frac{\beta}{2}(V_d - V_B - V_T)^2 - \frac{\beta}{2}(V_x - V_A - V_T)^2 = 0 \end{cases} \tag{4.20}$$

that is,

$$\Rightarrow \begin{cases} (V_s - V_A - V_T) = (V_x - V_B - V_T) \\ (V_d - V_B - V_T) = (V_x - V_A - V_T) \end{cases} \tag{4.21}$$

or

$$\Rightarrow V_x = \frac{1}{2}(V_s + V_d) \tag{4.22}$$

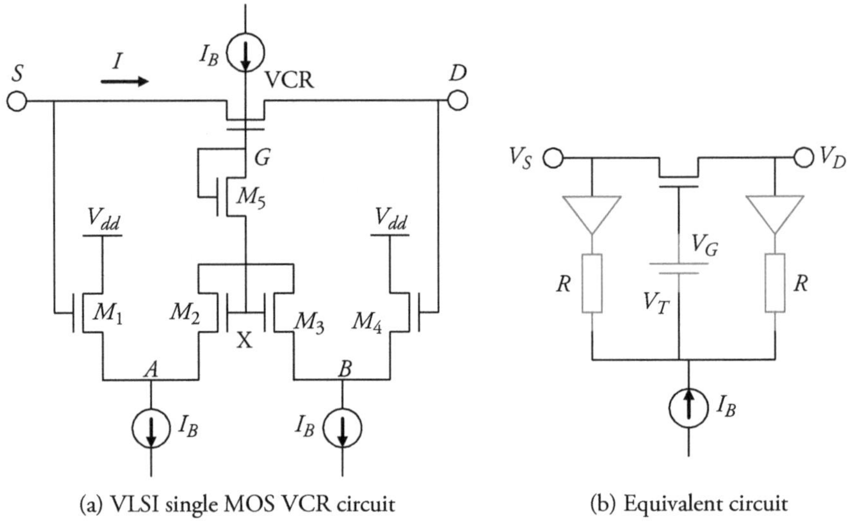

FIGURE 4.4 NMOS implementation of linear MOS VCR circuit.

The desired gate control voltage for linear VCR can be generated as follows:

$$\Rightarrow V_g = V_x + V_T + \sqrt{\frac{2I_B}{\beta_5}} = \frac{1}{2}(V_s + V_d) + V_T + \sqrt{\frac{2I_B}{\beta_5}} \tag{4.23}$$

By substituting Eq. (4.23) into Eq. (4.3), a linear conductance of the NMOS VCR can be achieved as follows:

$$\Rightarrow G \equiv \frac{1}{R} = \beta_{\text{VCR}}\sqrt{\frac{2I_B}{\beta_5}} \tag{4.24}$$

4.1.2 Two-Transistor Linear MOS VCR Circuit

Alternatively, linear MOS VCR can also be implemented using two identical MOS transistors as shown in Fig. 4.5. In this circuit structure, the gate voltages of the MOS transistors are generated by summing the control voltage and the two terminal voltages. The total current flowing through the MOS VCR is given as

$$\begin{aligned} I &= \beta\left(V_c + V_s - V_T - \frac{V_d + V_s}{2}\right)(V_d - V_s) + \beta\left(V_c + V_d - V_T - \frac{V_d + V_s}{2}\right)(V_d - V_s) \\ &= 2\beta(V_c - V_T)(V_d - V_s) \end{aligned} \tag{4.25}$$

Note that such a circuit structure provides a linear VCR with the equivalent conductance as follows:

$$\Rightarrow G = \frac{1}{R} = 2\beta(V_c - V_T) \tag{4.26}$$

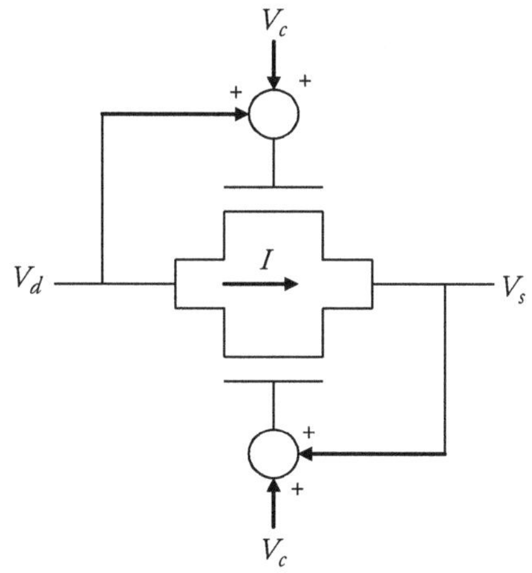

FIGURE 4.5 Two-transistor linear MOS VCR.

4.2 Differential Linear MOS VCR Structures

As described in Eq. (4.3), linear MOS VCR can be realized based on symmetrical terminal voltages applied to the MOS transistor. Such a concept can be used in the differential MOS VCR structures where the symmetry is maintained through the fully differential configuration of the signaling.

4.2.1 Two-Transistor Differential MOS VCR

A two-transistor differential linear MOS VCR structure is shown in Fig. 4.6, where two identical MOS transistors with the same gate and source terminal voltages are used. The drain voltages, on the other hand, are kept symmetrical with respect to the source (common-mode) voltages.

In the circuit in Fig. 4.6, the differential currents of the two-transistor VCR can be calculated as follows:

$$\begin{aligned} I_+ - I_- = 2I &= \beta\left(V_c - V_T - V_{CM} + \frac{V_i}{2}\right)V_i - \beta\left(V_c - V_T - V_{CM} + \frac{-V_i}{2}\right)(-V_i) \\ &= 2\beta(V_c - V_T - V_{CM})V_i \end{aligned} \tag{4.27}$$

Note that such a circuit structure provides a linear differential VCR with equivalent differential conductance as follows:

$$\Rightarrow G \equiv \frac{1}{R} \equiv \frac{I}{V_i} = \beta(V_c - V_T - V_{CM}) \tag{4.28}$$

Note also that even though the differential current is linear with respect to the differential input voltages in the two-transistor differential linear VCR, the current in each branch is generally nonlinear with respect to the terminal voltages.

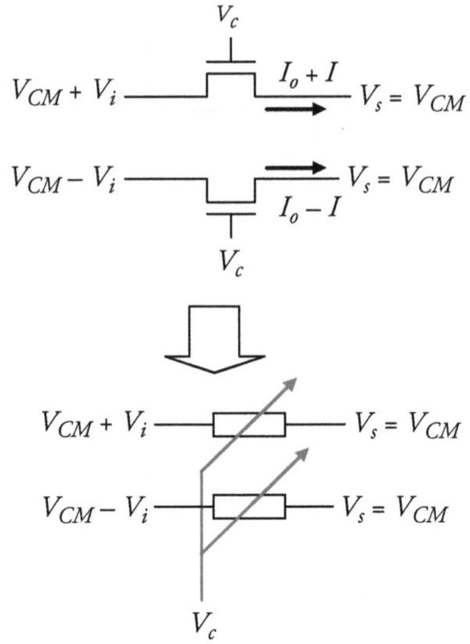

FIGURE 4.6 VLSI two-transistor differential linear MOS VCR circuit structure.

4.2.2 Four-Transistor Differential MOS VCR

The two-transistor differential MOS VCR suffers from symmetrical input signal constraints. Linearity degradation occurs if the input signal is not perfectly symmetrical with respect to the source terminal voltages. The second drawback of the two-transistor differential linear VCR is its dependency of differential conductance on the threshold voltage of the MOS transistors, which has large variations under the VLSI fabrication technology.

An improved differential linear MOS VCR employing four identical MOS transistors is shown in Fig. 4.7, where all transistors are in the linear operation mode with identical source terminal voltage V_{CM} (usually maintained through a feedback of fully differential OPAMP circuit). The input voltages are now free to use any values as soon as the devices are in linear mode.

For this four-transistor differential VCR, the differential current of the VCR is given as follows:

$$\begin{aligned} I_+ - I_- &= \beta\left(V_{c1} - V_T - \frac{V_{i+} + V_{\text{CM}}}{2}\right)(V_{i+} - V_{\text{CM}}) \\ &+\beta\left(V_{c2} - V_T - \frac{V_{i-} + V_{\text{CM}}}{2}\right)(V_{i-} - V_{\text{CM}}) \\ &-\beta\left(V_{c1} - V_T - \frac{V_{i-} + V_s}{2}\right)(V_{i-} - V_{\text{CM}}) \\ &-\beta\left(V_{c2} - V_T - \frac{V_{i+} + V_{\text{CM}}}{2}\right)(V_{i+} - V_{\text{CM}}) \\ &= \beta(V_{c1} - V_{c2})(V_{i+} - V_{i-}) \end{aligned} \tag{4.29}$$

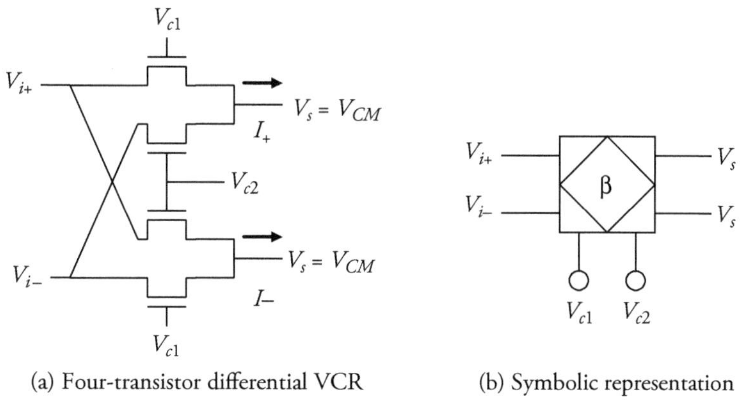

FIGURE 4.7 VLSI four-transistor differential linear MOS VCR circuit structure.

It can be seen that the circuit provides a linear differential VCR with the following equivalent differential conductance:

$$\Rightarrow G \equiv \frac{1}{R} \equiv \frac{I_+ - I_-}{(V_{d+} - V_s) - (V_{d-} - V_s)} = \beta(V_{c1} - V_{c2}) \tag{4.30}$$

Note that the differential resistance is independent of the threshold voltages of the devices.

4.3 Grounded Linear MOS VCR Structures

All the floating MOS VCR circuits discussed have been based on the devices in linear operations. Linear VCR circuits can also be realized using the device operated in the saturation region. Considering the *I-V* characteristic equation of the MOS device in saturation as

$$I = \frac{\beta}{2}(V_{gs} - V_T)^2 \tag{4.31}$$

generally, the current to the gate voltage relation is nonlinear. However, the difference of two currents for identical MOS devices with a different gate voltage can be designed to be linear under certain constrained conditions as follows:

$$\begin{aligned} &\Rightarrow I = I_2 - I_1 = \frac{\beta}{2}(V_{gs2} - V_T)^2 - \frac{\beta}{2}(V_{gs1} - V_T)^2 \\ &= \frac{\beta}{2}(V_{gs2} - V_{gs1})(V_{gs2} + V_{gs1} - 2V_T) \end{aligned} \tag{4.32}$$

4.3.1 Saturation MOS-Based Linear VCR

Figure 4.8 shows a linear grounded VLSI MOS VCR circuit structure based on MOS transistors in saturation where the two NMOS transistors are identical and are both biased in saturation mode.

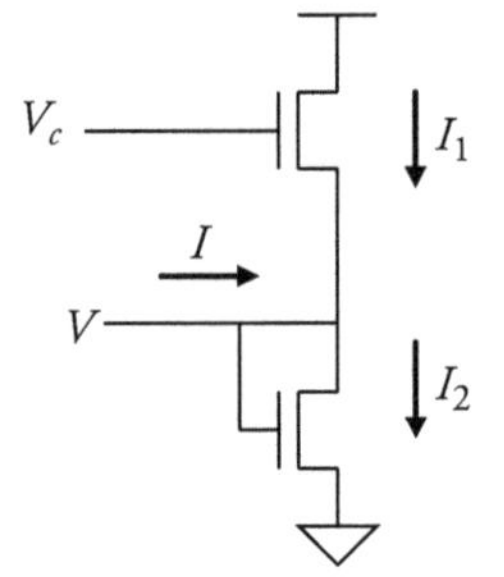

FIGURE 4.8 Linear grounded MOS VCR based on MOS in saturation.

The input current of the circuit can be expressed as

$$\begin{aligned} I = I_2 - I_1 &= \frac{\beta}{2}(V_{gs2} - V_T)^2 - \frac{\beta}{2}(V_{gs1} - V_T)^2 \\ &= \beta\left(V - \frac{V_C}{2}\right)(V_C - 2V_T) \end{aligned} \tag{4.33}$$

The equivalent conductance of the VCR is given as

$$\Rightarrow G \equiv \frac{1}{R} \equiv \frac{dI}{dV} = \beta(V_C - 2V_T) \tag{4.34}$$

4.3.2 Symmetrical MOS-Based Linear VCR

An alternative approach to construct a linear grounded MOS VCR is based on a pair of identical MOS transistors with one in saturation mode and one in linear mode as shown in Fig. 4.9.

The current flowing through the given MOS VCR can be expressed by summing the two MOS drain currents as follows:

$$I = I_2 + I_1 = \begin{cases} \dfrac{\beta}{2}V_T^2 + \beta(V_C - 2V_T)V & V > V_T \\ \beta(V_C - V_T)V & V < V_T \end{cases} \tag{4.35}$$

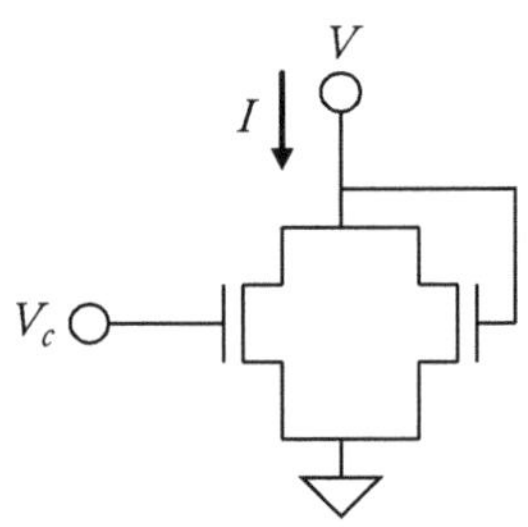

FIGURE 4.9 Linear grounded VLSI MOS VCR circuit–based saturation and linear MOS pair.

It can be seen that the circuit provides a segmented linear VCR with equivalent linear conductance as follows:

$$\Rightarrow G = \begin{cases} \beta(V_C - 2V_T) & V > V_T \\ \beta(V_C - V_T) & V < V_T \end{cases} \tag{4.36}$$

4.4 Distributed MOS-RC Circuits

In the double-poly process technologies, a resistor is usually made using the top poly layer, and the capacitor is formed between the two poly layers. Such a structure can be used to construct distributed RC network CT signal processing applications. Two major advantages of such circuit structure are sharp cutoff in the stop band compared with the lumped RC network and elimination of substrate power supply coupling and smaller area compared with the lumped RC network.

A distributed RC network with total resistance R and capacitance C can be modeled as a π network as shown in Figs. 4.10 and 4.11 with the parameter given as follows:

$$\begin{cases} G = \dfrac{\sqrt{sC/R}}{\sinh(\sqrt{sRC})} \\ P = \cosh(\sqrt{sRC}) \end{cases} \tag{4.37}$$

The transfer function of the network is given as

$$H(s) = \frac{\sqrt{sC/R}}{\sqrt{sC/R}\cosh(\sqrt{sRC}) + sC_o \sinh(\sqrt{sRC})} \tag{4.38}$$

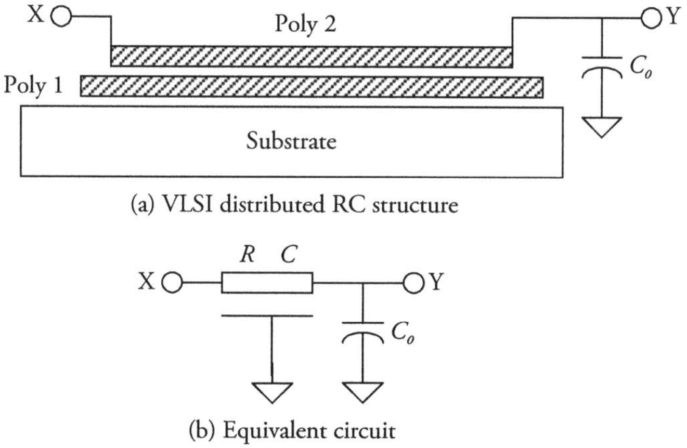

Figure 4.10 VLSI distributed RC circuit structure.

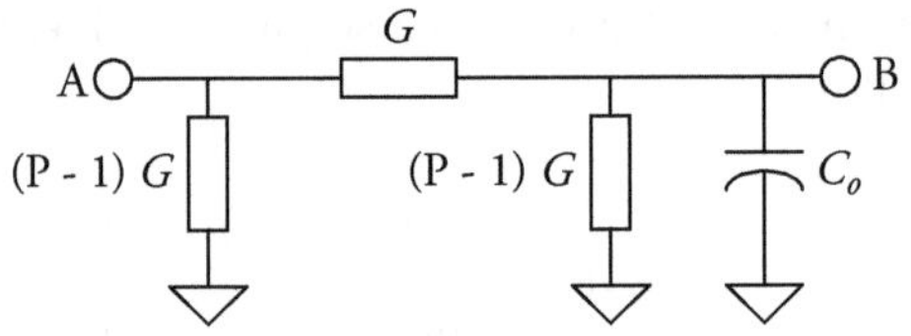

FIGURE 4.11 Equivalent circuit of VLSI distributed RC circuit structure.

An example of CT analog signal processing circuit using the distributed RC structure is shown in Fig. 4.12.

The distributed MOS-RC structure can also be constructed using MOS devices, as shown in Fig. 4.13, where M_0 serves as the distributed MOS-RC passive component, and the circuit provides a low-pass filter function with unit dc gain.

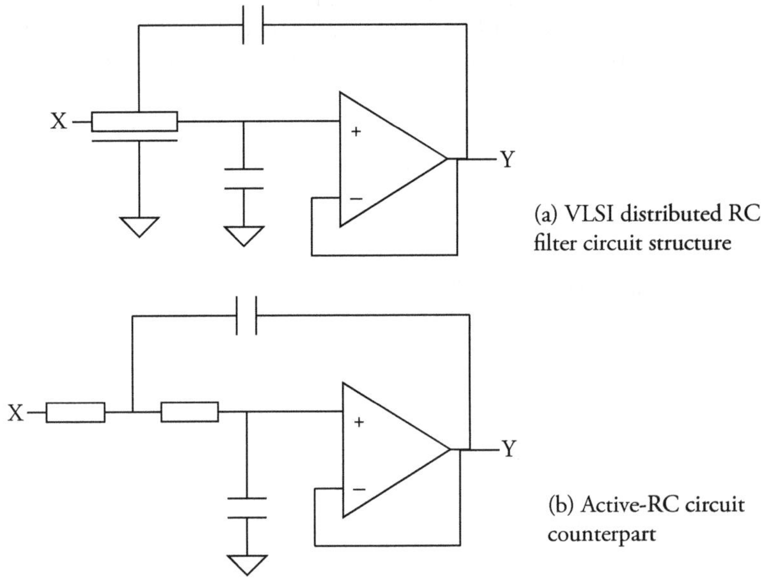

FIGURE 4.12 VLSI distributed RC filter structure.

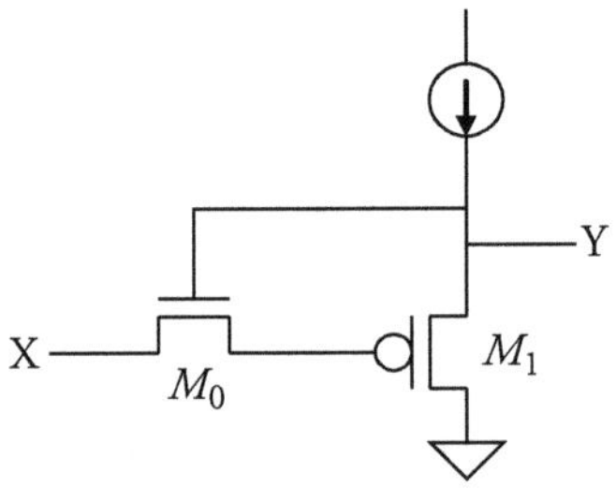

FIGURE 4.13 VLSI distributed MOS-RC filter structure.

4.5 Basic MOS-C Signal Processing Circuit Structures

The VCR circuits can be used to realize various analog signal processing functions.

4.5.1 MOS-C Scaling Circuit

Similar to the active-RC circuit, the scaling circuit can be constructed by replacing the linear resistors with the MOS VCR. The MOS-C single-transistor scaling circuit structure is shown in Fig. 4.14.

The transfer function of the block is given as follows:

$$H(s) = \frac{V_o(s)}{V_i(s)} = -\frac{\beta_1(V_{c1} - V_T)}{\beta_2(V_{c2} - V_T)} \tag{4.39}$$

Fully differential scaling circuit based on the two-transistor differential MOS VCR circuit is shown in Fig. 4.15.

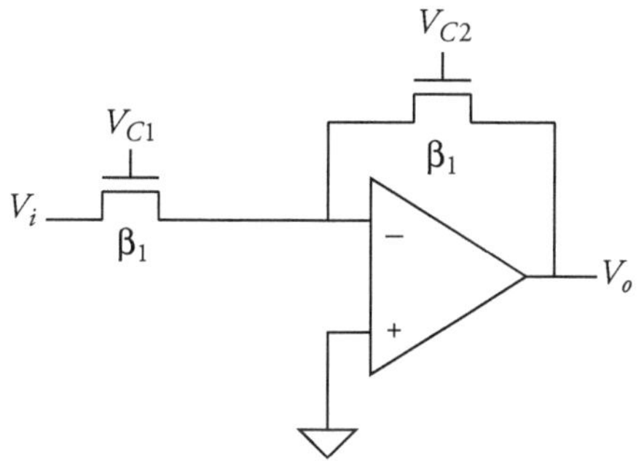

FIGURE 4.14 MOS-C scaling circuit based on single-transistor VCR circuit.

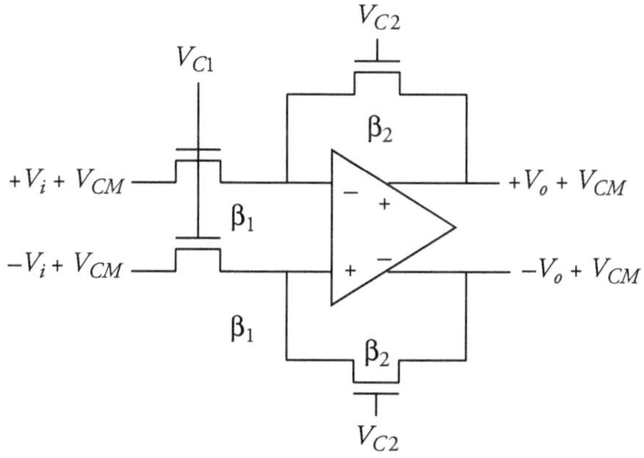

FIGURE 4.15 Two-transistor differential MOS-C scaling circuit.

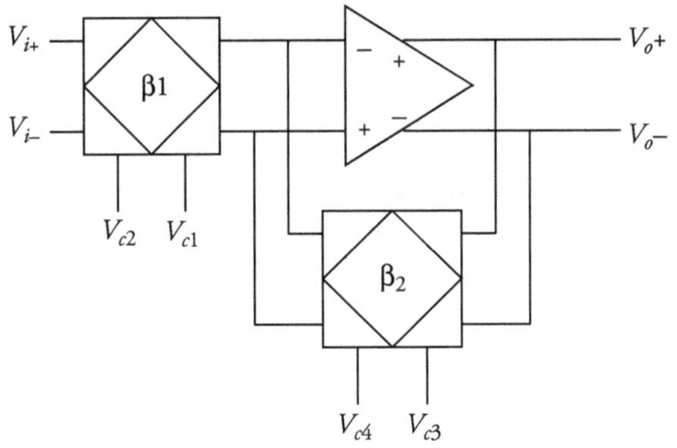

Figure 4.16 Four-transistor differential MOS-C scaling circuit.

The s-domain transfer function of the block is given as follows:

$$H(s) = \frac{V_o(s)}{V_i(s)} = -\frac{\beta_1(V_{c1} - V_T)}{\beta_2(V_{c2} - V_T)} \tag{4.40}$$

Similarly, an MOS-C scaling circuit can also be constructed using a four-transistor differential VCR circuit as shown in Fig. 4.16.

The transfer function of the circuit block is given as

$$H(s) = \frac{V_{o+}(s) - V_{o-}(s)}{V_{i+}(s) - V_{i-}(s)} = -\frac{\beta_1(V_{C2} - V_{C1})}{\beta_2(V_{C4} - V_{C3})} \tag{4.41}$$

where $V_{c4} > V_{c3}$ is required to ensure stability of the circuit.

4.5.2 MOS-C Adder Circuit

The single-transistor MOS VCR circuit–based adder circuit structure is shown in Fig. 4.17.

The transfer function of the circuit block is given as follows:

$$V_o(s) = -(V_{i1}(s) + V_{i2}(s)) \tag{4.42}$$

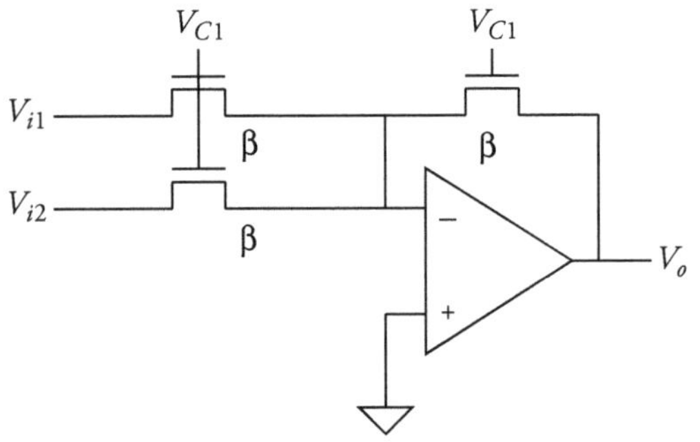

Figure 4.17 Single-transistor MOS VCR–based VLSI adder circuit.

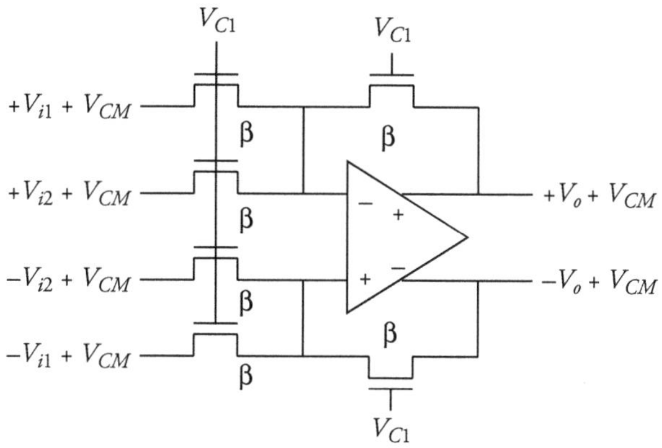

FIGURE 4.18 Two-transistor differential MOS VCR adder.

A differential adder based on the two-transistor differential MOS VCR circuit is shown in Fig. 4.18.

The transfer function of the adder is given as follows:

$$V_o(s) = -(V_{i1}(s) + V_{i2}(s)) \tag{4.43}$$

Figure 4.19 shows an MOS-C adder based on four-transistor differential VCR circuit.

The transfer function is given as

$$V_{o+}(s) - V_{o-}(s) = -[(V_{i1+}(s) - V_{i1-}(s)) + (V_{i2+}(s) - V_{i2-}(s))] \tag{4.44}$$

where $V_{c2} > V_{c1}$ is required to ensure stability of the circuit.

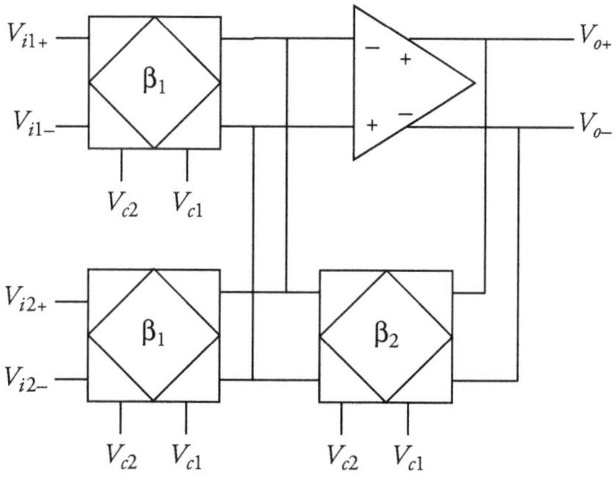

FIGURE 4.19 Four-transistor differential MOS VCR–based adder circuit.

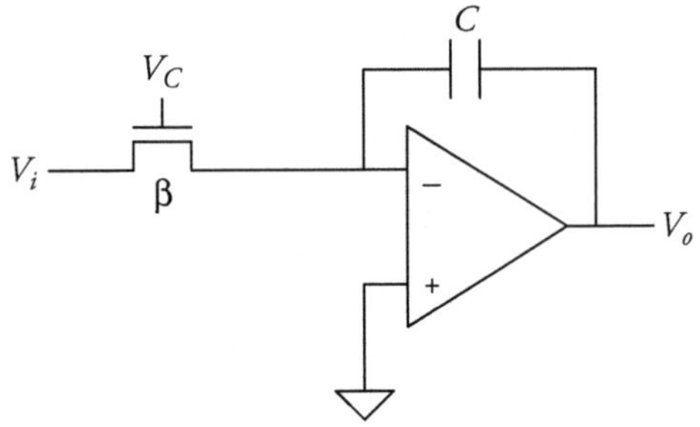

FIGURE 4.20 Single-transistor MOS-C–based integrator circuit.

4.5.3 MOS-C Integrator Circuit

VCR circuits can be used to realize various integration analog signal processing operations. Figure 4.20 shows a VLSI integrator circuit based on the MOS VCR.

The s-domain transfer function of the integrator is given as follows:

$$H(s) = \frac{V_o(s)}{V_i(s)} = -\frac{\beta(V_c - V_T)}{Cs} \tag{4.45}$$

Figure 4.21 shows the differential integrator based on two-transistor MOS VCR structure.

The s-domain transfer function of the integrator is given as follows:

$$H(s) = \frac{V_o(s)}{V_i(s)} = -\frac{\beta(V_c - V_T)}{Cs} \tag{4.46}$$

The integration can also be realized using four-transistor differential MOS VCR as shown in Fig. 4.22.

And the s-domain transfer function of the integrator is given as follows:

$$H(s) = \frac{V_{o+}(s) - V_{o-}(s)}{V_{i+}(s) - V_{i+}(s)} = -\frac{\beta(V_{C2} - V_{C1})}{Cs} \tag{4.47}$$

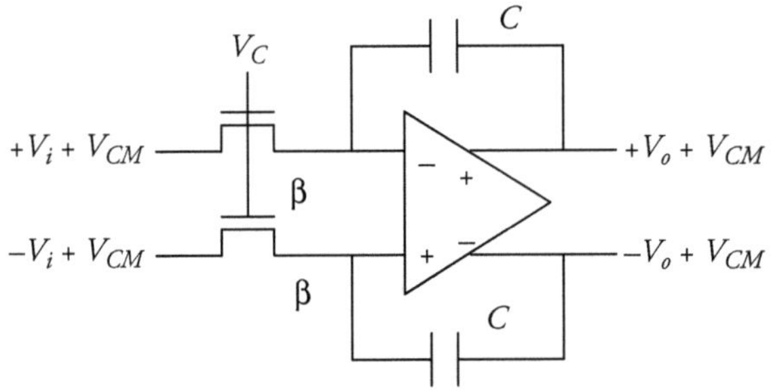

FIGURE 4.21 Two-transistor differential VCR–based MOS-C integrator circuit.

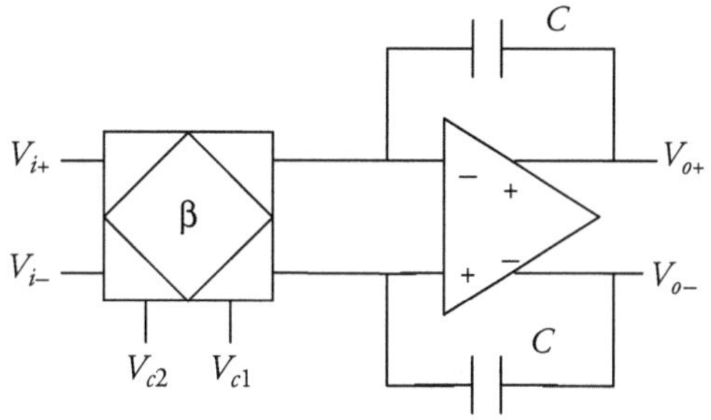

Figure 4.22 Four-transistor differential VCR-based MOS-C integrator circuit.

References

1. Y. Tsividis, M. Banu, J. Knoury, "Continuous-Time MOSFET-C Filters in VLSI," *IEEE J. Solid-State Circuits*, vol. SC-21, no. 1, pp. 15–30, February 1986.
2. M. Banu, Y. Tsividis, "Fully Integrated Active-RC Filters in MOS Technology," *IEEE J. Solid-State Circuits*, vol. SC-18, no. 6, pp. 644–651, December 1983.
3. M. Banu, Y. Tsividis, "An Elliptic Continuous-Time CMOS Filter with On-Chip Automatic Tuning," *IEEE J. Solid-State Circuits*, vol. SC-20, no. 6, pp. 1114–1121, December 1985.
4. M. Banu, Y. Tsividis, "Detailed Analysis of Nonidealities in MOS Fully Integrated Active-RC Filters Based on Balanced Networks," *IEE Proceedings, vol. 131. Pt. G.*, no. 5, pp. 190–196, October 1984.
5. J. M. Khoury, Y. Tsividis, "Analysis and Compensation of High-Frequency Effects in Integrated MOSFET-C Continuous-Time Filters," *IEEE Trans. Circuits Syst.*, vol. CAS-34, no. 8, pp. 862–875, August 1987.
6. A. Van Bezooijen, N. Ramalho, J. O. Voorman, "Balanced Integrator Filters at Video Frequencies," *Digest ESSCIRC'91*, pp. 1–4, September 1991.
7. Z. Czarnul, "Modification of Banu-Tsivids Continuous-Time Integrator Structure," *IEEE Trans. Circuits Syst.*, vol. CAS-33, no. 7, pp. 714–716, July 1986.
8. Z. Czarnul, "Novel MOS Resistive Circuit for Synthesis of Fully Integrated Continuous-Time Filters," *IEEE Trans. Circuits Syst.*, vol. CAS-33, no. 7, pp. 718–721, July 1986.
9. M. Ismail, S. V. Smith, R. G. Beale, "A New MOSFET-C Universal Filter Structure for VLSI," *IEEE J. Solid-State Circuits*, vol. SC-22, no. 1, pp. 183–194, February 1988.
10. A. Yoshizawa, Y. P. Tsividis, "Anti-Blocker Design Techniques for MOSFET-C Filters for Direct Conversion Receivers," *IEEE J. Solid-State Circuits*, vol. 37, no. 3, pp. 757–760, March 2002.
11. S. P. Singh, J. V. Hansom, J. Vlach, "A New Floating Resistor for CMOS Technology," *IEEE Trans. Circuits Syst.*, vol. 36, no. 9, pp. 1217–1220, September 1989.
12. H. P. Schmid, G. S. Moschytz, "Active-MOSFET-C Single-Amplifier Biquadratic Filters for Video Frequencies," *IEE Proceedings, Circuits, Devices and Systems*, vol. 147, no. 1, pp. 35–41, February 2000.
13. M. W. L. Cunha, S. N. Filho, M. C. Schneider, A. L. Dalcastagne, "Automatic Tuning of MOSFET-C Filters Using Digitally Programmable Current Attenuators," *ISCAS'97*, vol. 1, pp. 329–332, June 1997.

14. J. I. Osa, A. Carlosena, "MOSFET-C Sinusoidal Oscillator with Variable Frequency and Amplitude," *ISCAS'00*, vol. 2, pp. 725–728, May 2000.
15. B. K. Ahuja, "Implementation of Active Distributed RC Anti-Aliasing/Smoothing Filters," *IEEE J. Solid-State Circuits*, vol. SC-17, no. 6, pp. 1076–1080, December 1982.
16. S. Ramet, "A Low-Distortion Anti-Aliasing/Smoothing Filter for Sampled Data Integrated Circuits," *IEEE J. Solid-State Circuits*, vol. SC-23, no. 5, pp. 1267–1270, October 1988.
17. Y. P. Tsividis, "Minimal Transistor-Only Micropower Integrated VHF Active Filter," *El. Letters*, vol. 23, no. 15, pp. 777–778, July 1987.
18. L. J. Pu, Y. P. Tsividis, "Transistor-Only Frequency-Selective Circuits," *IEEE J. Solid-State Circuits*, vol. SC-25, no. 3, pp. 821–832, October 1990.

CHAPTER 5

Introduction to VLSI Gm-C Circuit Techniques

VLSI active-RC and MOS-C circuit techniques suffer from bandwidth limitation due mainly to the bandwidth limitation of the OPAMPs' circuits since their dominant pole frequencies are required to be much lower than the unity gain frequency to ensure circuit stability. The VLSI Gm-C circuit structures can offer enhanced operation bandwidth due in part to the fact that all internal nodes in the circuits have low impedance. In addition, the bias-dependent nature of the Gm circuit effectively provides the tuning necessary for accurate frequency operation of signal processing circuits based on VLSI Gm-C circuit structures. A major drawback of practical VLSI Gm-C circuits is their limited linearity due mainly to large signal swing at the input of the Gm cell to support the open-loop circuit operations. Consequently, the development of various linear Gm structures has been one of the main focuses in the VLSI Gm-C circuit design and developments.

5.1 Basic VLSI Gm Circuit Structures

An ideal transconductor (or Gm) circuit is a perfectly linear voltage-to-current converter that operates at large signals as shown in Fig. 5.1.

Note that a reference arrow at the output is used on all Gm symbols in this text to indicate the reference direction of the output current when the input voltage is selected to be positive. An ideal Gm circuit has the desired properties, including an infinite output impedance, a perfectly linear infinite input impedance, and an infinitely wide bandwidth.

The differential Gm circuits can also be defined as in Fig. 5.2.

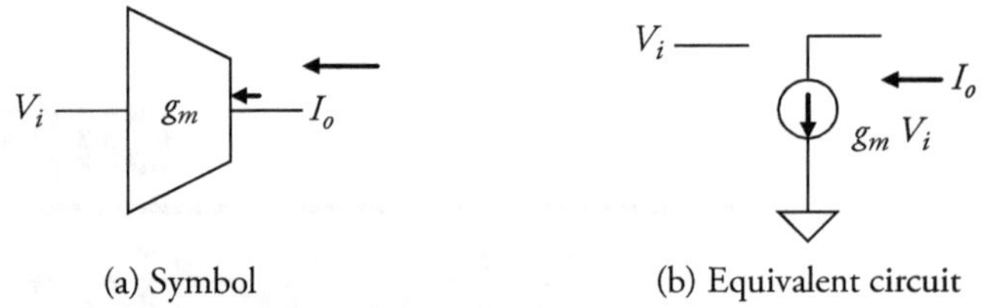

FIGURE 5.1 Basic Gm circuit model.

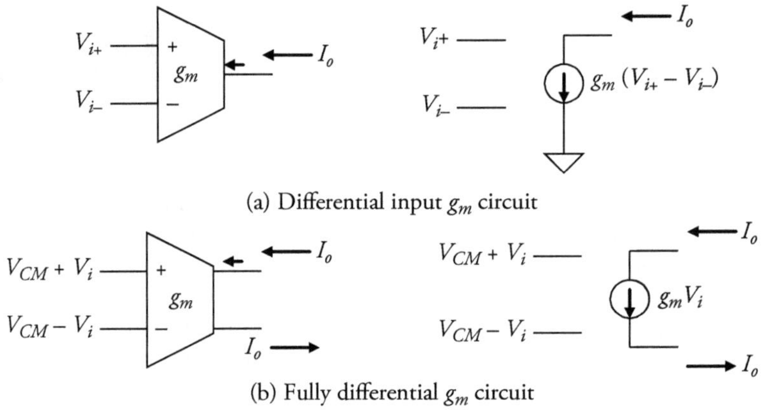

FIGURE 5.2 Differential Gm circuit model.

5.1.1 Basic MOS Gm Circuit Structure

Figure 5.3 shows the VLSI NMOS transistor and its small-signal equivalent circuit where the drain current is given either in the linear or in the saturation operation modes, respectively, as

$$I_o = \begin{cases} \beta\left(V_i - V_T - \frac{V_o}{2}\right)V_o & \text{Linear} \\ \frac{\beta}{2}(V_i - V_T)^2 & \text{Saturation} \end{cases} \tag{5.1}$$

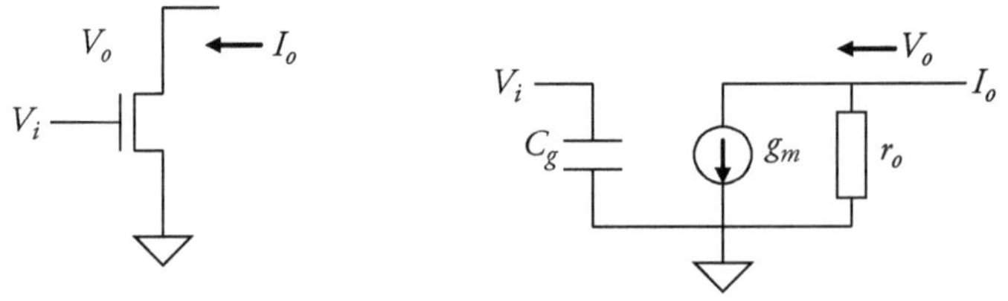

FIGURE 5.3 Basic VLSI MOS Gm circuit and its small-signal equivalent circuit.

The small-signal Gm values of the transistor in the two operation modes can be derived as follows:

$$\Rightarrow g_m = \frac{\partial I_o}{\partial V_i} = \begin{cases} \beta V_o & \text{Linear} \\ \sqrt{2\beta I_o} & \text{Saturation} \end{cases} \tag{5.2}$$

It can be seen that such a device can be used to approximate the ideal Gm circuit.

The basic VLSI Gm circuit structure based on the MOS transistor offers extremely wide bandwidth basically only limited by the f_T of the given manufacture process technology. Such Gm circuit structures have been widely used in various high-frequency applications such as RF circuit applications.

Gm circuits based on the basic VLSI MOS Gm structure suffer from several major performance limitations including poor linearity, low-output impedance, and, to a less degree, the input Miller capacitance effect. For applications that require better Gm performance, advanced circuit techniques are usually used to enhance the performances of the basic MOS Gm circuits. These circuit techniques include (1) the use of cascode circuit structures to improve the output impedance of the Gm circuit and to minimize the input capacitance effects contributed by the Miller effect, (2) the use of differential circuit structures to improve the signal dynamic range and to improve noise performance, (3) the use of negative impedance compensation technique to improve the Gm circuit output impedance, and (4) the use of various linearization circuit techniques to improve the Gm circuit linearity.

5.1.2 Cascode MOS Gm Circuit Structures

Figure 5.4 shows two widely used cascode MOS circuit structures for improving the output impedance of the MOS Gm circuits. One of the additional benefits of the cascode Gm circuit is the reduced Miller capacitance effects of the input devices that help to improve the bandwidth of the circuits.

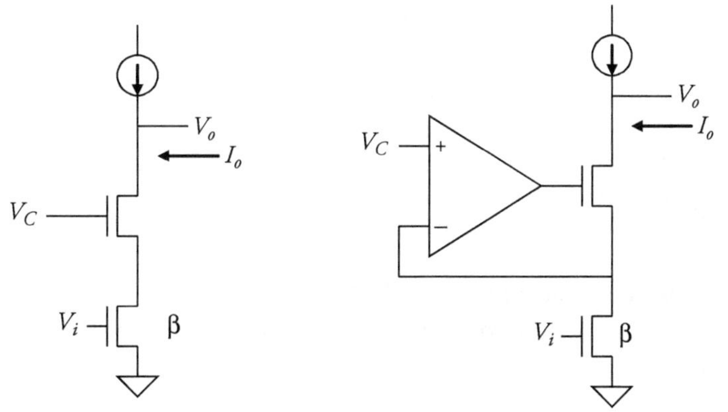

FIGURE 5.4 Cascode MOS Gm circuit structures.

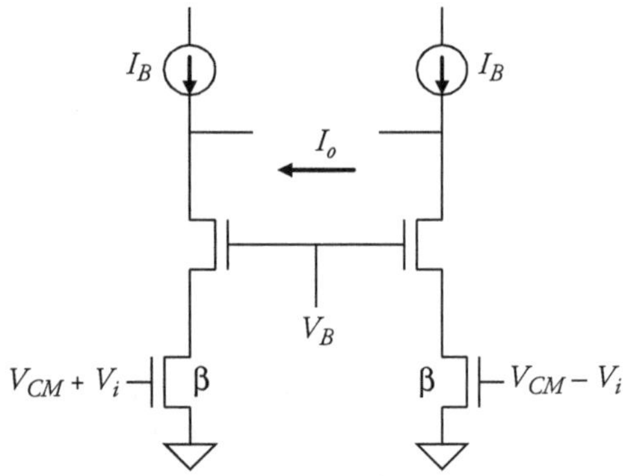

FIGURE 5.5 Fully differential VLSI MOS Gm circuit structure.

5.1.3 VLSI Fully Differential Gm Circuit Structures

Fully differential Gm circuit structures as shown in Fig. 5.5 are widely used in high-performance Gm circuits to achieve better signal-to-noise dynamic range and to reject common-mode circuit noise. This is based on the fact that the differential signaling configurations directly offer 6-dB signal dynamic range improvement for the same power supply voltage and that a significant portion of cross talk noise is common mode in nature.

5.2 Linear-Based MOS Gm Circuit Structures

The linear dependency of the drain current on the gate-to-source voltage for an MOS transistor in the linear mode as shown in Eq. (5.3) suggests the possibility of constructing linear Gm circuits using linear MOS transistors.

$$I = \beta\left(V_g - V_T - \frac{V_d + V_s}{2}\right)(V_d - V_s) \tag{5.3}$$

Linear-based VLSI MOS Gm circuit structures offer several advantages including their better linearity compared with saturation-based VLSI MOS Gm structures and their continuous tuning capability compared with passive-component–based Gm structures. On the other hand, the dependency of the current on the drain voltage in the above equation implies the key limitation of low-output impedance that requires certain circuit design techniques' output impedance enhancement.

5.2.1 Source-Grounded MOS Gm Circuit

For a source-grounded NMOS transistor in linear mode, the Gm value can be expressed as follows:

$$g_m = \frac{\partial I_o}{\partial V_i} = \beta V_o \tag{5.4}$$

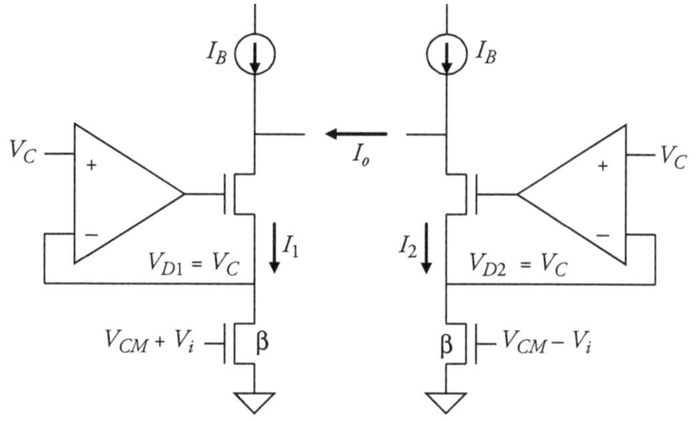

FIGURE 5.6 Differential linear MOS Gm circuit with regulated cascode bias.

Such a Gm circuit structure provides inherently linear behavior as the Gm is independent of the input voltage. However, the low-output impedance resulting from the dependency of the current on the output voltage is a fundamental issue in linear MOS–based Gm circuits. This effect can usually be minimized using a regulated cascode circuit as shown in Fig. 5.6, which significantly improves the output impedance of the Gm.

In this Gm circuit, the drain voltages of the two identical NMOS transistors are forced to a given control voltage V_C using the cascode circuits employing high-gain OPAMPs in negative feedback configuration. The differential Gm can be derived by calculating the difference of the above currents as follows:

$$\begin{cases} I_B + I_o = I_1 = \beta\left(V_i + V_{CM} - V_T - \frac{V_C}{2}\right)V_C \\ I_B - I_o = I_2 = \beta\left(-V_i + V_{CM} - V_T - \frac{V_C}{2}\right)V_C \end{cases} \tag{5.5}$$

$$\Rightarrow g_m \equiv \frac{I_o}{V_i} = \beta V_C \tag{5.6}$$

It can be seen that control voltage V_C can be used for the tuning for this Gm circuit.

5.2.2 Fixed-Bias Source Degeneration Gm Circuit

Figure 5.7 shows a fully differential Gm structure based on linear source degeneration resistor.

From Fig. 5.7, the small-signal voltage and current relation can be expressed as follows:

$$\begin{cases} I_o = g_{mi}(V_i - V_x) \\ I_o = V_x/(R/2) \end{cases} \tag{5.7}$$

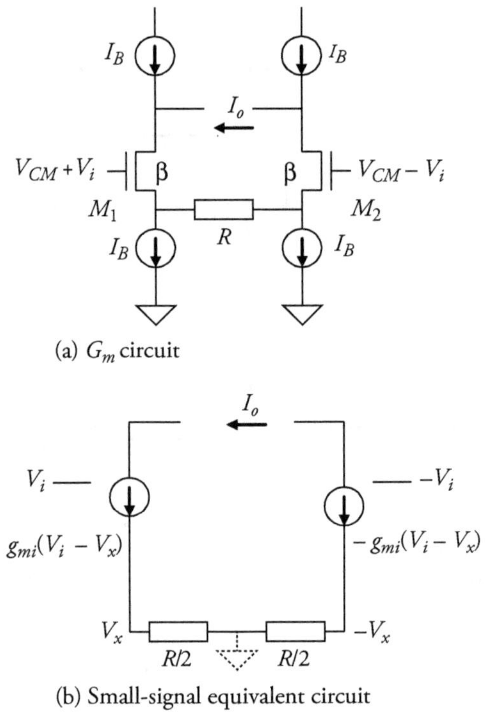

FIGURE 5.7 Differential Gm with linear source degeneration resistor.

The effective Gm of the circuit can be derived as follows:

$$\Rightarrow g_m \equiv \frac{I_o}{V_i} = \frac{g_{\mathrm{mi}}}{1 + g_{\mathrm{mi}}\frac{R}{2}} \tag{5.8}$$

The above Gm can be simplified to $g_m = 2/R$ if the circuit is in deep degeneration ($g_{mi}R >> 1$). However, this Gm circuit structure suffers from limitations in linearity at low degeneration level mainly due to input MOS pair Gm variation across the input voltage range and the poor tuning capability.

An alternative degeneration VLSI Gm circuit, which replaces the fixed resistor with an MOS VCR circuit, is shown in Fig. 5.8. Such a circuit structure offers continuous tuning capability.

By replacing the fixed resistance in Eq. (5.8), the Gm of the new Gm circuit structure can be expressed as follows:

$$\Rightarrow g_m \equiv \frac{I_o}{V_i} = \frac{g_{\mathrm{mi}}}{1 + g_{\mathrm{mi}}\frac{R_{\mathrm{VCR}}}{2}} \tag{5.9}$$

The dependency of the Gm value of the input MOS on the input signal contributes to the nonlinearity of the circuit. Such an effect can be eliminated using the MOS Gm circuit based on the fixed biasing as shown in Fig. 5.9. In this circuit, the current in the

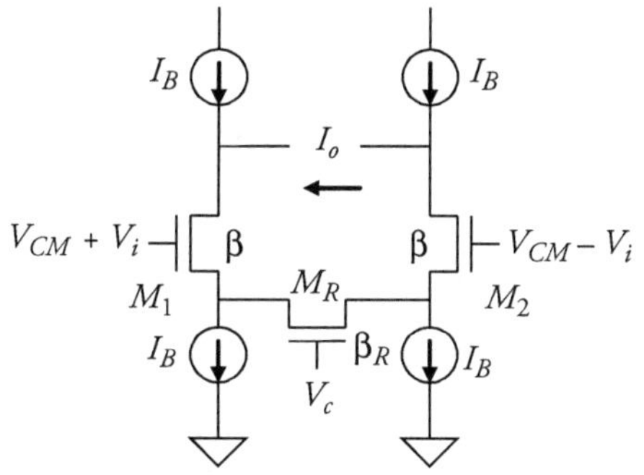

FIGURE 5.8 Simple differential Gm with fixed-bias linear MOS degeneration.

input MOS pair M_1 and M_2 are fixed, resulting in constant gate-to-source voltage drops independent of the input signal voltage. With the constant gate-to-source voltage drops on the input MOS pair, the Gm value of the Gm circuit can simply be expressed as follows:

$$g_m \equiv \frac{I_o}{V_i} = \frac{2}{R_{\text{VCR}}} = 2\beta_R(V_c - V_T) \tag{5.10}$$

An improved Gm circuit is shown in Fig. 5.10, where M_1, M_2, M_3, and M_4 are identical MOS transistors. The two linear MOS VCR circuits formed by M_{R1} and M_{R2} are also purposely matched.

By biasing the two MOS VCR circuits differently, this Gm circuit structure provides improved linearity with the Gm value given as

$$g_m \equiv \frac{I_o}{V_i} = \frac{2}{R_{\text{VCR}}} = 2\beta_R(V_{c1} - V_{c2}) \tag{5.11}$$

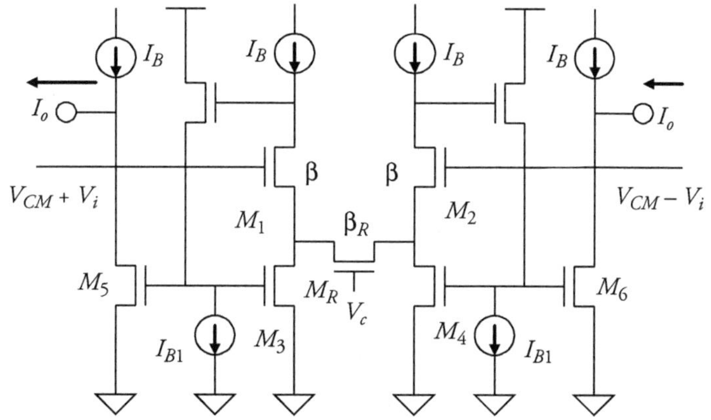

FIGURE 5.9 Improved MOS Gm circuit based on fixed-bias linear MOS degeneration.

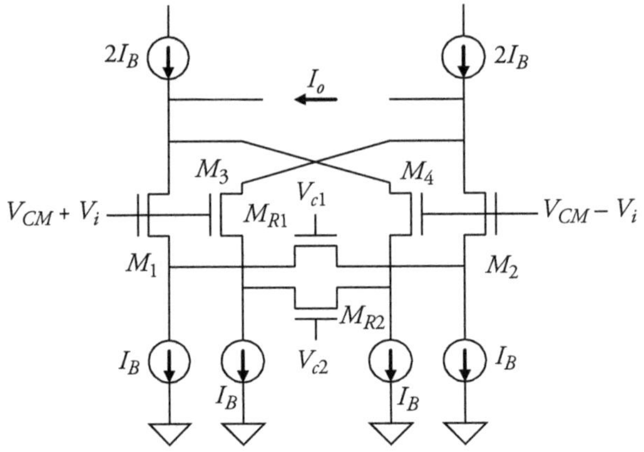

FIGURE 5.10 Improved fixed-bias source degeneration Gm structure.

5.2.3 Varying-Bias Source Degeneration Gm Circuit

An alternative improvement to the linearity of the source degeneration Gm structure is shown in Fig. 5.11, where the biasing of the source degeneration VCR circuit is adaptively controlled based on the input signal voltage levels. The effect of such adaptive biasing scheme is that the variations of the input pair Gm are effectively compensated in the adaptively tuned MOS VCR circuits.

In this Gm circuit, the currents passing through the input pair are given as

$$\begin{cases} I_B + I_o = \dfrac{1}{2}\beta_1(V_{CM} + V_i - V_A - V_T)^2 \\ I_B - I_o = \dfrac{1}{2}\beta_1(V_{CM} - V_i - V_B - V_T)^2 \end{cases} \tag{5.12}$$

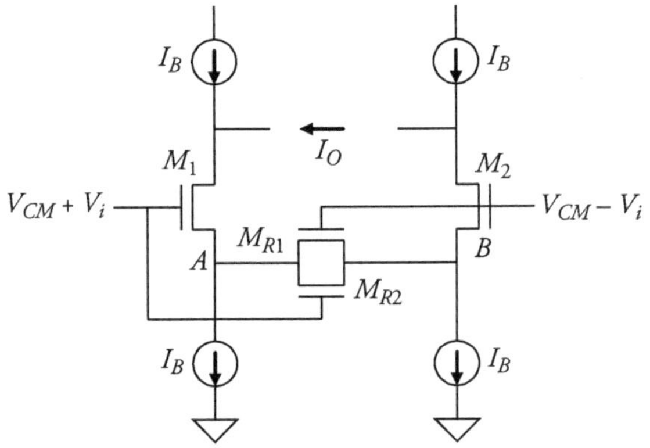

FIGURE 5.11 Varying-bias source degeneration Gm circuit.

The difference of the current flowing through the two input MOS transistors is given as follows:

$$\Rightarrow 2I_o = \frac{1}{2}\beta_1(V_{CM} + V_i - V_A - V_T)^2 - \frac{1}{2}\beta_1(V_{CM} - V_i - V_B - V_T)^2$$
$$= \beta_1\left(V_{CM} - V_T - \frac{(V_A + V_B)}{2}\right)(2V_i - (V_A - V_B)) \tag{5.13}$$

The total current flowing through the two MOS VCRs is given in the following equation:

$$\Rightarrow I_o = \beta_R\left[\begin{matrix}\left(V_{CM} + V_i - V_T - \dfrac{V_A + V_B}{2}\right)\\ +\left(V_{CM} - V_i - V_T - \dfrac{V_A + V_B}{2}\right)\end{matrix}\right](V_A - V_B)$$
$$= 2\beta_R\left(V_{CM} - V_T - \frac{V_A + V_B}{2}\right)(V_A - V_B) \tag{5.14}$$

By combining Eqs. (5.12) and (5.13), we have that

$$4\beta_R(V_A - V_B) = \beta_1[2V_i - (V_A - V_B)] \tag{5.15}$$

$$\Rightarrow (V_A - V_B) = \frac{\beta_1}{\beta_1 + 4\beta_R}2V_i \tag{5.16}$$

On the other hand, the total current flowing through the input MOS pair around the operation point ($V_i = 0$, $V_A = V_B$) is given as

$$2I_B = \frac{1}{2}\beta_1(V_{CM} + V_i - V_A - V_T)^2 + \frac{1}{2}\beta_1(V_{CM} - V_i - V_B - V_T)^2$$
$$\approx 2\frac{1}{2}\beta_1\left(V_{CM} - \frac{V_A + V_B}{2} - V_T\right)^2 \tag{5.17}$$

that is,

$$\Rightarrow \left(V_{CM} - \frac{V_A + V_B}{2} - V_T\right) \approx \sqrt{2I_B/\beta_1} \tag{5.18}$$

Combining Eqs. (5.14) and (5.18), we have the following:

$$\Rightarrow I_o \approx \frac{4\beta_1\beta_R}{\beta_1 + 4\beta_R}\sqrt{2I_B/\beta_1}V_i \tag{5.19}$$

The Gm of the varying-bias linear MOS degenerated Gm structure can finally be expressed as follows:

$$\Rightarrow g_m \equiv \frac{I_o}{V_i} \approx \frac{4\beta_1\beta_R}{\beta_1 + 4\beta_R}\sqrt{2I_B/\beta_1} \tag{5.20}$$

Note that the Gm value is proportional to the square root of the bias current I_B. Such a property can be used for the tuning of the Gm VLSI signal processing applications.

5.3 Saturation-Based MOS Gm Circuit Structures

MOS transistors in saturation offer inherently higher Gm circuit output impedance as compared with their linear MOS–based Gm circuit counterparts. However, an MOS transistor in saturation is inherently nonlinear:

$$I_d = \frac{\beta}{2}(V_{gs} - V_T)^2 \tag{5.21}$$

The above fact suggests that Gm circuit linearity enhancement should be the key design focus for various saturation-based MOS Gm circuit structures.

One of the commonly used methods to construct a linear saturation–based MOS Gm circuit is based on the difference of the currents of two or more saturation devices, as shown in Fig. 5.12, as follows:

$$\begin{aligned} I_o = I_{d1} - I_{d2} &= \frac{\beta}{2}(V_{gs1} - V_T)^2 - \frac{\beta}{2}(V_{gs2} - V_T)^2 \\ &= \frac{\beta}{2}(V_{gs1} + V_{gs2} - 2V_T)(V_{gs1} - V_{gs2}) \end{aligned} \tag{5.22}$$

This family of the Gm circuit structures is relayed on the following design method to linearize the circuit performance:

$$\begin{cases} V_i \propto (V_{gs1} - V_{gs2}) \\ \beta(V_{gs1} + V_{gs2} - 2V_T) = \text{const.} \end{cases} \tag{5.23}$$

5.3.1 Fixed Common-Mode Gm Circuit

A saturation-based MOS Gm structure is shown in Fig. 5.13, where the common-mode voltage V_{CM} of the differential inputs is used for the Gm tuning generation independent of the input signal.

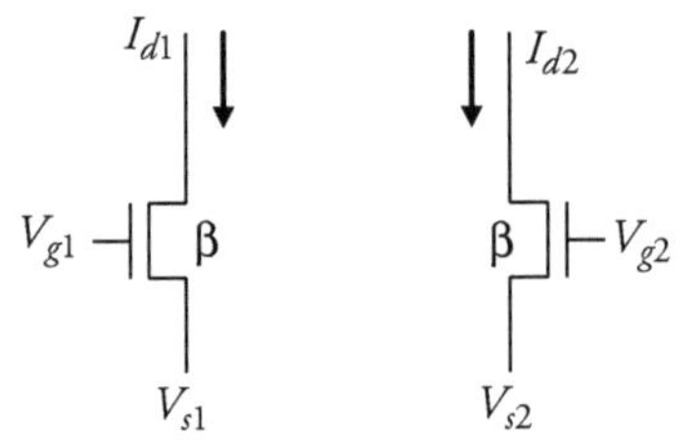

FIGURE 5.12 Differential MOS circuit structure.

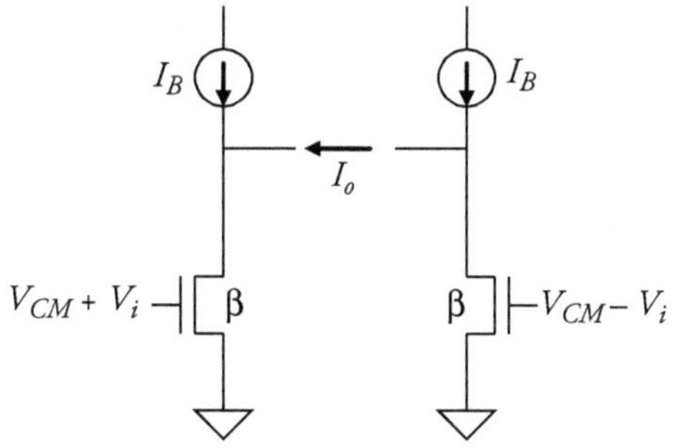

FIGURE 5.13 Fixed common-mode voltage-based Gm circuit.

For this VLSI Gm circuit structure, the drain currents of the MOS devices are given as follows:

$$\begin{cases} I_B + I_o = \dfrac{\beta}{2}(V_{CM} + V_i - V_T)^2 \\ I_B - I_o = \dfrac{\beta}{2}(V_{CM} - V_i - V_T)^2 \end{cases} \tag{5.24}$$

When we calculate the difference of the above currents, we have the following:

$$\Rightarrow I_o = \beta(V_{CM} - V_T)V_i \tag{5.25}$$

Note that this Gm structure is linear with the differential Gm value given as follows:

$$\Rightarrow g_m = \beta(V_{CM} - V_T) \tag{5.26}$$

5.3.2 Floating Source Gm Circuit

A linear saturation–based VLSI MOS Gm circuit can also be implemented based on the floating source voltage Gm circuit shown in Fig. 5.14.

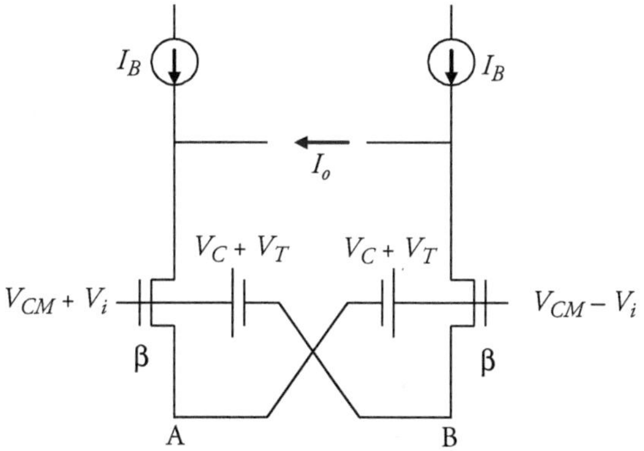

FIGURE 5.14 Floating source–based Gm circuit structure.

With the source voltages of the input MOS pair controlled adaptively with respect to the input signal voltages, the current-voltage relations of the MOS in saturation mode are given as

$$\begin{cases} I_B + I_o = \frac{\beta}{2}(V_{\text{CM}} + V_i - V_A - V_T)^2 = \frac{\beta}{2}(2V_i + V_C)^2 \\ I_B - I_o = \frac{\beta}{2}(V_{\text{CM}} - V_i - V_B - V_T)^2 = \frac{\beta}{2}(-2V_i + V_C)^2 \end{cases} \tag{5.27}$$

The differential output current of the circuit can be derived from the difference of the above currents as follows:

$$\Rightarrow I_o = \frac{\beta}{2} 2V_C 2V_i \tag{5.28}$$

The Gm value of this circuit is given as follows:

$$\Rightarrow g_m \equiv \frac{I_o}{V_i} = 2\beta V_C \tag{5.29}$$

Figure 5.15 shows the VLSI circuit implementation of the floating source saturation MOS Gm circuit architecture.

In this circuit implementation, the two NMOS transistors M_3 and M_4 are selected to be significantly larger ($N >> 1$) than the input MOS transistor pair M_1 and M_2. Under such conditions, the variations in the currents of transistor M_3 and M_4 is minimized, and it can approximately be neglected. The differential input signal voltages and the voltage across the input and the source nodes are approximately given as follows:

$$\begin{cases} \Delta V_1 \equiv (V_{\text{CM}} + V_i) - V_B \approx V_T + \sqrt{\frac{2I_B}{\beta}} \\ \Delta V_2 \equiv (V_{\text{CM}} - V_i) - V_A \approx V_T + \sqrt{\frac{2I_B}{\beta}} \end{cases} \tag{5.30}$$

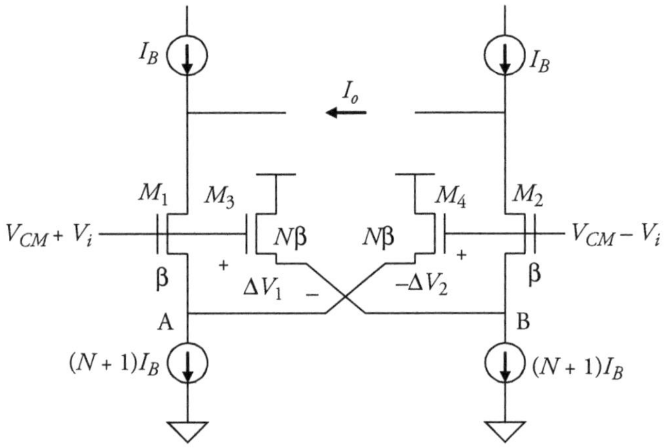

FIGURE 5.15 Implementation of floating source Gm architecture (I).

V_c and G_m values of the Gm structure can be expressed as

$$\Rightarrow \begin{cases} V_C = \sqrt{\dfrac{2I_B}{\beta}} \\ g_m \approx \beta V_C = \sqrt{2\beta I_B} \end{cases} \tag{5.31}$$

The above circuit can also be analyzed based on a more accurate analysis method with a more accurate Gm expression as follows:

$$g_m = \frac{N}{1+N}\sqrt{2\beta I_B} \tag{5.32}$$

5.3.3 Super-CMOS-Based Gm Circuit

One interesting super-CMOS device structure consisting of two saturation PMOS and NMOS devices is shown in Fig. 5.16.

The drain currents of the PMOS and NMOS transistors can be expressed as follows:

$$\begin{cases} I = I_p = \dfrac{\beta_p}{2}(V_A - V_{\text{gp}} + V_{\text{Tp}})^2 \\ I = I_n = \dfrac{\beta_n}{2}(V_{\text{gn}} - V_A - V_{\text{Tn}})^2 \end{cases} \tag{5.33}$$

Eliminating the node voltage V_A from the above equations, we have that

$$\Rightarrow I = \frac{1}{2}\left(\frac{\sqrt{\beta_p}\sqrt{\beta_n}}{\sqrt{\beta_p}+\sqrt{\beta_n}}\right)^2 (V_{\text{gn}} - V_{\text{gp}} - V_{\text{Tn}} + V_{\text{Tp}})^2 \tag{5.34}$$

The above equation can be further simplified by introducing a set of super-MOS device parameters as follows:

$$\begin{cases} \beta^{\text{eq}} \equiv \left[\sqrt{\beta_p}\sqrt{\beta_n}/(\sqrt{\beta_p}+\sqrt{\beta_n})\right]^2 \\ V_T^{\text{eq}} \equiv V_{\text{Tn}} - V_{\text{Tp}} \\ V_{\text{gs}}^{\text{eq}} = V_{\text{gn}} - V_{\text{gp}} \end{cases} \tag{5.35}$$

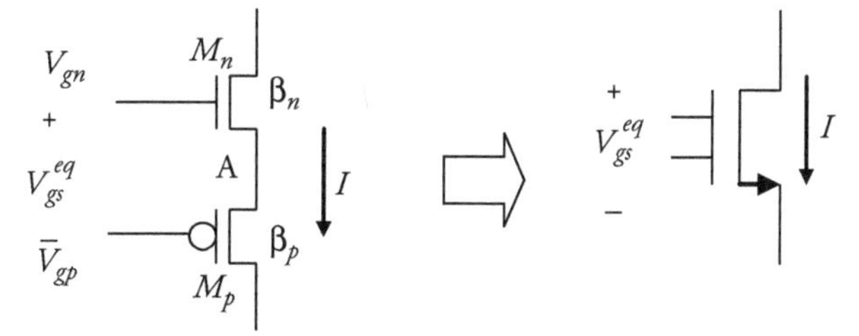

(a) Super MOS structure (b) Symbolic

FIGURE 5.16 Super-CMOS device structure.

The I-V characteristic equation of the super-CMOS device can be expressed in a way similar to a single MOS device in saturation as follows:

$$\Rightarrow I = \frac{\beta^{eq}}{2}(V_{gs}^{eq} - V_T^{eq})^2 \tag{5.36}$$

The super-CMOS device can be used to construct linear Gm circuits. A MOS Gm circuit–based circuit structure is shown in Fig. 5.17. Based on Eq. (5.36), the output current of the circuit can be calculated as follows:

$$\begin{aligned}\Rightarrow I_o = I_2 - I_1 &= \frac{\beta^{eq}}{2}\left(V_{gs2}^{eq} - V_T^{eq}\right)^2 - \frac{\beta^{eq}}{2}\left(V_{gs1}^{eq} - V_T^{eq}\right)^2 \\ &= \frac{\beta^{eq}}{2}\left(V_{gs2}^{eq} + V_{gs1}^{eq} - 2V_T^{eq}\right)\left(V_{gs2}^{eq} - V_{gs1}^{eq}\right) = 2\beta^{eq}\left(V_C - V_T^{eq}\right)V_i\end{aligned} \tag{5.37}$$

It can be seen that the above circuit works as a linear Gm circuit with the Gm value given as

$$\Rightarrow g_m = 2\beta^{eq}\left(V_C - V_T^{eq}\right) \tag{5.38}$$

The super-CMOS device can be used to implement the floating source Gm structure to construct a linear Gm circuit. One circuit implementation is shown in Fig. 5.18.

In this circuit, the drain currents of the two super-CMOS transistors are given as follows:

$$\begin{cases} I_B + I_o = \dfrac{\beta^{eq}}{2}\left(V_{CM} + V_i - V_A - V_T^{eq}\right)^2 = \dfrac{\beta^{eq}}{2}(2V_i + V_C)^2 \\ I_B - I_o = \dfrac{\beta^{eq}}{2}\left(V_{CM} - V_i - V_B - V_T^{eq}\right)^2 = \dfrac{\beta^{eq}}{2}(-2V_i + V_C)^2 \end{cases} \tag{5.39}$$

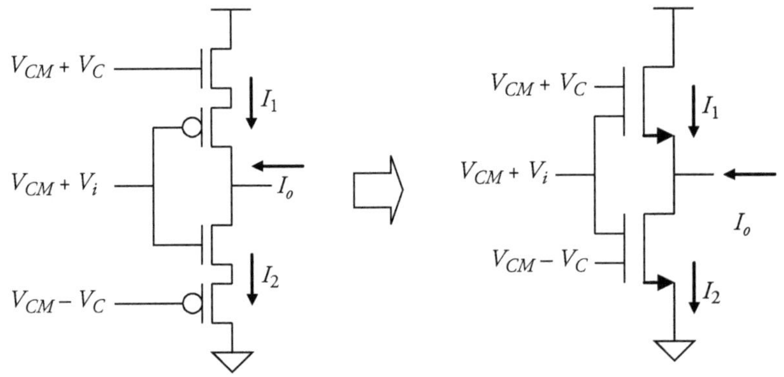

FIGURE 5.17 CMOS pair–based VLSI Gm circuit.

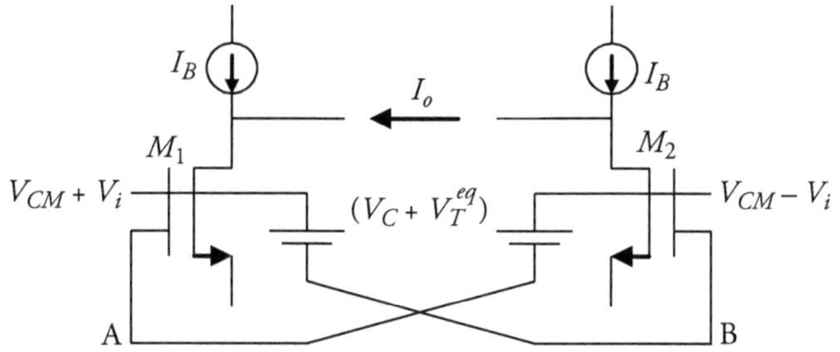

FIGURE 5.18 Super-CMOS device–based floating source Gm circuit.

The output current and the Gm can be calculated by taking the difference of the two currents.

$$\Rightarrow \begin{cases} I_o = 2\beta^{\text{eq}} V_C V_i \\ g_m \equiv \dfrac{I_o}{V_i} = 2\beta^{\text{eq}} V_C \end{cases} \tag{5.40}$$

A detailed VLSI implementation of the above Gm circuit structure is shown in Fig. 5.19, where M_1 and M_3, and M_2 and M_4 form the two input CMOS pair transistors. Considering the floating source control CMOS pair transistors formed by M_5, M_8, and M_6, M_7, we have the following equation:

$$\begin{cases} I_B = \dfrac{\beta^{\text{eq}}}{2}\left(V_{\text{CM}} + V_i - V_B - V_T^{\text{eq}}\right)^2 \\ I_B = \dfrac{\beta^{\text{eq}}}{2}\left(V_{\text{CM}} - V_i - V_A - V_T^{\text{eq}}\right)^2 \end{cases} \tag{5.41}$$

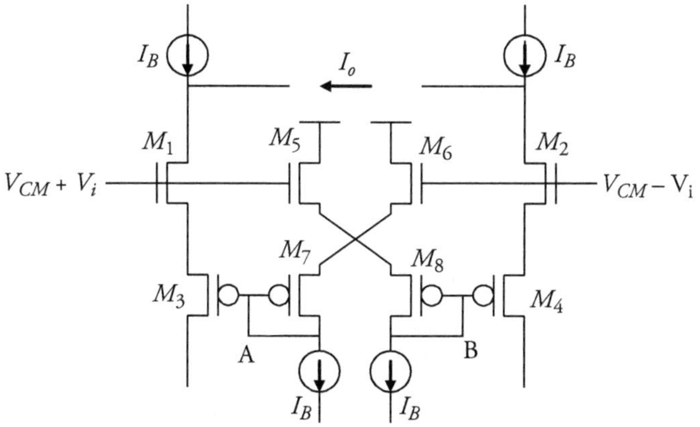

FIGURE 5.19 VLSI floating source Gm circuit–based super-CMOS devices.

The voltages across the transistors are given as follows:

$$\Rightarrow \begin{cases} V_{CM} + V_i - V_B = \sqrt{\dfrac{2I_B}{\beta^{eq}}} + V_T^{eq} \\ V_{CM} - V_i - V_A = \sqrt{\dfrac{2I_B}{\beta^{eq}}} + V_T^{eq} \end{cases} \tag{5.42}$$

Combining the equations above, we have that

$$\Rightarrow \begin{cases} V_C = \sqrt{\dfrac{2I_B}{\beta^{eq}}} \\ g_m \equiv \dfrac{I_o}{V_i} = 2\beta^{eq} V_C = 2\sqrt{2I_B \beta^{eq}} \end{cases} \tag{5.43}$$

5.3.4 Bias-Offset Cross Pairs–Based Gm Circuit

Figure 5.20 shows a saturation-based MOS Gm circuit based on the bias-offset circuit technique.

In this circuit, the secondary input MOS transistor pair M_3 and M_4 is bias offset by using the circuit consisting of M_5, M_6, M_7, and M_8 with respect to the primary input MOS transistor pair M_1 and M_2. To calculate the offset voltage ΔV of the bias in the circuit path formed by M_5 and M_7, we have the following equation:

$$I_1 = \frac{1}{2}\beta_1(\Delta V - V_T)^2 = \frac{1}{2}\beta_1(V_C - V_T)^2 \tag{5.44}$$

that is,

$$\Rightarrow \Delta V = V_C \tag{5.45}$$

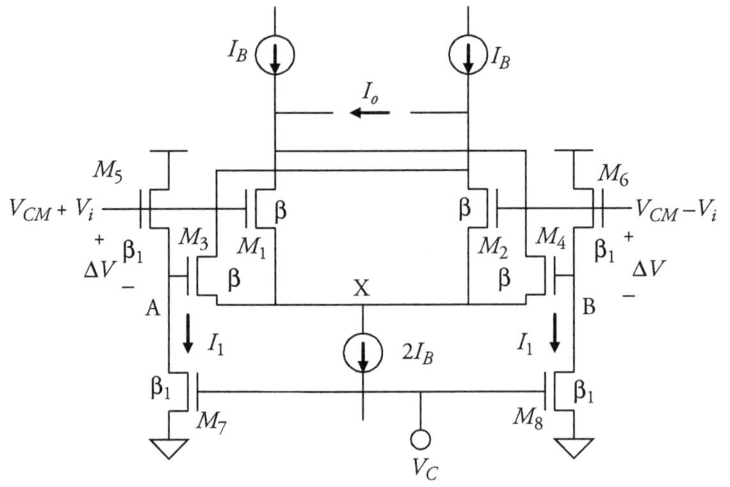

FIGURE 5.20 VLSI bias-offset cross pairs–based Gm circuit.

The total currents of the input MOS transistors are given as

$$\begin{cases} I_B + I_o = \frac{1}{2}\beta(V_{CM} + V_i - V_X - V_T)^2 + \frac{1}{2}\beta(V_{CM} - V_i - \Delta V - V_X - V_T)^2 \\ I_B - I_o = \frac{1}{2}\beta(V_{CM} - V_i - V_X - V_T)^2 + \frac{1}{2}\beta(V_{CM} + V_i - \Delta V - V_X - V_T)^2 \end{cases} \quad (5.46)$$

The output current can be calculated from the difference of the currents in the following equation:

$$\begin{aligned} &\Rightarrow I_o = \frac{1}{2}\beta(V_{CM} - V_X - V_T)(2V_i) + \frac{1}{2}\beta(V_{CM} - \Delta V - V_X - V_T)(-2V_i) \\ &= \beta \Delta V V_i \end{aligned} \quad (5.47)$$

The Gm of the given circuit can be expressed as follows:

$$\Rightarrow g_m \equiv \frac{I_o}{V_i} = \beta V_C \quad (5.48)$$

5.3.5 CMOS Inverter-Based Gm Circuit

As discussed in previous sections, a single MOS transistor–based Gm structure can offer extremely high-bandwidth operation. However, such a Gm circuit suffers from limitations including low-output impedance and poor linearity.

In Fig. 5.21a, a CMOS inverter with shorted input and output is used to generate the common-mode voltage reference of the circuit. Figure 5.21b shows the basic uncompensated inverter–based VLSI Gm circuit identical (or matched) to the common-mode generation circuit.

The drain currents of the MOS devices in the basic Gm circuit can be expressed as follows:

$$\begin{cases} I_n = \frac{\beta_n}{2}(V_i + V_{CM} - V_{TN})^2 \\ I_p = \frac{\beta_p}{2}(V_{CC} - V_i - V_{CM} + V_{TP})^2 \end{cases} \quad (5.49)$$

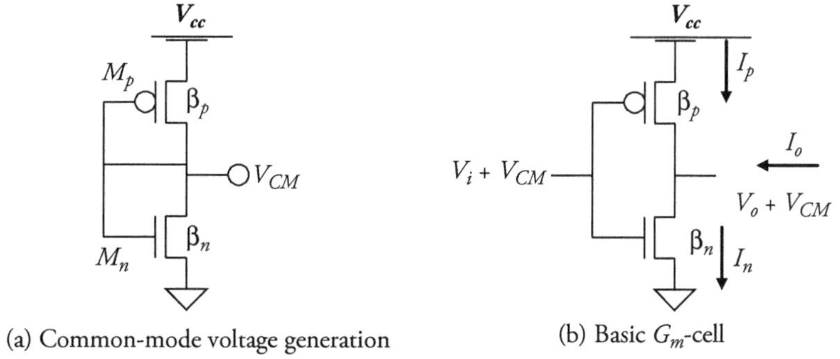

FIGURE 5.21 Basic inverter-based Gm circuit building blocks.

The output current of the CMOS inverter–based Gm structure can be calculated as

$$\begin{aligned}\Rightarrow I_o &= I_n - I_p \\ &= \frac{\beta_n}{2}(V_i + V_{\mathrm{CM}} - V_{\mathrm{TN}})^2 - \frac{\beta_p}{2}(V_{\mathrm{CC}} - V_i - V_{\mathrm{CM}} + V_{\mathrm{TP}})^2 \\ &= \frac{1}{2}\begin{bmatrix}\sqrt{\beta_n}(V_i + V_{\mathrm{CM}} - V_{\mathrm{TN}}) + \\ \sqrt{\beta_p}(V_{\mathrm{CC}} - V_i - V_{\mathrm{CM}} + V_{\mathrm{TP}})\end{bmatrix} \times \begin{bmatrix}\sqrt{\beta_n}(V_i + V_{\mathrm{CM}} - V_{\mathrm{TN}}) - \\ \sqrt{\beta_p}(V_{\mathrm{CC}} - V_i - V_{\mathrm{CM}} + V_{\mathrm{TP}})\end{bmatrix}\end{aligned} \tag{5.50}$$

The common-mode biasing condition can be calculated by letting $V_i = 0$ and $I_o = 0$ as

$$\sqrt{\beta_n}(V_{\mathrm{CM}} - V_{\mathrm{TN}}) - \sqrt{\beta_p}(V_{\mathrm{CC}} - V_{\mathrm{CM}} + V_{\mathrm{TP}}) = 0 \tag{5.51}$$

that is,

$$\Rightarrow V_{\mathrm{CM}} = \frac{\sqrt{\beta_p}(V_{\mathrm{CC}} + V_{\mathrm{TP}}) + \sqrt{\beta_n}V_{\mathrm{TN}}}{\sqrt{\beta_p} + \sqrt{\beta_n}} \tag{5.52}$$

By substituting Eq. (5.52) into Eq. (5.50), we have that

$$I_o = \frac{\left(\sqrt{\beta_n} + \sqrt{\beta_p}\right)}{2}\begin{pmatrix}\left(\sqrt{\beta_n} - \sqrt{\beta_p}\right)(V_i + V_{\mathrm{CM}}) \\ +\left(\sqrt{\beta_p}(V_{\mathrm{CC}} + V_{\mathrm{TP}}) - \sqrt{\beta_n}V_{\mathrm{TN}}\right)\end{pmatrix}V_i \tag{5.53}$$

Note that this uncompensated Gm circuit can provide a linear *I-V* relation if the PMOS and NMOS transistors are selected to be approximately matched.

$$\sqrt{\beta_n} \approx \sqrt{\beta_p} \approx \beta \tag{5.54}$$

Under such a matched condition, the output current and the Gm can be further simplified as follows:

$$\Rightarrow \begin{cases} I_o = \beta(V_{\mathrm{CC}} + V_{\mathrm{TP}} - V_{\mathrm{TN}})V_i \\ g_m = \beta(V_{\mathrm{CC}} + V_{\mathrm{TP}} - V_{\mathrm{TN}}) \end{cases} \tag{5.55}$$

Equation (5.55) implies the low-voltage operation capability of this Gm structure with the minimum supply given as

$$(V_{\mathrm{cc}})_{\mathrm{min}} = V_{\mathrm{TN}} + |V_{\mathrm{TP}}| \tag{5.56}$$

The basic uncompensated CMOS inverter–based Gm circuit can be used to construct a high-bandwidth fully differential VLSI Gm circuit as shown in Fig. 5.22.

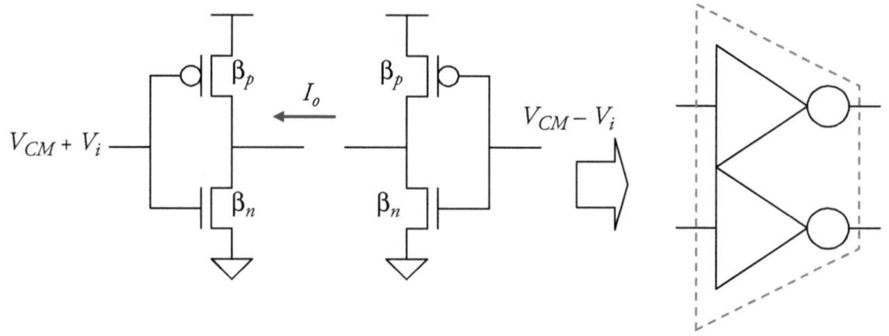

FIGURE 5.22 Basic fully differential inverter–based Gm structure.

The Gm value of above fully differential VLSI Gm structure with matched PMOS and NMOS is given as follows:

$$\Rightarrow g_m = \beta(V_{cc} + V_{TP} - V_{TN}) \tag{5.57}$$

The linear dependency of the Gm value on the V_{cc} suggests that the supply voltage can readily be used for the circuit tuning for such a Gm circuit structure.

Note that this fully differential Gm circuit can offer the similar attractive feature of very wide bandwidth operation and enhanced linearity similar to its basic Gm building circuit components. However, such a circuit also suffers from the limitation of relatively low-output impedance as the single MOS device without cascading. Fortunately, such a low-output impedance limitation can be improved using an output impedance enhancement technique based on impedance cancellation effect using positive feedback circuit effect as shown in Fig. 5.23.

In this VLSI Gm circuit, the g_{mi} and the r_{oi} are the g_m and the output impedance of the basic differential Gm structure, which are defined as follows:

$$\Rightarrow \begin{cases} g_m = \left(\dfrac{\partial I_o}{\partial V_i}\right)_{V_o} \\ r_o = \left(\dfrac{\partial I_o}{\partial V_o}\right)_{V_i}^{-1} \end{cases} \tag{5.58}$$

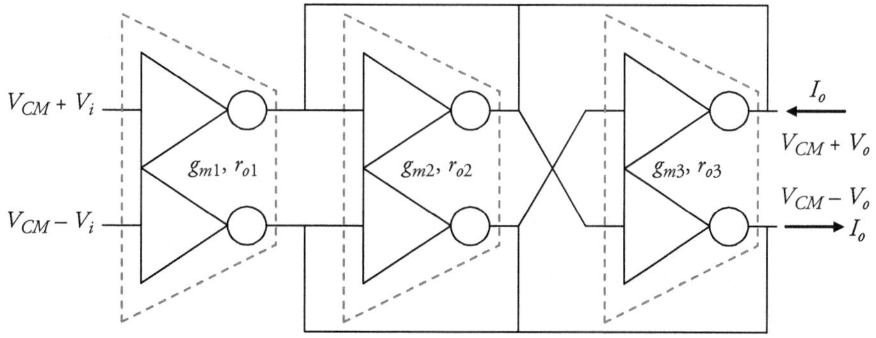

FIGURE 5.23 Output-enhanced inverter–based fully differential Gm circuit.

The differential output current of the output impedance–enhanced Gm circuit is given as

$$\Rightarrow g_{m1}V_i + (g_{m2} - g_{m3})V_o + \left(\frac{1}{r_{o1}} + \frac{1}{r_{o2}} + \frac{1}{r_{o3}}\right)V_o = I_o \tag{5.59}$$

The effective Gm value and output impedance of the improved fully differential Gm circuit can be expressed as

$$\begin{cases} g_m \equiv \left(\dfrac{\partial I_o}{\partial V_i}\right)_{Vo} = g_{m1} \\ r_o \equiv \left(\dfrac{\partial I_o}{\partial V_o}\right)_{Vi}^{-1} = \dfrac{1}{(g_{m2} - g_{m3}) + \left(\dfrac{1}{r_{o1}} + \dfrac{1}{r_{o2}} + \dfrac{1}{r_{o3}}\right)} \end{cases} \tag{5.60}$$

By properly selecting the g_{m2} and g_{m3} such that

$$\Rightarrow (g_{m2} - g_{m3}) + \left(\frac{1}{r_{o1}} + \frac{1}{r_{o2}} + \frac{1}{r_{o3}}\right) = 0 \tag{5.61}$$

the Gm value and the effective output impedance of the improved Gm circuit are given as

$$\Rightarrow \begin{cases} g_m = g_{m1} \\ r_o = \infty \end{cases} \tag{5.62}$$

It can be seen that the effective output impedance of the circuit can significantly be improved.

References

1. A. P. Nedungadi, R. L. Geiger, "High-Frequency Voltage-Controlled Continuous-Time Lowpass Filter Using Linearised CMOS Integrators," *El. Letters*, vol. 22, no. 14, pp. 729–731, July 3, 1986.
2. C. S. Park, R. Schaumann, "Design of a 4-MHz Analog Integrated CMOS Transconductance-C Bandpass Filter," *IEEE J. Solid-State Circuits*, vol. SC-23, no. 4, pp. 897–996, August 1988.
3. H. Khorramabadi, P. R. Gray, "High-Frequency CMOS Continuous-Time Filters," *IEEE J. Solid-State Circuits*, vol. SC-19, no. 6, pp. 939–948, December 1984.
4. J. Pennock, P. Frith, R. G. Barker, "CMOS Triode Transconductor Continuous-Time Filters," *IEEE CICC*, pp. 378–381, May 1986.
5. A. Kaiser, "A Micropower CMOS Continuous-Time Low-Pass Filter," *IEEE J. Solid-State Circuits*, vol. SC-24, no. 3, pp. 736–743, June 1989.
6. V. Gopinathan, Y. P. Tsividis, K.-S. Tan, R. K. Hester, "Design Considerations for High-Frequency Continuous-Time Filters and Implementation of an Antialiasing Filter for Digital Video," *IEEE J. Solid-State Circuits*, vol. SC-25, no. 6, pp. 1368–1378, December 1990.

7. Y.-T. Wang, A. A. Abidi, "CMOS Active Filter Design at Very High Frequencies," *IEEE J. Solid-State Circuits*, vol. SC-25, no. 6, pp. 1562–1574, December 1990.
8. F. Krummenacher, N. Joehl, "A 4-MHz CMOS Continuous-Time Filter with On-Chip Automatic Tuning," *IEEE J. Solid-State Circuits*, vol. SC-23, no. 3, pp. 750–758, June 1988.
9. F. Krummenacher, "Design Considerations in High-Frequency CMOS Gm Amplifier Capacitor (TAC) Filters," *Proc. IEEE ISCAS*, pp. 100–105, May 1989.
10. M. Steyaert, J. Silva-Martinez, W. Sansen, "High Performance OTA-R-C Continuous-Time Filters with Full CMOS Low Distortion Floating Resistors," *Proc. ESSCIRC'91*, pp. 5–8, September 1991.
11. F. Krummenacher, G. Van Ruymbeke, "Integrated Selectivity for Narrow-Band FM IF Systems," *IEEE J. Solid-State Circuits*, vol. SC-25, no. 3, pp. 757–760, June 1990.
12. J.-Y. Lee, C.-C. Tu, W.-H. Chen, "A 3V Linear Input Range Tunable CMOS Transconductor and Its Application to a 3.3V 1.1MHz Chebyshev Low-Pass Gm-C Filter for ADSL," Proceedings of the *IEEE CICC 2000*, pp. 387–390, May 2000.
13. A. Tajalli, M. Atarod, A. Adibi, "A 1.5-V Supply, Video Range Frequency, Gm-C Filter," *IEEE ISCAS'00*, vol. 2, pp. 148–151, May 2000.
14. C. Durdodt, A. Hanke, S. Heinen, U. Langmann, "Comparison of an Inductorless Low-IF and Zero-IF Receiver for Bluetooth," *MWSCAS-2002*, vol. 1, pp. I-563–566, August 2002.
15. E. A. M. Klumperink, B. Nauta, "Systematic Comparison of HF CMOS Transconductors," *IEEE Trans. on Circuits and Systems II: Analog and Digital Signal Processing*, vol. 50, no. 10, pp. 728–741, October 2003.
16. C.-C. Huang, K. A. I. Halonen, M. Ismail, V. Porra, A. Hyogo, "A Low-Voltage, Low-Power CMOS Fifth-Order Elliptic Gm-C Filter for Baseband Mobil, Wireless Communication," *IEEE Trans. on Circuits and Systems for Video Technology*, vol. 7, no. 4, pp. 584–593, August 1997.
17. S. Szczepanski, J. Jakusz, R. Schaumann, "A Linear Fully Balanced CMOS OTA for VHF Filtering Applications," *IEEE Trans. on Circuits and Systems II: Analog and Digital Signal Processing*, vol. 44, no. 3, pp. 174–187, March 1997.
18. P. Bowron, K. A. Mezher, A. A. Muhieddine, "Dynamic-Range Comparison of Second-Order Active-RC and OTA-C Filters," *Proceedings, the 37th Midwest Symposium on Circuits and Systems*, vol. 2, pp. 776–779, August 1994.
19. S. D'Amico, A. Baschirotto, "An Active-Gm-RC Continuous-Time Biquad Low-Pass Cell," *ICECS*, vol. 2, pp. 551–554, December 2003.
20. P. V. Ananda Mohan, "Novel OTA-C Filter Structures Using Grounded Capacitors," *ISCAS'91*, vol. 3, pp. 1347–1350, June 1991.
21. P. Bowron, H. M. Dahir, "Generalization and Evaluation of Second-Order Analogue OTA-C Filters," *IEE Colloquium on Electronic Filters*, pp. 4/1–4/4, June 1989.
22. Z. Ni, D. Gu, "Multiloop OTA-C Filter Realization with Chainlike Feedback Structure," *Proceedings, 1991 International Conference on Circuits and Systems*, vol. 2, pp. 684–687, June 1991.
23. C.-C. Hsu, W.-S. Feng, "Testable Design of Multiple-stage OTA-C Filters," *IEEE Transactions on Instrumentation and Measurement*, vol. 49, no. 5, pp. 929–934, October 2000.
24. A. C. M. de Queiroz, L. P. Caloba, "OTA-C Filters Derived from Unbalanced Lattice Passive Structures," *ISCAS'93*, vol. 4, pp. 2256–2256, May 1993.

25. Z. Shujiang, "Designing OTA-C Filters by SFG Approach," *ISCAS'93*, vol. 4, pp. 2248–2251, May 1993.
26. C.-M. Chang, B. M. Al-Hashimi, "Analytical Synthesis of Current-Mode High-Order OTA-C Filters," *IEEE Trans. on Circuits and Systems I: Fundamental Theory and Applications*, vol. 50, no. 9, pp. 1188–1192, September 2003.
27. F. Galvez-Durand, L. Caloba, A. C. M. de Queiroz, "Stability of High-Pass and Band-Reject Analog Continuous-Time OTA-C Filters," *Proceedings of the 40th Midwest Symposium on Circuits and Systems*, vol. 1, pp. 64–67, August 1997.
28. H. Nevarez-Lozano, J. A. Hill, E. Sanchez-Sinencio, "Frequency Limitations of Continuous-Time OTA-C Filters," *IEEE International Symposium on Circuits and Systems*, vol. 3, pp. 2169–2172, June 1988.
29. P. V. A. Mohan, "Generation of OTA-C Filter Structures from Active RC Filter Structures," *IEEE Trans. on Circuits and Systems*, vol. 37, no. 5, pp. 656–660, May 1990.
30. K.-J. Lee, K.-S. Huang, W.-C. Wang, "Concurrent Test Method for OTA-C Filters," *Electronics Letters*, vol. 33, no. 1, January 1997.
31. A. Wyszynski, R. Schaumann, "Using Multiple-Input Transconductors to Reduce Number of Components in OTA-C Filter Design," *Electronics Letters*, vol. 28, no. 3, January 1992.
32. H. W. Su, Y. Sun, "Performance Comparison of Continuous-Time IFLF and LF Low-Pass All-Pole OTA-C Filters," *IEEE APCCAS'00*, pp. 271–274, December 2000.
33. R. Wilcock, B. M. Al-Hashimi, "Application of Group Delay Equalization in Testing Fully-Balanced OTA-C Filters," *ISCAS'02*, vol. 4, pp. IV-643–IV-646, May 2002.
34. K. A. Mezher, P. Bowron, "Noise-Flow-Graph Analysis of OTA-C Filters," *ISCAS'01*, vol. 1, pp. 695–698, May 2001.
35. R. Lojacono, M. Salerno, F. Sargeni, "A novel Architecture of Digitally Programmable Continuous-time OTA-C Filters," *Proceedings, 37th Midwest Symposium on Circuits and Systems*, vol. 2, pp. 1103–1106, August 1994.
36. J. E. Cousseau, L. P. Caloba, "Frequency Bounds on OTA-C Filters," *ISCAS'93*, vol. 4, pp. 2252–2255, May 1993.
37. Q. Q. Han, "Fully Integrated Continuous Time OTA-C Filter with Leapfrog Structure," Proceedings, *1991 International Conference on Circuits and Systems*, vol. 2, pp. 696–698, June 1991.
38. A. C. M. de Queiroz, L. P. Caloba, "Some Practical Problems in OTA-C Filters Related with Parasitic Capacitances," *ISCAS'90*, vol. 3, pp. 2279–2282, May 1990.
39. Y. Sun, J. K. Fidler, "General Multiple Loop Feedback OTA-C Filter Architectures," *IEE Third One-day Colloquium on Analog Signal Processing*, pp. 3/1–3/6, November 1996.
40. Y. Sun, J. K. Fidler, "Some Design Methods of OTA-C and CCII-RC Filters," *IEE Colloquium on Digital and Analogue Filters and Filtering Systems*, pp. 7/1–7/8, November 1993.
41. A. C. M. de Queiroz, L. P. Caloba, "Reactive Loop Elimination and the Use of Special Approximations in the Design of OTA-C Filters," *Proceedings of the 38th Midwest Symposium on Circuits and Systems*, vol. 2, pp. 1341–1344, August 1995.
42. C. A. Barbargires, "Explicit Design of General High-Order FLF OTA-C Filters," *Electronics Letters*, vol. 35, no. 16, pp. 1289–1290, August 1999.
43. Y. Sun, "Second-Order OTA-C Filters Derived from Nawrocki-Klein Biquad," *Electronics Letters*, vol. 34, no. 15, pp. 1449–1450, July 1998.

44. D. Chen, H. Chen, F. Qi, Q. Gao, S. Qin, "A 5th Order LP Butterworth OTA-C Filter with Automatic Tuning System," *Proceedings, 5th International Conference on ASIC*, vol. 1, pp. 518–521, October 2003.
45. L. Caloba, F. Galvez-Durandde, A. C. M. Queiroz, "OTA-C Filters with Finite Zeros and without Floating Capacitors," *IEEE 39th Midwest Symposium on Circuits and Systems*, vol. 2, pp. 925–928, August 1996.
46. B. Pankiewicz, M. Wojcikowski, S. Szczepanski, Y. Sun, "A CMOS Field Programmable Analog Array and Its Application in Continuous-Time OTA-C Filter Design," *ISCAS'01*, vol. 1, pp. 5–8, May 2001.
47. J. Silva-Martinez, "Design Issues for UHF OTA-C Filter Realizations," *2001 Southwest Symposium on Mixed-Signal Design*, pp. 93–98, 25–27, February 2001.
48. A. B. Bhattacharyya, S. Dey, "Sub-Circuit Analysis for Power Supply Rejection Ratio in Regulated Cascode Operational Gm amplifiers and Filters," *Proceedings, Twelfth International Conference on VLSI Design*, pp. 164–168, January 1999.
49. Y. Sun, K. Fidler, "Fully-Balanced Structures of Continuous-Time MLF OTA-C Filters," *1998 IEEE International Conference on Electronics, Circuits and Systems*, vol. 1, pp. 157–160, September 1998.
50. B. Pankiewicz, M. Wojcikowski, S. Szczepanski, Y. Sun, "A Field Programmable Analog Array for CMOS Continuous-Time OTA-C Filter Applications," *IEEE Journal of Solid-State Circuits*, vol. 37, no. 2, pp. 125–136, February 2002.
51. P. Bowron, K. A. Mezher, A. A. Muhieddine, "Dynamic-Range Comparison of Second-Order Active-RC and OTA-C Filters," *Proceedings of the 37th Midwest Symposium on Circuits and Systems*, vol. 2, pp. 776–779, August 1994.
52. F. Rezzi, V. Pisati, R. Castello, R. Alini, "Novel Linearization Circuit for BiCMOS Transconductors Used in High Frequency OTA-C Filters," *ISCAS'93*, vol. 2, pp. 1132–1135, May 1993.
53. L. P. Caloba, F. Galvez-Durand, A. C. M. de Queiroz, "Parasitic Effects in OTA-C Filters with Finite Zeros," *Proceedings of the 34th Midwest Symposium on Circuits and Systems*, vol. 2, pp. 1049–1052, May 1991.
54. Y. Sun, J. K. Fidler, "Performance Analysis of Multiple Loop Feedback OTA-C Filters," *IEE Colloquium on Digital and Analogue Filters and Filtering Systems*, pp. 9/1–9/7, November 1994.
55. K.-J. Lee, W.-C. Wang, K.-S. Huang, "A Current-Mode Testable Design of Operational Gm Amplifier-Capacitor Filters," *IEEE Trans. on Circuits and Systems II: Analog and Digital Signal Processing*, vol. 46, no. 4, pp. 401–413, April 1999.
56. Y. Sun, J. K. Fidler, "OTA-C Realization of General High-Order Transfer Functions," *Electronics Letters*, vol. 29, no. 12, pp. 1057–1058, June 1993.
57. T. Inoue, F. Ueno, Y. Aramaki, O. Matsumoto, M. Suefuji, "A Design of CMOS OTA's Using Simple Linearizing Techniques and Their Application to High-Frequency Continuous-Time Filters," *IEEE International Symposium on Circuits and Systems*, vol. 3, pp. 1741–1744, June 1991.
58. P. M. Van Peteghem, B. J. Haby, H. M. Fossati, G. L. Rice, "A Very-Linear CMOS Gm Stage for OTA-C Filters," *Proceedings of the IEEE 1989 Custom Integrated Circuits Conference*, pp. 25.1.1/1–25.1.1/4, May 1989.
59. Y. Sun, J. K. Fidler, "High-Order Current-Mode Continuous-Time Multiple Output OTA Capacitor Filters," *IEE 15th SARAGA Colloquium on Digital and Analogue Filters and Filtering Systems*, pp. 8/1–8/6, November 1995.

60. E. Hayahara, M. Enomoto, "Deriving Technique of Equivalent Forms from General Filter Circuits and Its Application," *IEEE Trans. on Circuits and Systems I: Fundamental Theory and Applications,* vol. 46, no. 9, pp. 1037–1041, September 1999.
61. Y. Sun, J. K. Fidler, "Structure Generation and Design of Multiple loop Feedback OTA-Grounded Capacitor Filters," *IEEE Trans. on Circuits and Systems I: Fundamental Theory and Applications,* vol. 44, no. 1, pp. 1–11, January 1997.
62. A. Wyszynski, R. Schaumann, S. Szczepanski, P. Van Halen, "Design of a 2.7-GHz Linear OTA and a 250-MHz Elliptic Filter in Bipolar Transistor-Array Technology," *IEEE Trans. on Circuits and Systems II: Analog and Digital Signal Processing,* vol. 40, no. 1, pp. 19–31, January 1993.
63. J. Ramirez-Angulo, E. Sanchez-Sinencio, "Active Compensation of Operational Gm Amplifier Filters Using Partial Positive Feedback," *IEEE Journal of Solid-State Circuits,* vol. 25, no. 4, pp. 1024–1028, August 1990.
64. E. Sanchez-Sinencio, J. Silva-Martinez, "CMOS Gm Amplifiers, Architectures and Active Filters: A Tutorial," *IEE Proceedings Circuits, Devices and Systems,* vol. 147, no. 1, pp. 3–12, February 2000.
65. Y. Sun, "OTA-C Filter Design Using Inductor Substitution and Bruton Transformation Methods," *Electronics Letters,* vol. 34, no. 22, pp. 2082–2083, October 1998.
66. Y. Sun, J. K. Fidler, "Synthesis and Performance Analysis of Universal Minimum Component Integrator-Based IFLF OTA-grounded Capacitor Filter," *IEE Proceedings Circuits, Devices and Systems,* vol. 143, no. 2, pp. 107–114, April 1996.
67. J. A. De Lima, F. M. Alcaide, "On Designing Linearly Tunable High-Q OTA-C Filters with Low Sensitivity," *ISCAS'04,* vol. 1, pp. 21–24, May 2004.
68. M. Jeong, S. Takagi, Z. Czarnul, N. Fujii, "Design of a Novel Linear 3-Input CMOS OTA and Its Application to Filter Realization," *IEEE Asia Pacific Conference on Circuits and Systems,* pp. 121–124, November 1996.
69. A. Lewinski, J. Silva-Martinez, "OTA linearity Enhancement Technique for High Frequency Applications with IM3 Below -65dB," *Proceedings of the IEEE 2003 Custom Integrated Circuits Conference,* pp. 9–12, September 2003.
70. B. A. Minch, "Synthesis of Multiple-Input Translinear Element Log-Domain Filters," *ISCAS'99,* vol. 2, pp. 697–700, June 1999.
71. S. Solis-Bustos, J. Silva-Martinez, "A 4 Hz Low-Pass Continuous-Time Filter," *1998 IEEE International Conference on Electronics, Circuits and Systems,* vol. 1, pp. 169–172, September 1998.
72. J. E. Cousseau, L. P. Caloba, "Frequency Bounds on OTA-C Filters," *ISCAS'93,* pp. 2252–2255, May 1993.
73. A. C. M. de Queiroz, L. P. Caloba, "OTA-C Filters Derived from Unbalanced Lattice Passive Structures," *ISCAS'93,* pp. 2256–2256C, May 1993.
74. S. Zhao, "Designing OTA-C Filters by SFG Approach," *ISCAS'93,* pp. 2248–2251, May 1993.
75. F. Rezzi, V. Pisati, R. Castello, R. Mini, "Novel Linearization Circuit for BiCMOS Transconductors Used in High Frequency," *ISCAS'93,* pp. 1132–1135, May 1993.
76. L. P. Caloba, A. C. M. de Queiroz, "Constant-Resistance All-Pass OTA-C Filters," *Proceedings of the 33rd Midwest Symposium on Circuits and Systems,* pp. 488–491, August 1990.
77. L. P. Caloba, A. C. M. de Queiroz, E. Sanchez-Sinencio, "Signal-Flow Graph OTA-C Band-Pass and Band-Reject Integrated Filters," *IEEE International Symposium on Circuits and Systems,* vol. 3, pp. 1624–1627, May 1989.

78. F. Baruqui, L. P. Caloba, "Stability and Active Compensation of OTA-C Continuous Time High Frequency Filters," *Proceedings of the 32nd Midwest Symposium on Circuits and Systems,* vol. 1, pp. 169–172, August 1989.
79. Y.-S. Hwang, S.-I. Liu, D. S. Wu, Y.-P. Wu, "Table-Based Linear Transformation Filters Using OTA-C Techniques," *Electronics Letters,* vol. 30, no. 24, pp. 2021–2022, November 1994.
80. B. Pankiewicz, M. Solecki, S. Szczepanski, "A Fully Differential CMOS OTA for Continuous-Time Filter Applications," Proceedings, *ICCSC'02,* pp. 42–45, June 2002.
81. E. Ibaragi, S. Nishioka, A. Hyogo, K. Sekine, "A CMOS OTA Free from Second Order Effects with a High Input Resistance Gm Control Terminal," *ISCAS'01,* vol. 1, pp. 300–303, May 2001.
82. J. Silva-Martinez, J. S. Suner, "Very Low Frequency IC Filters," *Proceedings of the 38th Midwest Symposium on Circuits and Systems,* vol. 2, pp. 1325–1328, August 1995.
83. D. Y. Kim, S. W. Choi, J. C. Ahn, N. Fujii, "The Design and Comparison of Elliptic Filters with an OTA-C Structure," *Proceedings of the 33rd Midwest Symposium on Circuits and Systems,* vol. 1, pp. 484–487, August 1990.
84. J. Ramirez-Angulo, E. Sanchez-Sinencio, "Active Compensation of Operational Gm Amplifier Filters Using Partial Positive Feedback," *Proceedings of the 32nd Midwest Symposium on Circuits and Systems,* vol. 2, pp. 1206–1210, August 1989.
85. C.-C. Hung, K. Halonen, M. Ismail, V. Porra, "Micropower CMOS GM-C Filters for Speech Signal Processing," *ISCAS'97*, vol. 3, pp. 1972–1975, June 1997.
86. C.-C. Hung, K. A. I. Halonen, M. Ismail, V. Porra, A. Hyogo, "A Low-Voltage, Low-Power CMOS Fifth-order Elliptic GM-C Filter for Baseband Mobile, Wireless Communication," *IEEE Trans. on Circuits and Systems for Video Technology,* vol. 7, no. 4, pp. 584–593, August 1997.
87. C.-C. Hung, K. Halonen, V. Porra, M. Ismail, "Low-Voltage, Micropower Weak-Inversion CMOS GM-C Filter," *ICECS'96*, vol. 2, pp. 1178–1181, October 1996.
88. J. Shin, S. Min, S. Kim, J. Choi, S. Lee, H. Park, J. Kim, "3V Baseband Gm-C Filters for Wireless Transceiver Applications," *ISCAS'03,* vol. 1, pp. I-457–I-460, May 2003.
89. T. Itakura, T. Ueno, H. Tanimoto, A. Yasuda, R. Fujimoto, T. Arai, H. Kokatsu, "A 2.7V, 200-kHz, 49-dBm, Stopband-IIP3, Low-Noise, Fully Balanced Gm-C Filter IC," *IEEE Journal of Solid-State Circuits,* vol. 34, no. 8, pp. 1155–1159, August 1999.
90. C.-C. Hung, K. Halonen, V. Porra, M. Ismail, "Low-Voltage CMOS GM-C Filter with Rail-to-Rail Common-Mode Voltage," *IEEE 39th Midwest Symposium on Circuits and Systems*, vol. 2, pp. 921–924, August 1996.
91. C. Yoo, W. Kim, "A ±1.5 V, 4 MHz Low-Pass Gm-C Filter in CMOS," *Proceedings of the ASP-DAC'98,* pp. 341–342, February 1998.
92. J. Mahattanakul, C. Toumazou, "Modular Log-Domain Filters Based upon Linear Gm-C Filter Synthesis," *IEEE Trans. on Circuits and Systems I: Fundamental Theory and Applications,* vol. 46, no. 12, pp. 1421–1430, December 1999.
93. L. Ramezani, "An Adjustable Bandwidth Analog CMOS Gm-C Filter," *ICECS'03,* vol. 2, pp. 420–422, December 2003.
94. S. Mehrmanesh, H. A. Aslanzadeh, M. B. Vahidfar, M. Atarodi, "A 1.8V High Dynamic Range CMOS Gm-c Filter for Portable Video Systems," *The 14th International Conference on Microelectronics*, pp. 38–41, December 2002.
95. A. Tajalli, M. Atarodi, A. Adibi, "A 1.5-V Supply, Video Range Frequency, Gm-C Filter," *ISCAS'00,* vol. 2, pp. 148–151, May 2000.

96. J.-Y. Lee, C.-C. Tu, W.-H. Chen, "A 3 V Linear Input Range Tunable CMOS Transconductor and Its Application to a 3.3 V 1.1 MHz Chebyshev Low-Pass Gm-C Filter for ADSL," *CICC'00*, pp. 387–390, May 2000.
97. C. Yoo, K. Jung, J.-W. Lee, W. Kim, "A 15 MHz, 2.6 mW, Sixth-Order Bandpass Gm-C Filter in CMOS," *Proceedings of the 40th Midwest Symposium on Circuits and Systems*, vol. 1, pp. 316–319, August 1997.
98. S. Mehrmanesh, M. B. Vahidfar, H. A. Aslanzadeh, M. Atarodi, "An Ultra Low-Voltage Gm-C Filter for Video Applications," *ISCAS'03*, vol. 1, pp. I-561–I-564, May 2003.
99. J. Lee, K. Hayatleh, F. J. Lidgey, J. Drew, "Tunable Linear Gm Cell for Gm-C Filter Applications," *ISCAS'02*, vol. 4, pp. IV-253–IV-256, May 2002.
100. A. Hassan, K. Sharaf, H. El-Ghitani, H. F. Ragaie, "A New Low Power Transconductor for Gm-C Filters," *ISCAS'02*, vol. 4, pp. IV-249–IV-252, May 2002.
101. A. Gharbiya, M. Syrzycki, "Highly Linear, Tunable, Pseudo Differential Transconductor Circuit for the Design of Gm-C Filters," *IEEE CCECE 2002*, vol. 1, pp. 521–526, May 2002.
102. O. Naess, Y. Berg, "Tunable Floating-Gate Low-Voltage Transconductor," *ISCAS'02*, vol. 4, pp. IV-663–IV-666, May 2002.
103. A. C. Carusone, D. A. Johns," A 5th Order Gm-C Filter in 0.25um CMOS with Digitally Programmable Poles and Zeroes," *ISCAS'02*, vol. 4, pp. IV-635–IV-638, May 2002.
104. E. Rodriguez-Villegas, A. Yufera, A. Rueda, "A 1.25-V Micropower Gm-C Filter Based on FGMOS Transistors Operating in Weak Inversion," *IEEE Journal of Solid-State Circuits*, vol. 39, no. 1, pp. 100–111, January 2004.
105. Y. Palaskas, Y. Tsividis, "Dynamic Range Optimization of Weakly nonlinear, Fully Balanced, Gm-C Filters with Power Dissipation Constraints," *IEEE Trans. on Circuits and Systems II: Analog and Digital Signal Processing*, vol. 50, no. 10, pp. 714–727, October 2003.
106. F. Munoz, A. Torralba, R. G. Carvajal, J. Ramirez-Angulo, "Two New VHF Tunable CMOS Low-Voltage Linear Transconductors and Their Application to HF GM-C Filter Design," *ISCAS'00*, vol. 5, pp. 173–176, May 2000.
107. A. Tajalli, S. Mojtaba Atarodi, "Design Considerations for a 1.5-V, 10.7-MHz Bandpass Gm-C Filter in a 0.6um Standard CMOS Technology," *ISCAS'03*, vol. 1, pp. I-521–I-524, May 2003.
108. A. Hassan, K. Sharaf, H. El-Ghitani, H. F. Ragai, "The Design and Implementation of a Bandpass GM-C Filter for Bluetooth," *45th Midwest Symposium on Circuits and Systems*, vol. 2, pp. II-629–II-632, August 2002.
109. C. Yoo, S. W. Lee, W. Kim, "A ±1.5-V, 4-MHz CMOS Continuous-time Filter with a Single-Integrator Based Tuning," *IEEE Journal of Solid-State Circuits*, vol. 33, no. 1, pp. 18–27, January 1998.
110. F. Rezzi, I. Bietti, M. Cazzaniga, R. Castello, "A 70-mW Seventh-Order Filter with 7-50 MHz Cutoff Frequency and Programmable Boost and Group Delay Equalization," vol. 32, no. 12, pp. 1987–1999, December 1997.
111. S. D'Amico, A. Baschirotto, "Low-Power Compact Gm-C Filters Structures," *ISCAS'04*, vol. 1, pp. 873–876, May 2004.
112. O. Naess, Y. Berg, "Tunable Ultra-Low Voltage Gm Amplifier and GmC Filter," *ISCAS'00*, vol. 2, pp. 709–712, May 2000.

113. J. van der Tang, D. Kasperkovitz, A. Bretveld, "A 65 mW, 0.4-2.3 GHz Bandpass Filter for Satellite Receivers," *CICC'00*, pp. 383–386, May 2000.
114. G. J. Allbutt, D. G. Nairn, "Tunable Capacitors for Digital CMOS Compatible High Frequency Analog Filters," *ISCAS'96*, vol. 1, pp. 89–92, May 1996.
115. M. Ugajin, A. Yamagishi, J. Kodate, M. Harada, T. Tsukahara, "A 1-V CMOS/SOI Bluetooth RF Transceiver for Compact Mobile Applications," 2003 Symposium on VLSI Circuits 2003, pp. 123–126, June 2003.
116. W. Jendernalik, S. Szczepanski, "A CMOS OTA-C Channel-Select Filter for Mobile Receiver," *Proceedings, ICCSC'02*, pp. 50–53, June 2002.
117. S. Mehrmanesh, M. Atarodi, "A High Dynamic Range CMOS Variable Gain Filter for ADSL," *ISCAS'02*, vol. 4, pp. IV-257–IV-260, May 2002.
118. E. Raisanen-Ruotsalainen, K. Lasanen, M. Sijander, J. Kostamovaara, "A Low-Power 5.1.2 kHz CMOS Gm-C Bandpass Filter with On-chip Center Frequency Tuning," *ISCAS'02*, vol. 4, pp. IV-651–IV-654, May 2002.
119. C. Dualibe, P. Jespers, M. Verleysen, "Embedded Fuzzy Control for Automatic Channel Equalization after Digital Transmissions," *ISCAS'01*, vol. 3, pp. 173–176, May 2001.

CHAPTER 6

Introduction to VLSI Continuous-Time Current-Mode Circuit Techniques

VLSI voltage-mode signal processing circuits—such as active-RC, MOS-C, and Gm-C circuits—offer a variety of integrated signal processing circuit solutions. However, in some VLSI continuous-time analog signal processing application cases, signals are represented by their current values instead of the typically used voltage values. In these applications, VLSI continuous-time current-mode (CTI) circuits provide an alternative approach to the VLSI voltage-mode signal processing operations. Two commonly used types of VLSI CTI circuits covered in this chapter are the MOS current-mirror-based CTI circuits and the second-generation current conveyor (CCII)–based CTI circuits. Current-mirror-based CTI circuits inherently have very few internal nodes and, hence, are less sensitive to the secondary parasitic poles effects. Current-mirror circuits also have potential advantages over classic voltage-mode active-RC, MOS-C, and Gm-C circuit techniques in terms of lower power consumption, lower operating supply voltage, higher bandwidth, smaller layout area, and full compatibility to digital VLSI CMOS process technologies. The CCII-based CTI circuits, on the other hand, offer the flexible analog signal processing capabilities comparable to the traditional OPAMP circuit–based voltage-mode signal processing circuits. Current-mode subcircuits have been widely used in various VLSI analog circuits at subcircuit design levels, such as the current biasing circuits in the VLSI OPAMPs, comparators, transmitter, and other analog circuits. Major drawbacks of the VLSI CTI circuits are limitations in linearity, device mismatch, and finite output impedance.

6.1 Basic MOS CTI Circuit Structures

The basic MOS current-mirror circuit is shown in Fig. 6.1.

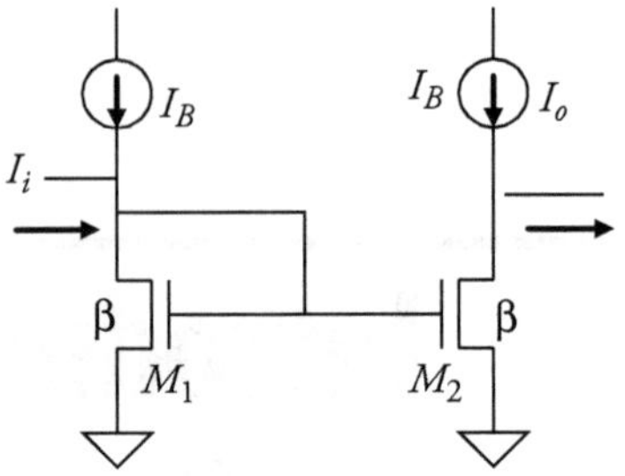

FIGURE 6.1 Basic MOS current-mirror circuit structure.

When the two identical NMOS devices are properly biased at saturation modes, the low-frequency drain currents of the devices ignoring the output resistance effects can be expressed as

$$\begin{cases} I_{D1} = I_B + I_i = \frac{1}{2}\beta(V_g - V_T)^2 \\ I_{D2} = I_B - I_o = \frac{1}{2}\beta(V_g - V_T)^2 \end{cases} \tag{6.1}$$

The input-to-output current transfer function can be derived as

$$\Rightarrow I_o = -I_i \tag{6.2}$$

It can be seen that such a circuit structure offers a current-mirroring function.

6.1.1 Basic CTI Scaling Circuit

The basic MOS current-mirror-based scaling circuit is shown in Fig. 6.2.

When the two matched (with ratio of α) NMOS devices are properly biased at saturation modes, such circuits offer a current-mode scaling operation with the scaling factor determined by the layout ratio factor α as

$$\Rightarrow I_o = -\alpha I_i \tag{6.3}$$

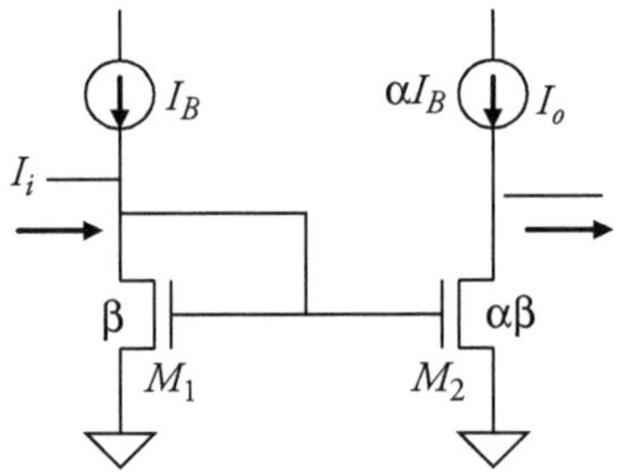

FIGURE 6.2 Basic MOS current-mirror circuit structure.

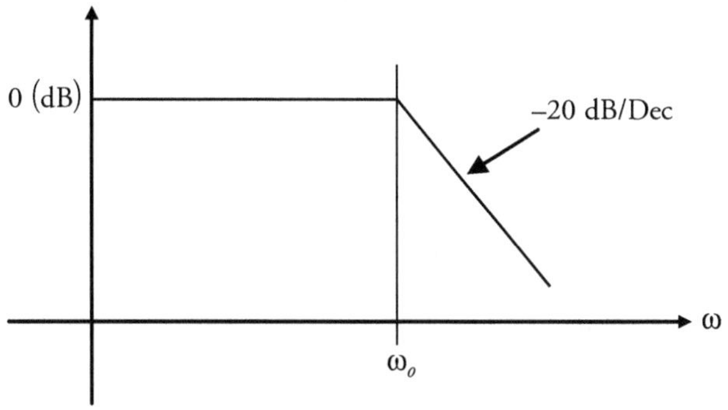

FIGURE 6.3 Typical frequency response of basic MOS current-mirror circuit.

6.1.2 Basic CTI Integrator Circuit

When the parasitic capacitor in a practical current-mirror circuit is included, the basic current-mirror circuit offers approximately −20 dB/Dec attenuation for signal frequency beyond its cutoff frequency (ω_o). Such an effect can be modeled using the signal transfer function shown in Fig. 6.3 as

$$H(s)=\frac{I_o}{I_i}=-\frac{1}{\left(\frac{s}{\omega_o}\right)+1} \tag{6.4}$$

The frequency response can be modeled by including the pole contributed by the input impedance of the mirror and the gate and drain parasitic capacitances presented at the input node as shown in Fig. 6.4.

The s-domain transfer function of the current-mirror circuit can be effectively expressed using the device impedance and the parasitic capacitance in the following equation:

$$H(s)=\frac{I_o}{I_i}=-\frac{1}{\left(\frac{s}{g_m/C}\right)+1} \tag{6.5}$$

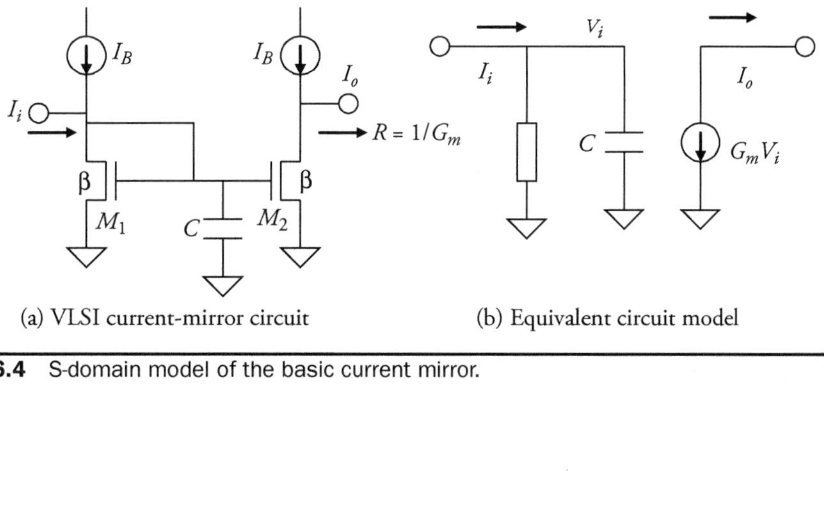

FIGURE 6.4 S-domain model of the basic current mirror.

Note that such a circuit is equivalent to a lossy integrator with the cutoff frequency given by the effective pole of the above transfer function as follows:

$$\Rightarrow \omega_o = \frac{g_m}{C} \tag{6.6}$$

C is the parasitic capacitances at the input node, mainly contributed by the gate capacitance, the drain-diffusion capacitance of the input MOS, and the routing capacitance.

For the deep submicron VLSI process technologies, the pole frequency ω_o can be as high as hundreds of megahertz or even gigahertz, implying the possibility of extremely high-bandwidth operation of the current-mirror-based VLSI CTI circuits. Since such frequency range is usually too high for the typical audio- or video-frequency band signal processing applications, the basic current-mirror-based integrator circuit has to be properly "degenerated" for low-frequency applications. In such cases, an explicit resistor or capacitor can be added into the basic current-mirror circuit to significantly lower the cutoff frequency of the CTI integrator to the desired frequency as shown in Fig. 6.5. Such a method also offers better control of the cutoff frequency of the integrator in the practical design, and it can be used for the frequency tuning.

When the added capacitor and resistor are much higher than the parasitic capacitance and the input impedance of the MOS device, the transfer function of the reduced bandwidth CTI lossy integrator is given as

$$H(s) = \frac{I_o}{I_i} = -\frac{1}{\text{RCs}+1} \tag{6.7}$$

6.1.3 Basic CTI Lossless Integrator Circuit

The CTI lossless integrator can be constructed from the lossy integrator in the same way as shown in Fig. 6.6.

The s-domain transfer function of the constructed lossless integrator is given as follows:

$$H(s) = \frac{I_o}{I_i} = -\frac{1}{\text{RCs}} \tag{6.8}$$

It can be seen that the time constant is simply the RC time constant of the lossy integrator.

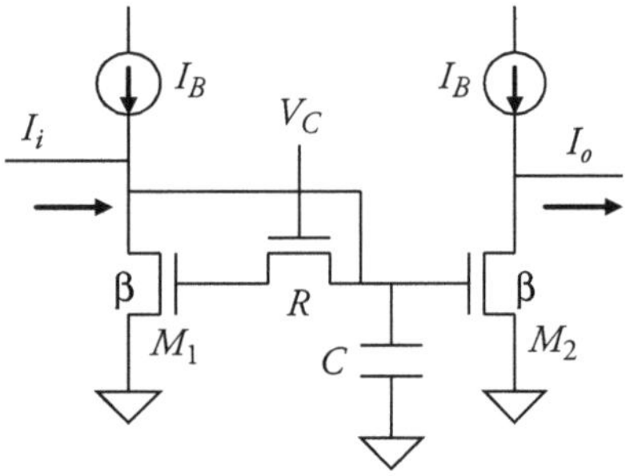

Figure 6.5 Reduced bandwidth CTI lossy integrator.

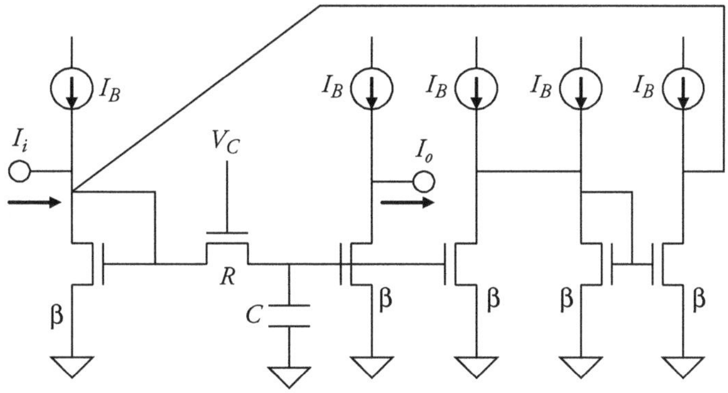

FIGURE 6.6 VLSI current-mirror-based CTI lossless integrator circuit.

6.1.4 Fully Differential CTI Integrator Circuit

One of the major performance limitations of the single-ended current-mode circuits is that the signal paths for the positive and negative inputs are different. The inverting stage may introduce a parasitic nondominant pole, which degrades the quality factor of the integrator and the signal processing circuit response. Intuitively, this suggests that the fully balanced structure will have identical paths for both positive and negative inputs, and hence, the circuit bandwidth can be further extended. A fully differential current-mirror circuit is shown in Fig. 6.7.

For the circuit in Fig. 6.7, we have the following:

$$\begin{cases} I_{o+} = -(I_{i+} + I_{o-})\left(\dfrac{1}{1+s/\omega_o}\right) \\ I_{o-} = -(I_{i-} + I_{o+})\left(\dfrac{1}{1+s/\omega_o}\right) \end{cases} \tag{6.9}$$

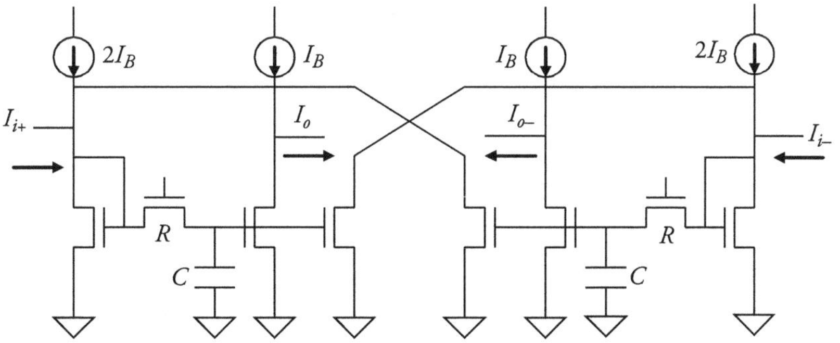

FIGURE 6.7 Current-mirror-based balanced VLSI CTI lossless integrator circuit.

Equation (6.9) can be rewritten as

$$\Rightarrow \begin{pmatrix} 1+\dfrac{s}{\omega_o} & 1 \\ 1 & 1+\dfrac{s}{\omega_o} \end{pmatrix} \begin{pmatrix} I_{o+} \\ I_{o-} \end{pmatrix} = \begin{pmatrix} I_{i-} \\ I_{i+} \end{pmatrix} \tag{6.10}$$

or

$$\Rightarrow \begin{pmatrix} I_x \\ I_y \end{pmatrix} = \frac{\begin{pmatrix} 1+\dfrac{s}{\omega_o} & -1 \\ -1 & 1+\dfrac{s}{\omega_o} \end{pmatrix}}{\left(1+\dfrac{s}{\omega_o}\right)^2 - 1} \begin{pmatrix} I_{i-} \\ I_{i+} \end{pmatrix} \tag{6.11}$$

The above matrix equation can be separated into the differential- and common-mode current signals as

$$\Rightarrow \begin{cases} I_{o+} - I_{o-} = -\dfrac{1}{s/\omega_o}(I_{i+} - I_{i-}) \\ I_{o+} + I_{o-} = \dfrac{1}{s/\omega_o + 2}(I_{i+} + I_{i-}) \end{cases} \tag{6.12}$$

Note that such a circuit can operate in two modes. The circuit offers a lossy integration function to the common-mode signal. However, the circuit offers a lossless CT integration operation for differential-mode current inputs.

6.2 Introduction to Current Conveyor Circuits

The second type of current-mode signal processing circuit is based on the current conveyor circuits. Similar to the OPAMP circuits in the voltage-mode signal processing systems, the current conveyors may serve as robust building blocks for various current-mode signal processing systems.

6.2.1 Second-Generation Conveyor

The second-generation current conveyor (CCII) circuits shown in Fig. 6.8 are defined as three-terminal networks with their input and output equations as

$$\begin{cases} I_Y = 0 \\ V_X = V_Y \\ I_Z = I_X \end{cases} \tag{6.13}$$

Based on the above definition, a CCII circuit is a three-terminal circuit where terminal Y is a high-impedance voltage-input port, terminal X is a low-impedance voltage-follower I/O port, and terminal Z is a high-impedance current-follower (inverter) output port.

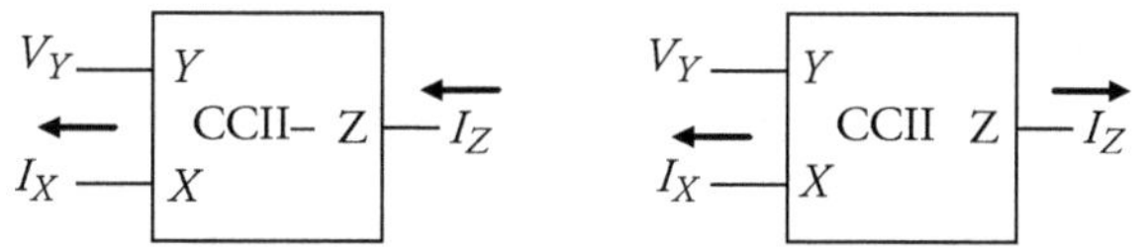

FIGURE 6.8 Positive and negative CCII structures.

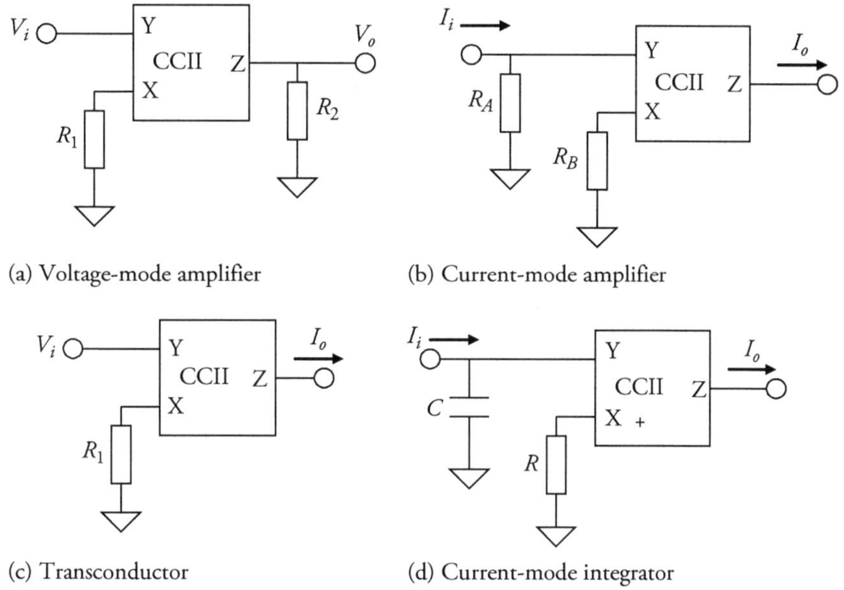

FIGURE 6.9 CCII-based CTI analog signal processing circuits.

The circuit operates as follows: An input voltage at the voltage-input port Y introduces a same voltage at the voltage-follower terminal X. The terminal voltage at X introduces a current if there is a path to ground from X. This current from X will be mirrored to terminal Z. The positive and negative CCII circuits differ from each other on the mirroring current direction of terminal Y to terminal Z as shown in the figures.

6.2.2 CCII-Based Signal Processing Circuit Structures

Using the CCII circuits, various analog signal processing functions shown in Fig. 6.9 can be easily implemented.

6.3 CMOS CCII Circuit Structures

Most historically implemented CCII circuits have been based on bipolar technology. However, as CMOS is the preferred VLSI technology in mixed-signal system, CMOS CCII circuits are very attractive, although CMOS-based CCII implementations suffer from certain limitations caused by the fundamental CMOS characteristics such as the body effect, the threshold-voltage mismatch between NMOS and PMOS, and the lower gate transconductance.

6.3.1 CMOS Single-End CCII Circuits

Fundamentally, the CCII is a combination of two different building blocks—a voltage follower, and a current follower (CCII–) or a current inverter (CCII+). The basic MOS transistor configurations of these functions are the source follower, the common-gate stage, and the current-mirror stage. By combining these functions together, the basic CCII circuit structures can be constructed.

Several VLSI circuit realizations of single-ended CCII circuits are shown in Figs. 6.10, 6.11, and 6.12.

6.3.2 CMOS Fully Differential CCII Circuits

MOS CCII circuits can also be implemented in fully differential configurations. Such circuits offer advantages of high-signal dynamic range, smaller offset, better noise, and power supply rejection. Figure 6.13 shows some basic and improved differential VLSI CCII circuit structures.

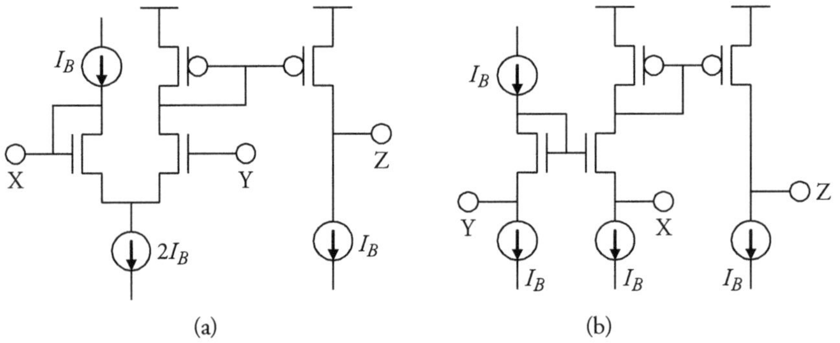

FIGURE 6.10 CMOS CCII+ (current inverter) circuits structures (I).

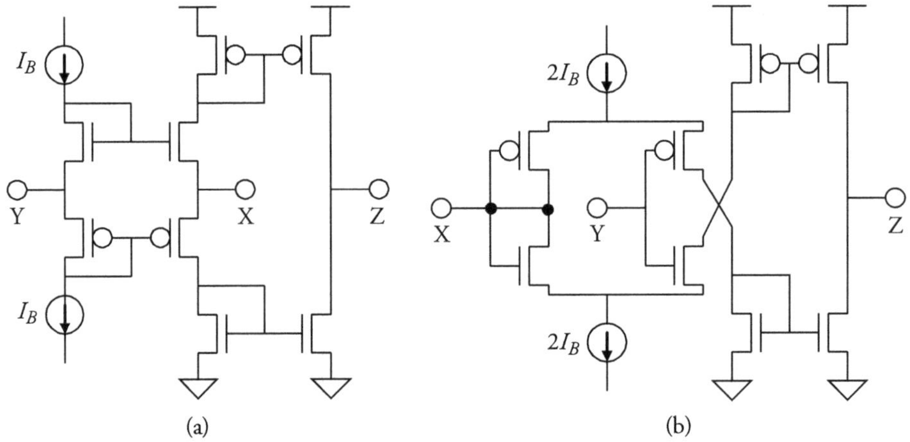

FIGURE 6.11 CMOS CCII+ (current inverter) circuits structures (II).

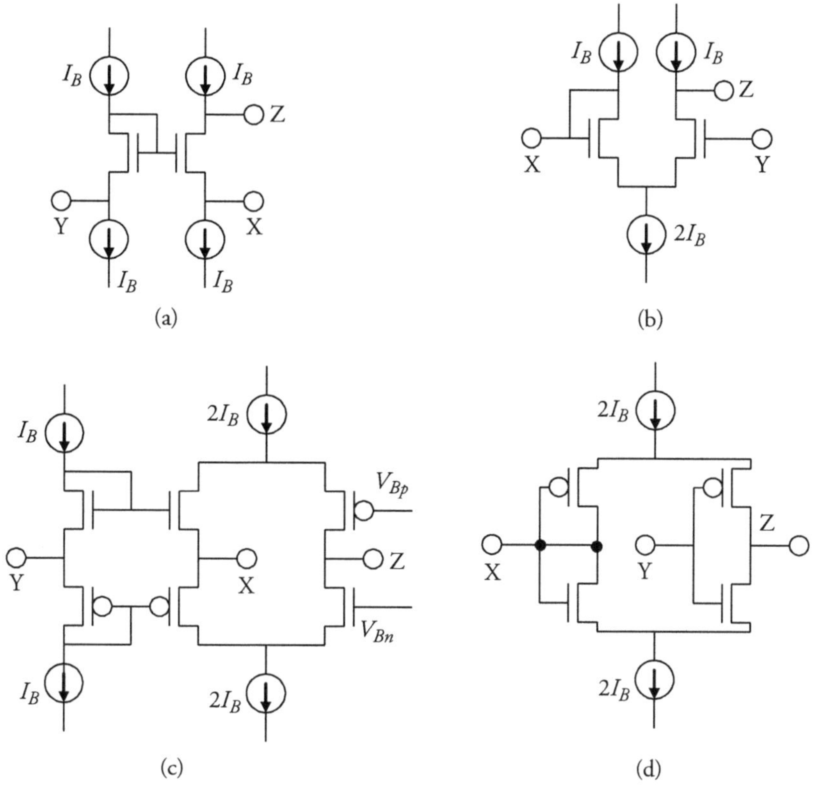

FIGURE 6.12 CMOS CCII- (current-follower) circuit structures.

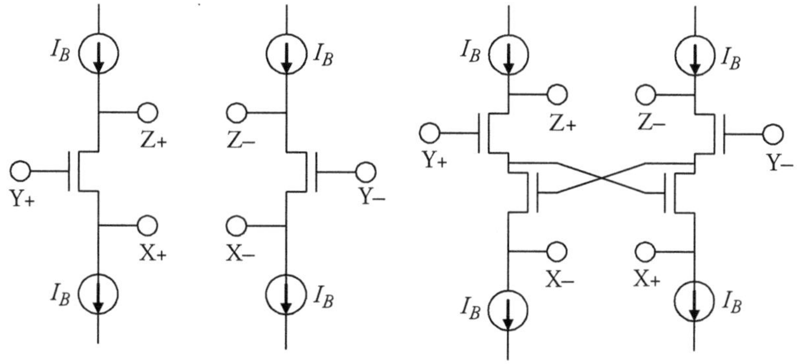

(a) Basic differential CMOS CCII (b) Improved circuit

FIGURE 6.13 CMOS differential CCII circuit structures.

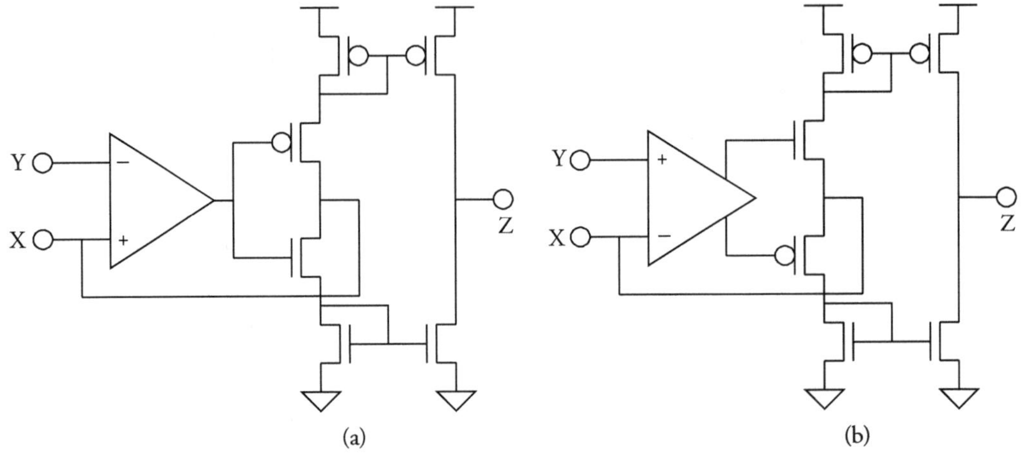

Figure 6.14 VLSI CCII circuit structure based on OPAMP circuits with feedback.

6.3.3 Feedback-Based VLSI CCII Circuits

Feedback can typically be introduced to improve CCII performance of unity-gain amplifier with respect to impedance levels and gain precision. Many options exist for utilizing feedback to improve the CCII performance, including the use of local feedback to obtain a low *X*-input resistance and the global feedback to improve the overall performance of the voltage and/or current buffers.

Figure 6.14 shows CCII circuit structures, which are based on the feedback structures utilizing voltage-mode OPAMP circuits.

Major drawbacks of practical VLSI CCII circuit using feedback OPAMP circuits are the limited operation bandwidth, potential stability problem, and reduced slew rate.

References

1. S.-S. Lee, R. H. Zele, D. J. Allstot, G. Liang, "CMOS Continuous-Time Current-Mode Filters for High-Frequency Applications," *IEEE J. Solid-State Circuits*, vol. 28, no. 3, pp. 323–329, June 1993.
2. V. C. Vincence, C. G. Montoror, M. C. Schneider, "A High-Swing MOS Cascode Bias Circuit for Operation at any Current Level," *IEEE ISCAS'00*, May 28, 2000.
3. C. Thamviriyakul, A. Worapishet, C. Tourmazou, "A Low-Voltage Linear CMOS First-Order Two-Pair for Integrated CMA Filters," *IEEE ISCAS'97*, June 1997.
4. T. Dostal, K. Vrba, I. Lattenberg, "Conveyor-Based Cauer Filters," *APCCAS'02*, vol. 2, pp. 415–418, October 2002.
5. J. Tingleff, C. Toumazou, "Integrated Current Mode Wave Active Filters Based on Lossy Integrators [CMOS chip]," *IEEE Trans. on Circuits and Systems I: Fundamental Theory and Applications*, vol. 42, no. 5, pp. 237–244, May 1995.
6. H. F. A. Hamed, "Low Voltage Low Power Highly Linear CCIIs and Its Applications," *Proceedings of the 15th International Conference on Microelectronics*, pp. 59–62, December 2003.
7. H. F. A. Hamed, "A low Voltage Digitally Programmable Current-Mode Filter," *Proceedings of the 15th International Conference on Microelectronics*, pp. 413–416, December 2003.

CHAPTER 7

VLSI Discrete-Time Signal Processing System Fundamentals

Electrical signals, in general, can be classified into the digital and analog signal families. The analog signals, in turn, can be further classified into the continuous-time (CT) signals and discrete-time (DT), or sampled-data, signals. CT signals are continuous functions both in their signal values (voltage, current, etc.) and in their time references. The DT signals are continuous only in their values while these signals are defined discretely in the time references. The DT circuits occupy the intermediate positions between analog and digital signal processing circuits. DT signals are usually generated from the CT forms through the sampling operation. Due to their discrete-time nature, DT circuits are usually characterized by their z-domain transfer functions based on the z-transform. The z-domain transfer functions are in fact closely related to the s-domain transfer functions of the CT signal processing circuits. Such a relationship allows the approximation of a given DT signal processing circuit using a corresponding CT signal processing circuit through various s-to-z transformation techniques. Similar to the CT circuits, signal flow graph (SFG) and block diagram representations are useful methods for the specifying and implementing of the VLSI DT processing circuits.

7.1 Discrete-Time Signals and Sampling Process

Analog signals can be represented in the CT signal and the DT (or sampled-data) signal forms. Both signal forms are widely used in the VLSI signal processing circuits for various signal processing applications.

7.1.1 Discrete-Time Signals

Unlike the CT signals discussed in previous chapters where the signal is defined continuously in both the values (voltage, current, etc.) and time references as shown in Fig. 7.1a, the DT signals are defined under DT references as shown in Fig. 7.1b.

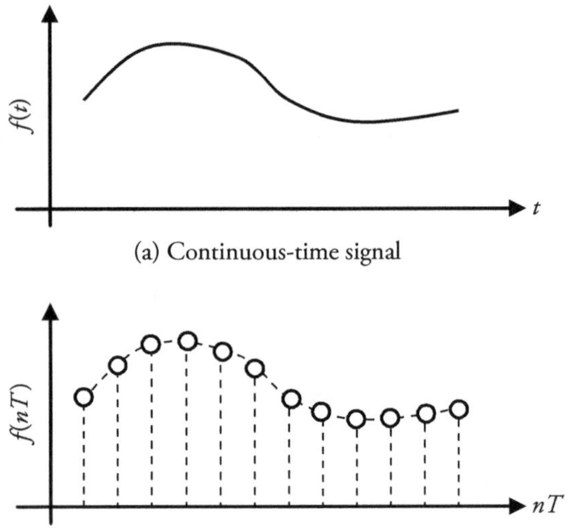

(a) Continuous-time signal

(b) Discrete-time signal

FIGURE 7.1 The CT and DT signals.

Some important aspects of the VLSI DT-signal representations are as follows:

- Mathematically, a DT signal only exists at the specific time points; values on other time points are not defined (or do not exist).
- Electrically, the values of a VLSI DT signal do exist continuously in the time-domain, on both the defined or undefined time points.
- However, only these signal values under definition are used to represent the DT signal. These undefined values, on the other hand, are purposely ignored in the VLSI DT signal processing operations.
- These timing resources outside the DT-signal definition are usually used for specific VLSI DT circuit operations, such as switching operation, circuit settling, and autozeroing.
- Different from CT signal processing circuits where noise at all-time points can contribute to the noise content of the signal, the DT signal processing circuits offer an attractive feature for time-domain noise-filtering capability since the noise at the undefined time points can be effectively "filtered" out from the signal.

7.1.2 VLSI Discrete-Time Signals

The finite processing speed of the practical VLSI signal processing circuits, supported by specific VLSI technology and circuit structure, result in certain constraints for practical VLSI DT signals. Such constraints can be described using the setup time and the hold time of the sampled circuits. These setup-time and hold-time constraints require the input signal of a VLSI sampled-data circuit to be valid for a certain time duration before the effective sampling-time points (setup-time requirement), and it continues to be valid for a certain time duration after the effective sampling points (hold-time requirement) for proper circuit operation as shown in Figs. 7.2 and 7.3.

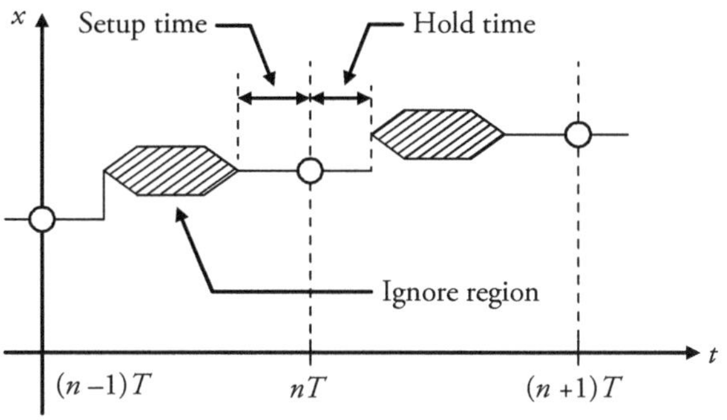

FIGURE 7.2 VLSI sampled-data signal.

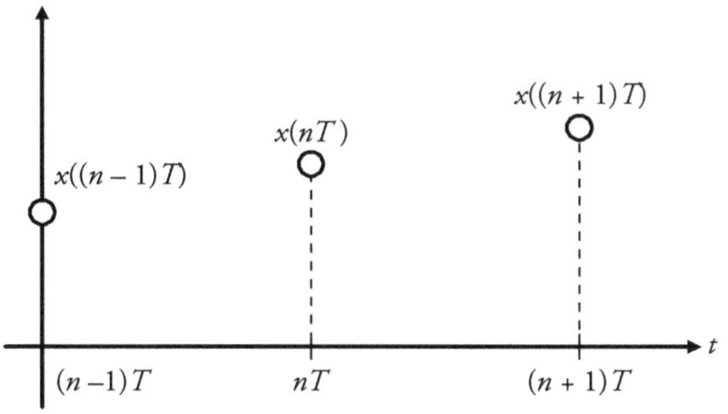

FIGURE 7.3 Mathematical sampled-data signal.

7.1.3 Sampling of Continuous-Time Signals

DT signals can be generated from the corresponding CT signals through the sampling operation. Sampling of the signal can usually be viewed as a data-compression process, which is used to approximately represent all values of the signal inside a sampling interval using a single value. However, the validity of such a data-compression process relies on the fact that the signal inside each sampling period is relatively unchanged. In other words, the sampling frequency is much higher than the signal frequency. Mathematically, it can be derived so that the sampling frequency should be at least twice of the frequency bandwidth of the CT signal to ensure that the compressed information is properly recovered.

The above sampling frequency requirement is usually stated as the Nyquist theorem in many signal processing textbooks. An ideal sampling process shown in Fig. 7.4 can be mathematically modeled using a sampling function consisting of a train of unit impulse functions (or delta functions) as

$$f^*(t) = f(t)\cdot P_\delta(t) \equiv f(t)\cdot \sum_{n=-\infty}^{\infty} \delta(t-nT) = \sum_{n=-\infty}^{\infty} f(nT)\cdot\delta(t-nT) \tag{7.1}$$

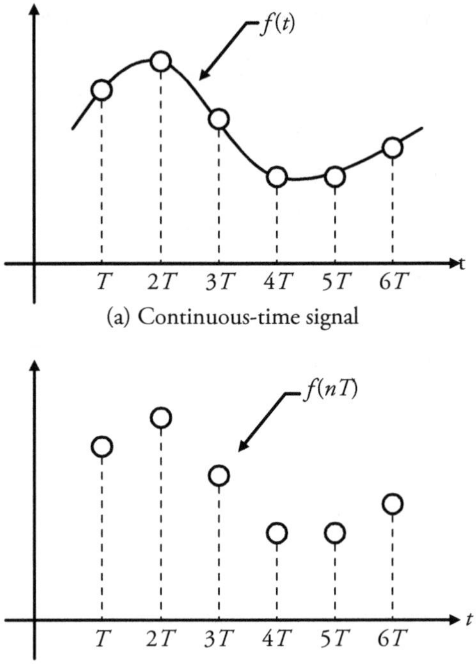

Figure 7.4 Sampling operation of continuous-time signal.

where

$$f(nT) = f(t)\,|_{t=nT} \tag{7.2}$$

and

$$\begin{cases} \delta(t) = 0 & t \neq 0 \\ \delta(t) = \infty & t = 0 \\ \int_{-\infty}^{\infty} \delta(t)dt = 1 \end{cases} \tag{7.3}$$

One of the simplest VLSI sampling circuit implementations is shown in Fig. 7.5. Such a CMOS circuit offers the sample/hold (S/H) circuit operations described as follows:

- When the sampling control clock ϕ is high, the output tracks the input signal.
- When the sampling control clock ϕ is low, the output holds the sampled input-signal value.

The sampled-data signal generated from such a sample/hold circuit is shown in Fig. 7.6.

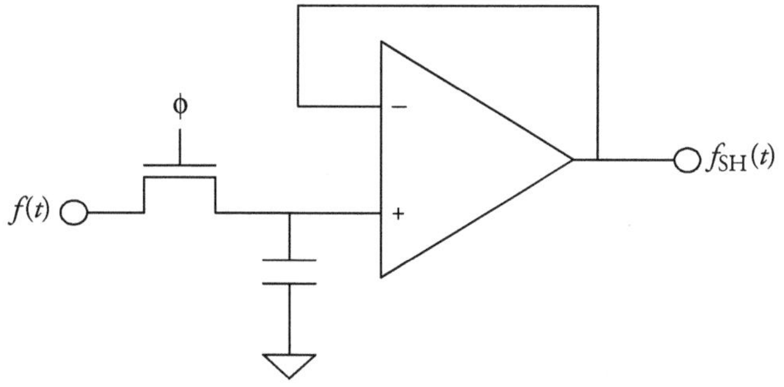

Figure 7.5 VLSI sample/hold circuit structure.

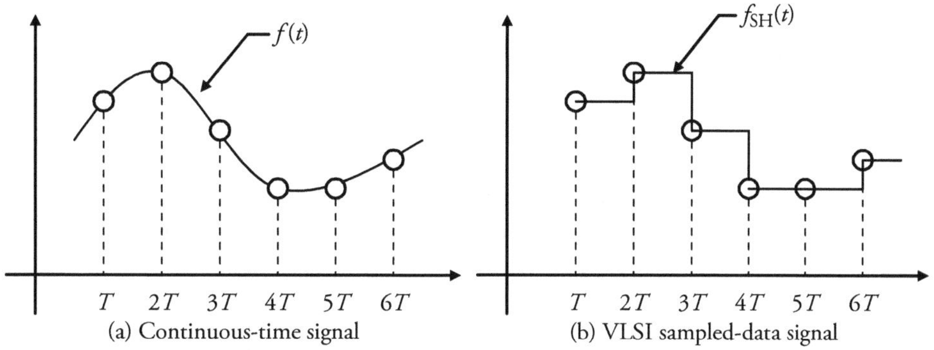

Figure 7.6 Sampling of continuous-time signal based on VLSI sample/hold circuit.

The advantage of sampled-data signal generation based on a VLSI sample/hold circuit is that it offers the maximized stable period of the sampled signal required by the VLSI sampled-data signal processing operation.

Mathematically, a sample/hold operation can be expressed as

$$f_{S/H}(t) = \sum_{n=0}^{\infty} f(nT)[u(t-nT) - u(t-nT-T)] \tag{7.4}$$

where

$$u(t) \equiv \begin{cases} 0 & t<0 \\ 1 & t \geq 0 \end{cases} \tag{7.5}$$

7.2 Discrete-Time System Responses

Figure 7.7 shows an example DT circuit block diagram based on the linear model of second-order DT sigma-delta modulator circuit. Such a circuit is constructed based on three basic building elements, including the unit delay element (delay by one clock period), the adder, and the scaler.

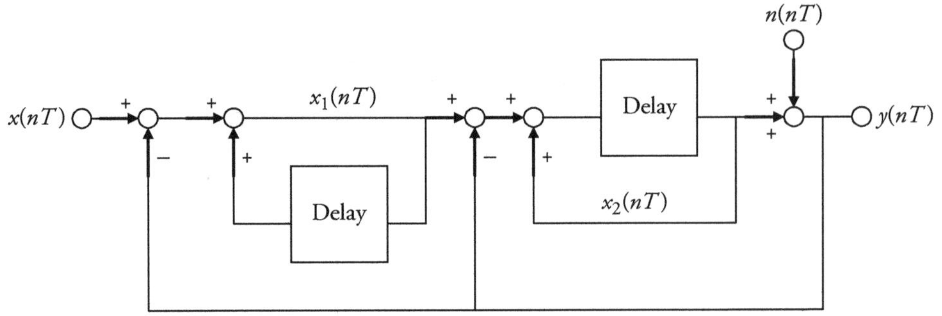

FIGURE 7.7 Second-order discrete-time sigma-delta modulator circuit.

To solve for the responses of such DT circuits for a DT signal $x(nT)$ and noise $n(nT)$ inputs, we first have to construct the time-domain equations based on the introduced system's state variable x_1, x_2, and properties of the basic building elements as

$$\begin{cases} x_1(nT) = x(nT) - y(nT) + x_1(nT - T) \\ x_2(nT) = x_1(nT - T) - y(nT - T) + x_2(nT - T) \\ y(nT) = n(nT) + x_2(nT) \end{cases} \tag{7.6}$$

Note that the general solution of such DT circuits to a general DT signal input usually is not trivial. However, such a solution can be highly simplified by making the constraint to the input signal form as follows:

$$\begin{cases} x(nT) = Xe^{jn\omega T} \\ n(nT) = Ne^{jn\omega T} \end{cases} \tag{7.7}$$

The steady-state system responses of the system for such an input form can easily be derived as

$$\Rightarrow \begin{cases} X_1 e^{jn\omega T} = Xe^{jn\omega T} - Ye^{jn\omega T} + X_1 e^{jn\omega T} e^{-j\omega T} \\ X_2 e^{jn\omega T} = (X_1 e^{jn\omega T} - Ye^{jn\omega T} + X_2 e^{j\omega nT}) e^{-j\omega T} \\ Ye^{jn\omega T} = Ne^{jn\omega T} + X_2 e^{jn\omega T} \end{cases} \tag{7.8}$$

Solving for Y with respect to input X and N from Eq. (7.8), we have the following:

$$\Rightarrow y(nT) = Ye^{jn\omega T} = [Xe^{-j\omega T} + (1 - e^{-j\omega T})^2 N] e^{jn\omega T} \tag{7.9}$$

From Eq. (7.9), we can get the input- to output-signal transfer function of the given circuit as

$$\Rightarrow \frac{Y}{X} = e^{-j\omega T} \tag{7.10}$$

Similarly, the noise-transfer function is given as follows:

$$\Rightarrow \frac{Y}{N} = (1 - e^{-j\omega T})^2 = \left(2\sin(\frac{\omega T}{2})\right)^2 e^{-j\omega T} \qquad (7.11)$$

Note that such a circuit provides a high-pass noise-shaping characteristic.

7.2.1 Z-Transform

The above linear sampled-data equation solution technique implies a generic DT circuit-response solution that is based on the linear combination of the special DT sinusoidal signals. This concept leads to the z-transform–based DT system analysis method that is very similar to the Laplace transform–based method for the CT circuit introduced in the earlier chapters.

In fact, by applying the Laplace transform to the sampled signal shown in Eq. (7.6), we have the following:

$$\Rightarrow F^*(s) = \int_0^{\infty}\left(\sum_{n=-\infty}^{\infty} f(nT)\cdot\delta(t-nT)\right)e^{-st}\,dt = \sum_{n=-\infty}^{\infty} f(nT)e^{-snT} \qquad (7.12)$$

It should be noted that the frequency-domain variable for the sampled signals always come in the form of *exp*(sT). It is convenient to introduce a frequency-domain variable z and the corresponding function in z as follows:

$$\begin{cases} z \equiv e^{sT} \\ F(z) \equiv F^*(s)\big|_{z=e^{sT}} \end{cases} \qquad (7.13)$$

By using parameter z, we may now introduce the (two-side) z-transform of a DT signal $f(nT)$ as

$$\Rightarrow \begin{cases} Z(f(nT)) \equiv F(z) = \sum_{n=-\infty}^{\infty} f(nT)z^{-n} \\ f(nt) \equiv Z^{-1}(F(z)) \end{cases} \qquad (7.14)$$

$F(z)$ is known as the z-transform of the time-domain DT signal $f(nT)$. $f(nT)$ can be calculated by the inverse z-transform of $F(z)$.

For many applications, we may limit the DT signals to the positive (one-side) time points by multiplying the DT signal with the step function $u(t)$, which forces the DT signal to zero value for all negative time points:

$$f(nT) \rightarrow f(nT)\cdot u(t) \qquad (7.15)$$

In such cases, the z-transform can be simplified to the one-side (or simply called z-transform) form as follows:

$$\Rightarrow \begin{cases} F(z) = \sum_{n=0}^{\infty} f(nT)z^{-n} \\ f(nt) \equiv Z^{-1}(F(z)) \end{cases} \qquad (7.16)$$

Note that in applications where the DT signals do not have zero values for the negative time points, we may force them to zero by including the initial condition at $t = 0$.

7.2.2 Solution to DT Circuit Based on Z-Transform

We may now solve for the steady-state response of the DT sigma-delta modulator circuit given in Fig. 7.7 with the help of the z-transform. By applying the z-transform to Eq. (7.6), we have the following equations:

$$\begin{cases} X_1(z) = X(z) - Y(z) + z^{-1}X_1(z) + x_1(nT)\,|_{n=0} \\ X_2(z) = (X_1(z) - Y(z) + X_2(z))z^{-1} + (x_1(nT) - y(nT) + x_2(nT))\,|_{n=0} \\ Y(z) = N(z) + X_2(z) \end{cases} \tag{7.17}$$

The output $Y(z)$ as the function of inputs can be solved by eliminating X_1 and X_2 as

$$\Rightarrow Y(z) = z^{-1}X(z) + (1 - z^{-1})^2 N(z) + Y_0(z) \tag{7.18}$$

where

$$Y_0(z) \equiv x_1(nT)\,|_{n=0} + (1 - z^{-1})(x_1(nT) - y(nT) + x_2(nT))\,|_{n=0} \tag{7.19}$$

which represents the initial states (or the history) of the system. It approaches zero after some clock periods that can be ignored in studying the system's steady-state responses. The system response for a general input $x(nT)$ and $n(nT)$ can now be computed by taking inverse z-transform of the z-domain transfer function. The steady-state responses (or transfer functions) to the input signal x and noise n can now be computed directly from Eqs. (7.17), (7.18), and (7.19) as follows:

$$\Rightarrow \begin{cases} \dfrac{Y(z)}{X(z)}\Big|_{N=0} = z^{-1} \\ \dfrac{Y(z)}{N(z)}\Big|_{X=0} = (1 - z^{-1})^2 \end{cases} \tag{7.20}$$

Based on the above equations, we can see that the system's steady-state response to the sinusoidal signal of frequency ω can be computed directly by using the system's z-domain transfer function and letting

$$z = e^{j\omega T} \tag{7.21}$$

The inherent relationship between Laplace transform and z-transform can further be seen in the following discussions.

7.2.3 Expression of VLSI Sampling Operation

By taking the Laplace transform for the practical sample/hold circuit, we have the following:

$$\begin{aligned} \Rightarrow F_{S/H}(s) &= \sum_{n=0}^{\infty} f(nT)L[u(t - nT) - u(t - nT - T)] \\ &= \left(\frac{1 - e^{-sT}}{s}\right)\sum_{n=0}^{\infty} f(nT)e^{-sTn} = \left(\frac{1 - e^{-sT}}{s}\right)F^*(s) \end{aligned} \tag{7.22}$$

It consists of two terms with the first term representing the nonideal sampling operation and the second term representing the ideal sampling function.

At high oversampling ratio (OSR >> 1)

$$sT << 1 \tag{7.23}$$

we have the following equation:

$$\Rightarrow \frac{1-e^{-sT}}{s} \approx \frac{1-(1-sT)}{s} = T \tag{7.24}$$

and that

$$\Rightarrow F_{S/H}(s)|_{sT\to 0} = T \cdot F^*(s) \tag{7.25}$$

that is, the S/H function approaches the ideal sampling when the sampling frequency is much higher than the signal frequency.

7.3 Z-Domain Transfer Function and System Responses

With the introduction of the z-transform of the sampled-data signals, we are now ready to introduce the z-domain transfer function of a linear sampled-data system.

7.3.1 Z-Domain Transfer Function

For a general DT signal processing system, the input and the output are usually related through a linear-difference equation as

$$\sum_{k=0}^{n} b_k y(nT-kT) = \sum_{k=0}^{m} a_k x(nT-kT) \tag{7.26}$$

By applying the z-transform to both sides and ignoring the initial condition of the system, we have that

$$\Rightarrow \sum_{k=0}^{n} b_k z^{-k} Y(z) = \sum_{k=0}^{m} a_k z^{-k} X(z) \tag{7.27}$$

or

$$\Rightarrow H(z) \equiv \frac{Y(z)}{X(z)} = \frac{\sum_{k=0}^{m} a_k z^{-k}}{\sum_{k=0}^{n} b_k z^{-k}} \tag{7.28}$$

The above equation provides the general z-domain transfer function of the given system. Furthermore, from the last section, we can also obtain the system's response to the sinusoidal signal of frequency ω as follows:

$$H(\omega) = H(z)|_{z=e^{j\omega T}} \tag{7.29}$$

The transfer function in Eq. (7.28) can also be expressed in the complex zero-pole form as follows:

$$H(z)=\frac{\sum_{k=0}^{m} a_k z^{-k}}{\sum_{k=0}^{n} b_k z^{-k}}=\frac{a_m \prod_{k=1}^{m}\left(z^{-1}-\frac{1}{z_k}\right)}{b_n \prod_{k=1}^{n}\left(z^{-1}-\frac{1}{p_k}\right)} \tag{7.30}$$

It can also be expressed in the form of cascaded first- and second-order sections with all real coefficients as follows:

$$H(z)=\frac{\sum_{k=0}^{m} a_k z^{-k}}{\sum_{k=0}^{n} b_k z^{-k}}=\prod_k \left(\frac{c_{0k}+c_{1k}z^{-1}+c_{2k}z^{-2}}{1-2r\cos\theta_k z+r^2 z^{-2}}\right)\prod_k \left(\frac{d_{0k}+d_{1k}z^{-1}}{1-p_k z^{-1}}\right) \tag{7.31}$$

Note that the transfer function of a DT system is not unique and may have many other forms. We are going to see later that each form represents a system implementation.

7.3.2 Gain and Phase Responses of the DT Systems

The gain and phase responses of the DT circuit are given as

$$\begin{cases} \text{Gain} = |H(z)|_{z=e^{j\omega T}} \\ \theta = \angle H(z)\,|_{z=e^{j\omega T}} \end{cases} \tag{7.32}$$

Similar to the s-domain system, Bode plots can be generated from Eq. (7.32).

7.3.3 Time-Domain Response and Stability

From Eq. (7.30), we have that

$$Y(z)=\frac{a_m \prod_{k=1}^{m}\left(z^{-1}-\frac{1}{z_k}\right)}{b_n \prod_{k=1}^{n}\left(z^{-1}-\frac{1}{p_k}\right)}X(z)=\sum_{k=1}^{n}\frac{A_k}{(1-p_k z^{-1})}+F(z) \tag{7.33}$$

where A_k is constant, and the $F(z)$ term is the z-domain function contributed by the input signal. Taking the inverse z-transform of Eq. (7.33), we have the following equations:

$$\Rightarrow y(mT)=Z^{-1}(y(z))=\sum_{k=1}^{n} A_k p_k^m + z^{-1}(F(z)) \tag{7.34}$$

In order for the output to be finite for any finite input signal, it requires that

$$|p_k| \le 1 \tag{7.35}$$

that is, for a DT system to be stable, the pole of the z-domain transfer function needs to be located inside the unity circle.

7.4 Z-Domain Block Diagram and Signal Flow Graph

As discussed in the previous sections, the transfer function of a linear system provides the behavioral representation of that system, that is, the system's response to a given input signal of various frequencies. However, a major drawback of this type of representation is that it does not provide the topology or the implementation information of the system; therefore, it cannot be directly used in system implementation.

For the linear system design, we are mainly interested in the implementation of the system based on the system's transfer function or other specifications and VLSI circuit building components. Theoretically, there are various ways to select the basic building elements for the VLSI system's implementation. One useful method is to select the unit delay, the adder, and the scaler (or gain element) as the basic circuit-delay elements for the DT system implementation. In principle, many linear DT systems or subsystems can be implemented by this method. The block diagram and SFG representations of the three basic building elements are shown in Figs. 7.8 to 7.10.

- The unit delay element
- The adder
- The scaler

7.4.1 Construction of Z-Domain SFG

The generic transfer function of a linear DT system can be expressed as follows:

$$H(z) \equiv \frac{Y}{X} = \frac{b_n z^{-n} + b_{n-1} z^{-(n-1)} + \cdots + b_1 z^{-1} + b_0}{a_n z^{-n} + a_{n-1} z^{-(n-1)} + \cdots + a_1 z^{-1} + 1} \tag{7.36}$$

When the unit delay, the adder, and the scaler are selected as the basic building blocks, we have that

$$\begin{aligned} &\Rightarrow (a_n z^{-n} + a_{n-1} z^{-(n-1)} + \cdots + a_1 z^{-1} + 1)Y \\ &= (b_n z^{-n} + b_{n-1} z^{-(n-1)} + \cdots + b_1 z^{-1} + b_0)X \end{aligned} \tag{7.37}$$

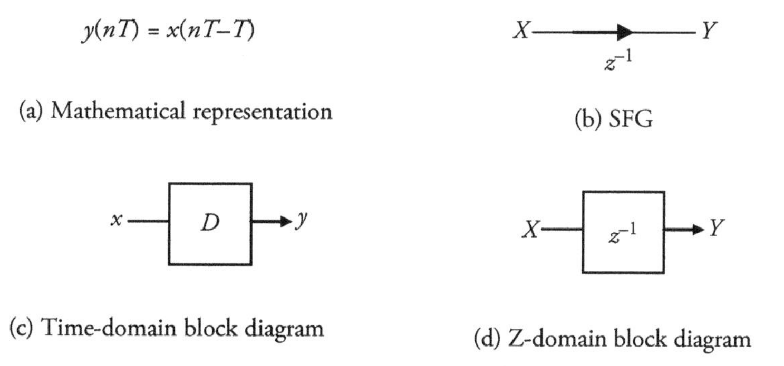

FIGURE 7.8 Representation of the unit delay operation.

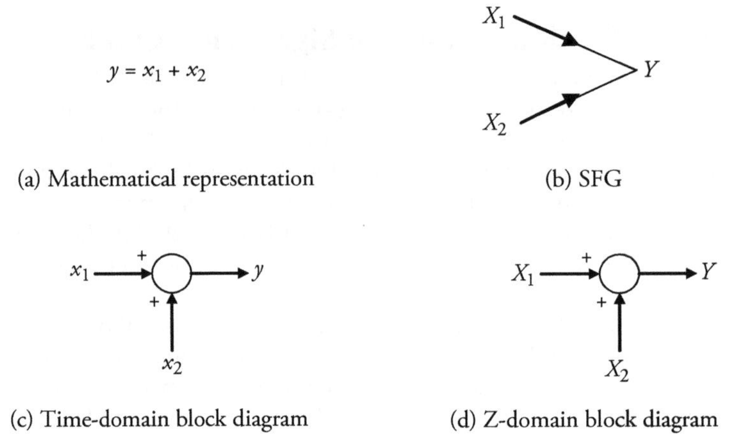

FIGURE 7.9 Representation of the addition operation.

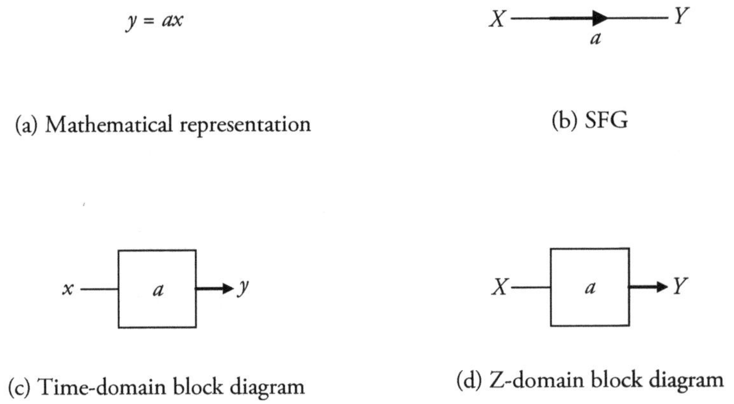

FIGURE 7.10 Representation of the scaling operation.

or

$$\Rightarrow Y = (b_n X - a_n Y)z^{-n} + (b_{n-1}X - a_{n-1}Y)z^{-(n-1)} + \cdots + (b_1 X - a_1 Y)z^{-1} + b_0 X \tag{7.38}$$

The above equation can be directly translated into the SFG as shown in Fig. 7.11.

Alternatively, we may manipulate the transfer function by introducing a variable U as

$$\frac{Y}{X} = \frac{b_n z^{-n} + b_{n-1}z^{-(n-1)} + \cdots + b_1 z^{-1} + b_0}{a_n z^{-n} + a_{n-1}z^{-(n-1)} + \cdots + a_1 z^{-1} + 1} \cdot \frac{U}{U} \tag{7.39}$$

or

$$\Rightarrow \begin{cases} U = X - (a_n z^{-n} + a_{n-1}z^{-(n-1)} + \cdots + a_1 z^{-1})U \\ Y = (b_n z^{-n} + b_{n-1}z^{-(n-1)} + \cdots + b_1 z^{-1} + b_0)U \end{cases} \tag{7.40}$$

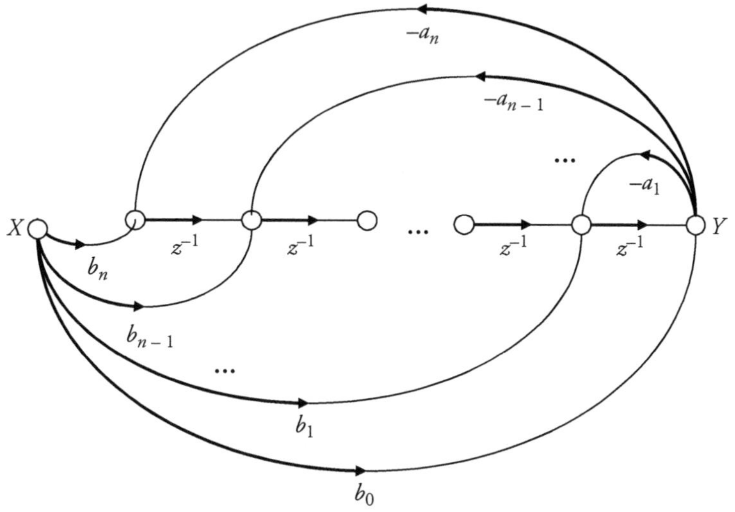

FIGURE 7.11 SFG of the general DT transfer function.

The above equations can be converted into an alternative SFG representation as shown in Fig. 7.12.

It can be seen that the SFG representation of the linear DT system is not unique. In fact, there are other SFG representations that can be derived based on the selection of the basic circuit building block for the system implementation.

There is one commonly used method in practical DT system implementation. In this method, the system transfer function is first partitioned into the product of the first-

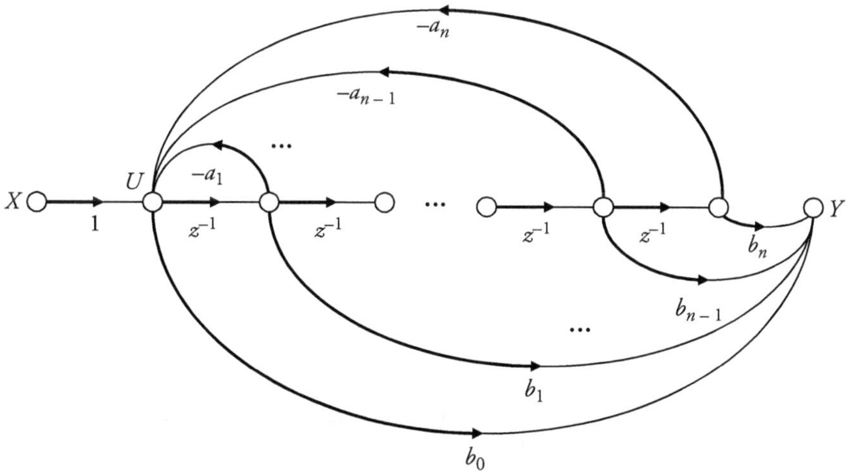

FIGURE 7.12 Alternative SFG of the general DT transfer function.

and second-order terms. Then the system can be implemented using cascading of these highly reusable first- and second-order subblocks:

$$H(s) = \frac{\sum_{k=0}^{m} a_k z^k}{\sum_{k=0}^{n} b_k z^k} = \prod_k \left(\frac{A_k z^2 + B_k z + C_k}{z^2 + \frac{z\omega_k}{Q_k} + \omega_k^2} \right) \prod_k \left(\frac{D_k z + E_k}{z + p_k} \right) \tag{7.41}$$

7.5 S-to-Z Transformations

Although VLSI DT circuits can be designed directly in z-domain, sometimes it is convenient to design it through the CT transfer function by taking advantage of the large collection of available CT circuit building blocks developed over the years. These CT circuit functions can be used with minor modifications for the DT circuit implementations through well-developed transformation methods.

7.5.1 Standard S-to-Z Transform

For the two linear system configurations shown in Fig. 7.13, if both of them have the same responses for the same delta function input, the DT transfer function $H(z)$ is known as the standard s-to-z transform (or impulse invariant transform) of CT transfer function $H_A(s)$.

It can be proven mathematically that for a circuit with a single pole, we have that

$$H_A(s) = \frac{A}{s+p} \Rightarrow H(z) = \frac{A}{1 - e^{-pT} z^{-1}} \tag{7.42}$$

For a circuit with multiple poles, we have that

$$H_A(s) = \frac{A}{(s+p)^n} \Rightarrow H(z) = (-1)^{n-1} \frac{d^{n-1}}{dp^{n-1}} \left[\frac{A}{(n-1)!} \frac{1}{1 - e^{-pT} z^{-1}} \right] \tag{7.43}$$

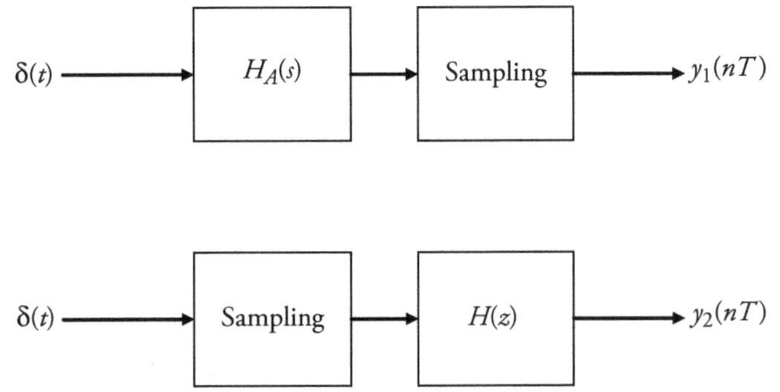

FIGURE 7.13 Standard s-to-z transformation.

7.5.2 Matched-Z Transform

The standard transformation requires the partial-fraction expansion of the given s-domain transfer function in order to perform the s-to-z mapping. The matched-z transform, on the other hand, simplifies the calculation process by using the transformation as follows:

$$H_A(s) = \frac{A}{(s+p)^n} \Rightarrow H(z) = \frac{A}{(1-e^{-pT}z^{-1})^n} \tag{7.44}$$

7.5.3 Forward Euler Integration

Various numerical methods, which are suitable for VLSI signal processing circuit implementation, have been developed in past years. In the forward Euler integration method, a mapping between the s- and z-domain transfer functions is formed by the approximation of the CT signal integration with a forward Euler numerical integration.

For the CT integration operation shown below

$$y(t) = \int^{t} x(t)dt \tag{7.45}$$

$$\frac{Y(s)}{X(s)} = \frac{1}{s} \tag{7.46}$$

the output and input relation can be approximated using the forward Euler numerical integration method as shown in Fig. 7.14.

$$y(nT+T) = \int_{t=nT}^{nT+T} x(t)dt \approx x(nT)\cdot T + y(nT) \tag{7.47}$$

By taking the z-transform of the equation, we have the following:

$$\Rightarrow \frac{Y(z)}{X(z)} = \frac{Tz^{-1}}{1-z^{-1}} \tag{7.48}$$

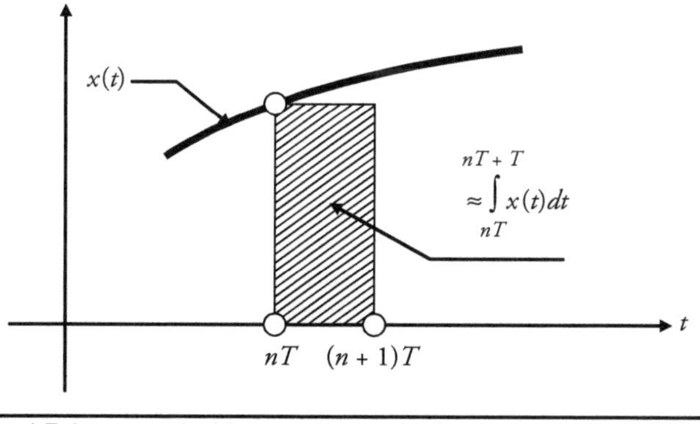

FIGURE 7.14 Forward Euler numerical integration method.

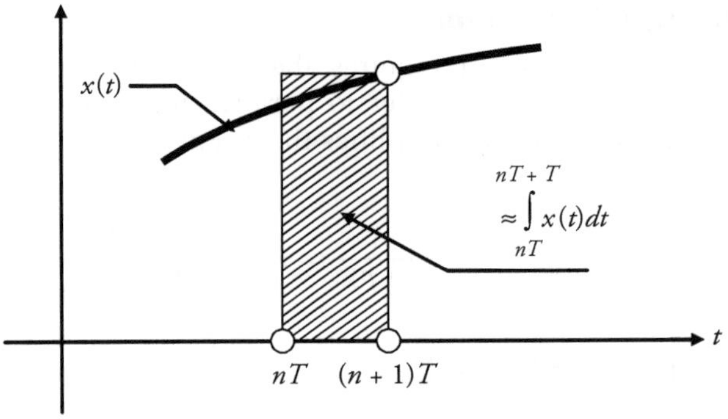

Figure 7.15 Backward Euler numerical integration method.

We can get the numerical-based s-to-z transformation based on the forward Euler numerical integration method as follows:

$$\Rightarrow \frac{1}{s} \approx \frac{Tz^{-1}}{1-z^{-1}} \tag{7.49}$$

7.5.4 Backward Euler Integration

An alternative numerical s-to-z transformation can be developed based on the backward Euler numerical integration method as shown in Fig. 7.15.

In this method, the CT integration is approximated by the backward Euler integration rule as follows:

$$y(nT+T) = \int_{t=nT}^{nT+T} x(t)dt \approx x(nT+T)\cdot T + y(nT) \tag{7.50}$$

The z-domain transfer function of the integrator can be derived by taking the z-transform of Eq. (7.50) as follows:

$$\Rightarrow \frac{Y(z)}{X(z)} = \frac{T}{1-z^{-1}} \tag{7.51}$$

The numerical s-to-z transformation based on this method is given as

$$\Rightarrow \frac{1}{s} \approx \frac{T}{1-z^{-1}} \tag{7.52}$$

7.5.5 Bilinear Transformation

The forward and backward Euler integration can be combined to form the bilinear numerical integration as shown in Fig. 7.16.

The numerical integration based on the bilinear integration rule can be derived based on the bilinear numerical integration method as follows:

$$y(nT+T) = \int_{t=nT}^{nT+T} x(t)dt \approx \frac{x(nT+T)+x(nT)}{2}\cdot T + y(nT) \tag{7.53}$$

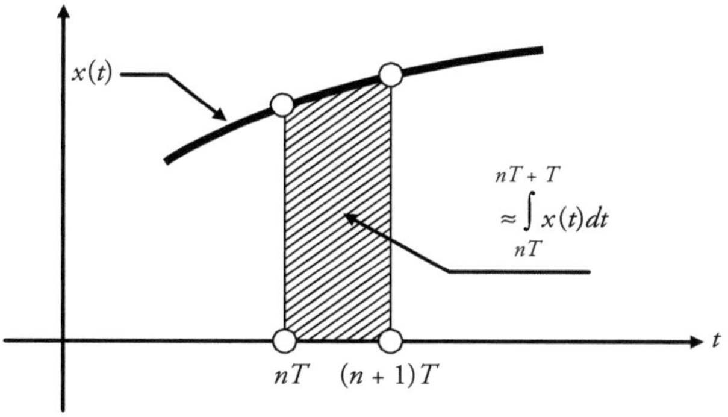

FIGURE 7.16 Bilinear integration method.

Numerical s-to-z transformation based on the bilinear integration can be derived by taking the z-transform of Eq. (7.53) as

$$\Rightarrow \frac{Y(z)}{X(z)} = \frac{T}{2}\frac{1+z^{-1}}{1-z^{-1}} \tag{7.54}$$

or

$$\Rightarrow \frac{1}{s} \approx \frac{T}{2}\frac{1+z^{-1}}{1-z^{-1}} \tag{7.55}$$

7.5.6 LDI Transformation

Alternatively, numerical integration can also be realized using the middle-point value in the integration approximation as shown in Fig. 7.17.

$$y(nT+T) = \int_{t=nT}^{nT+T} x(t)dt \approx x\left(nT + \frac{T}{2}\right)\cdot T + y(nT) \tag{7.56}$$

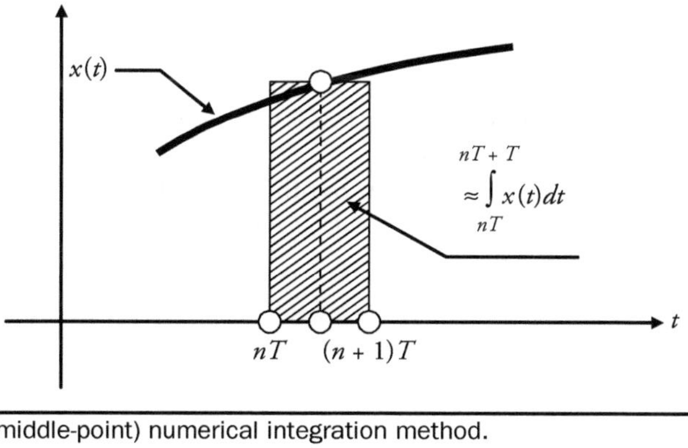

FIGURE 7.17 LDI (middle-point) numerical integration method.

Such a method leads to the midpoint or LDI numerical integration–based s-to-z transformation as

$$\Rightarrow \frac{Y(z)}{X(z)} = \frac{Tz^{-1/2}}{1-z^{-1}} \tag{7.57}$$

and

$$\Rightarrow \frac{1}{s} = \frac{Tz^{-1/2}}{1-z^{-1}} \tag{7.58}$$

7.5.7 Multistep Transformation

To minimize the effect of frequency warping and/or to provide closer approximation of analog signal processing, the multistep integration method may be used. A multistep integration formula is a numerical method that yields $y((n+1)T)$ based on the value of $y(t)$ and $x(t)$ sampled at $k+1$ previous time points and $x((n+1)T)$.

$$y((n+1)T) = y(nT) + T\sum_{i=-1}^{l} c_i x((n-i)T) \Big/ \sum_{i=-1}^{l} c_i \tag{7.59}$$

The coefficients a_i and b_i are both real. Such integration in a formula yields the following numerical s-to-z transformation:

$$\Rightarrow \frac{1}{s} \approx \frac{T}{\sum_{i=0}^{l} c_l} \frac{c_0 + c_1 z^{-1} + c_2 z^{-2} + \ldots + c_l z^{-l}}{1-z^{-1}} \tag{7.60}$$

References

1. M. A. Al-Alaoui, "Novel Stable Higher Order s-to-z Transforms," *IEEE Trans. on Circuits and Systems I: Fundamental Theory and Applications*, vol. 48, no. 11, pp. 1326–1329, November 2001.
2. F. Larsen, M. Ismail, "LDI Z-domain Images of Lossless LC Ladders," *IEEE International Symposium on Circuits and Systems*, vol. 1, pp. 144–147, June 1991.
3. M. Lee, G. Temes, C. Chang, M. Ghaderi, "Bilinear Switched-Capacitor Ladder Filters," *IEEE Trans. on Circuits and Systems*, vol. 28, no. 8, pp. 811–822, August 1981.
4. G. J. Smolka, "Switched-Capacitor Bandpass Filter Based on Simulation of Canonical Reactance Sections," *IEEE J. of Solid-State Circuits*, vol. 19, no. 3, pp. 360–367, June 1984.
5. P. V. A. Mohan, "Analysis of Discrete-Time Filters," *IEEE Trans. on Education*, vol. 32, no. 3, pp. 398–401, August 1989.
6. M. Ismail, F. Larsen, "Z-domain Synthesis of Discrete-Time Reactance Functions," *IEEE Trans. on Circuits and Systems I: Fundamental Theory and Applications*, vol. 39, no. 4, pp. 316–324, April 1992.

CHAPTER 8

Introduction to VLSI Switched-Capacitor Circuit Techniques

VLSI-integrated continuous-time (CT) signal processing circuits—including the active-RC, MOS-C, Gm-C, and CTI circuit structures—suffer from several inherent limitations, such as the poor absolute frequency accuracies (e.g., time constant) mainly resulting from the PVT variations of component parameter with the VLSI fabrication processes and application conditions. Consequently, these VLSI circuits usually require certain tuning circuits for achieving targeted application specifications. These VLSI circuit types also suffer from the high-cost penalty associated with the realizations of large on-chip time-constant circuits for low-frequency (e.g., audio band) signal processing applications. The VLSI switched-capacitor (SC) circuits, which are based on the discrete-time (DT) signal processing operation, were first developed to design accurate analog signal processing circuits economically at voice-band frequencies in fully integrated form. CMOS technologies were selected for VLSI SC signal processing circuit implementations to achieve low-power, compact-designed, and fully compatible VLSI digital circuits. SC circuits can be used to simulate resistors with very good accuracy by rapidly charging and discharging the capacitor across two terminal voltages. By using this scheme, the RC constants of the signal processing circuits only depend on the clock frequency and the ratio of capacitor values. All of this can be controlled very accurately in VLSI circuit implementations. In addition, the SC circuits are particularly suitable for VLSI circuit realization of high-value on-chip resistances, which effectively solves the fundamental problem of applying VLSI analog signal processing circuits in most low-frequency signal processing applications.

8.1 Basic VLSI Switched-Capacitor Circuits

The very basic SC circuit that is based on two-phase nonoverlap clocks is shown in Fig. 8.1.

During the charging phase, electrical charge is transferred from the terminal V_a (supply) to the capacitor with the total charge transferred in this phase given as

$$\Delta q = CV_a - q_o \tag{8.1}$$

where q_o is assumed to be the initial charge stored in the capacitor.

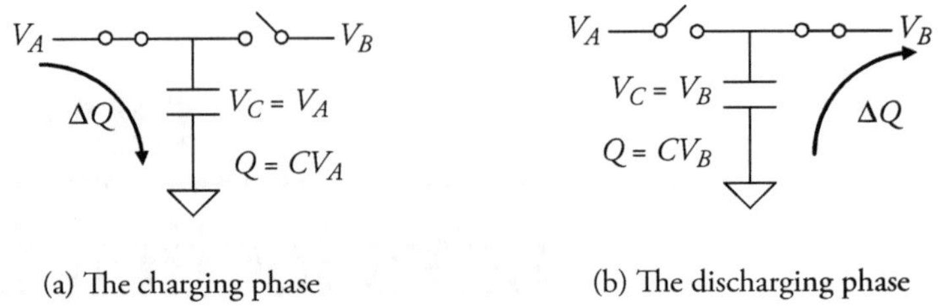

FIGURE 8.1 The basic switched-capacitor circuit operation.

During the discharging phase, the charge is transferred from the capacitor to terminal V_B (load) with the total electrical charge transferred in this phase given as follows:

$$\Delta q = C(V_a - V_b) \tag{8.2}$$

If a periodic control clock is used to control the above operation under the following conditions—the frequency of the signal is much lower than that of the clock frequency, and the time constant of the system is much greater than the period of the clock—then such an SC circuit is equivalent to a linear resistor connected across the two terminals, and the equivalent resistance is given as follows:

$$\Rightarrow R \equiv \frac{(V_a - V_b)}{I} = \frac{(V_a - V_b)}{\frac{\Delta q}{T}} = \frac{(V_a - V_b)}{\frac{C}{T}(V_a - V_b)} = \frac{T}{C} \tag{8.3}$$

The significances of the SC-simulated resistor shown in Eq. (8.3) are as follows:

- Large-value resistors can be practically realized using the VLSI processes. Based on Eq. (8.3), a high-value resistor (e.g., 1 MΩ) can be easily implemented with a capacitor of reasonable value (e.g., 1 pF) and clock frequency (e.g., 1 MHz) in VLSI form.
- Accuracy of the resistance implemented by the SC method is improved compared with the direct VLSI resistor implementation since both clock frequency and capacitance can be controlled much accurately than resistance.
- The RC time constant of the on-chip VLSI circuit under SC implementation is highly improved since it only depends on the clock period and the ratio of capacitances; both of them can be controlled very accurately under PVT conditions.
- Since an on-chip capacitor typically has better linearity than a resistor in VLSI, an SC-implemented resistor will also improve the linearity of the signal processing circuits.

By replacing the resistor in the active-RC integrator shown in Fig. 8.2 using an SC resistor, the SC integrator can be constructed.

The basic SC circuit structure suffers from the fundamental accuracy limitation due to the parasitic capacitance effects of the VLSI implementation where the parasitic capacitance from the upper plate of the VLSI capacitor contributes to the SC value

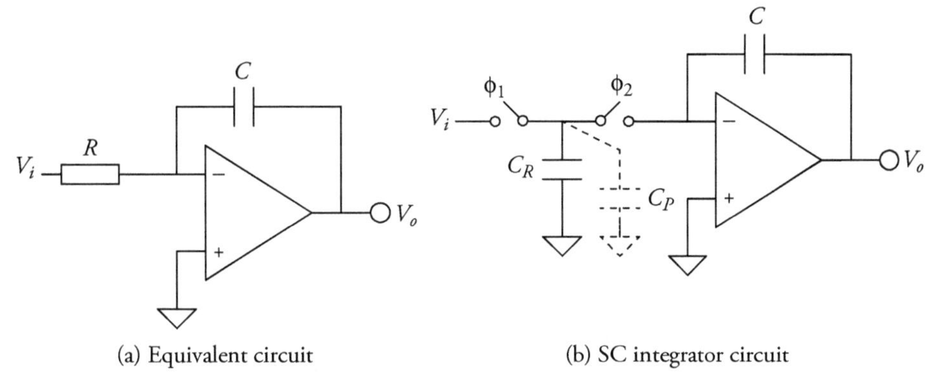

FIGURE 8.2 Basic VLSI SC integrator circuit structure.

uncertainty. Such a problem can effectively be solved using the so-called parasitic-insensitive SC circuit structures as shown in Fig. 8.3.

In the circuit shown in Fig. 8.3a, the capacitor is first discharged with both terminals of the capacitor shorted to ground during phase 1 ($\phi_1 = 1$). During phase 2 ($\phi_2 = 1$), the capacitor is charged to the input voltage. It can be seen that the parasitic capacitor effect (C_{pa} and C_{pb}) can be eliminated from the resistance in Eq. (8.3) since the parasitic capacitor is connected to ground and does not contribute to the current flow from terminal V_a to terminal V_b during ϕ_2 operation. Since what we have here is a direct current path from

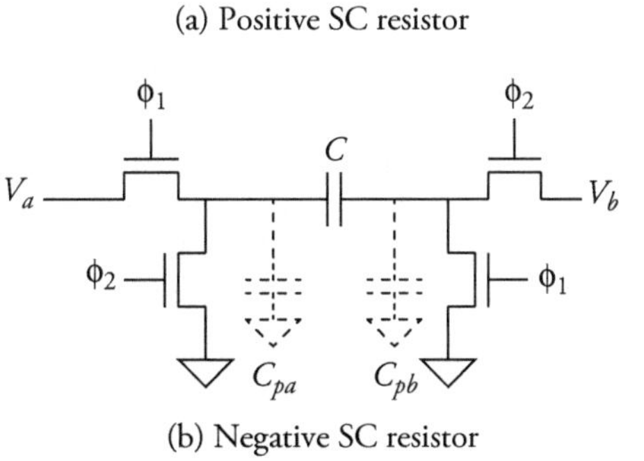

FIGURE 8.3 Parasitic-insensitive SC resistor circuit structures.

the input to the output, no delay effect occurs between the input and output signals during this phase:

$$\begin{cases} R = \dfrac{T}{C} \\ \Delta Q\left(nT + \dfrac{T}{2}\right) = V_a\left(nT + \dfrac{T}{2}\right) \end{cases} \tag{8.4}$$

In the alternative SC circuit shown in Fig. 8.3b, the capacitor C and its parasitic capacitor C_{pa} are first charged to the input signal voltage in phase 1 ($\phi_1 = 1$). During phase 2 ($\phi_2 = 1$), the two capacitors are discharged to ground. Similar to the previous circuit, only the capacitor C contributes to the discharged current to terminal V_b, and the parasitic capacitor effect is also eliminated. Since the direction of the discharge current path in such cases is opposite to that in the previous circuit, a negative SC resistance with a half-clock-period delay between the input and output signals is effectively realized.

$$\begin{cases} R = -\dfrac{T}{C} \\ \Delta Q\left(nT + \dfrac{T}{2}\right) = V_a\left(nT + \dfrac{T}{2}\right) \end{cases} \tag{8.5}$$

8.1.1 Charge Redistribution Analysis Method

VLSI SC circuits are based on the principle of electrical charge redistribution among the capacitors within the SC circuits where the OPAMP or OTA circuits merely serve as the helpers for the necessary environment for the charge redistribution operations (isolation, gain, and stabilization of the capacitor terminal voltages). The charge redistribution method provides an effective way of analyzing the SC circuit at all operation frequencies down to the Nyquist rate.

Note that the direct mapping of switched capacitors to resistors introduced in the previous section is only valid under the oversampling conditions. Under the oversampling condition, the charge redistribution method and the direct mapping method are equivalent to each other.

8.1.2 Noninverted Switched-Capacitor Integrator

For the noninverted SC integrator circuit shown in Fig. 8.4, the equivalent circuit during the ϕ_1 phase ($\phi_1 = 1$) is shown in Fig. 8.5. In this phase, the switched capacitor C_R is charged to the input-signal voltage value, and the integration capacitor C holds the charge accumulated in the previous clock cycles.

At the beginning of this phase, the voltages across the two capacitors are given as follows:

$$\begin{cases} V_o(t) = V_o(nT) \\ V_{CR}(t) = V_i(nT) \end{cases} \tag{8.6}$$

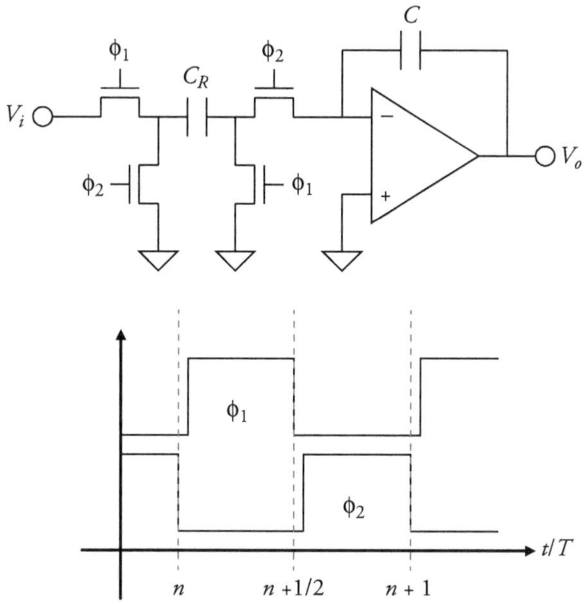

FIGURE 8.4 Noninverted SC integrator circuit.

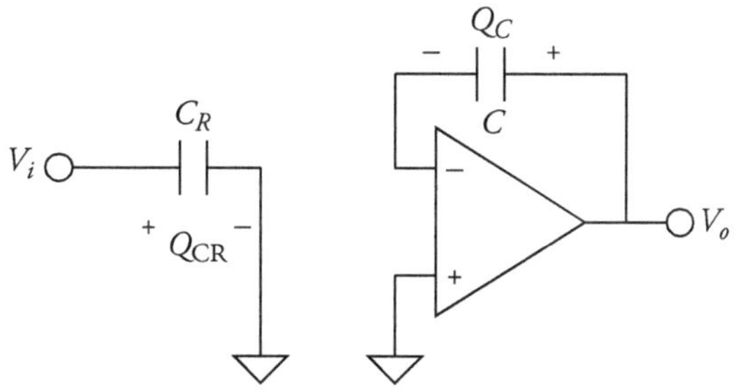

FIGURE 8.5 Equivalent SC integrator circuit during the ϕ_1 phase.

At the end of the ϕ_1 phase, these voltages can be expressed in the following equations:

$$\begin{cases} V_o(nT + T/2) = V_o(nT) \\ V_{CR}(nT + T/2) = V_i(nT + T/2) \end{cases} \tag{8.7}$$

The electrical charges stored on the two capacitors can be derived as follows:

$$\begin{cases} Q_C(nT + T/2) = C \cdot V_o(nT) \\ Q_{CR}(nT + T/2) = C_R \cdot V_i(nT + T/2) \end{cases} \tag{8.8}$$

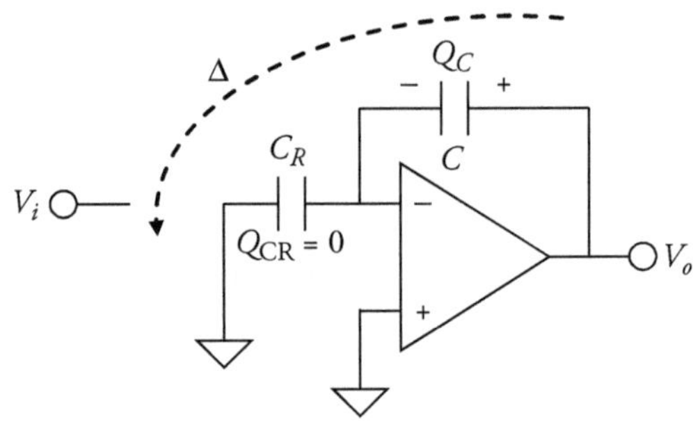

FIGURE 8.6 SC integrator circuit during the ϕ_2 phase.

During the ϕ_2 phase ($nT + T/2 < t < nT + T$), the equivalent circuit of the SC integrator is shown in Fig. 8.6.

Note that when the SC circuit configuration is switched from the circuit shown in Fig. 8.5 to the circuit shown in Fig. 8.6, electrical charge redistribution occurs. For example, the switched capacitor is usually charged to a nonzero voltage at the end of a ϕ_1 phase, while at the end of the ϕ_2 phase, it is required to be discharged to the zero voltage based on the fact that the two terminals of C_R will be forced to the same ground voltage during the ϕ_2 phase because of the OPAMP circuit in a negative feedback connection. Since the OPAMP circuit has the infinite input impedance, the above discharge process of C_R within ϕ_2 can only be realized through the path across the integration capacitor C. As a result, the integration capacitor C is charged with the same amount of electrical charge when the switched capacitor is discharged. In other words, the electrical charge is redistributed between the two capacitors in the circuit.

At the end of the ϕ_2 phase, the charges on the two capacitors can be expressed as follows:

$$\begin{cases} Q_C(t) = Q_C(nT) + Q_{CR}(nT + T/2) \\ Q_{CR}(nT + T) = 0 \end{cases} \tag{8.9}$$

The output voltage is then given as follows:

$$V_o(nT + T) = V_o(nT) + \frac{C_R}{C} V_i(nT + T/2) \tag{8.10}$$

Assuming that the input signal is gated by the opposite phase (ϕ_2 phase), we can further derive the DT input and output transfer functions of such an SC integrator as

$$V_o(nT + T) = V_o(nT) + \frac{C_R}{C} V_i(nT + T/2) = V_o(nT) + \frac{C_R}{C} V_i(nT) \tag{8.11}$$

The z-domain transfer function can also be derived by applying the z-transform as follows:

$$\Rightarrow \frac{V_o(z)}{V_i(z)} = \frac{C_R}{C} \frac{z^{-1}}{1 - z^{-1}} \tag{8.12}$$

8.1.3 Inverted Switched-Capacitor Integrator

The charge redistribution technique can also be used to analyze the operation of the inverted SC integrator circuit as shown in Fig. 8.7.

For the ϕ_1 phase ($nT < t < nT+T/2$), this SC integrator is equivalent to the circuit shown in Fig. 8.8.

The voltages applied to the two capacitors can be expressed as follows:

$$\begin{cases} V_o(t) = V_o(nT) \\ V_{CR}(t) = 0 \end{cases} \tag{8.13}$$

The charges on the two capacitors at the end of this phase can be derived as

$$\begin{cases} Q_C(nT+T/2) = C \cdot V_o(nT) \\ Q_{CR}(nT+T/2) = 0 \end{cases} \tag{8.14}$$

During the ϕ_2 phase ($nT+T/2 < t < nT+T$), the SC integrator is equivalent to the circuit shown in Fig. 8.9.

Similarly, charge redistribution occurs when the SC circuit is switched between the ϕ_1 and ϕ_2 configurations. At the end of the ϕ_2 phase, C_R is charged from zero to the input voltage through the path across the integration capacitor C as follows:

$$\begin{cases} Q_C(t) = Q_C(nT) - Q_{CR}(nT+T) \\ Q_{CR}(nT+T) = C_R \cdot V_i(nT+T) \end{cases} \tag{8.15}$$

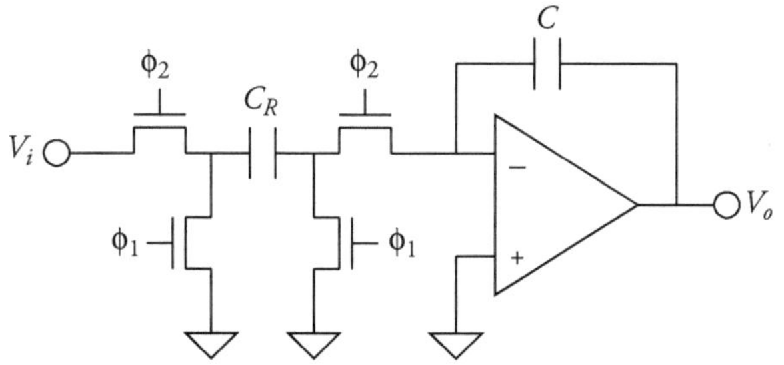

FIGURE 8.7 Inverted SC integrator circuit.

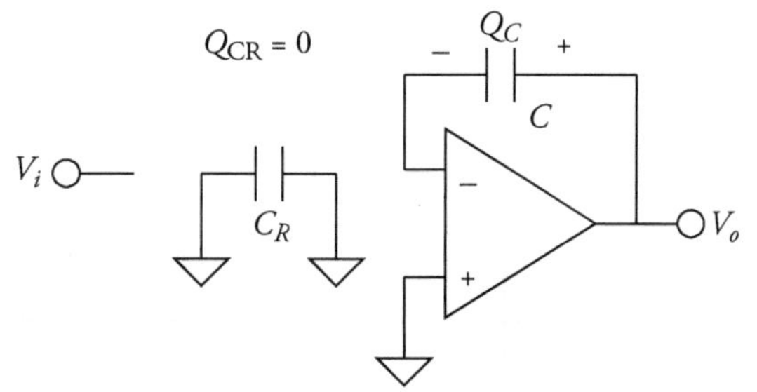

FIGURE 8.8 Equivalent SC integrator circuit during the ϕ_1 phase.

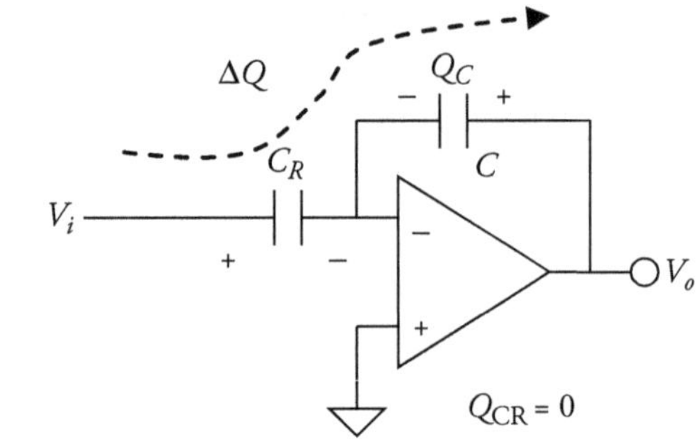

Figure 8.9 SC integrator circuit during the ϕ_2 phase.

Consequently, the DT input-to-output voltage transfer function can be derived as follows:

$$V_o(nT+T) = V_o(nT) - \frac{C_R}{C} V_i(nT+T) \tag{8.16}$$

The z-domain transfer function can further be derived as

$$\Rightarrow \frac{V_o(z)}{V_i(z)} = -\frac{C_R}{C}\frac{1}{1-z^{-1}} \tag{8.17}$$

8.1.4 Doubly Pumped Switched-Capacitor Integrator

Figure 8.10 shows a doubly pumped SC integrator circuit structure. The equivalent circuit of such an integrator for the ϕ_1 phase ($nT < t < nT+T/2$) is shown in Fig. 8.11.

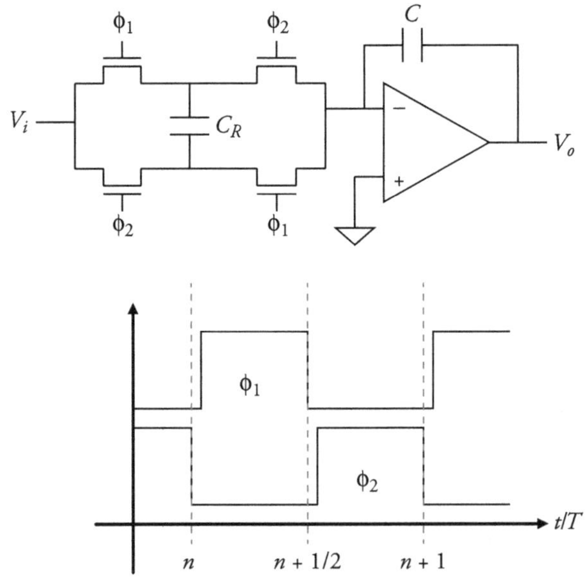

Figure 8.10 Doubly pumped SC integrator circuit structure.

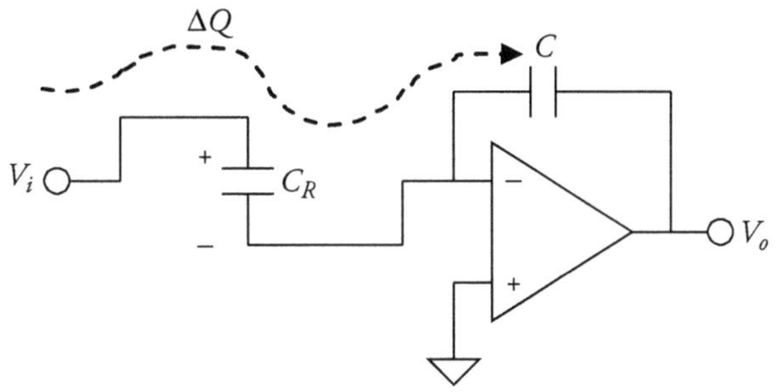

Figure 8.11 Doubly pumped SC integrator circuit at ϕ_1 phase.

The electrical charges stored on the two capacitors at the beginning of the ϕ_1 phase are given as follows:

$$\begin{cases} Q_C(nT) = C \cdot V_o(nT) \\ Q_{CR}(nT) = -C_R \cdot V_i(nT) \end{cases} \tag{8.18}$$

After the charge redistribution process during the ϕ_1 phase, the electrical charges stored on the two capacitors at the end of the ϕ_1 phase are given as

$$\begin{cases} Q_C(nT+T/2) = C \cdot V_o(nT) - C_R \cdot [V_i(nT+T/2) + V_i(nT)] \\ Q_{CR}(nT+T/2) = C_R \cdot V_i(nT+T/2) \end{cases} \tag{8.19}$$

The equivalent circuit of the SC integrator during the ϕ_2 phase is shown in Fig. 8.12.

The charge redistribution process occurs when the circuit is switched from the ϕ_1 to ϕ_2 circuit configurations. At the end of the ϕ_2 phase, the electrical charges stored on the two capacitors are given as follows:

$$\begin{cases} Q_C(nT+T) = C \cdot V_o(nT+T/2) - C_R \cdot [V_i(nT+T) + V_i(nT+T/2)] \\ Q_{CR}(nT+T) = -C_R \cdot V_i(nT+T) \end{cases} \tag{8.20}$$

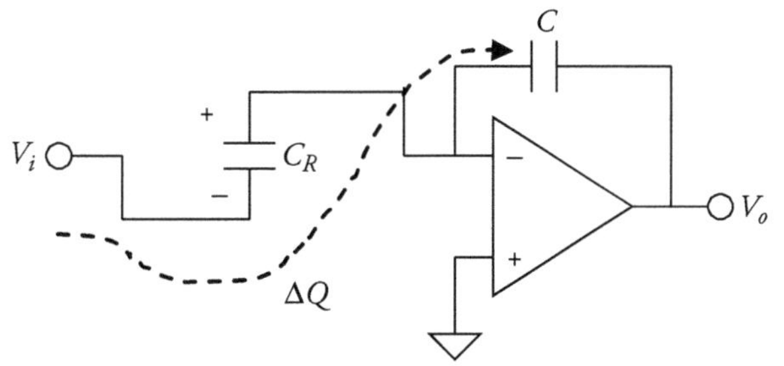

Figure 8.12 Doubly pumped SC integrator circuit.

Consequently, the DT input-to-output voltage transfer function of this SC integrator can be derived as

$$V_o(nT+T) = V_o(nT) - \frac{C_R}{C}[V_i(nT) + 2V_i(nT+T/2) + V_i(nT+T)] \tag{8.21}$$

The z-domain transfer function can be further derived as follows:

$$\Rightarrow \frac{V_o(z)}{V_i(z)} = -\frac{C_R}{C}\frac{z^{-1}+2z^{-1/2}+1}{1-z^{-1}} = -\frac{C_R}{C}\frac{1+z^{-1/2}}{1-z^{-1/2}} \tag{8.22}$$

8.2 Basic VLSI SC Circuit Elements

A typical VLSI SC circuit consists of the following elements: VLSI MOS switch circuits, clock-generation circuits, VLSI capacitors, and VLSI OPAMP (OTA) circuits.

8.2.1 VLSI Analog Switch Circuit Structures

The desired properties of the VLSI SC switch circuits include high *off* resistance and low *on* resistance, minimized switching delay between the ϕ_1 and ϕ_2 circuit configurations, minimized influence between switch control and signals, and low-signal voltage offset. Note that switches based on VLSI MOS transmission gates can meet or approximate the above functions. Practical MOS switches are realized using either NMOS and/or PMOS transistors, and they are controlled through the voltage signals on the gates. The basic SC switch in the CMOS process technology uses a single-NMOS (or PMOS) transistor as shown in Fig. 8.13.

Based on the MOS transistor *I-V* characteristic equations in the linear and cutoff modes, we have the following:

$$I = \begin{cases} 0 & V_{gs} - V_T < 0 \\ \beta\left(V_g - V_T - \frac{V_d + V_s}{2}\right)(V_d - V_s) & V_{gs} - V_T \geq 0 \end{cases} \tag{8.23}$$

The effective switch *off* and *on* resistances of the basic MOS switch circuit can be derived as follows:

$$\Rightarrow R = \begin{cases} \infty & V_{gs} - V_T < 0 \\ \dfrac{1}{\beta\left(V_g - V_T - \frac{V_d + V_s}{2}\right)} & V_{gs} - V_T \geq 0 \end{cases} \tag{8.24}$$

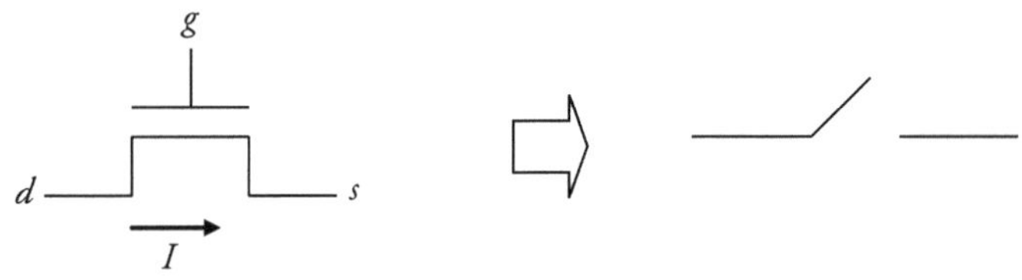

Figure 8.13 Basic MOS switch circuit.

Note that a typical *off* resistance of the MOS switch is typically in the order of 10^{10}–10^{12} Ω and the *on* resistance is in the order of 0.5–10 kΩ. These values are usually PVT dependent.

Note that the MOS pass gate–based switch offers zero dc offset in the switch operation. However, the parasitic capacitances between the gate and source and the drain may introduce the dc offset and the charge injection (or clock feedthrough, CFT). The total voltage error in a practical circuit caused by the switch charge injection is typically in the millivolt range. Practically speaking, a smaller switch-transistor size, a slower switching-edge rate, and/or a small resistance of the voltage source may be used to minimize the charge injection effect. Among various charge injection minimization techniques, the half-sized dummy transistor as shown in Fig. 8.14, using inversed switching clock polarity, is commonly used in design to compensate for the charge injection (or CFT) from the main switching transistor.

The *on* resistance of a VLSI analog switch employing a single-NMOS (or -PMOS) device strongly depends on the signal level. For an NMOS switch, a high-level signal may significantly degrade the *on* resistance that impacts the speed of the switch. On the other hand, a low-level signal will degrade the switch speed in the same way for a PMOS switch. An improved VLSI SC switch circuit structure employing both NMOS and PMOS transistors is shown in Fig. 8.15.

Such switch structures offer advantages over the basic NMOS or PMOS switches by providing reduced *on* resistance and charge injections. In practical design, the half-sized dummy transistors as shown in Fig. 8.15b can also be used in such structures for

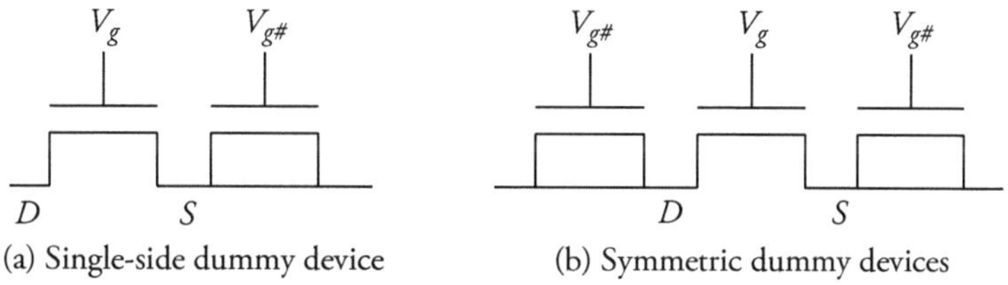

FIGURE 8.14 MOS switch with dummy transistors for charge injection reduction.

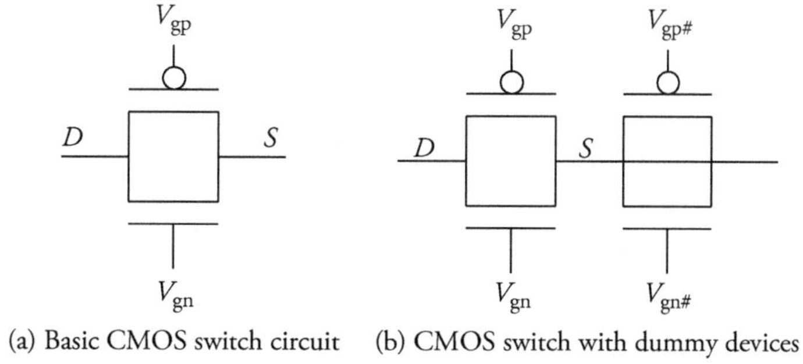

FIGURE 8.15 CMOS switch circuit structures.

further charge injection cancellation. Based on the *I-V* equation in the MOS device in linear mode given as

$$I = \beta_n \left(V_{\text{gn}} - V_{\text{Tn}} - \frac{V_d + V_s}{2} \right)(V_d - V_s)$$
$$+ \beta_p \left(-V_{\text{gp}} + V_{\text{Tn}} + \frac{V_d + V_s}{2} \right)(V_d - V_s) \tag{8.25}$$

the effective *on* resistance of the switch can be derived as

$$\Rightarrow R = \frac{1}{\beta_n \left(V_{\text{gn}} - V_{\text{Tn}} - \frac{V_d + V_s}{2} \right) + \beta_p \left(-V_{\text{gp}} + V_{\text{Tp}} + \frac{V_d + V_s}{2} \right)} \tag{8.26}$$

8.2.2 Nonoverlap Clock-Generation Circuit

A commonly used two-phase nonoverlap clock-generation circuit structure is shown in Fig. 8.16, where the nonoverlap regions of the two-phase clock are controlled by the delay time of the buffer used.

8.2.3 VLSI OTA Circuit Structures

OPAMP (or OTA) circuits in SC circuits are only required to drive capacitive loads. Consequently, the OPAMP circuit in VLSI SC circuits can typically be optimized for gain, stability, and power dissipation with higher output impedance. OTAs are commonly used in many practical SC circuit design applications.

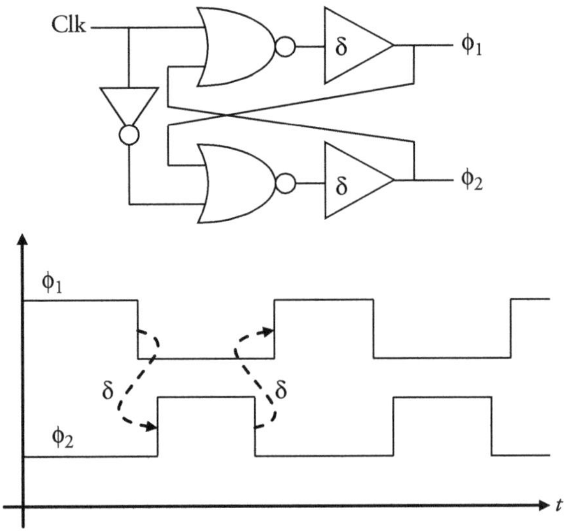

Figure 8.16 VLSI two-phase nonoverlap clock-generation circuit.

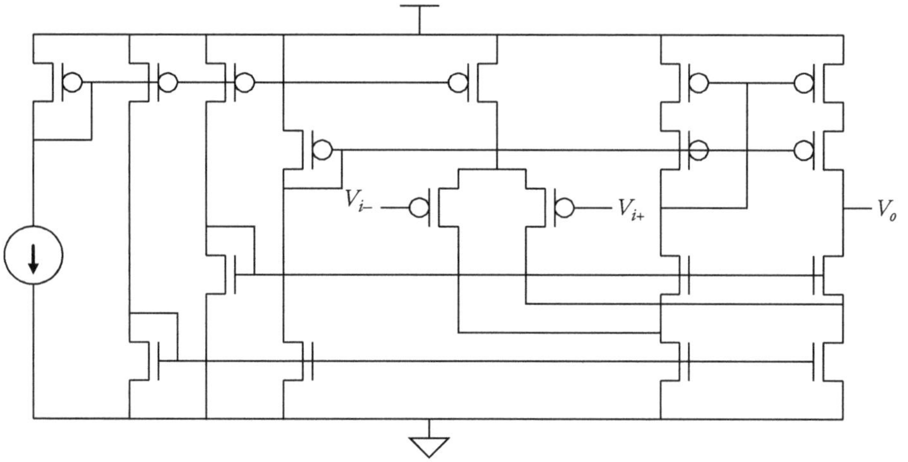

FIGURE 8.17 VLSI folded-cascode OTA circuit structure.

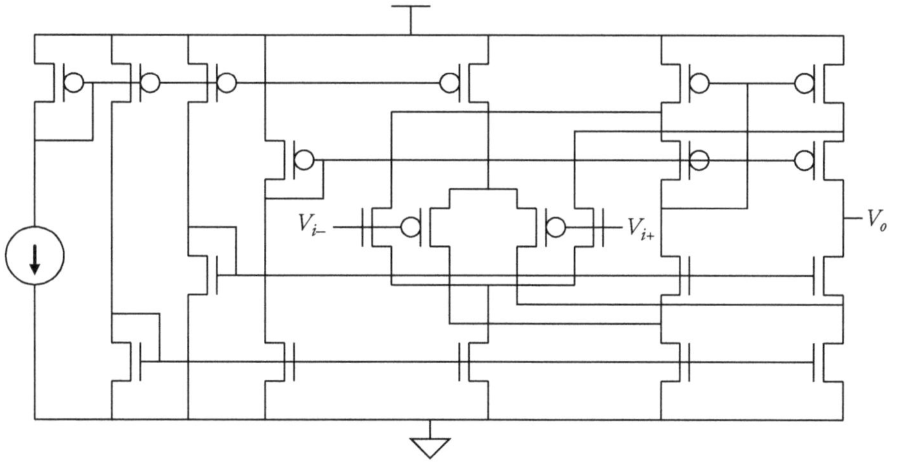

FIGURE 8.18 VLSI rail-to-rail input folded-cascode OTA circuit structure.

Figures 8.17 and 8.18 are two commonly used VLSI folded-cascode OTA circuit structures. These circuits offer the features of higher-gain and high-output impedance, which are suitable for capacitive load in the SC circuit applications.

In practical SC circuit design, fully differential OTA circuits, such as the one shown in Fig. 8.19, are typically used for better noise immunity.

8.3 VLSI Sample/Hold Circuit Structures

The sample/hold (S/H) circuits are essential building circuit blocks of DT systems where the S/H circuits are typically used to avoid frequency response distortion caused by CT signal feedthrough effect (see Fig. 8.20).

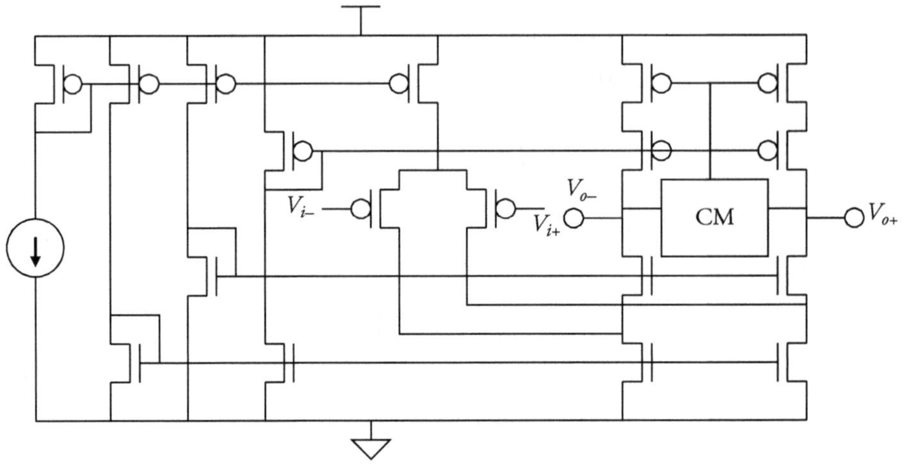

FIGURE 8.19 VLSI fully differential folded-cascode OTA circuit structure.

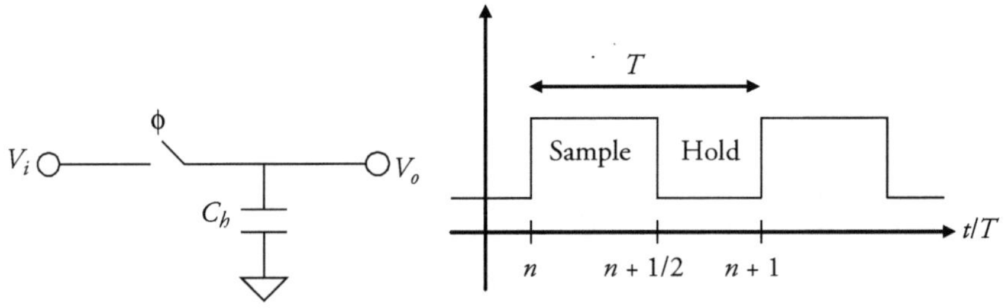

FIGURE 8.20 Basic sample/hold circuit.

The S/H circuit works as follows: the output signal tracks the input signal during the sample phase, and the output holds the previous sampled input signal value during the hold phase.

$$\begin{cases} V_o(t) = V_i(t) & nT < t < nT + T_s \\ V_o(t) = V_i(nT + T_s) & nT + T_s < t < nT + T \end{cases} \tag{8.27}$$

Practical S/H circuits suffer from various nonideal effects as shown in Fig. 8.21. These nonideal effects are characterized by parameters such as the acquisition time, slew rate, aperture uncertainty time, hold-mode drop rate, voltage offset, and CFT.

8.3.1 VLSI Buffered Sample/Hold Circuit

A buffered VLSI S/H circuit employing two OPAMP circuits is shown in Fig. 8.22, where the first OPAMP circuit serves as the input buffer and the second one as the output buffer. Such an S/H circuit structure offers better isolation at the input and

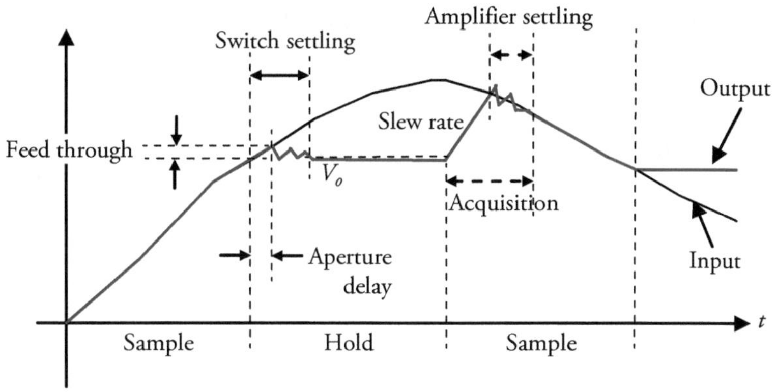

FIGURE 8.21 Nonideal effects of practical sample/hold circuits.

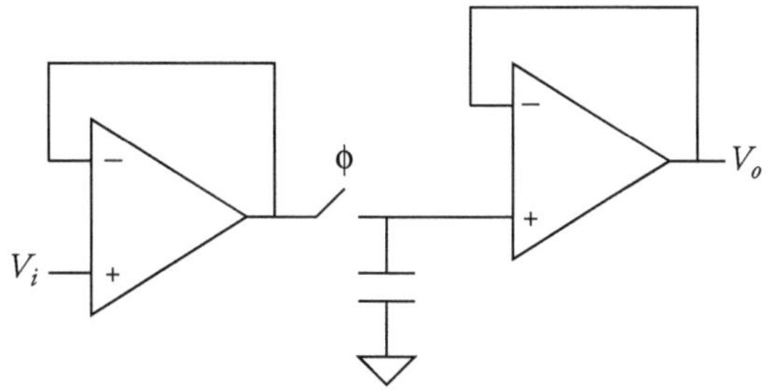

FIGURE 8.22 The buffered VLSI S/H circuit structure.

enhanced driving capability at the output. However, such a circuit suffers from the limitation of high offset error contributed from both OPAMP circuits.

An improved VLSI buffered S/H circuit structure is shown in Fig. 8.23. Such an S/H circuit structure effectively eliminates the offset voltage of the output OPAMP circuit by connecting the sample capacitor within its feedback loop.

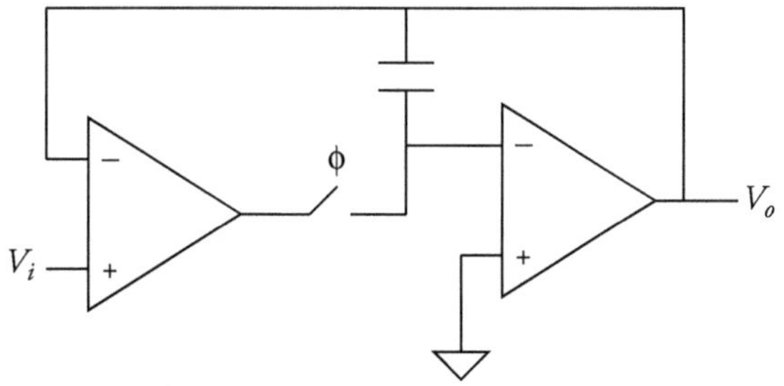

FIGURE 8.23 Improved of VLSI S/H circuit structure.

8.3.2 Correlated Double Sampling Techniques

Offset error in VLSI S/H circuits can be minimized using the autozero methods based on the correlated double sampling (CDS) techniques. In the CDS method, the offset voltage of the sampler circuit is sampled first and stored in the first clock phase, and then this stored offset voltage is subtracted from the sampled signal value that occurs in the second clock phase. In fact, the CDS techniques are also effective in reducing other nonideal effects, such as the low-frequency noise of the devices in the sampler and the power supply noise. Figure 8.24 shows a VLSI SC S/H circuit structure that employs a CDS-based autozero technique.

In the SC circuit in Fig. 8.24, the nonideal OPAMP circuit with offset and input reference noise is modeled using an effective offset voltage source V_{off} at the offset-free ideal OPAMP circuit input. Such a circuit works as follows:

- During the sampling phase (ϕ_1), this offset voltage at the OPAMP input is sampled and stored in the sampling capacitor C_h. The node voltage V_x and the electrical charge stored on the sampling capacitor C_h at the end of the ϕ_1 phase are given as follows:

$$\begin{cases} V_x\left(nT+\dfrac{T}{2}\right)=V_{\text{offset}}\left(nT+\dfrac{T}{2}\right) \\ Q_h\left(nT+\dfrac{T}{2}\right)=\left[V_i\left(nT+\dfrac{T}{2}\right)-V_{\text{offset}}\left(n\mathrm{T}+\dfrac{T}{2}\right)\right]C_h \end{cases} \tag{8.28}$$

- During the hold phase (ϕ_2 phase), node voltage on V_x keeps tracking the effective offset voltage because of the OPAMP circuit in feedback operation, and the charge stored on sampling capacitor C_h remains the same as the sampling phase. As a result, the node voltages at the end of the ϕ_2 phase are given as follows:

$$\begin{cases} V_x(nT+T)=V_{\text{offset}}(nT+T) \\ V_o(nT+T)=\left[V_i\left(nT+\dfrac{T}{2}\right)-V_{\text{offset}}\left(n+\dfrac{T}{2}\right)\right]+V_{\text{offset}}(nT+T) \end{cases} \tag{8.29}$$

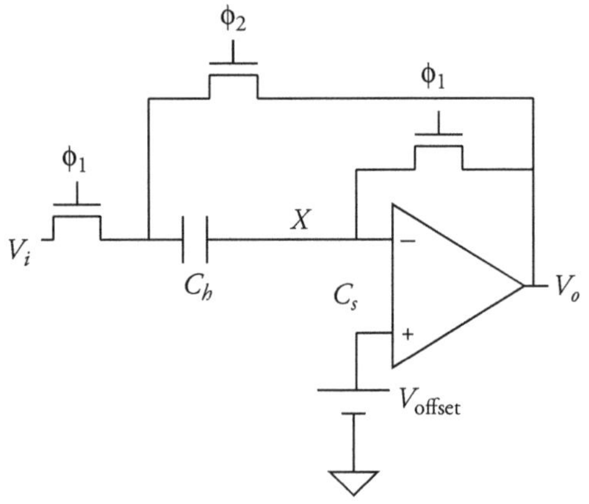

FIGURE 8.24 The VLSI SC CDS S/H circuit structure.

Note the following for such a CDS-based S/H circuit:

- The time-independent offset (dc offset) can be totally eliminated from the sampled signal.

$$\begin{cases} V_x(nT+T) = V_{\text{offset}}\left(nT+\frac{T}{2}\right) \\ V_o(nT+T) = V_i\left(nT+\frac{T}{2}\right) \end{cases} \tag{8.30}$$

- The time-dependent offset (ac offset or other input refereed noise), CDS circuit provides a high-pass offset shaping similar to a first-order sigma-delta modulator.

$$V_o(nT+T) = V_i\left(nT+\frac{T}{2}\right) + \left[V_{\text{offset}}(nT+T) - V_{\text{offset}}\left(nT+\frac{T}{2}\right)\right] \tag{8.31}$$

$$\begin{cases} H_{\text{offset}}(z) = (1-z^{-1/2}) \\ |H_{\text{offset}}(z)|_{z=\exp(j\omega T)} = \left|2\sin\left(\frac{\omega T}{4}\right)\right| \end{cases} \tag{8.32}$$

Consequently, all low-frequency noise sources within the OPAMP circuit, such as the $1/f$ noise and power supply noise, are attenuated.

An alternative SC CDS circuit technique based on the unity-gain buffer circuit configuration is shown in Fig. 8.25.

Such a circuit works as follows:

- During the ϕ_1 phase (sampling phase), the input signal and the input-referred offset is sampled.

$$\begin{cases} Q_{h1}\left(nT+\frac{T}{2}\right) = V_i\left(nT+\frac{T}{2}\right)C_{h1} \\ Q_{h2}\left(nT+\frac{T}{2}\right) = \left(-V_{\text{offset}}\left(nT+\frac{T}{2}\right)\right)C_{h2} \end{cases} \tag{8.33}$$

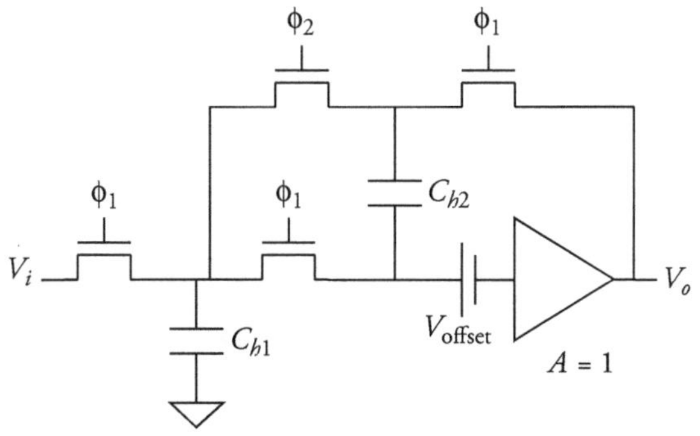

FIGURE 8.25 VLSI CDS S/H circuit based on unity-gain buffer circuit structure.

- During the ϕ_2 phase (hold phase), the offset is subtracted from the sampled input signal.

$$\begin{aligned} V_o(nT+T) &= Q_{h1}/C_{h1} - Q_{h2}/C_{h2} - V_{\text{offset}} \\ &= V_i\left(nT+\frac{T}{2}\right) + \left[V_{\text{offset}}\left(nT+\frac{T}{2}\right) - V_{\text{offset}}(nT+T)\right] \end{aligned} \tag{8.34}$$

8.3.3 Nonunity-Gain VLSI SC S/H Circuit

Figure 8.26 shows a VLSI SC S/H circuit structure that offers nonunity-gain circuit operation.

This circuit works as follows:

- At the end of the sample phase, C_1 is charged, and C_2 is discharged. The charge stored on the two capacitors are given as follows:

$$\begin{cases} Q_{C1} = \left(V_i\left(nT+\frac{T}{2}\right) - V_{\text{offset}}\left(nT+\frac{T}{2}\right)\right)C_1 \\ Q_{C2} = \left(-V_{\text{offset}}\left(nT+\frac{T}{2}\right)\right)C_2 \end{cases} \tag{8.35}$$

- At the end of the hold phase, C_1 is discharged, and C_2 is charged, and the output voltage can be derived as follows:

$$V_o(nT+T) = \frac{C_1}{C_2}V_i\left(nT+\frac{T}{2}\right) + \frac{C_1+C_2}{C_2}\left[V_{\text{offset}}(nT+T) - V_{\text{offset}}\left(nT+\frac{T}{2}\right)\right] \tag{8.36}$$

Note that the input signal can be amplified with the desired gain and that the offset of the S/H circuit can be eliminated.

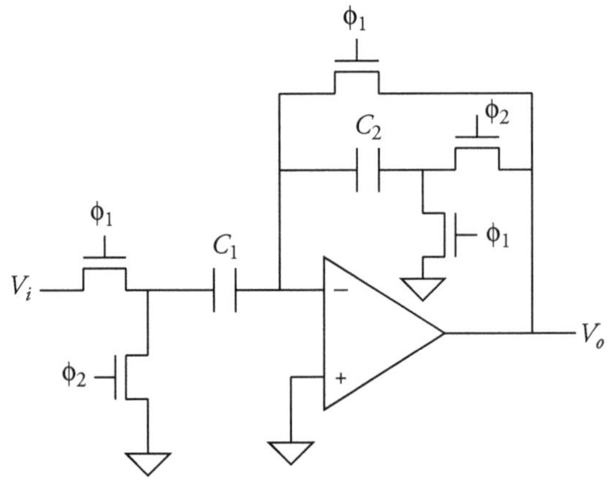

FIGURE 8.26 Nonunity-gain VLSI SC S/H circuit structure.

8.3.4 Charge Injection Compensation Techniques

Charge injection of the MOS switch in SC circuits may result to signal processing errors (typically a few millivolts). A VLSI SC charge injection–compensated circuit structure is shown in Fig. 8.27, where dummy capacitors and switches are added in the positive input terminal of the OPAMP. Such a structure matches the total capacitance and switch devices connected to the OPAMP positive and negative input terminals. Consequently, charged injections on the positive and negative terms of the OPAMP circuit are compensated in the output signal.

An alternative way to reduce the charge injection in the SC circuit employs the so-called bottom-plate sampling (BPS) technique. This technique is based on the fact that if the offset or charge injection noise is made time independent, it can easily be eliminated in the SC circuit based on certain circuit techniques.

The BPS circuit technique focuses on eliminating the signal-dependent charge injection noise.

Figure 8.28 shows a VLSI SC sampling circuit structure employing the BPS technique.

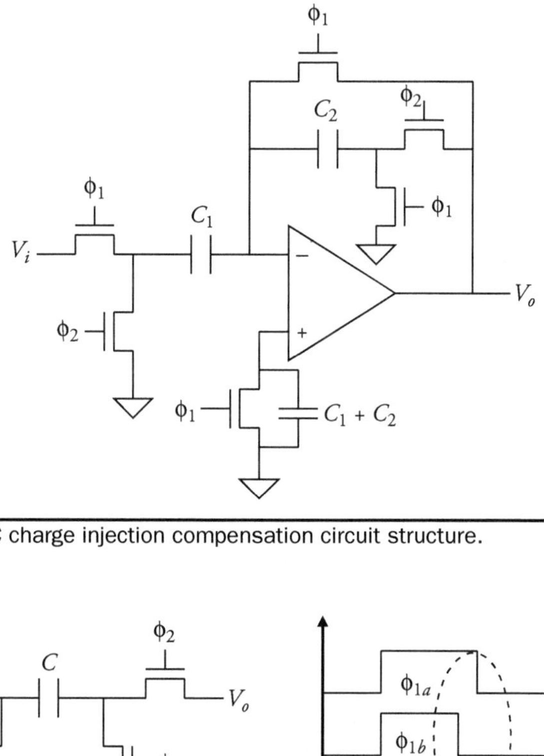

FIGURE 8.27 VLSI SC charge injection compensation circuit structure.

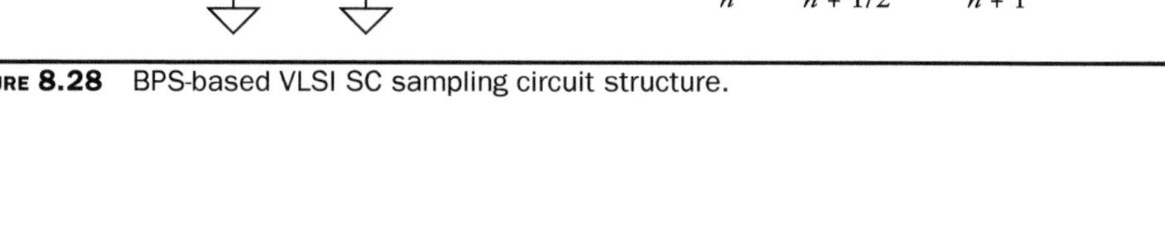

FIGURE 8.28 BPS-based VLSI SC sampling circuit structure.

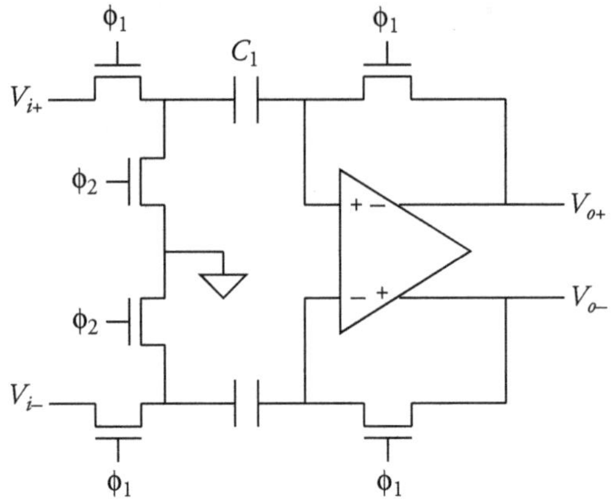

Figure 8.29 VLSI fully differential SC S/H circuit structure.

In this BPS-based SC circuit, the switch connected to the bottom plate of the sampling capacitor is turned off earlier than the switch connected to the top (input) plate. As a result, the sampling of the signal in this phase is effectively controlled by the bottom-plate switch.

Note that the bottom plate of the sampling capacitor during the sampling phase is always connected to the ground node, independent of the input signal level. As a result, the charge injection from the sampling phase is signal independent.

8.3.5 Fully Differential SC S/H Circuit

Figure 8.29 shows a VLSI fully differential SC S/H circuit structure. Such a circuit offers advantages over the single-end S/H circuit structures like better noise immunity, less charge injection, and high signal dynamic range; and it is commonly used in high-performance SC circuits. In addition, the CDS and BPS techniques can also be used in the fully differential VLSI SC S/H circuit structures for further performance enhancements.

References

1. G. M. Jacobs, D. J. Allstot, R. W. Brodersen, P. Gary, "Design Techniques for MOS Switched Capacitor Ladder Filters," *IEEE Trans. on Circuits and Systems*, vol. CAS-25, no. 12, pp. 1014–1021, December 1978.
2. K. Martin, A. S. Sedra, "Exact Design of Switched-Capacitor Bandpass Filter Using Coupled-Biquad Structures," *IEEE Trans. on Circuits and Systems*, vol. CAS-27, no. 6, pp. 469–475, December 1980.
3. M. S. Lee, G. C. Tempes, C. Chang, M. B. Ghaderi, "Bilinear Switched-Capacitor Ladder Filters," *IEEE Trans. on Circuits and Systems*, vol. CAS-28, no. 8, pp. 881–821, December 1981.
4. F. Krummenacher, "Micropower Switched-Capacitor Biquadratic Cell," *IEEE J. Solid-State Circuits*, vol. SC-17, no. 3, pp. 507–512, June 1982.

5. R. Gregorian, K. W. Martin, G. C. Tempes, "Switched-Capacitor Circuit Design," *Proc. of the IEEE*, vol. 71, no. 8, pp. 941–966, August 1983.
6. R. B. Datar, A. S. Sedra, "Exact Design of Stray-Insensitive Switched-Capacitor Ladder Filters," *IEEE Trans. on Circuits and Systems*, vol. CAS-30, no. 12, pp. 888–898, December 1983.
7. R. K. Henderson, J. I. Sewell, "Matrix Methods for Switched-Capacitor Filter Design," *ISCAS'88*, vol. 2, pp. 1021–1024, June 1988.
8. K. Chen, S. Eriksson, "Z-domain Synthesis of Switched-Capacitor LDI Ladder Filters," *ISCAS'88*, vol. 2, pp. 1013–1016, June 1988.
9. U.-K. Moon, "CMOS High-Frequency Switched-Capacitor Filters for Telecommunication Applications," *IEEE J. Solid-State Circuits*, vol. 35, no. 2, pp. 212–220. February 2000.
10. E. M. Saad, "Switched Capacitor Circuits and Their Applications, an Overview," *Proceedings, Radio Science Conference 2002*, pp. 51–59, March 2002.
11. A. S. Korotkov, "Synthesis of Narrow-band Coupled Resonator SC-Filters," *MELECON 98*, vol. 1, pp. 554–557, May 1998.
12. W. Aloisi, G. Giustolisi, G. Palumbo, "Exploiting the High-Frequency Performance of Low-Voltage Low-Power SC Filters," *IEEE Trans. on Circuits and Systems II: Express Briefs*, vol. 55, no. 2, pp. 77–84, February 2004.
13. L. Yue, J. I. Sewell, "A Comparison Study of SC Biquads in the Realization of SC Filters," *IEE Colloquium on Digital and Analogue Filters and Filtering Systems*, pp. 101/1–101/9, November 1993.
14. J. S. Pereira, A. Petraglia, "Low-Sensitivity Direct-form IIR SC Filters with Improved Phase Linearity," *ISCAS'00*, vol. 3, pp. 169–172, May 2000.
15. K. Nakayama, M. Zhiqiang, G. Yamamoto, "A Method to Minimize Total Capacitance in Cascade Realization of SC Filters," *Proc. of the 32nd Midwest Symposium on Circuits and Systems*, vol. 2, pp. 968–972, August 1989.
16. A. Handkiewicz, "Two-Dimensional SC Filters with Circularly Symmetric Response," *European Conference on Circuit Theory and Design*, pp. 252–255, September 1989.
17. M. D. Lutovac, D. Tosic, D. Novakovic, "Programmable Low-Pass/High-Pass SC-Filters," *MELECON 98*, vol. 1, May 1998.
18. Y. Lu, J. I. Sewell, "A Comparison Study of SC Biquads in the Realization of SC Filters," *ISCAS '94*, vol. 5, pp. 711–714, June 1994.
19. M. Biey, F. Montecchi, A. Premoli, "Performances of Bandpass Cascaded SC Filters with Reduced Pole-Qs," *IEEE International Symposium on Circuits and Systems*, vol. 2, pp. 1253–1256, June 1988.
20. A. T. Younis, R. E. Massara, "On the Design of Optimal Switched Capacitor Filters Based on the Use of Lossy Frequency-Dependent Negative Resistance (FDNR) Prototype Structures," *IEE 1988 Saraga Colloquium on Electronic Filters*, pp. 2/1–2/6, May 1988.
21. E. Sanchez-Sinencio, R. Geiger, J. Silva-Martinez, "Tradeoffs Between Passive Sensitivity, Output Voltage Swing, and Total Capacitance in Biquadratic SC Filters," *IEEE Trans. on Circuits and Systems*, vol. 31, no. 11, pp. 984–987, November 1984.
22. A. S. Korotkov, H. Hauer, R. Unbehauen, "Practical Design of Element Simulation Type SC Filters," *IEEE Trans. on Circuits and Systems I: Fundamental Theory and Applications*, vol. 47, no. 6, pp. 934–940, June 2000.

23. H. Baher, "Microelectronic Switched Capacitor Filters," *IEEE Circuits and Devices Magazine*, vol. 7, no. 1, pp. 33–36, January 1991.
24. J. J. F. Rijns, H. Wallinga, "Spectral Analysis of Double-Sampling Switched-Capacitor Filters," vol. 38, no. 11, pp. 1269–1279, November 1991.
25. G. W. Roberts, A. S. Sedra, "Synthesizing Switched-Current Filters by Transposing the SFG of Switched-Capacitor Filter Circuits," *IEEE Trans. on Circuits and Systems*, vol. 38, no. 3, pp. 337–340, March 1991.
26. J. Silva-Martinez, E. Sanchez-Sinencio, "Strategic SC Filter Design Based on a Comparative Study of Various S-to-z Mappings," *IEEE Trans. on Circuits and Systems*, vol. 36, no. 11, pp. 1465–1472, November 1989.
27. M. D. Lutovac, D. Novakovic, I. Markoski, "Selective SC-filters with Low Passive Sensitivity," *Electronics Letters*, vol. 33, no. 8, pp. 674–675, April 1997.
28. B. Nowrouzian, "Theory and Design of LDI Lattice Digital and Switched-Capacitor Filters," *IEE Proceedings, Circuits, Devices and Systems*, vol. 139, no. 4, pp. 517–526, August 1992.
29. A. Baschirotto, D. Bijno, R. Castello, F. Montecchi, "A 1 V 1.2 μW 4th Order Bandpass Switched-OPAMP SC Filter for a Cardiac Pacer Sensing Stage," *ISCAS'00 Geneva*, vol. 3, pp. 173–176, May 2000.
30. J. S. Pereira, A. Petraglia, M. F. Quelhas, "Approximating Linear Phase with IIR SC Filters," *ISCAS'02*, vol. 4, pp. IV-623–IV-626, May 2002.
31. J. L. Ausin, G. Torelli, J. F. Duque-Carrillo, J. M. Carrillo, P. Merchan, "Programmable Time-Multiplexed SC Filters without Dynamic Range Degradation," *IEEE International Conference on Electronics, Circuits and Systems*, vol. 1, pp. 373–376, September 1998.
32. P. Filoramo, G. Giustolisi, G. Palmisano, G. Palumbo, "An Approach to the Design of Low-Voltage SC Filters," *ISCAS'98*, vol. 1, pp. 265–268, June 1998.
33. J. Goette, A. Kaelin, W. Guggenbuhl, "Criteria and Methods for the Selection of Optimized Capacitor Values in SC-Filters," *ISCAS'89*, vol. 3, pp. 1696–1700, May 1989.
34. J. Silva-Martinez, E. Sanchez-Sinencio, "A Sparing SC Filter Design Approach with Reduced Transmission Zeros," *ISCAS'88*, vol. 2, pp. 1721–1724, June 1988.
35. B. G. Lofmark, "SC-filter Structures with Reduced Requirements on Op-amp Bandwidth and Settling Time," *ISCAS'88*, vol. 2, pp. 1473–1477, June 1988.
36. E. S. Sinencio, J. S. Martinez, R. Geiger, "Biquadratic SC Filters with Small GB Effects," *IEEE Trans. on Circuits and Systems*, vol. 31, no. 10, pp. 876–884, October 1984.
37. P. Van Peteghem, W. Sansen, "Power Consumption Versus Filter Topology in SC Filters," *IEEE Trans. on Circuits and Systems*, vol. 33, no. 2, pp. 150–157, February 1986.
38. A. S. Korotkov, K. H. Feistel, R. Unbehauen, "Synthesis of Switched-Capacitor Parametric Bandpass Filters," *IEEE Trans. on Circuits and Systems I: Fundamental Theory and Applications*, vol. 46, no. 4, pp. 484–490, April 1999.
39. D. Vazquez, A. Rueda, J. L. Huertas, E. Peralias, "A High-Q Bandpass Fully Differential SC Filter with Enhanced Testability," *IEEE J. of Solid-State Circuits*, vol. 33, no. 7, pp. 976–986, July 1998.
40. A. Baschirotto, R. Castello, "A 1-V 1.8-MHz CMOS Switched-OPAMP SC Filter with Rail-to-Rail Output Swing," *IEEE J. of Solid-State Circuits*, vol. 32, no. 12, pp. 1979–1986, December 1997.

41. B. Maundy, E. I. El-Masry, "A Switched Capacitor Bidirectional Associative Memory," *IEEE Trans. on Circuits and Systems*, vol. 37, no. 12, pp. 1568–1572, December 1990.
42. R. Castello, A. G. Grassi, S. Donati, "A 500-nA Sixth-Order Bandpass SC Filter," *IEEE J. of Solid-State Circuits*, vol. 25, no. 3, pp. 669–676, June 1990.
43. P. M. Van Peteghem, "On the Relationship between PSRR and Clock Feed through in SC Filters," *IEEE J. of Solid-State Circuits*, vol. 23, no. 4, pp. 997–1004, August 1988.
44. P. Filoramo, G. Giustolisi, G. Palmisano, G. Palumbo, "Approach to the Design of Low-Voltage SC Filters," *IEE Proceedings, Circuits, Devices and Systems*, vol. 147, no. 3, pp. 196–200, June 2000.
45. M. C. Schneider, C. Galup-Montoro, J. C. M. Bermudez, "Explicit Formula for Harmonic Distortion in SC Filters with Weakly Nonlinear Capacitors," *IEE Proceedings Circuits, Devices and Systems*, vol. 141, no. 6, pp. 505–509, December 1994.
46. A. Dabrowski, U. Menzi, G. S. Moschytz, "Design of Switched-Capacitor FIR Filters with Application to a Low-Power MFSK Receiver," *IEE Proceedings, Circuits, Devices and Systems*, vol. 139, no. 4, pp. 450–466, August 1992.
47. M. F. Quelhas, A. Petraglia, A. P. Baruqui, "Power Consumption Estimation in SC Filters," *IEEE International Symposium on Industrial Electronics*, vol. 1, pp. 298–300, June 2003.
48. A. Baschirotto, R. Castello, "A 1V 1.8MHz CMOS Switched-OPAMP SC Filter with Rail-to-Rail Output Swing," *44th ISSCC*, pp. 58–59, February 1997.
49. A. S. Korotkov, "Predistortion Technique for Element Simulation Type SC-Filters," *MELECON '96*, vol. 3, pp. 1259–1262, May 1996.
50. J. L. Ausin, R. Perez-Aloe, J. F. Duque-Carrillo, G. Torelli, E. Sanchez-Sinencio, "High-Selectivity SC Filters with Continuous Digital Q-Factor Programmability," *ISCAS'02*, vol. 4, pp. IV-631–IV-634, May 2002.
51. R. Harjani, B. Vinnakota, "Digital Detection of Analog Parametric Faults in SC Filters," *Proceedings, 36th Design Automation Conference*, pp. 772–777, June 1999.
52. U. Kleine, T. Pasch, "Low-Voltage Medium Q-SC Filters for High Frequency Communication Applications," *1998 IEEE International Conference on Electronics, Circuits and Systems*, vol. 2, pp. 39–42, September 1998.
53. U. Marschner, W.-J. Fischer, E.-G. Kranz, "Low Expense Architectures for a Dynamic Spectrum Analyzer Based on SC-Filters," *ISCAS'98*, vol. 2, pp. 276–279, June 1998.
54. A. Baschirotto, F. Montecchi, R. Castello, "A 150 M sample/s 20 mW BiCMOS Switched-Capacitor Biquad Using Precise Gain Op Amps," *42nd ISSCC*, pp. 212–213, February 1995.
55. G. H. Wang, K. Watanabe, "Current-Mode Switched-Capacitor Circuit Synthesis by C-invariant Dual Transformation," *ISCAS'91*, vol. 3, pp. 1553–1556, June 1991.
56. X. F. Wania, D. A. Johns, A. S. Sedra, "Programmable Multiplexed Switched-Capacitor Filters," *Proceedings of the 33rd Midwest Symposium on Circuits and Systems*, vol. 2, pp. 973–976, August 1990.
57. A Baschirotto, R Castello, F. Montecchi, "Finite Gain Compensation Techniques for High-Q Bandpass SC Filters," *ISCAS'90*, vol. 4, pp. 23813–2816, May 1990.
58. R. Geiger, E. Sanchez-Sinencio, "Operational Amplifier Gain-Bandwidth Product Effects on the Performance of Switched-Capacitor Networks," *IEEE Trans. on Circuits and Systems*, vol. 29, no. 2, pp. 96–106, February 1982.

59. R. Castello, P. Gray, "Performance Limitations in Switched-Capacitor Filters," *IEEE Trans. on Circuits and Systems*, vol. 32, no. 9, pp. 865–876, September 1985.
60. D. Bruckmann, U. Kleine, "Novel Voltage Inverter Switches with Minimum Sensitivity Properties," *IEEE Trans. on Circuits and Systems*, vol. 32, no. 7, pp. 723–726, July 1985.
61. P. Li, J. I. Sewell, "The LUD Approach to Switched-Capacitor Filter Design," *IEEE Trans. on Circuits and Systems*, vol. 34, no. 12, pp. 1611–1614, December 1987.
62. P. M. Van Peterghem, W. M. C. Sansen, "Power Consumption versus Filter Topology in SC Filters," *IEEE J. of Solid-State Circuits*, vol. 21, no. 1, pp. 40–47, February 1986.
63. P. Landau, D. Michel, D. Melnik, "A Reduced Capacitor Spread Algorithm for Elliptic Bandpass SC Filters," *IEEE J. of Solid-State Circuits*, vol. 22, no. 4, pp. 624–626, August 1987.
64. W. M. C. Sansen, H. Qiuting, K. A. I. Halonen, "Transient Analysis of Charge-Transfer in SC Filters-gain Error and Distortion," *IEEE J. of Solid-State Circuits*, vol. 22, no. 2, pp. 268–276, April 1987.
65. J. L. Ausin, J. F. Duque-Carrillo, G. Torelli, E. Sanchez-Sinencio, "Switched-Capacitor Circuits with Periodical Nonuniform Individual Sampling," *IEEE Trans. on Circuits and Systems II: Analog and Digital Signal Processing*, vol. 50, no. 8, pp. 404–414, August 2003.
66. L. Lentola, G. M. Corelazzo, E. Malavasi, A. Baschirotto, "Design of SC Filters for Video Applications," *IEEE Trans. on Circuits and Systems for Video Technology*, vol. 10, no. 1, pp. 14–22, February 2000.
67. B. Vinnakota, R. Harjani, "DFT for Digital Detection of Analog Parametric Faults in SC Filters," *IEEE Trans. on Computer-Aided Design of Integrated Circuits and System*, vol. 19, no. 7, pp. 789–798, July 2000.
68. R. P. Martins, J. E. Franca, "Cascade Switched-Capacitor IIR Decimating Filters," *IEEE Trans. on Circuits and Systems I: Fundamental Theory and Applications*, vol. 42, no. 7, pp. 367–376, July 1995.
69. R. P. Martins, J. E. Franca, F. Maloberti, "An Optimum CMOS Switched-Capacitor Anti-Aliasing Decimating Filter," *IEEE J. of Solid-State Circuits*, vol. 28, no. 9, pp. 962–970, September 1993.
70. J. L. Huertas, A. Rueda, D. Vazquez, "Testable Switched-Capacitor Filters," *IEEE J. of Solid-State Circuits*, vol. 28, no. 7, pp. 719–724, July 1993.
71. A. Baschirotto, R. Castello, F. Montecchi, "IIR Double-Sampled Switched-Capacitor Decimators for High-Frequency Applications," *IEEE Trans. on Circuits and Systems I: Fundamental Theory and Applications*, vol. 39, no. 4, pp. 300–304, April 1992.
72. P. Mehta, M. Darwish, T. Thomson, "Switched-Capacitor Filters," *IEEE Trans. on Power Electronics*, vol. 5, no. 3, pp. 331–336, July 1990.
73. C.-Y. Wu, T.-C. Yu, S.-S. Chang, "New Monolithic Switched-Capacitor Differentiators with Good Noise Rejection," *IEEE J. of Solid-State Circuits*, vol. 24, no. 1, pp. 177–180, February 1989.
74. J. C. M. Bermudez, M. C. Schneider, C. G. Montoro, "Compatibility of Switched Capacitor Filters with VLSI Processes," *IEE Proceedings G, Circuits, Devices and Systems*, vol. 139, no. 4, pp. 413–418, August 1992.
75. E. I. El-Masry, E. G. Hix, "Novel Switched-Capacitor Integrator for High-Frequency Applications," *IEE Proceedings G, Circuits, Devices and Systems*, vol. 136, no. 5, pp. 263–267, October 1989.

76. D. Novakovic, B. Loncar, P. Osmokrovic, "Efficient High-Pass Low-Sensitive Selective SC-Filters," *22nd International Conference on Microelectronics*, 2000, vol. 2, pp. 739–742, May 2000.
77. I. Seo, R. M. Fox, "Low-Power Switched-Capacitor Filters Using Charge Transfer Integrators," *ISCAS'03*, vol. 1, pp. I-609–I-612, May 2003.
78. K. Hajek, J. Sedlacek, "General Multiple LC Prototype Filter Solutions and Optimization," *9th International Conference on Electronics, Circuits and Systems*, vol. 1, pp. 165–168, September 2002.
79. J. M. Canive, A. Petraglia, "On the Testability of SC Filters Based on Allpass Sections," *ISCAS'01*, vol. 1, pp. 65–68, May 2001.
80. A. Baschirotto, "IIR Double-Sampled Switched-Capacitor Building Blocks for High-Frequency Decimators," *ISCAS'91*, vol. 3, pp. 1673–1676, June 1991.
81. B. Nowrouzian, "A Novel Approach to the Exact Design of LDI Symmetrical Digital and Switched-Capacitor Filters," *Proceedings of the 33rd Midwest Symposium on Circuits and Systems*, vol. 2, pp. 967–972, August 1990.
82. R. E. Massara, A. T. Younis, "An Efficient Design Method for Optimal MOS Integrated Circuit Switched-Capacitor LDI Ladder Filters," *Proceedings of the 33rd Midwest Symposium on Circuits and Systems*, 1990, vol. 2, pp. 956–959, August 1990.
83. R. Raut, B. B. Bhattacharrya, S. M. Faruque, "An Application of Systolic Array Design Architecture to Switched Capacitor Filter Circuits," *ICASSP-89*, vol. 4, pp. 2401–2404, May 1989.
84. M. Nalecz, J. J. Mulawka, "Parasitic-Compensated Building Blocks for Active Switched-Capacitor Filters," *ISCAS'88*, vol. 2, pp. 1483–1486, June 1988.
85. A. M. Davis, H. P. Nguyen, "Exact Synthesis of SC Filters Using Forward Difference Integrators," *ISCAS'88*, vol. 2, pp. 1009–1012, June 1988.
86. S. Setty, C. Toumazou, K. Manetakis, P. Cheung, R. Spence, "Ideas and Concepts for the Automation of Analogue Filters Using Qualitative Reasoning," *IEE 15th SARAGA Colloquium on Digital and Analogue Filters and Filtering Systems*, pp. 7/1–7/9, November 1995.
87. H. Shafeeu, A. Betts, J. Taylor, "Approaches to Ultra-Narrow-Band Analogue IC Filter Design Using Switched-Capacitors," *IEE Colloquium on Advances in Analogue VLSI*, pp. 2/1–2/9, May 1991.
88. G. Temes, H. Orchard, M. ahanbegloo, "Switched-Capacitor Filter Design Using the Bilinear z-Transform," *IEEE Transactions on Circuits and Systems*, vol. 25, no. 12, pp. 1039–1044, December 1978.
89. M. Lee, C. Chang, "Switched-Capacitor Filters Using the LDI and Bilinear Transformations," *IEEE Trans. on Circuits and Systems*, vol. 28, no. 4, pp. 265–270, April 1981.
90. E. El-Masry, "Strays-Insensitive State-Space Switched-Capacitor Filters," *IEEE Trans. on Circuits and Systems*, vol. 30, no. 7, pp. 474–488, July 1983.
91. J. Garcia-Vazquez, E. Sanchez-Sinencio, "Finite Gain-Bandwidth Product Effects on a Pair of Pseudo-N-Path SC Filters," *IEEE Trans. on Circuits and Systems*, vol. 31, no. 6, pp. 583–584, June 1984.
92. C. Campbell, K. Reineck, "A Pole/Zero Prewarping Procedure in SC Filter Design," *IEEE Trans. on Circuits and Systems*, vol. 31, no. 9, pp. 821–825, September 1984.
93. G. Fischer, G. Moschytz, "SC Filters for High-Frequencies with Compensation for Finite-Gain Amplifiers," *IEEE Trans. on Circuits and Systems*, vol. 32, no. 10, pp. 1050–1056, October 1985.

94. T. Inoue, F. Ueno, "On Robustness of the Stability of Multiphase Switched-Capacitor Two-Ports," *IEEE Trans. on Circuits and Systems*, vol. 32, no. 6, pp. 522–529, June 1985.
95. J. Da Franca, "A Single-Path Frequency-Translated Switched-Capacitor Bandpass Filter System," *IEEE Trans. on Circuits and Systems*, vol. 32, no. 9, pp. 938–944, September 1985.
96. A. Davis, "Flow Graph Synthesis of Darlington – Cauer Switched-Capacitor Filters," *IEEE Trans. on Circuits and Systems*, vol. 32, no. 7, pp. 727–732, July 1985.
97. M. H. Fino, J. E. Franca, A. Steiger-Garcao, "Automatic Symbolic Analysis of Switched-Capacitor Filtering Networks Using Signal Flow Graphs," *IEEE Trans. on Computer-Aided Design of Integrated Circuits and Systems*, vol. 14, no. 7, pp. 858–867, July 1995.
98. J. Crols, M. Steyaert, "Switched-OPAMP: An Approach to Realize Full CMOS Switched-Capacitor Circuits at Very Low Power Supply Voltages," *IEEE J. of Solid-State Circuits*, vol. 29, no. 8, pp. 936–942, August 1994.
99. C.-Y. Wu, P.-H. Lu, M.-K. Tsai, "Design Techniques for High-frequency CMOS Switched-Capacitor Filters Using Non-OPAMP-Based Unity-Gain Amplifiers," *IEEE J. of Solid-State Circuits*, vol. 26, no. 10, pp. 1460–1466, October 1991.
100. A. Handkiewicz, "Two-Dimensional Switched Capacitor Filter Design System for Real-Time Image Processing," *IEEE Trans. on Circuits and Systems for Video Technology*, vol. 1, no. 3, pp. 241–246, September 1991.
101. K. Nakayama, Y. Takahashi, Y. Sato, Y. Nukada, "An Adaptive SC Line Equalizer System for Four-Wire Full-Duplex and Multirate Digital Transmission," *IEEE Transactions on Circuits and Systems*, vol. 35, no. 9, pp. 1073–1081, September 1988.
102. A. E. Salama, "Factored State-Space Stray-Insensitive Switched-Capacitor Filters," *IEE Proceedings, Circuits, Devices and Systems*, vol. 143, no. 6, pp. 325–330, December 1996.
103. T. F. Yong, M. J. Long, "Systematic Method for Design of Switched Capacitor Filters," *Electronics Letters*, vol. 28, no. 3, pp. 309–310, January 1992.
104. Li Qiang, H. Yifeng, X. Ke, M. Hao, "A Novel Offset Compensation biquad Switched-Capacitor Filter Design," *5th International Conference on ASIC*, vol. 1, no. 21–24, pp. 643–646, October 2003.
105. J. L. Ausin, J. F. Duque-Carillo, G. Torelli, E. Sanchez-Sinencio, F. Maloberti, "Periodical Nonuniform Individually Sampled Switched-capacitor Circuits," *ISCAS'00*, vol. 5, pp. 449–452, May 2000.
106. H. Matsumoto, Z. Tang, O. Ishizuka, "A Buffer-Based Switched-Capacitor Integrator with Reduced Capacitance Ratio," *Proceedings of the 33rd Midwest Symposium on Circuits and Systems*, vol. 2, pp. 719–720, August 1990.
107. J. A. Hegt, "Signal Flow Graph Based Synthesis of Strays-Insensitive Switched-Capacitor Filters," *ISCAS'90*, vol. 3, pp. 2177–2180, May 1990.
108. I. M. Sahawaneh, A. A. Sakla, E. I. El-Masry, "Comparison between Cascade and Parallel Switched Capacitor Structures," *Proceedings of the 32nd Midwest Symposium on Circuits and Systems*, vol. 2, pp. 973–976, August 1989.
109. A. K. Betts, J. T. Taylor, D. G. Haigh, "Synthesis Method for Switched-Capacitor FIR Decimators and Interpolators," *ISCAS'88*, vol. 3, pp. 2463–2466, June 1988.
110. Q. Huang, "A Novel Technique for the Reduction of Capacitance Spread in High Q SC Circuits," *ISCAS'88*, vol. 2, pp. 1249–1252, June 1988.

111. C. Kurth, G. Moschytz, "Two-Port Analysis of Switched-Capacitor Networks Using Four-Port Equivalent Circuits in the Z-domain," *IEEE Trans. on Circuits and Systems*, vol. 26, no. 3, pp. 166–180, March 1979.
112. U. Brugger, D. von Grunigen, G. Moschytz, "A Comprehensive Procedure for the Design of Cascaded Switched-Capacitor Filters," *IEEE Trans. on Circuits and Systems*, vol. 28, no. 8, pp. 803–810, August 1981.
113. K. Martin, A. Sedra, "Effects of the Op Amp Finite Gain and Bandwidth on the Performance of Switched-Capacitor Filters," *IEEE Transactions on Circuits and Systems*, vol. 28, no. 8, pp. 822–829, August 1981.
114. S. Wong, C. Salama, "A Switched Differential Op-amp with Low Offset and Reduced 1/f Noise," *IEEE Trans. on Circuits and Systems*, vol. 33, no. 11, pp. 1119–1127, November 1986.
115. F. A. P. Baruqui, A. Petraglia, J. E. Franca, S. K. Mitra, "Efficient Design of Integrated Switched-Capacitor Decimation Filters," *IEEE Trans. on Circuits and Systems II: Analog and Digital Signal Processing*, vol. 47, no. 11, pp. 1314–1318, November 2000.
116. B. Raahemi, A. Opal, "Group Delay and Group Delay Sensitivity of Periodically Switched Linear Networks," *IEEE Trans. on Circuits and Systems I: Fundamental Theory and Applications*, vol. 47, no. 1, pp. 96–104, January 2000.
117. A. Petraglia, "Random Capacitance Ratio Error Effects in the Frequency Response of Switched-capacitor Filters," *IEEE-CAS Region 8 Workshop on Analog and Mixed IC Design*, pp. 112–116, September 1996.
118. K. van Hartingsveldt, P. Quinn, A. van Roermund, "A 10.7MHz CMOS SC Radio IF Filter with variable Gain and a Q of 55," *ISSCC'00*, pp. 152–153, 452, February 2000.
119. M. M. Amourah, R. L. Geiger, "Gain and Bandwidth Boosting Techniques for High-Speed Operational Amplifiers," *ISCAS'01*, vol. 4, pp. 674–677, May 2001.
120. A. Hossein Nejad-Malayeri, T. Manku, "A 270MHz, 1.8V Fully Differential CMOS Operational Amplifier for Switched Capacitor Channel Select Filters in Wide-band Wireless Applications," *ISCAS'01*, vol. 1, pp. 659–662, May 2001.
121. B. Manai, P. Loumeau, "A Wide-band CMOS Switched Capacitor Filter for UMTS Direct Conversion Receiver," *MWSCAS'01*, vol. 1, pp. 377–380, August 2001.
122. Z. Kaelin, G. S. Moschytz, "Exact Design of Arbitrary Parasitic-Insensitive Elliptic SC-Ladder Filters in the z-domain," ISCAS'88, vol. 3, pp. 2485–2488, June 1988.
123. T. S. Fiez, B. Lee, D. J. Allstot, "CMOS Switched-Current Biquadratic Filters," *ISCAS'90*, vol. 3, pp. 2300–2303, May 1990.
124. C.-Y. Wu, T.-C. Yu, "New Forward- and Backward-Mapping SC Differentiator and Their Applications in the Design of Biquad SCF," *Proceedings of Technical Papers VLSI Technology, Systems and Applications*, pp. 169–173, May 1989.
125. M. J. Svihura, B. Nowrouzian, "A New Approach to the Design of Bilinear-LDI Switched-Capacitor Filters Having Low Passband Sensitivity," *Proceeding, IEEE Pacific Rim Conference on Communications, Computers and Signal Processing*, pp. 193–197, June 1989.
126. R. E. Vallee, E. I. El-Masry, "High Performance CMOS Operational-Amplifier," *ISCAS'89*, vol. 2, pp. 1475–1478, May 1989.
127. S. Michael, "Stray Insensitive Programmable Switched-Capacitor Filter," *Proceedings of the 32nd Midwest Symposium on Circuits and Systems*, vol. 2, pp. 1189–1192, August 1989.

CHAPTER 9

Introduction to VLSI Switched-Current Circuit Techniques

VLSI switched-capacitor circuits suffer from several fundamental design constraints such as requiring high-gain OPAMP or OTA circuits and linear capacitors. Such constraints make them difficult to be implemented in certain VLSI process technologies that do not offer these circuit structures. VLSI switched-current (SI) circuits were developed as an alternative discrete-time (DT) signal processing circuits that can be directly realized using the fully digital VLSI process technologies. The basic VLSI SI circuit structure is based on the switched MOS current-mirror circuits that use the nonlinear MOS capacitors to store charge. The SI circuit does not require highly linear capacitors. They also do not need the OPAMP or OTA circuits. As a result, SI circuits are fully digital CMOS process technology compatible. Similar to the current-mirror-based VLSI CTI circuits, VLSI SI circuits are current-mode signal processing circuits that employ the current for signal presentation and processing. Fundamental limitations of SI circuits are the switch charge injection, final output impedance, and mismatch effects of the MOS transistors in the current-mirror circuits.

9.1 Basic VLSI Switched-Current Circuit

Figure 9.1 shows a basic VLSI SI sample/hold (S/H) circuit structure. Such a circuit is based on a switched MOS current mirror employing MOS switch (SW) device, supporting the signal sampling and hold operations.

The basic SI circuit works as follows:

- During the sampling phase ($\phi = 1$), the sampling switch is in the *on* state, and the SI circuit works as a current mirror. In this phase, the output current signal tracks the input current signal. The input and output current signal relation is fully determined by the size ratio of the two input and output MOS devices as

$$i_o(t) = -\frac{\beta_2}{\beta_1} i_i(t) \quad \left(n < \frac{t}{T} < n + \frac{1}{2}\right) \tag{9.1}$$

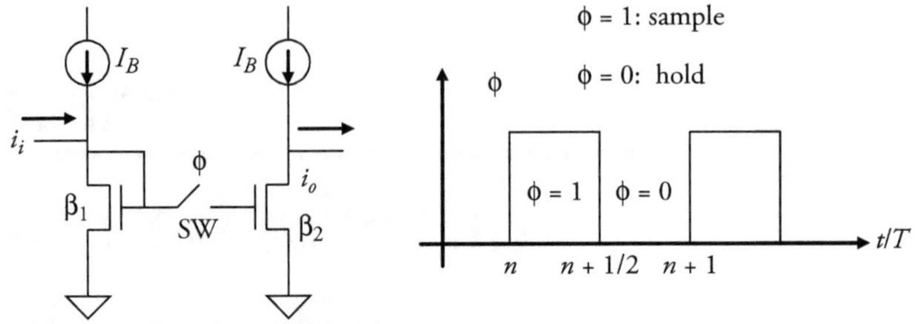

FIGURE 9.1 Basic VLSI SI circuit structure.

- During the hold phase ($\phi = 0$), the sampling switch is in the *off* state, and the sampled voltage value on the gate of the M_2 device is held throughout the entire hold phase. Consequently, the output current in this phase holds the previously sampled input current value as follows:

$$i_o(t) = -\frac{\beta_2}{\beta_1} i_i(t)|_{\frac{t}{T}=n+1/2} \quad \left(n+\frac{1}{2} \leq \frac{t}{T} < n+1\right) \tag{9.2}$$

9.2 Basic VLSI Switched-Current Signal Processing Elements

A set of VLSI signal processing circuits can be constructed based on the SI circuit structures. These circuits can be used to realize the basic S/H, the delay, the addition, and the scaling operations necessary to supporting various VLSI analog signal processing applications.

9.2.1 Switched-Current Unity Delay Circuit

The SI unity delay circuit is the most basic circuit block in the sampled-data signal processing operation where the output signal and the input signal are simply related in the time and z-domain by the following expressions:

$$i_o(nT) = i_i(nT - T) \tag{9.3}$$

and

$$\frac{I_o}{I_i} = z^{-1} \tag{9.4}$$

The VLSI SI unity delay circuit can be constructed from two identical SI S/H circuits, controlled by nonoverlap clocks as shown in Fig. 9.2.

For practicality, it is sometimes convenient to use a half-cycle delay SI circuit in DT signal processing operation such as the LDI integration. Such a signal processing operation can be expressed by the following equations:

$$i_o(nT) = i_i(nT - T/2) \tag{9.5}$$

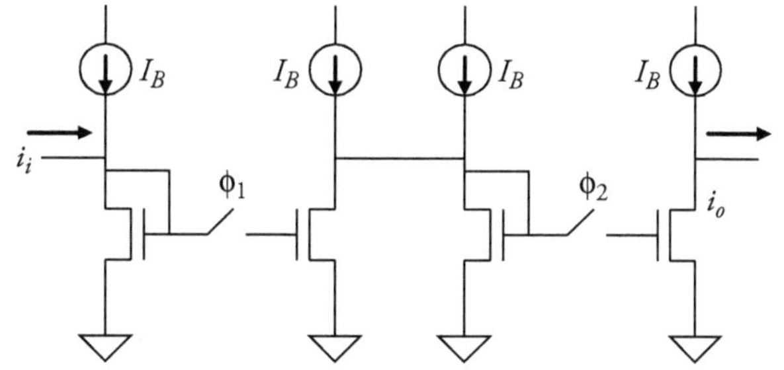

FIGURE 9.2 Basic VLSI SI unity delay circuit structure.

and

$$\frac{I_o}{I_i} = z^{-1/2} \tag{9.6}$$

In this case, the basic VLSI SI S/H circuit of the same device size can directly be used (see Fig. 9.3). However, one thing should be kept in mind that this circuit will require the input of the DT signal to be clocked by an opposite clock phase for proper operation due to the transparent nature of the S/H circuit at the sample phase.

9.2.2 Switched-Current Adder Circuit

Due to the current nature of the SI signals, addition operation in the SI circuit can be directly realized by wired addition similar to the CTI circuit implementation.

Note that in practical VLSI SI circuit implementation, it is desired that the addition operation to be realized is combined with other signal processing operation (such as delay and scaling operations) to minimize the device count required for overall signal processing operation.

The time and z-domain operation of the combined addition with half-cycle delay operation can be expressed as follows:

$$i_o(nT) = -[i_{i1}(nT - T/2) + i_{i2}(nT - T/2)] \tag{9.7}$$

$$I_o = -z^{-1/2}(I_{i1} + I_{i2}) \tag{9.8}$$

It can be seen that the SI circuit shown in Fig. 9.4 can realize such an operation.

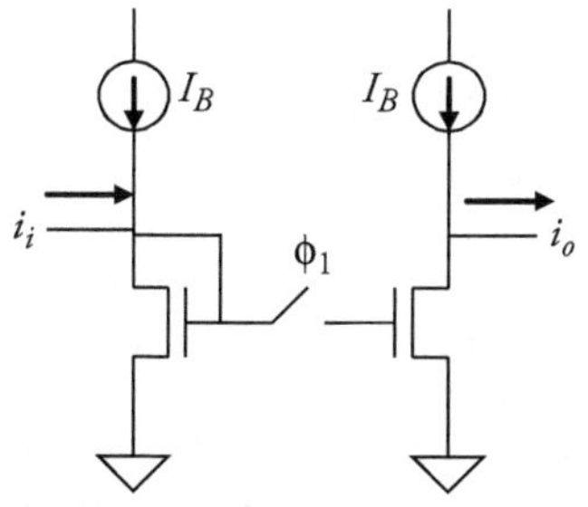

FIGURE 9.3 Basic VLSI SI half-cycle delay circuit structure.

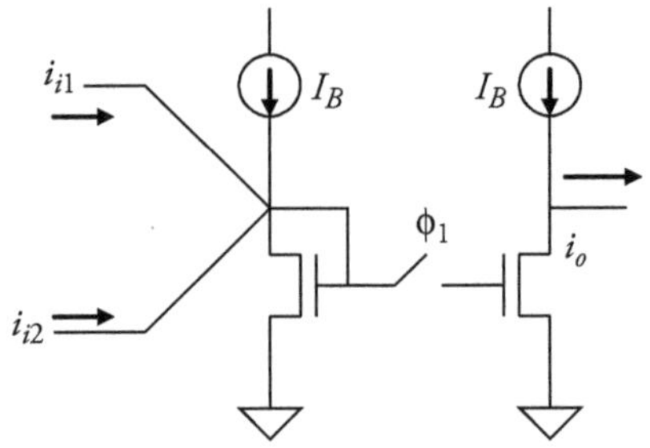

FIGURE 9.4 VLSI SI addition circuit structure with half-cycle delay.

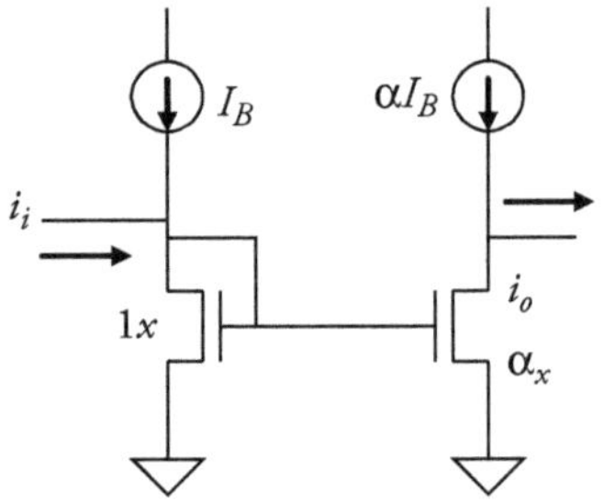

FIGURE 9.5 VLSI SI/CTI scaling circuit structure.

9.2.3 Switched-Current Scaling Circuit

Scaling operation in the VLSI SI circuit as expressed in Eq. (9.9) can be directly realized through the proper sizing of the input and output devices in the basic VLSI current-mirror circuit (see Fig. 9.5). By selecting the output device and the bias current to be α times of that of the input device, we have the following:

$$i_o(nT) = -\alpha[i_i(nT)] \tag{9.9}$$

9.3 VLSI Dynamic Switched-Current-Mirror Structures

The basic (or first-generation) VLSI SI S/H circuit structures suffer from accuracy limitation due mainly to the nonideal effects of the current mirror such as device parameter mismatch, finite output impedance, and switching charge injection. In practical VLSI SI circuit design, dynamic S/H (also known as the second-generation current-mirror) circuit structures can be employed to solve the device mismatch problem. The basic dynamic current S/H circuit is shown in Fig. 9.6, which consists of a single MOS transistor that is used for both the input and the output operations controlled by three clock phases.

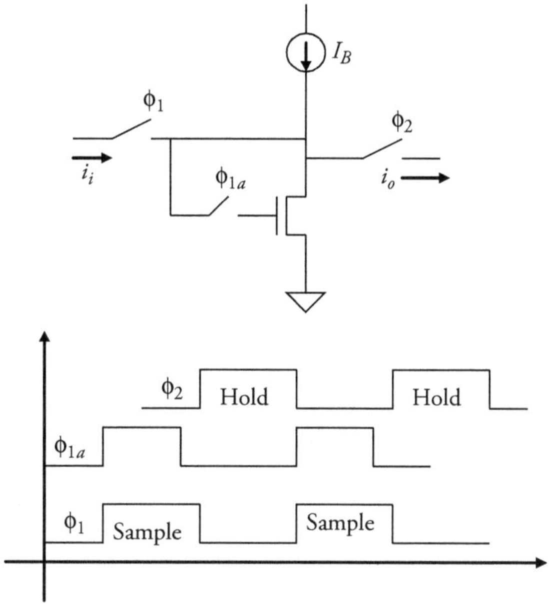

FIGURE 9.6 Basic VLSI dynamic SI S/H circuit structure.

Such a dynamic SI S/H circuit works as follows:

- During the sample phase, the SI S/H circuit is connected as an MOS diode circuit as shown in Fig. 9.7. The gate voltage of the input MOS device tracks the input current signal that can be derived from the saturation mode MOS device *V-I* characteristic equation as follows:

$$V_g(t) = V_T + \sqrt{\frac{2(I_i(t) + I_B)}{\beta}} \quad (nT < t < nT + T/2) \tag{9.10}$$

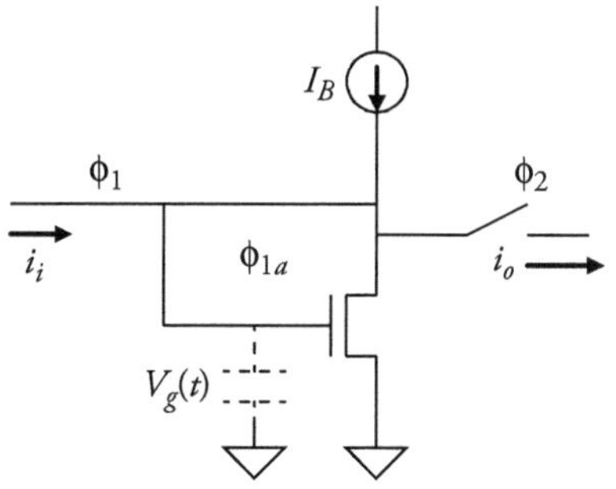

FIGURE 9.7 Equivalent SI dynamic S/H circuit during the sample phase.

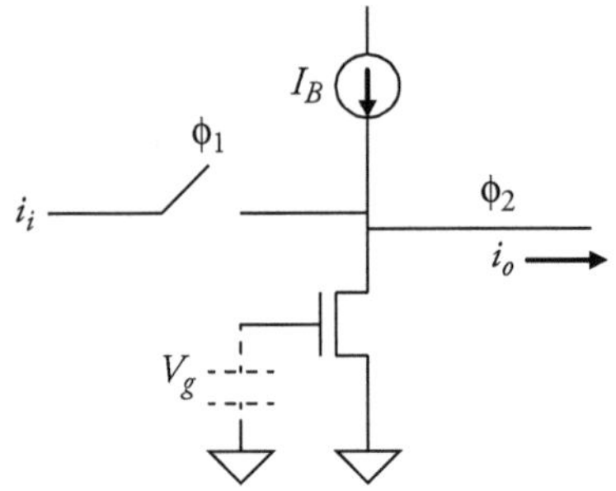

FIGURE 9.8 Equivalent SI dynamic sample/hold circuit during the hold phase.

This voltage is stored on the parasitic gate capacitor of the input MOS device. Note that during this phase, the output current is always zero, that is, not available to external circuit.

- During the hold phase, the sampled charge trapped on the parasitic gate capacitor of the input MOS device will continue being trapped during this phase as shown in Fig. 9.8. As a result, the drain current of the input MOS device will maintain the same value just before the sampling switch to the gate of the input MOS device is turned off.

$$i_o(t) = -i_i(t)\,|_{\frac{t}{T}=n+1/2} \left(nT + \frac{T}{2} \le t < nT + T\right) \tag{9.11}$$

Note that since both sample and hold circuits use the same MOS device in a time-sharing fashion, the size mismatch between the sample and hold devices can be effectively eliminated.

9.4 Switched-Current Charge Injection Compensation Techniques

Similar to the VLSI SC circuits, VLSI SI circuits suffer from the charge injection effects resulting from the switching device in the SI circuits. Charge injection has been one of the fundamental limitations in practical VLSI SI circuit structures, and a large number of charge injection compensation techniques have been developed in the past years.

9.4.1 Charge Injection Attenuation Circuits

A VLSI SI-circuit charge injection attenuation technique is shown in Fig. 9.9. This technique relies on adaptively controlling the capacitance of the sampling charge storage node such that a larger capacitance occurs at the end of the sample phase based on the Miller capacitor enhancement techniques.

This circuit operates as follows:

- During the sample phase, switches S_{1a}, S_{1b}, and S_{1c} are turned off in sequence. Charge injection occurs when switch S_{1a} is turned off. However, since the other side of added Miller capacitor C_M is connected to the drain of M_2 through switch

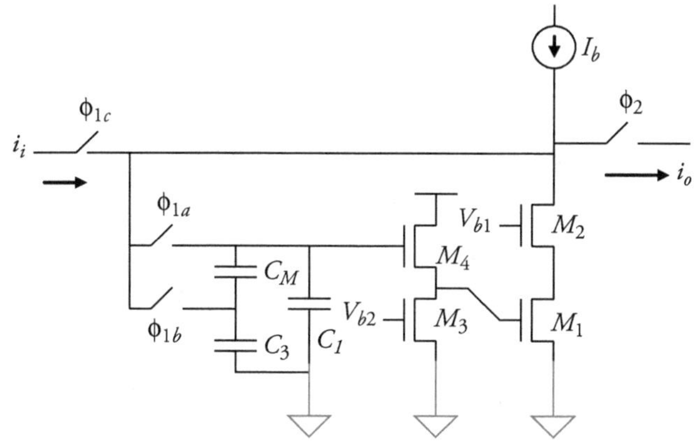

Figure 9.9 VLSI SI charge injection attenuation technique.

S_{1b}, such a connection results in significantly higher effective sampling gate capacitance because of the Miller effect.

- The node capacitance looked from S_{1a} is enlarged in this case by the loop gain A, and it is approximately given as follows:

$$C_{\text{eff}} = C_1 + (1 + A)C_M \tag{9.12}$$

- Due to the large capacitance presented at the sampling node, the charge injection due to switch S_{1a} is highly attenuated.
- When S_{1b} is turned off, charge may still inject to the storage node through C_M. Such a charge injection effect can be minimized using a smaller C_M value.

An alternative VLSI SI charge attenuation circuit structure based on the dynamic Miller effect compensation method is shown in Fig. 9.10. Such a circuit has a controllable loop gain. The loop gain of the circuit is set to the low-gain state during the sample

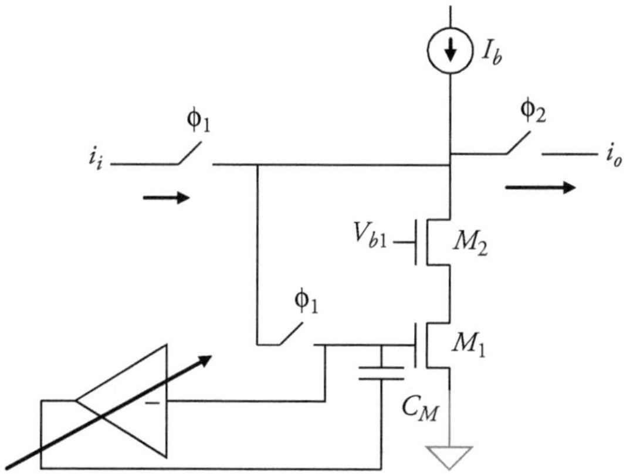

Figure 9.10 Alternative VLSI charge injection attenuation SI circuit structure.

phase and the hold phases. The loop is set to the high-gain state during the sample-to-hold transitions. The Miller effect enhanced by the high-gain feedback loop helps to minimize the charge injection effect.

9.4.2 Charge Injection Cancellation Circuits

The SI charge injection cancellation circuit technique based on the replica charge injection cancellation effect is shown in Fig. 9.11. As the dummy path is selected to be identical to the main signal path, the charge injections from the main signal path and the dummy path are expected to be equal and will cancel each other in the main signal path. The above replica circuit technique can also be implemented using the dynamic SI circuit as shown in Fig. 9.12.

An alternative SI circuit is shown in Fig. 9.13. In this circuit, the two switching transistors are designed to be identical, and one of the output branches is double sized. The output current is taken as the difference of the two output ports.

The two output currents are scaled by a factor of 2. On the other hand, the charge injection–induced current errors are scaled by a factor of 1. Consequently, the difference of the current from the two output ports eliminates the charge injection–induced error.

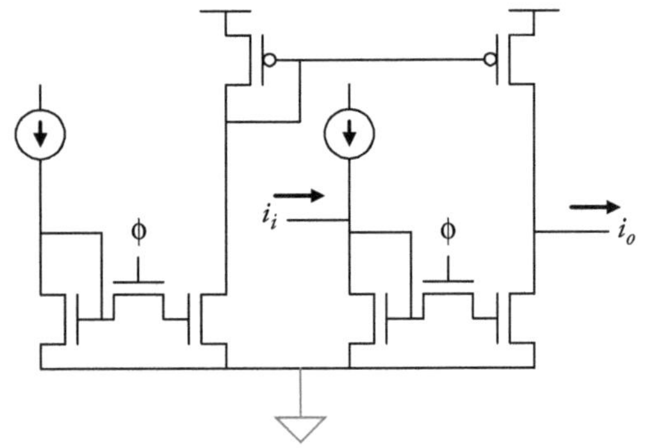

Figure 9.11 VLSI replica-based SI charge injection cancellation circuit structure.

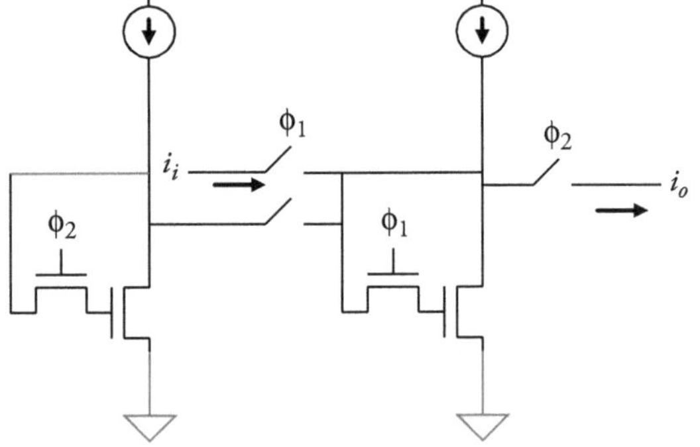

Figure 9.12 VLSI replica-based dynamic SI charge injection cancellation circuit structure.

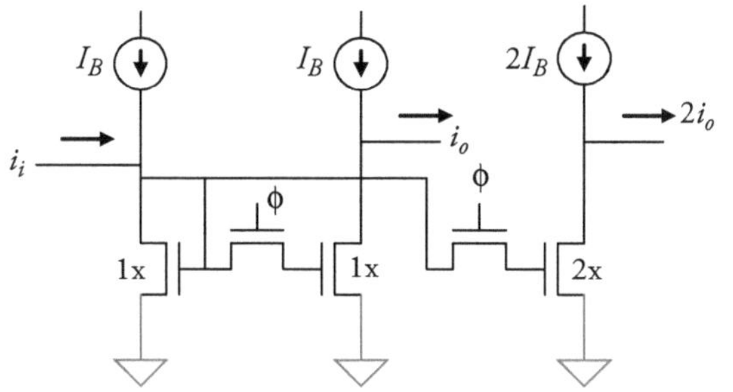

FIGURE 9.13 Alternative VLSI replica-based SI charge injection cancellation circuit.

9.4.3 Algorithmic Charge Cancellation Circuits

Algorithmic charge injection cancellation techniques based on the iteration sampling method for charge injection compensation in sequential steps is shown in Fig. 9.14. Such a circuit employs a two-step VLSI SI algorithmic charge injection cancellation technique.

This circuit consists of two current S/H elements, including the coarse current S/H element controlled by clock ϕ_{1a} and the fine current S/H circuit elements controlled by clocks ϕ_{1b} and ϕ_{1c}. The circuit works as follows:

- During the operation, the input signal is first sampled and stored in a coarse current mirror, which has relaxed current resolution and is tolerant to high charge injection error.
- Then the current difference between the input current and the sampled coarse current is sampled and held in the fine current mirror.
- Since the fine current mirror only needs to handle a small current signal range, the error introduced by fine current-mirror sampling charge injection weakly depends on the signal level and may be cancelled relatively easily.

The algorithmic charge injection schemes can also be extended to multistep approaches as shown in Fig. 9.15.

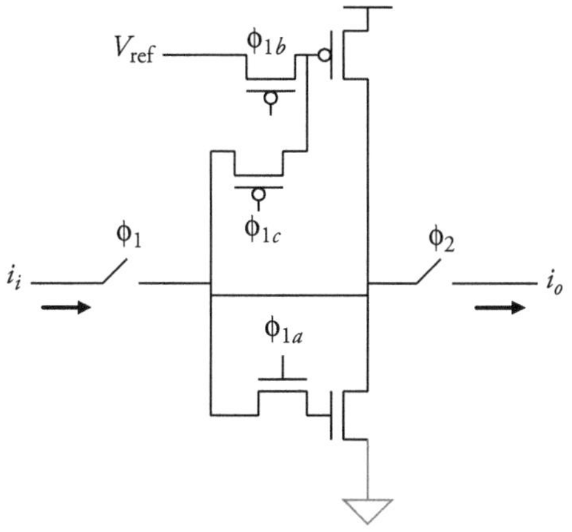

FIGURE 9.14 Two-step switched-current circuit structure.

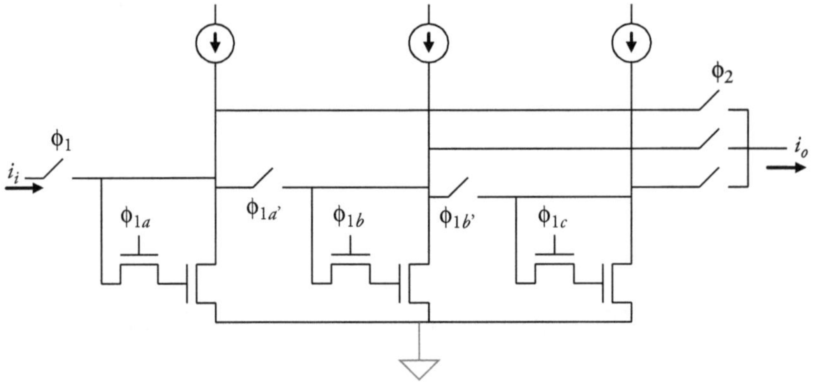

FIGURE 9.15 Multistep switched-current circuit structure.

9.4.4 Full-Differential VLSI SI Circuits

A differential SI circuit structure is shown in Fig. 9.16. Such an SI circuit structure can also be used to minimize the signal-independent charge injection. Such a circuit technique is based on the fact that the charge injection mainly appears as the common-mode error and can be minimized by the common-mode rejection ratio (CMRR).

However, the signal-dependent charge injection, which appears as the differential-mode error, cannot be effectively eliminated by this method.

9.4.5 Controlled Clocking and Input Techniques

Charge injection effects may also be minimized by employing either adaptive controlling of the clocking voltage level with respect to the input signal level or through keeping the input signal voltage (not the current) level constant with the operation in negative feedback. Under such conditions, the charge injection can be made signal independent and, therefore, can be effectively minimized or eliminated using other circuit techniques. A zero-switching SI circuit implementation based on such a design concept is shown in Fig. 9.17.

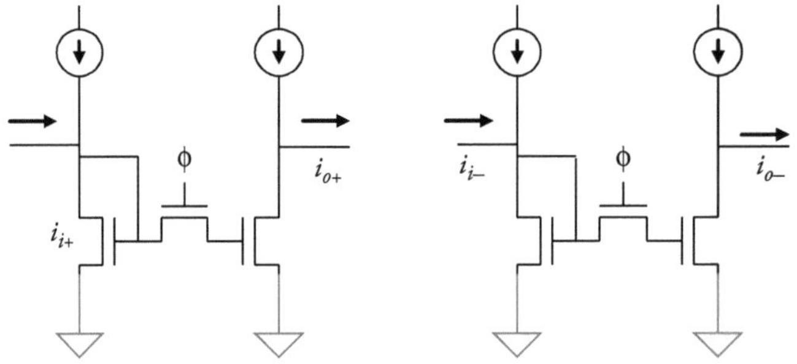

FIGURE 9.16 Differential switched-current circuit structure.

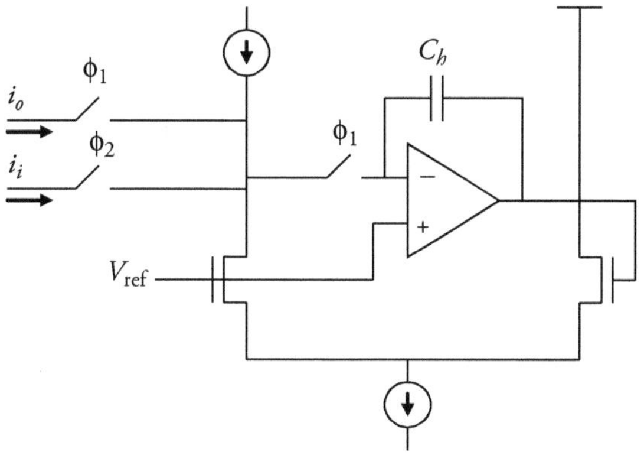

FIGURE 9.17 Zero-switching circuit structure.

References

1. A. Ng, J. I. Sewell, "Switched-Current Techniques under Pressure: High Order and Wideband Filter Design," *IEE Colloquium on Analog Signal Processing*, pp. 4/1–4/11, October 1998.
2. Y. Lu, J. I. Sewell, "First or Second Generation SI Cells? A Comparison of Sensitivity from an SI Filter System Viewpoint," *IEE Colloquium on Digital and Analogue Filters and Filtering Systems*, pp. 11/1–11/5, November 1994.
3. J. A. Barby, "Switched-current Filter Models for Frequency Analysis in the Continuous-Time Domain," *ISCAS'93*, vol. 2, pp. 1427–1430, May 1993.
4. Y. Lu, J. I. Sewell, "Multirate SC and SI Filter System Design by XFILT," *ISCAS'95*, vol. 3, pp. 2257–2260, May 1995.
5. Z. Q. Shang, J. I. Sewell, "Efficient Analysis of Some Nonlinearities in SC and SI Filter Networks," *IEE Colloquium on Digital and Analogue Filters and Filtering Systems*, pp. 10/1–10/5, November 1994.
6. G. W. Roberts, A. S. Sedra, "Synthesizing Switched-Current Filters by Transposing the SFG of Switched-Capacitor Filter Circuits," *IEEE Trans. on Circuits and Systems*, vol. 38, no. 3, pp. 337–340, March 1991.
7. B. Jonsson, S. Eriksson, "A Low Voltage Wave SI Filter Implementation using Improved Delay Elements," *ISCAS'94*, vol. 5, pp. 305–308, June 1994.
8. M. Goldenberg, R. Croman, T. S. Fiez, "Accurate SI Filters using RGC Integrators," *IEEE J. Solid-State Circuits*, vol. 29, no. 11, pp. 1388–1395, November 1994.
9. T. S. Fiez, D. J. Allstot, "CMOS Switched-Current Ladder Filters," *IEEE J. Solid-State Circuits*, vol. 25, no. 6, pp. 1360–1367, December 1990.
10. B. Jonsson, S. Eriksson, "Current-Mode N-port Adaptors for Wave Si Filters," *Electronics Letters*, vol. 29, no. 10, pp. 925–926, May 1993.
11. M. H. Fino, L. J. Mourao, "SymbSI-A Program for the Symbolic Signal flow Graph Generation of Switched Current Circuits," *IEEE International Conference on Electronics, Circuits and Systems*, vol. 3, pp. 211–214, September 1998.
12. M. Helfenstein, A. Muralt, G. S. Moschytz, "Direct Analysis of Multiphase Switched-Current Networks using Signal-flow Graphs," *ISCAS'95*, vol. 2, pp. 1476–1479, May 1995.

13. M. Helfenstein, A. Muralt, G. Fischer, P. Zbinden, D. Pfaff, F. Frey, G. S. Moschytzl, "SC and SI Filters in Baseband Applications: A Comparison," *ISCAS'97*, vol. 1, pp. 297–300, June 1997.
14. A. Handkiewicz, P. Sniatala, M. Lukowiak, "Low-Voltage High-Performance Switched Current Memory Cell," *Proceedings, Tenth Annual IEEE International ASIC Conference and Exhibit*, pp. 12–16, September 1997.
15. J. L. Chan, S. S. Chung, "Universal Switched-Current Integrator Blocks for SI Filter Design," *Proceedings of the ASP-DAC '99*, vol. 1, pp. 261–264, January 1999.
16. L. J. Mourao, M. H. Fino, "Automatic Symbolic Evaluation of Nonideal Effects in SI Circuit Behavior," *Third International Workshop on Design of Mixed-Mode Integrated Circuits and Applications*, pp. 40–43, July 1999.
17. J. Schechtman, A. C. M. De Queiroz, L. P. Caloba, "A Practical Implementation Scheme for Component Simulation SI Filters," *Proceedings of the 38th Midwest Circuits and Systems*, vol. 1, pp. 174–177, August 1995.
18. J. B. Hughes, K. W. Moulding, "A Switched-Current Double Sampling Bilinear z-transform Filter Technique," *ISCAS'94*, vol. 5, pp. 293–296, June 1994.
19. N. C. Battersby, C. Toumazou, "Towards High Frequency Switched-current Filters in CMOS and GaAs Technology," *ISCAS'93*, vol. 2, pp. 1239–1242, May 1993.
20. T. Fiez, D. Allstot, "A CMOS Switched-Current Filter Technique," *Digest of Technical Papers, 37th ISSCC*, pp. 206–207, February 1990.
21. G. K. Balachandran, P. E. Allen, "Switched-Current Circuits in Digital CMOS Technology with Low Charge-injection Errors," *IEEE J. Solid-State Circuits*, vol. 37, no. 10, pp. 1271–1281, October 2002.
22. B. Raahemi, A. Opal, "Group Delay and Group Delay Sensitivity of Periodically Switched Linear Networks," *IEEE Trans. on Circuits and Systems I: Fundamental Theory and Applications*, vol. 47, no. 1, pp. 96–104, January 2000.
23. T-HKuo, S-Y Lee, K-D Chen, "Offset Current Cancellation Based on a Multiple-Path Feedback Compensation (MPFC) Technique for Switched-Current Circuits and Systems," *IEEE Trans. on Circuits and Systems II: Analog and Digital Signal Processing*, vol. 44, no. 4, pp. 299–309, April 1997.
24. A. Yufera, A. Rueda, "Studying The Effects of Mismatching and Clock-feed Through in Switched-Current Filters using Behavioral Simulation," *IEEE Trans. on Circuits and Systems II: Analog and Digital Signal Processing*, vol. 44, no. 12, pp. 1058–1067, December 1997.
25. A. Worapishet, R. Sitdhikorn, B. J. Hughes, "Architecture for Low-voltage Switched-current Complex Bandpass Filters," *Electronics Letters*, vol. 38, no. 12, pp. 535–536, June 2002.
26. A. E. J. Ng, J. I. Sewell, "Ladder Decompositions for wideband SI Filter Applications," *IEE Proceedings, Circuits, Devices and Systems*, vol. 145, no. 5, pp. 306–313, October 1998.
27. T.-H. Kuo, S.-Y. Lee D.-J. Lu, T. Jih, J.-J. Tsaur, "3.3 V Mixed-mode IC Design using Switched-current Techniques for Speech Applications," *IEE Proceedings Circuits, Devices and Systems*, vol. 144, no. 6, pp. 367–374, December 1997.
28. K.-D. Chen, T.-H. Kuo, S.-.Y Lee, "A Novel Offset Current Cancellation Technique for Switched-current Circuits and Systems," *ISCAS'96*, vol. 1, pp. 417–420, May 1996.
29. A. Handkiewicz, M. Kropidlowski M. Lukowiak, "Computer Tools for Switched-Current Filter Design," *MWSCAS-2002*, vol. 2, pp. II-501–II-504, August 2002.

30. Q. Gao, S. Qin, X. Jia, Y. Song, "Computer Aided Design of Switched-current Filters," *4th International Conference on ASIC*, pp. 94–97, October 2001.
31. D. L. Shlemin, "A Design Technique for High Q-factor SI Filters," *APEIE-98*, vol. 1, pp. 142–152, September 1998.
32. I. Yusim, G. Ionis, K. Suyama, "Simulator for Switched-current Integrated Circuits," *Proceedings of the IEEE 1998 Custom Integrated Circuits Conference*, pp. 459–462, May 1998.
33. A. Yufera, A. Rueda, J. L Huertas, "Switched-Current Wave Analog Filters," *ISCAS'92*, vol. 2, pp. 859–862, May 1992.
34. T. S. Fiez, B. Lee, D. J. Allstot, "CMOS Switched-Current Biquadratic Filters," *IEEE International Symposium on Circuits and Systems*, vol. 3, pp. 2300–2303, May 1990.
35. J. B. Hughes, K. W. Moulding, J. Richardson, J. Bennett, W. Redman-White, M. Bracey, R. S. Soin, "Automated Design of Switched-current Filters," *IEEE J. Solid-State Circuits*, vol. 31, no. 7, pp. 898–907, July 1996.
36. T. S. Fiez, G. Liang, D. J. Allstot, "Switched-current Circuit Design Issues," *IEEE J. Solid-State Circuits*, vol. 26, no. 3, pp. 192–202, March 1991.
37. J. B. Hughes, "Top-down Design of A Switched-Current Video Filter," *IEE Proceedings, Circuits, Devices and Systems*, vol. 147, no. 1, pp. 73–81, February 2000.
38. A. Handkiewicz, M. Kropidlowski, M. Lukowiak, M. Bartkowiak, "Switched-Current Filter Design for Image Processing Systems," *13th Annual IEEE International ASIC/SOC Conference*, pp. 8–12, September 2000.
39. R. T. Goncalves, S. Noceti Filho, M. C. Schneider, C. Galup-Montoro, "Digitally Programmable Switched Current Filters," *ISCAS'96*, vol. 1, pp. 258–261, May 1996.
40. A. D. Remenyuk, E. V. Astrova, L. I. Korovin, I. G. Lang, V. B. Shuman, "Diffraction Effects in Macroporous Silicon with Through Pores," *Proceedings of 2003 5th International Conference on Transparent Optical Networks*, vol. 1, pp. 273–275, July 2003.
41. X. Li, X. Zeng, D. Zhou, X. Ling, "Behavioral Modeling of Analog Circuits by Wavelet Collocation Method," *ACM IEEE International Conference on Computer Aided Design*, pp. 65–69, November 2001.
42. Y. Lu, J. I. Sewell, "A Systematic Approach for Ladder Based Switched Current Filter Design," *ISCAS'95*, vol. 3, pp. 2253–2256, May 1995.
43. R. T. Goncalves, S. N. Filho, M. C. Schneider, C. Galup-Montoro, "Programmable Switched Current Filters using MOSFET-only Current Dividers," *Proceedings of the 38th Midwest Circuits and Systems*, vol. 2, pp. 1046–1049, August 1995.
44. A. Yufera, A. Rueda, J. L. Huertas, "A Study of the Sensitivity of Switched-current Wave Analog Filters to Mismatching and Clock-Feedthrough Errors," *ISCAS'94*, vol. 5, pp. 317–320, June 1994.

CHAPTER 10

Frequency-Scaling and -Transformation Techniques

VLSI analog filters are frequency-domain signal processing circuits that are typically specified and implemented in the frequency-domain. The behaviors of analog filters are usually characterized by their steady-state gain and phase responses, which are commonly presented by the Bode plots. Although in certain design applications, they may also be specified by their time-domain parameters such as their group delays. Frequency-scaling techniques allow the filter frequency responses, such as the gain and phase responses, to be scaled proportionally in the frequency-domain to meet the given design specifications by keeping the shape of the filter's response curves or the filter type unchanged. Frequency-transformation techniques, on the other hand, provide useful ways to convert the filter frequency responses among various filter types such as low-pass, high-pass, band-pass, band-stop, and complex types. Frequency-scaling and -transformation techniques are commonly used in the filter circuit design, design analysis, and circuit implementation. They provide systematic paths for the reuse of well-behaved filter architectures and VLSI circuit structures previously developed.

10.1 Low-Pass Prototype Filters

For the basic passive LRC filter circuit shown in Fig. 10.1, the s-domain transfer function is given as

$$H(s) \equiv \frac{V_o(s)}{V_i(s)} = \frac{1}{LCs^2 + RCs + 1} \tag{10.1}$$

The Bode plots of such a transfer function depend on the inductor, resistor, and capacitor device parameters. Theoretically speaking, there are infinite combinations for the parameters even for such a simple low-pass filter circuit. Mathematically, we may unify

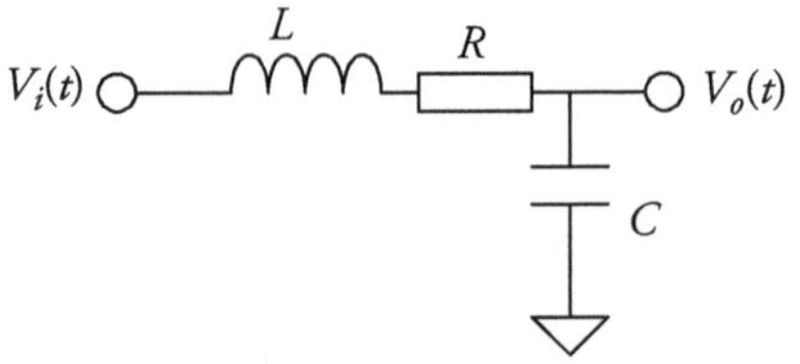

Figure 10.1 Passive LRC network.

all possible transfer functions of this filter by introducing three basic filter performance parameters as follows:

$$\begin{cases} \omega_o \equiv \sqrt{\dfrac{1}{LC}} \\ Q \equiv \dfrac{1}{R}\sqrt{\dfrac{L}{C}} \\ S = j\Omega \equiv \dfrac{s}{\omega_o} \end{cases} \tag{10.2}$$

The parameters ω_o and Q represent the cutoff frequency and the quality factor of the filter, respectively. S represents the normalized frequency parameter.

The transfer function of the normalized filter can be expressed as follows:

$$H(S) \equiv H\left(\frac{s}{\omega_o}\right) = \frac{1}{\left(\dfrac{s}{\omega_o}\right)^2 + \dfrac{1}{Q}\left(\dfrac{s}{\omega_o}\right) + 1} = \frac{1}{S^2 + \dfrac{1}{Q}S + 1} \tag{10.3}$$

Note that the above approach provides an effective way to compress the required information for specifying a given filter circuit structure. The significant aspects of the above mathematic manipulations are that (a) it provides an effective way to unify the responses of the filter that are based on the same circuit topology, (b) it allows to effectively reuse the existing design to meet different design specifications, and (c) it provides an effective way to analyze filter response under various manufacture processes and environment variations.

Based on the above example, we may now formally introduce the prototype-filter concept and the frequency-scaling technique. The normalized filter as described in Fig. 10.2 is also called a prototype filter. In particular, a low-pass prototype filter is a filter that has the following properties:

- It is a low-pass filter with a normalized characteristic frequency at (or near) $\Omega = 1$.
- It is implemented using circuit elements with dimensionless values.

It can be seen that the LRC circuit structure shown in Fig. 10.3 is a low-pass prototype filter.

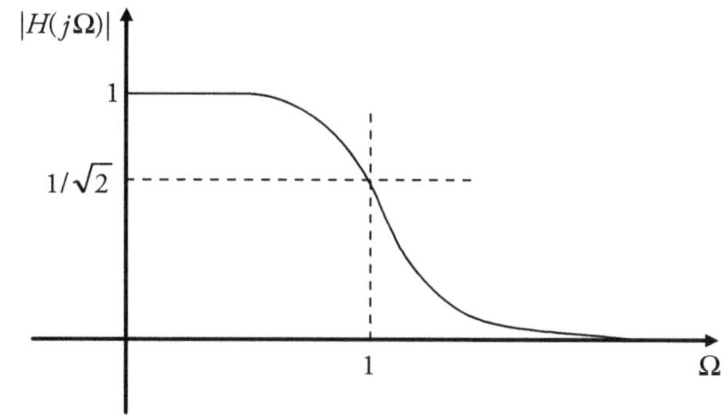

FIGURE 10.2 Example of low-pass prototype-filter response.

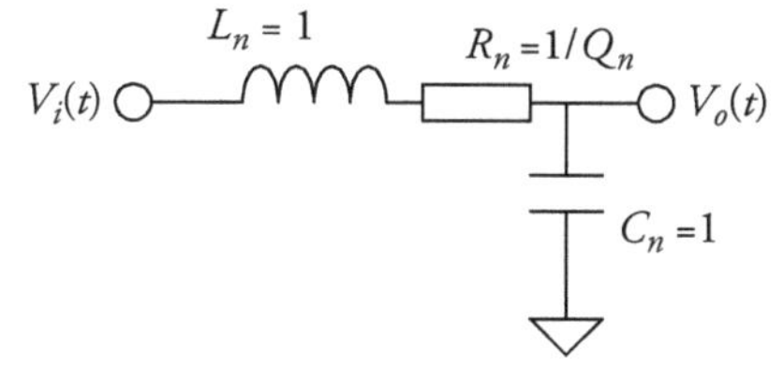

FIGURE 10.3 Example of low-pass prototype filter.

Similarly, we may also define other prototype-filter types as the high-pass prototype, the band-pass prototype filters, etc. However, as can be seen in the following discussion, these filter prototypes can usually be derived from low-pass prototype filters based on frequency-transformation techniques.

10.2 Frequency-Scaling Techniques

If we select a new set of circuit component parameters for the filter shown in Fig. 10.3 as

$$\begin{cases} R = R_s R_n \\ C = \dfrac{1}{R_s \omega_s} C_n \\ L = \dfrac{R_s}{\omega_s} L_n \end{cases} \tag{10.4}$$

where R_s and ω_s are filter design constants, we can see that the new filter will have the same quality factor as the prototype filter independent of R_s and ω_s, implying the similar shapes of the frequency response curves of both filters:

$$Q \equiv \frac{1}{R}\sqrt{\frac{L}{C}} = Q_n \tag{10.5}$$

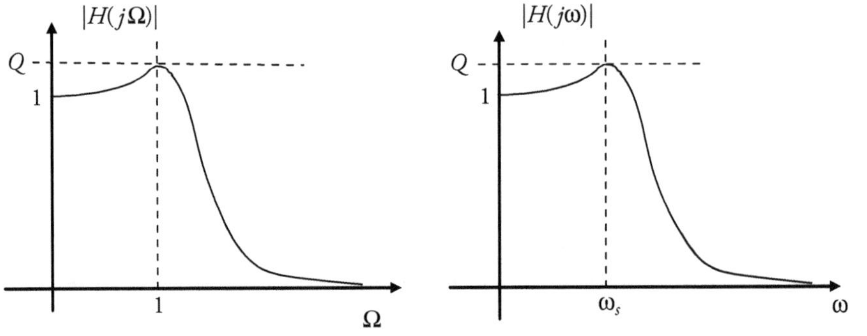

Figure 10.4 Frequency-scaling operation.

Note that the characteristic frequency of the given filter is proportionally scaled by ω_s times to a different value as

$$\omega_o = \sqrt{\frac{1}{LC}} = \omega_s \sqrt{\frac{1}{L_n C_n}} = \omega_s \tag{10.6}$$

We can see that the new filter constructed from the prototype filter by the scaling rule is a filter that has the same shape of response curves as the prototype filter, and the frequency responses are scaled proportionally by ω_s times as shown in Fig. 10.4.

10.2.1 Passive Filters

The above frequency-scaling rule can be extended to general filter circuits consisting of a large number of passive inductor, resistor, and capacitor circuit components. For a passive prototype-filter circuit consisting of a set of k resistors $\{R_n^i; i = 1, 2, 3, ..., k\}$, k capacitors $\{C_n^i; i = 1, 2, 3, ..., m\}$ and n inductors $\{L_n^i; i = 1, 2, 3, ..., n\}$, if the components are scaled following the transformations below.

$$\begin{cases} R^i = R_s R_n^i \quad i = 1, 2, 3, ..., k \\ C^i = \dfrac{1}{R_s \omega_s} C_n^i \quad i = 1, 2, 3, ..., m \\ L^i = \dfrac{R_s}{\omega_s} L_n^i \quad i = 1, 2, 3, ..., n \end{cases} \tag{10.7}$$

then the new filter will have frequency responses, which are proportionally scaled in the frequency-domain by ω_s times. In the above equations, the two scaling constants R_s and ω_s are known as the resistance and the frequency-scaling factors.

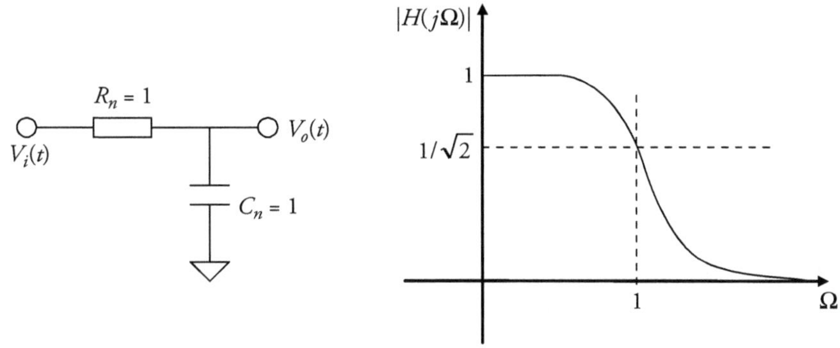

Figure 10.5 Low-pass RC prototype-filter circuit and response.

Taking the low-pass filter shown in Fig. 10.5 as an example and assuming a targeted cutoff frequency of 100 kHz, the transfer function of the prototype filter is given as follows:

$$H(S) = \frac{1}{1+S} \tag{10.8}$$

Applying the frequency scaling and selecting the scaling factors R_s = 10 kΩ and $\omega_s = 2\pi \times 100$ kHz, we can get the parameters of the passive circuit elements of the filter that meets our design specifications:

$$\begin{cases} R = R_s R_n = 10\ \text{k}\Omega \\ C = \dfrac{1}{R_s \omega_s} C_n = 160\ \text{pF} \end{cases} \tag{10.9}$$

The new circuit and the frequency response are shown in Fig. 10.6.

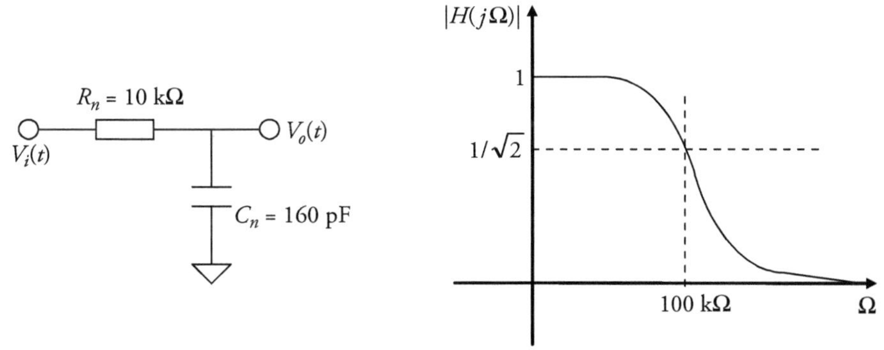

Figure 10.6 Low-pass RC prototype-filter circuit and response.

Note that we may also alternatively select R_s = 100 kΩ and $\omega_s = 2\pi \times 100$ kHz to obtain a different filter implementation, which also meets the given specifications.

$$\begin{cases} R = R_s R_n = 100\ \text{k}\Omega \\ C = \dfrac{1}{R_s\omega_s} C_n = 16\ \text{pF} \end{cases} \tag{10.10}$$

10.2.2 VLSI Active-RC Filters

For the VLSI active-RC filter circuits, the filter responses are mainly determined by the passive circuit components of the filter. Consequently, the scaling rules for the active-RC filters are the same as the passive LRC filters without the inductors:

$$\begin{cases} R^i = R_s R_n^i & i = 1, 2, 3, \ldots, k \\ C^i = \dfrac{1}{R_s\omega_s} C_n^i & i = 1, 2, 3, \ldots, m \end{cases} \tag{10.11}$$

Note that the driving capability of the OPAMP circuit in a practical design requires verification after the scaling to ensure the proper filter circuit operation even though it does not appear in the scaling equations.

10.2.3 VLSI MOS-C Filters

In VLSI MOS-C filter circuits, the MOS VCRs and capacitors are scaled in the frequency-scaling process. Similar to active-RC filters, OPAMP circuits do not appear in the scaling equations. The effective resistance of the MOS VCR and the capacitors in the MOS-C filters need to be scaled similar to the resistors and capacitors in active-RC filters, respectively, as follows:

$$\begin{cases} (R^i)_{\text{eff}} = R_s (R_n^i)_{\text{eff}} & i = 1, 2, 3, \ldots, k \\ C^i = \dfrac{1}{R_s\omega_s} C_n^i & i = 1, 2, 3, \ldots, m \end{cases} \tag{10.12}$$

10.2.4 VLSI Gm-C Filters

In VLSI Gm-C filter circuits, the capacitor and the Gm-devices are scaled in the frequency-scaling process. Frequency-scaling equations for a Gm-C filter consisting of *k* Gm-cells and *m* capacitors is given as

$$\begin{cases} g_m^i = \dfrac{1}{R_s} g_{mn}^i & i = 1, 2, 3, \ldots, k \\ C^i = \dfrac{1}{R_s\omega_s} C_n^i & i = 1, 2, 3, \ldots, m \end{cases} \tag{10.13}$$

where g_{mn}^i and g_m^i are the g_m values of the prototype and the scaled Gm-C filters.

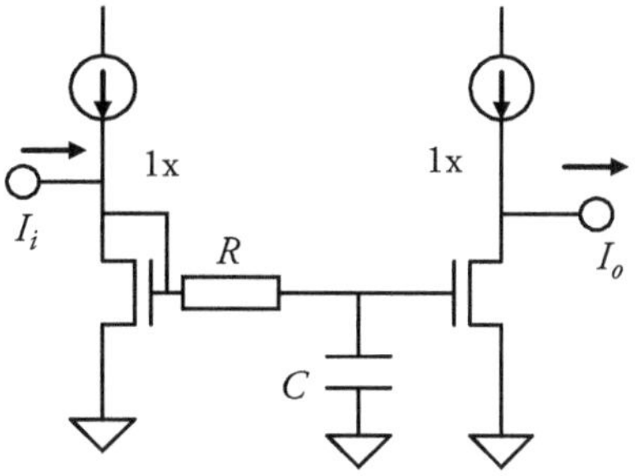

FIGURE 10.7 CTI integrator with RC bandwidth reduction.

10.2.5 VLSI CTI Filters

In current-mirror–based VLSI CTI filter circuits with resistor and capacitor bandwidth reduction as shown in Fig. 10.7, the resistor and capacitor components are scaled according to the following equations:

$$\begin{cases} R^i = R_s R_n^i & i = 1, 2, 3, \ldots, k \\ C^i = \dfrac{1}{R_s \omega_s} C_n^i & i = 1, 2, 3, \ldots, m \end{cases} \tag{10.14}$$

For high-bandwidth CTI filters constructed directly based on the current-mirror parasitic capacitors, frequency scaling is complicated as the filter structures may need to be adjusted when filters are scaled. For the CTI filter based on the current conveyors, frequency scaling is similar to that of the active-RC filters, and it can be realized directly using previous scaling equations.

10.2.6 VLSI SC Filters

In the switched-capacitor (SC) filter circuit, the capacitance and the switch clock period appear in the filter transfer function; therefore, they are required to be scaled in the frequency-scaling process. Under the frequency-scaling process, the conventional capacitors are scaled the same way as in passive filters. The switched capacitors, on the other hand, will be scaled similar to the conductors as follows:

$$\begin{cases} \left(\dfrac{C_{SC}}{T}\right)^i = \dfrac{1}{R_s}\left(\dfrac{C_{SC}}{T}\right)_n^i & i = 1, 2, 3, \ldots, k \\ C^i = \dfrac{1}{R_s \omega_s} C_n^i & i = 1, 2, 3, \ldots, m \end{cases} \tag{10.15}$$

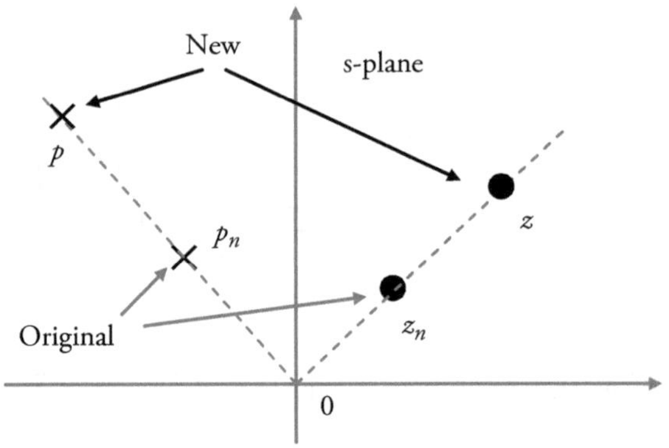

FIGURE 10.8 Poles and zeros of the filter circuit under frequency-scaling process.

10.2.7 VLSI SI Filters

For the VLSI switched-current (SI) filter circuits based on the current mirrors and switches, only the control clock period is required to be scaled in the frequency-scaling operations:

$$T = \frac{1}{\omega_s} T_n \tag{10.16}$$

10.2.8 Filter Pole/Zero under Frequency Scaling

The frequency-scaling operation corresponds to the scaling of all filter poles and zeros simultaneously by a real constant specified by the frequency-scaling factor ω_s as shown in Eq. (10.17) and Fig. 10.8:

$$\begin{cases} p^i = \omega_s p_n^i \\ z^i = \omega_s z_n^i \end{cases} \tag{10.17}$$

10.3 Frequency-Transformation Techniques

In VLSI filter design practice, other filter response types—such as the high pass, low pass, band pass, band stop—are often required and can be generated from the low-pass filter response employing the frequency-transformation operations.

10.3.1 LP-to-HP Transformation

Under the low-pass-to-high-pass (LP-to-HP) prototype transformation process, the low-pass prototype filter is transformed to a high-pass prototype filter that has the same characteristic frequency as shown in Fig. 10.9 based on the following frequency-transformation rule:

$$\begin{cases} S \rightarrow \frac{1}{s} \quad \text{or} \\ \Omega \rightarrow -\frac{1}{\omega} \end{cases} \tag{10.18}$$

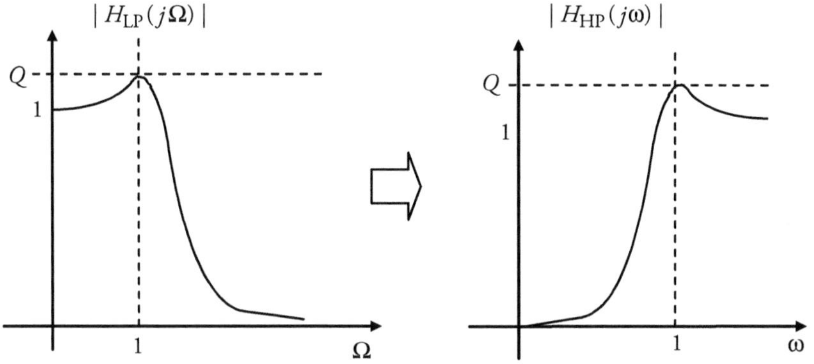

FIGURE 10.9 LP-to-HP prototype transformation.

The resulting filter transfer function has a high-pass response with the same cutoff frequency.

$$H_{HP}(s) = H_{LP}(S)|_{S=\frac{1}{s}} \tag{10.19}$$

Note that under the LP-to-HP transformation, each pole of the filter located at p_i is converted into a pole at $1/p_i$ and an added zero at the origin as shown in Fig. 10.10 as follows:

$$\frac{1}{S-p_i} \rightarrow \frac{1}{\frac{1}{s}-p_i} = -\frac{1}{p_i} \cdot \frac{s}{s-\frac{1}{p_i}} \tag{10.20}$$

Similarly, each zero in the transfer function will result in a zero at $1/z_i$ and an added pole at the origin as shown in Fig. 10.11 as follows:

$$S - z_i \rightarrow \frac{1}{s} - z_i = -z_i \cdot \frac{s-1/z_i}{s} \tag{10.21}$$

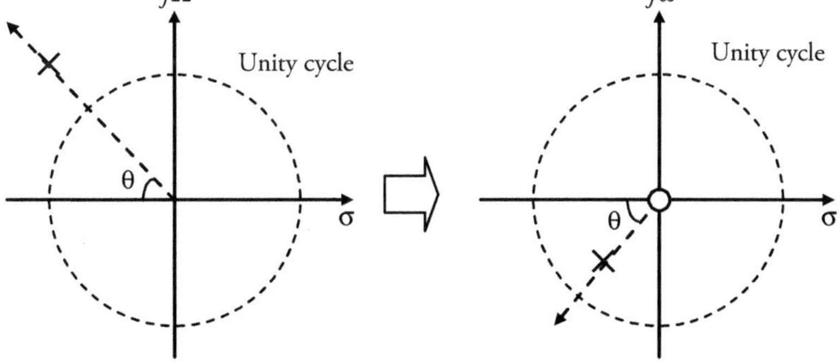

FIGURE 10.10 Transformation of pole in the LP-to-HP transformation.

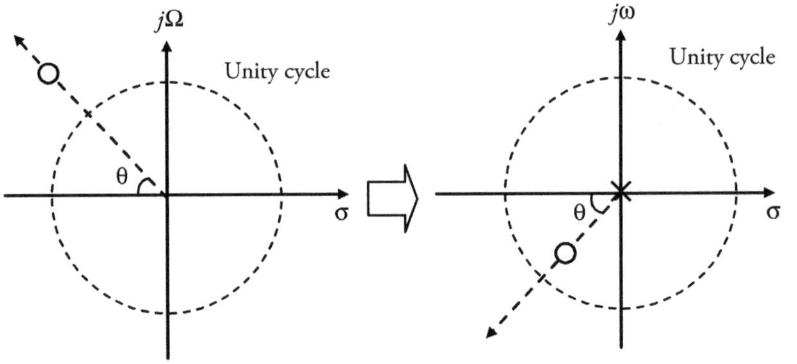

FIGURE 10.11 Transformation of zero in the LP-to-HP transformation.

In the VLSI filter circuit realization, the frequency transformation shown in Eq. (10.18) can in fact be directly realized by swapping the inductors and capacitors in the filter circuit as shown in Fig. 10.12.

For the low-pass prototype filter shown as Fig. 10.13, we can obtain a high-pass prototype filter by swapping the capacitor with the inductor in the circuit.

The Bode plots of the original low-pass prototype and the yielded high-pass prototype filters are shown in Fig. 10.14.

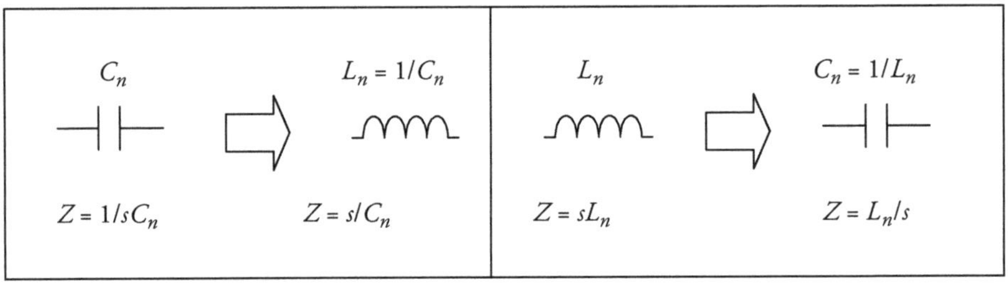

FIGURE 10.12 Transformation of capacitors and inductors in the LP-to-HP transformation.

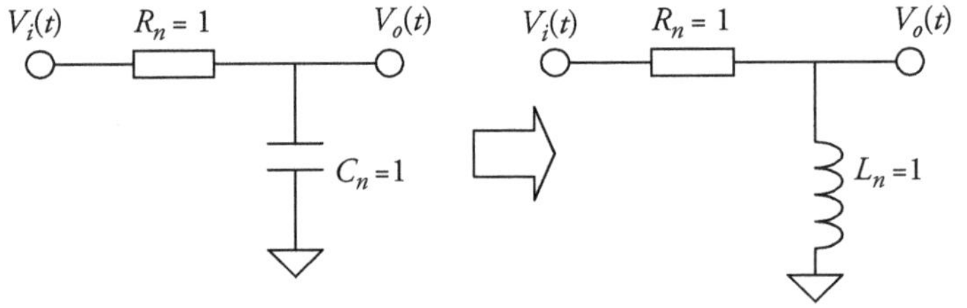

FIGURE 10.13 Realization of LP-to-HP transformation by capacitors and inductors swapping.

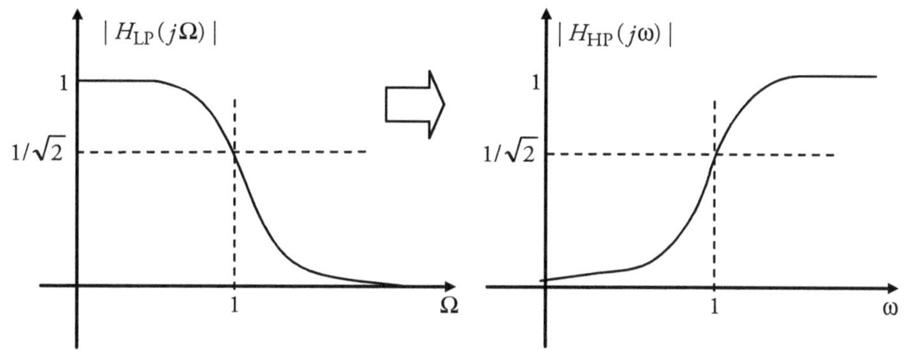

FIGURE 10.14 Bode plots transformation for the LP-to-HP transformation example.

10.3.2 LP-to-BP Transformation

Under the low-pass-to-band-pass (LP-to-BP) prototype transformation shown in Fig. 10.15, the low-pass prototype-filter circuit is converted into a band-pass prototype filter with the unity center frequency and specified bandwidth of Q (= $1/B$) based on the following transformation:

$$\begin{cases} S \to Q\left(\dfrac{1}{s}+s\right) & \text{or} \\ \Omega \to \dfrac{1}{B}\dfrac{\omega^2-1}{\omega} & \end{cases} \tag{10.22}$$

The new transfer function after the LP-to-BP transformation is given as follows:

$$H_{BP}(s) = H_{LP}(S)\big|_{S=Q\left(\frac{1}{s}+s\right)} \tag{10.23}$$

Note that under the LP-to-BP transformation, each pole p_i of the low-pass prototype filter is transformed into a pair of poles and a single zero (at the origin) as

$$\frac{1}{S-p_i} \to \frac{1}{Q\left(s+\dfrac{1}{s}\right)-p_i} = \frac{s/Q}{s^2-\dfrac{p_i}{Q}s+1} \tag{10.24}$$

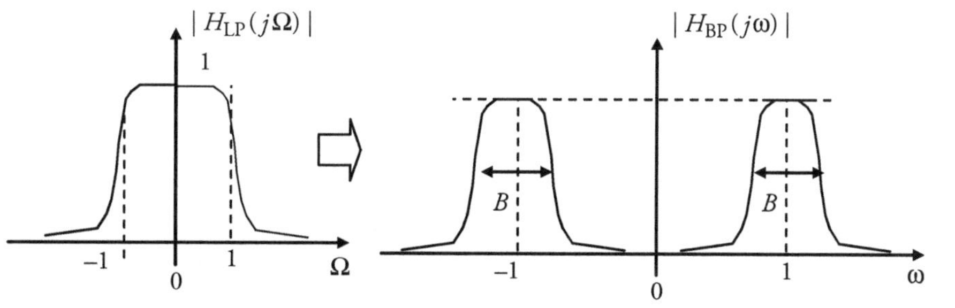

FIGURE 10.15 LP-to-BP prototype transformation.

Similarly, each zero is transformed into a pair of zeros and a single pole (at the origin) as follows:

$$S - z_i \rightarrow Q\left(s + \frac{1}{s}\right) - z_i = \frac{s^2 - \frac{z_i}{Q}s + 1}{s/Q} \tag{10.25}$$

For example, for the real pole in the low-pass filter transfer function, we have the following:

$$p_i = -|p_i| \tag{10.26}$$

The new poles after transformation are given as follows:

$$p_\pm = -\frac{|p_i|}{2Q} \pm \sqrt{\left(\frac{|p_i|}{2Q}\right)^2 - 1} \quad \text{for} \left(\frac{|p_i|}{2Q}\right)^2 \geq 1 \tag{10.27}$$

$$p_\pm = -\frac{|p_i|}{2Q} \pm j\sqrt{1 - \left(\frac{|p_i|}{2Q}\right)^2} \quad \text{for} \left(\frac{|p_i|}{2Q}\right)^2 < 1 \tag{10.28}$$

The root loci of the mapped poles are shown in Fig. 10.16.

For real-zero cases, each zero is transformed into a single pole at the origin and two zeros at the location.

$$z_\pm = -\frac{z_i}{2Q} \pm \sqrt{\left(\frac{z_i}{2Q}\right)^2 - 1} \quad \text{for} \left(\frac{z_i}{2Q}\right)^2 \geq 1 \tag{10.29}$$

$$z_\pm = -\frac{z_i}{2Q} \pm j\sqrt{1 - \left(\frac{z_i}{2Q}\right)^2} \quad \text{for} \left(\frac{z_i}{2Q}\right)^2 < 1 \tag{10.30}$$

The root loci of the mapped poles are shown in Fig. 10.17.

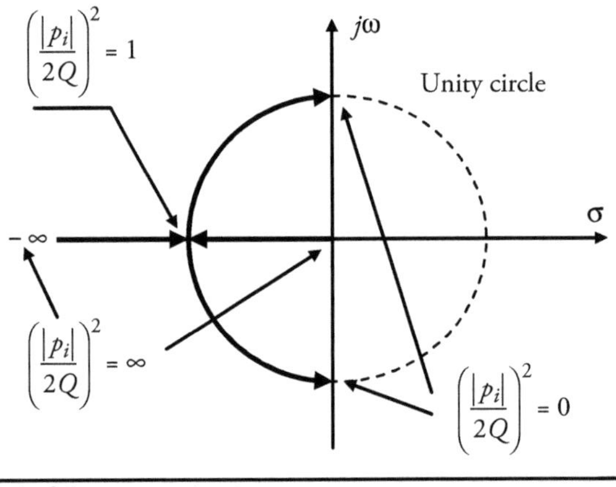

FIGURE 10.16 Root loci of the mapped poles.

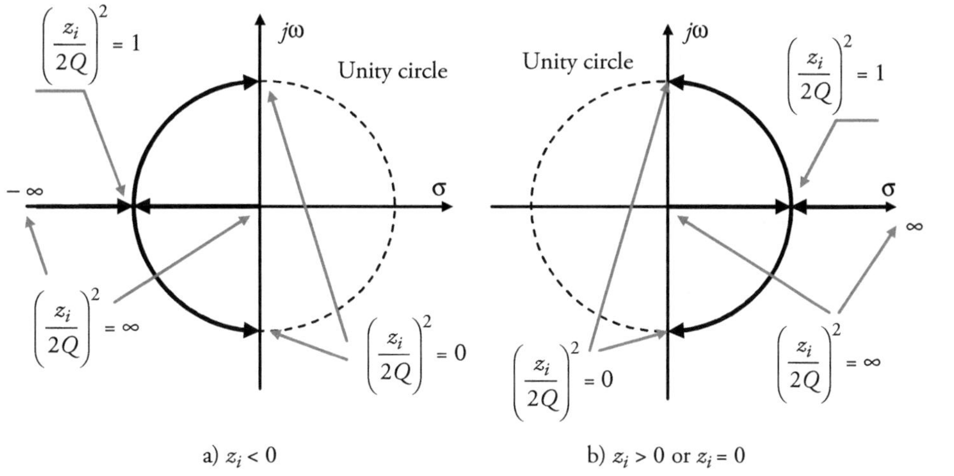

FIGURE 10.17 Root loci of the mapped zeros.

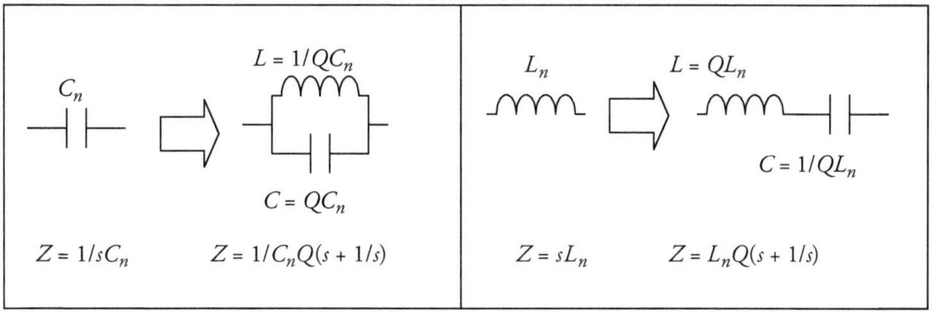

FIGURE 10.18 Transformation of capacitors and inductors in the LP-to-BP transformation.

In the VLSI circuit realization, the LP-to-BP transform can be directly implemented by converting the capacitor into a parallel capacitor/inductor pair and an inductor into a serial capacitor/inductor pair as shown in Fig. 10.18.

10.3.3 LP-to-BS Transformation

The low-pass-to-band-stop (LP-to-BS) transformation of the prototype-filter transfer function shown in Fig. 10.19 can be realized using the following transformation:

$$\begin{cases} S \rightarrow \dfrac{1}{Q\left(\dfrac{1}{s}+s\right)} \quad \text{or} \\ \Omega \rightarrow \dfrac{1}{B}\dfrac{\omega}{\omega^2 - 1} \end{cases} \tag{10.31}$$

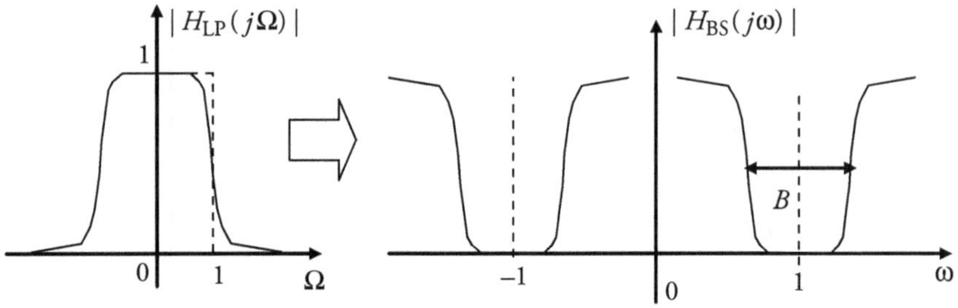

FIGURE 10.19 LP-to-BS prototype transformation.

where $Q = 1/B$. The resulting transfer function after the LP-to-BS transformation is given as follows:

$$H_{BP}(s) = H_{LP}(S)|_{S=\frac{1}{Q\left(\frac{1}{s}+s\right)}} \tag{10.32}$$

It can be seen that each pole p_i in the low-pass prototype filter is converted into a pair of poles and a pair of zeros (at $\omega = \pm 1$).

$$\frac{1}{S-p_i} \rightarrow \frac{1}{\left(Q\left(s+\frac{1}{s}\right)\right)^{-1} - p_i} = -\frac{(s^2+1)/p_i}{s^2 - \frac{1}{Qp_i}s + 1} \tag{10.33}$$

On the other hand, each zero is transformed into a pair of zeros and a pair of poles (at $\omega = \pm 1$).

$$S - z_i \rightarrow \frac{1}{Q\left(s+\frac{1}{s}\right)} - z_i = -z_i \frac{s^2 - \frac{1}{Qz_i}s + 1}{s^2+1} \tag{10.34}$$

The new poles for the real pole filter after transformation are given as

$$p_i = -|p_i| \tag{10.35}$$

$$p_\pm = -\frac{1}{2Q|p_i|} \pm \sqrt{\left(\frac{1}{2Q|p_i|}\right)^2 - 1} \quad \text{for} \left(\frac{1}{2Q|p_i|}\right)^2 \geq 1 \tag{10.36}$$

and

$$p_\pm = -\frac{1}{2Q|p_i|} \pm j\sqrt{1 - \left(\frac{1}{2Q|p_i|}\right)^2} \quad \text{for} \left(\frac{1}{2Q|p_i|}\right)^2 < 1 \tag{10.37}$$

The root loci of the mapped poles are shown in Fig. 10.20.

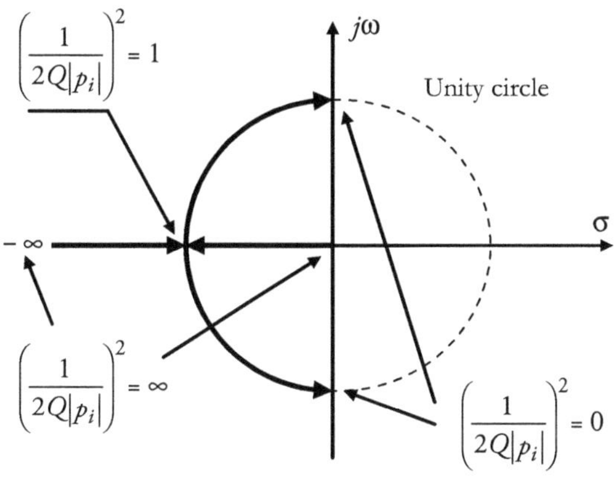

FIGURE 10.20 Root loci of the mapped poles.

Note that each zero is transformed into a single pole at the origin and a pair of zeros at

$$z_{\pm} = -\frac{1}{2Qz_i} \pm \sqrt{\left(\frac{1}{2Qz_i}\right)^2 - 1} \quad \text{for} \left(\frac{1}{2Qz_i}\right)^2 \geq 1 \tag{10.38}$$

and

$$z_{\pm} = -\frac{1}{2Qz_i} \pm j\sqrt{1 - \left(\frac{1}{2Qz_i}\right)^2} \quad \text{for} \left(\frac{1}{2Qz_i}\right)^2 < 1 \tag{10.39}$$

The root loci of the mapped poles are shown in Fig. 10.21.

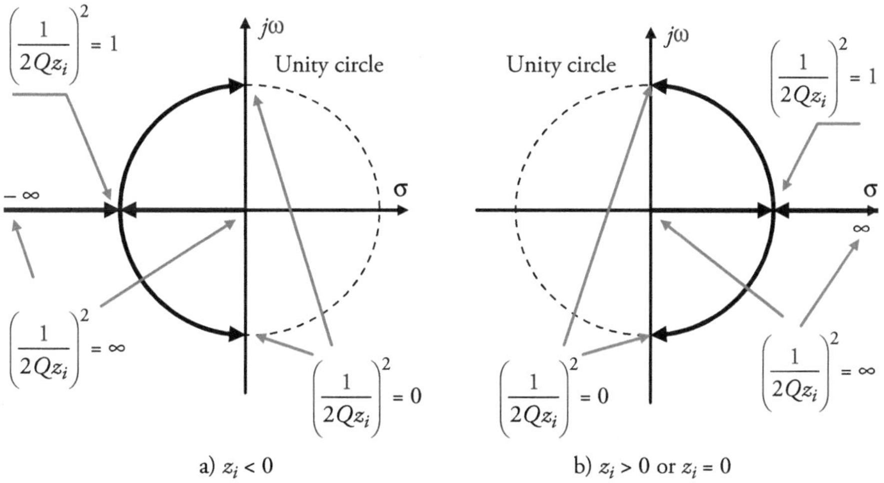

FIGURE 10.21 Root loci of the mapped zeros.

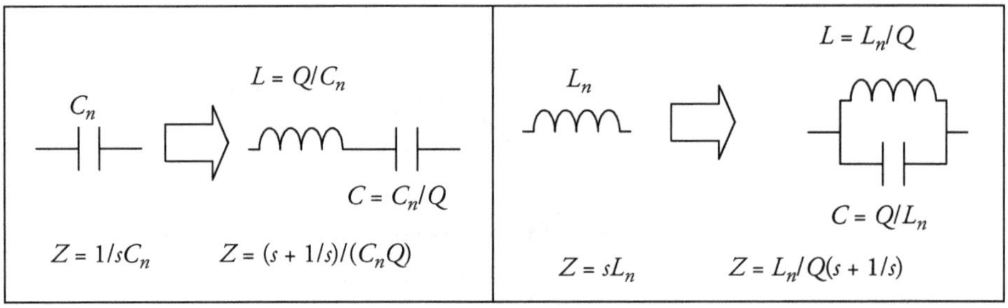

FIGURE 10.22 Transformation of capacitors and inductors in the LP-to-BS transformation.

The LP-to-BS transform can be realized in VLSI circuit implementation by transforming the capacitors into a capacitor/inductor pair in series and the inductors into a capacitor/inductor pair in parallel as shown in Fig. 10.22.

10.3.4 Hilbert Transform

Under the Hilbert transform shown in Fig. 10.23, the responses of a filter is shifted linearly to the frequency-domain of a given frequency ω_o without changing the shape of its response curves.

$$\begin{cases} S \to s - j\omega_o \text{ or} \\ \Omega \to \omega - \omega_o \end{cases} \tag{10.40}$$

The resulting transfer function of the filter after the LP-to-Hilbert transform is given as follows:

$$H_{\text{real}}(s) = H_{\text{Hilbert}}(S)\,|_{S=s-j\omega_o} \tag{10.41}$$

Under the Hilbert transform, the poles p_i and zeros z_i of the filter are transformed in the following ways:

$$\begin{cases} p_i \to p_i - j\omega_o \\ z_i \to z_i - j\omega_o \end{cases} \tag{10.42}$$

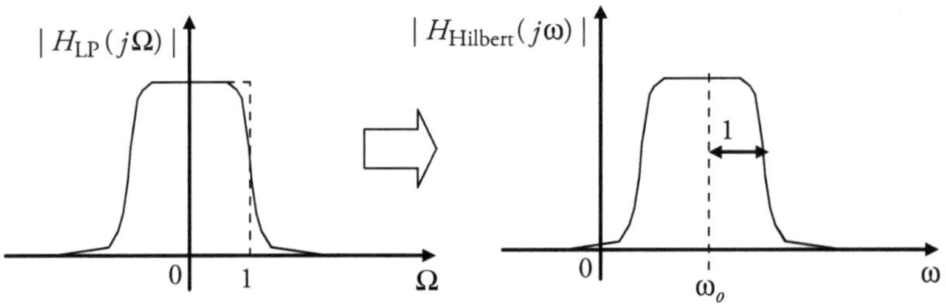

FIGURE 10.23 LP-to-Hilbert prototype transformation.

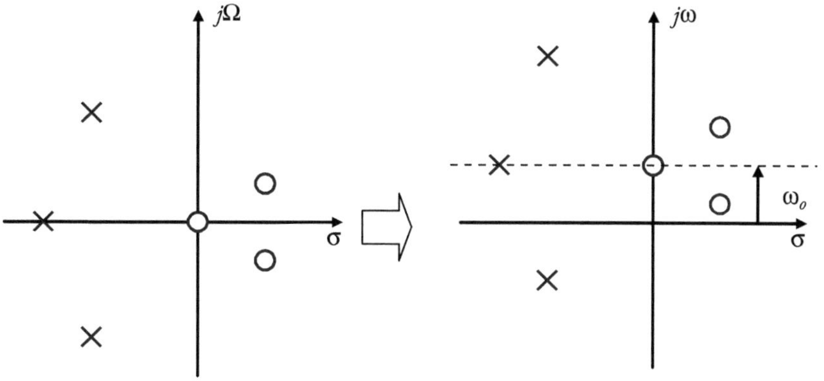

FIGURE 10.24 Transformation of poles and zeros under the Hilbert transform.

It can be seen that all poles and zeros are shifted in the $j\omega$ axis direction by ω_o during the Hilbert transform (see Fig. 10.24). Such a transformation converts a real filter into a complex filter.

References

1. N. Fujii, S.K. Mitra, "An Active RC Filter Capable of Easy Lowpass-Highpass Transformation," *ISCAS*, vol. 3, pp. 2403–2406, June 1988.
2. D. Jurisic, N. Mijat, G. S. Moschytz, "Low-Power High-Order Band-Pass Active-RC Allpole Filters Using a "Lossy" LP-BP Transformation," *9th International Conference on Electronics, Circuits and Systems*, vol. 1, pp. 209–213, September 2002.

CHAPTER 11

Design of VLSI Active-RC Filters

VLSI active-RC filters are commonly used in various communication system applications. Such filter circuit techniques offer advantages of low distortion benefited from the highly linear VLSI active-RC circuit components. VLSI active-RC filters also offer other advantages such as their design simplicity that can be directly mapped from most well-developed classic discrete filter circuit structures. The major limitations of VLSI active-RC filter structures are their operation frequency constraint limited by the OPAMP circuits and poor absolute frequency responses caused by the RC time constant variations due to the variations in VLSI manufactured circuit components. As a result, VLSI active-RC filter circuits are common choice for audio- or video-band signal processing applications and have limited applications in the RF circuit applications because of bandwidth limitation. Active-RC filter circuits usually have slightly higher power consumption compared with their passive filter circuit counterparts mainly resulting from the overhead of the OPAMP circuits.

Typical active-RC filter design methods include the SFG-based methods, the cascade-based methods, the component-simulation–based methods, and the ladder-simulation–based methods. The first two methods will be discussed in detail in this chapter, and the other methods will be covered in later chapters.

11.1 SFG-Based VLSI Active-RC Filter Design Methods

The SFG-based filter methods are based on the design concept that there is an easy mathematic path to construct the SFG of a given signal processing system based on the s-domain transfer function and that there is a one-to-one correlation between the SFG representation and the circuit implementation based on a targeted VLSI manufacture process technology. The SFG-based VLSI filter design approach provides a systematic method for VLSI circuit implementation for a specific signal processing system transfer function through the well-developed mathematic manipulation process. Such a method also provides the flexibility to construct the SFG based on different basic building blocks suitable for different VLSI process technologies. One of the most popular ways of VLSI active-RC filter design is based on the VLSI lossless integrator circuits. For a general s-domain transfer function of specific response as

$$\frac{Y}{X} = \frac{b_n s^{-n} + b_{n-1} s^{-(n-1)} + \cdots + b_1 s^{-1} + b_0}{a_n s^{-n} + a_{n-1} s^{-(n-1)} + \cdots + a_1 s^{-1} + 1} \tag{11.1}$$

we can construct a lossless integrator–based SFG through the following manipulation process:

$$\Rightarrow Y(a_n s^{-n} + a_{n-1} s^{-(n-1)} + \cdots + a_1 s^{-1} + 1) = (b_n s^{-n} + b_{n-1} s^{-(n-1)} + \cdots + b_1 s^{-1} + b_0)X \tag{11.2}$$

$$\Rightarrow Y = (b_n X - a_n Y)s^{-n} + (b_{n-1} X - a_{n-1} Y)s^{-(n-1)} + \cdots + (b_1 X - a_1 Y)s^{-1} + b_0 X \tag{11.3}$$

The SFG corresponding to the above equation is shown in Fig. 11.1. Note that the noninverted lossless integrators are used here to demonstrate the SFG-based filter design approach for their simplicity. However, in many filter design cases such as the single-ended active-RC filter designs, the inverted lossless integrators rather than the noninverted integrators are more convenient in the VLSI filter implementations.

Alternatively, we may manipulate the filter transfer function based on a different path by introducing a state variable U as

$$\frac{Y}{X} = \frac{b_n s^{-n} + b_{n-1} s^{-(n-1)} + \cdots + b_1 s^{-1} + b_0}{a_n s^{-n} + a_{n-1} s^{-(n-1)} + \cdots + a_1 s^{-1} + 1} \cdot \frac{U}{U} \tag{11.4}$$

$$\Rightarrow \begin{cases} U = X - (a_n s^{-n} + a_{n-1} s^{-(n-1)} + \cdots + a_1 s^{-1})U \\ Y = (b_n s^{-n} + b_{n-1} s^{-(n-1)} + \cdots + b_1 s^{-1} + b_0)U \end{cases} \tag{11.5}$$

The alternative SFG expression based on such a method is given in Fig. 11.2.

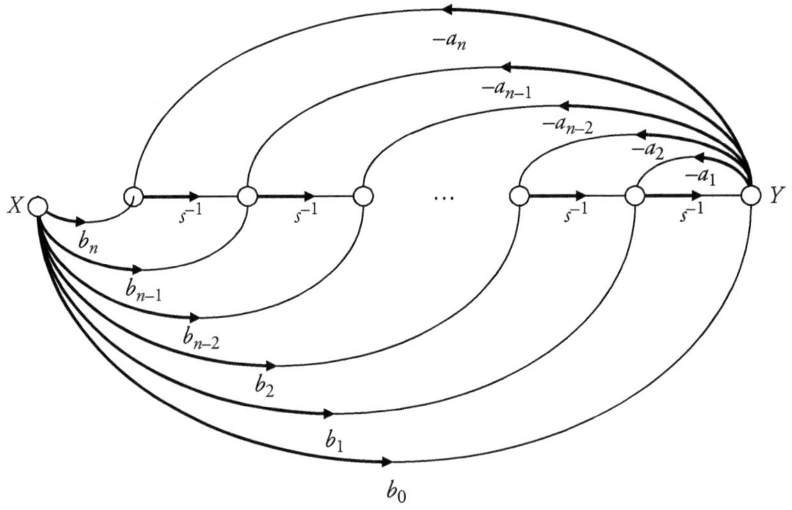

FIGURE 11.1 SFG structure based on lossless integrator.

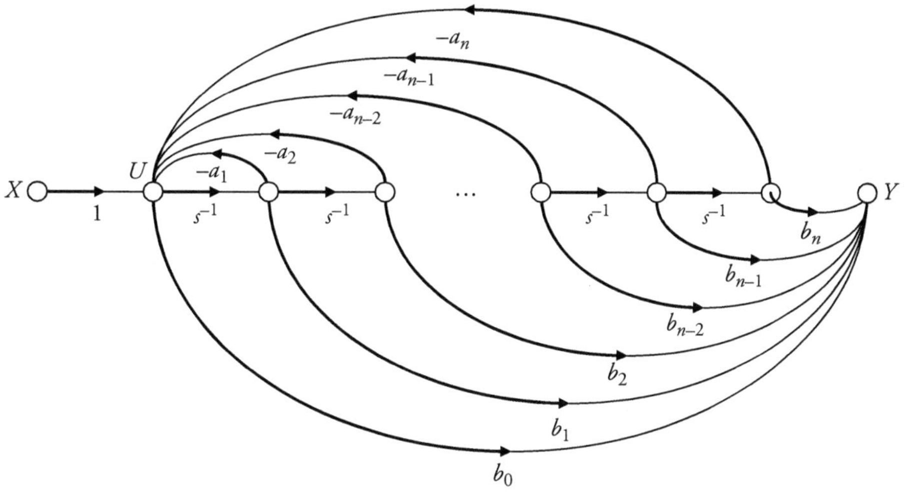

FIGURE 11.2 Alternative SFG to the general transfer function based on lossless integrator.

Note that there are, in fact, many paths of constructing SFG. Even from the same filter transfer function, there are different ways of expressing the input and output relation. From the discussion above, it can be seen that any given filter transfer function can be implemented using the SFG approach based on three basic building operations, including the integration with weighted inputs, the weighted addition, and the scaling element.

11.1.1 Active-RC Circuit Realization of Integration

One of the popular basic building elements of the SFG-based filter structure is the lossless integration. In a more effective form (see Fig. 11.3), the integration with addition of multiple weighted inputs is more suitable for practical VLSI filter circuit realizations.

$$Y = \frac{1}{s}(\alpha_1 X_1 + \alpha_2 X_2 + \cdots + \alpha_n X_n) \tag{11.6}$$

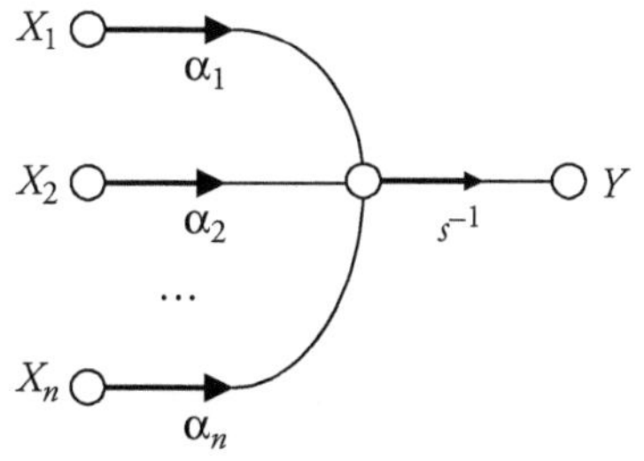

FIGURE 11.3 The transfer function and SFG of integration with weighted inputs.

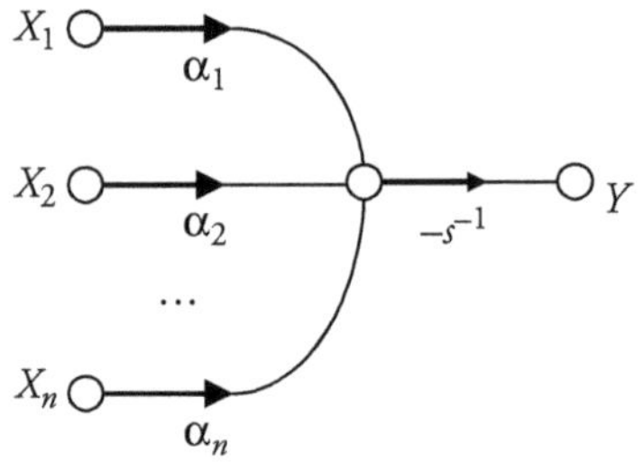

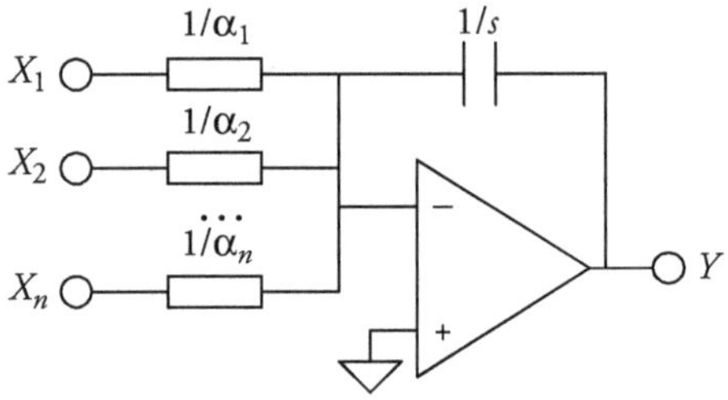

FIGURE 11.4 SFG and active-RC circuit realization of integration with weighted inputs.

For single-ended circuit implementation, the inverted integration is preferred and effective in the VLSI circuit implementation, considering the negative feedback configuration of the OPAMPs in the active-RC circuits. Figure 11.4 shows the SFG element and the active-RC circuit realization of the integration transfer function.

$$Y = -\frac{1}{s}(\alpha_1 X_1 + \alpha_2 X_2 + \cdots + \alpha_n X_n) \tag{11.7}$$

11.1.2 Active-RC Circuit Realization of Addition

Similar to the integration, a weighted addition shown in Fig. 11.5 is commonly used for realization of the VLSI active-RC filter.

$$Y = (\alpha_1 X_1 + \alpha_2 X_2 + \cdots + \alpha_n X_n) \tag{11.8}$$

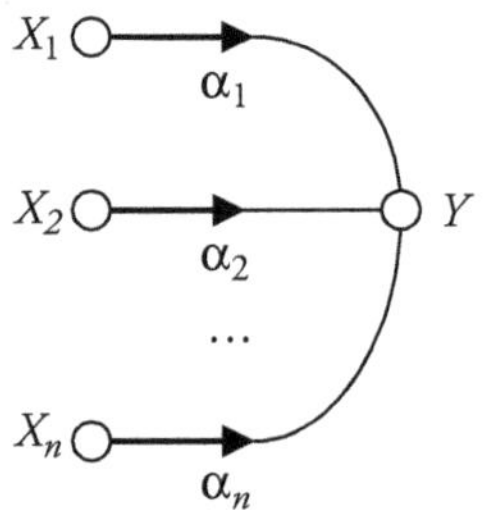

FIGURE 11.5 Transfer function and SFG of addition with weighted inputs.

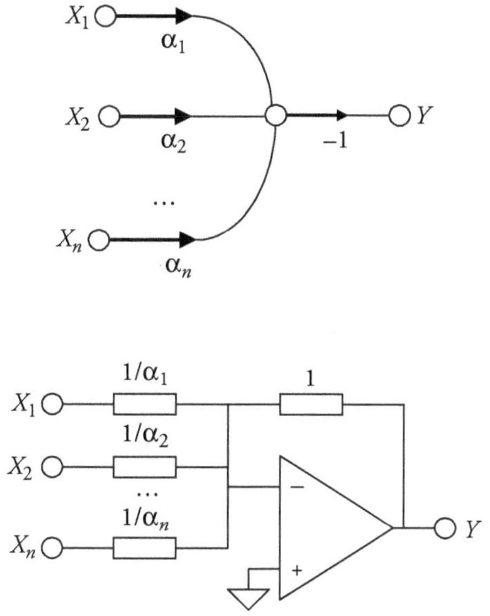

FIGURE 11.6 SFG and active-RC circuit of weighted addition.

Similar to the VLSI active-RC filter circuit implementations, the inverted addition can also be realized in the active-RC circuit for high implementation efficiency. The SFG and the active-RC circuit implementation of the inverted weighted addition are shown in Fig. 11.6.

$$Y = -(\alpha_1 X_1 + \alpha_2 X_2 + \cdots + \alpha_n X_n) \tag{11.9}$$

11.1.3 Active-RC Circuit Realization of Scaling

The scaling element for the active-RC filter implementation is shown in Fig. 11.7.

$$Y = -\alpha X \tag{11.10}$$

The SFG and the active-RC circuit implementation are shown in Fig. 11.8.

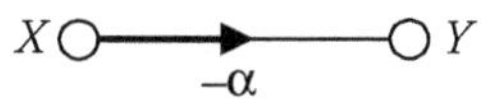

FIGURE 11.7 Transfer function and SFG of scaling.

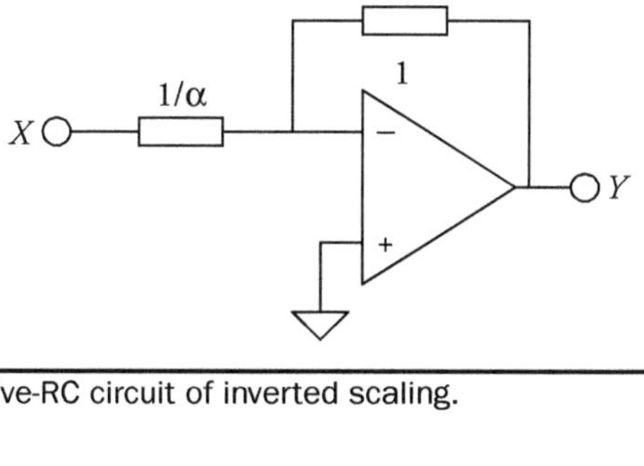

FIGURE 11.8 SFG and active-RC circuit of inverted scaling.

Note that both integration with the addition of weighted inputs and weighted addition readily provide the scaling operations. Therefore, dedicated scaling operations should only be used in VLSI filter design when the operation cannot be realized with other signal processing operations.

11.1.4 Design Example

The following example is used to demonstrate the design procedure of the SFG-based filter design method. Assume that we want to design a VLSI active-RC filter with the following transfer function and with $\omega_o = 2\pi \times 100$ kΩ and $Q = 10$:

$$\frac{Y}{X} = \frac{1}{\left(\frac{s}{\omega_o}\right)^2 + \frac{1}{Q}\left(\frac{s}{\omega_o}\right) + 1} \tag{11.11}$$

We may start the design process with the transfer function of a normalized (or prototype) filter as follows:

$$\frac{Y}{X} = \frac{1}{S^2 + \frac{1}{Q}S + 1} \tag{11.12}$$

By rewriting the above transfer function to a form that can be expressed using lossless integrators as

$$\begin{aligned} &\left(S^2 + \frac{1}{Q}S + 1\right)Y = X \Rightarrow \left(1 + \frac{1}{Q}S^{-1} + S^{-2}\right)Y = S^{-2}X \\ &\Rightarrow Y = S^{-2}X - \left(\frac{1}{Q}S^{-1} + S^{-2}\right)Y \\ &\Rightarrow Y = (-S^{-1})^2 X + \frac{1}{Q}(-S^{-1})Y - (-S^{-1})^2 Y \end{aligned} \tag{11.13}$$

we can construct an SFG corresponding to the above prototype filter (see Fig. 11.9).

The SFG in Fig. 11.9 can be further modified to a form that is more suitable for the VLSI active-RC circuit implementation as shown in Fig. 11.10.

Finally, this SFG can be directly mapped to the active-RC circuit using the basic building blocks (see Fig. 11.11).

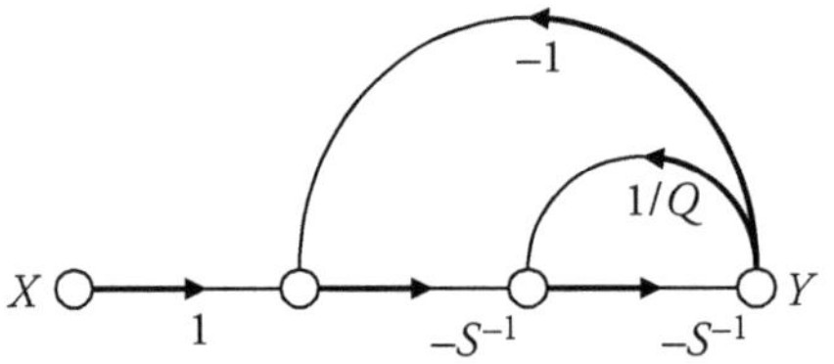

FIGURE 11.9 SFG representation of the given prototype filter.

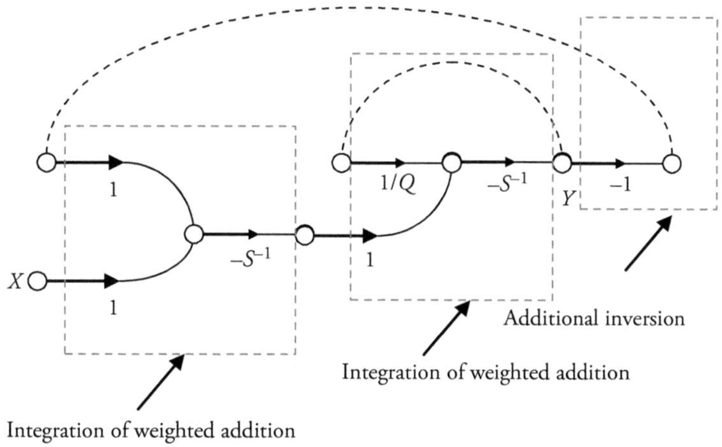

FIGURE 11.10 Alternative SFG of the example suitable for active-RC realization.

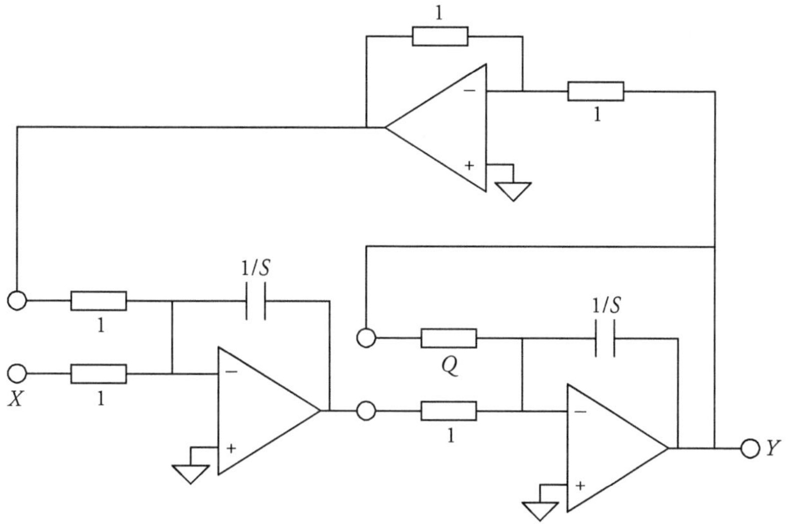

FIGURE 11.11 Active-RC circuits of the prototype filter.

By applying the frequency-scaling technique to the above prototype filter with R_s = 10 kΩ, we can get the final active-RC filter component parameters, which meet our design specifications (see Fig. 11.12).

$$\begin{aligned} R_s &= 10\ \text{k}\Omega \\ \omega_s &= 2\pi \cdot 100\ \text{k}\Omega \\ R &= R_n \cdot R_s = 10 \cdot R_n(\text{k}\Omega) \\ C &= \frac{C_n}{R_s\omega_s} = 159 \cdot C_n(\text{pF}) \end{aligned} \tag{11.14}$$

Finally, we obtain the VLSI filter circuit.

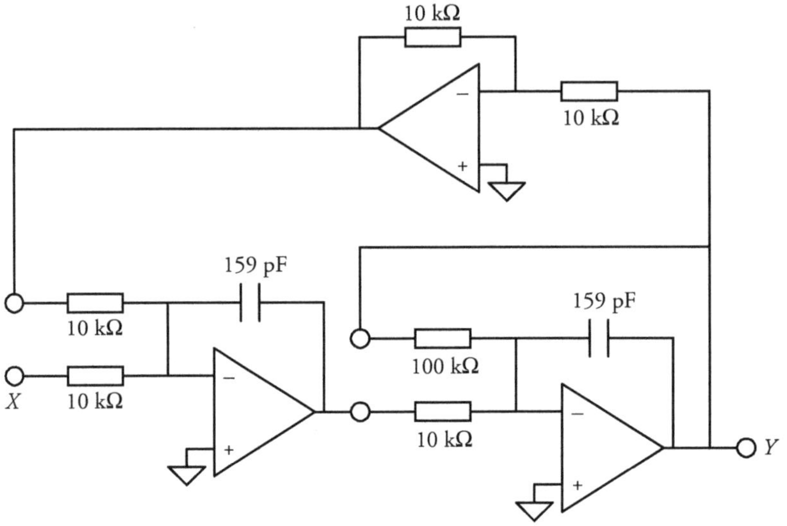

FIGURE 11.12 Final realization VLSI active-RC filter for the given example.

11.2 Cascade-Based VLSI Active-RC Filter Design Methods

Transfer function of a given filter can also be realized based on the multiplication of low-order transfer functions (typically first- and second-filter sections) as shown in Fig. 11.13. In this case, the filter can be implemented through cascading of low-order filter circuit sections.

$$H(s) = H_1(s) \cdot H_2(s) \cdot \cdots \cdot H_n(s) \tag{11.15}$$

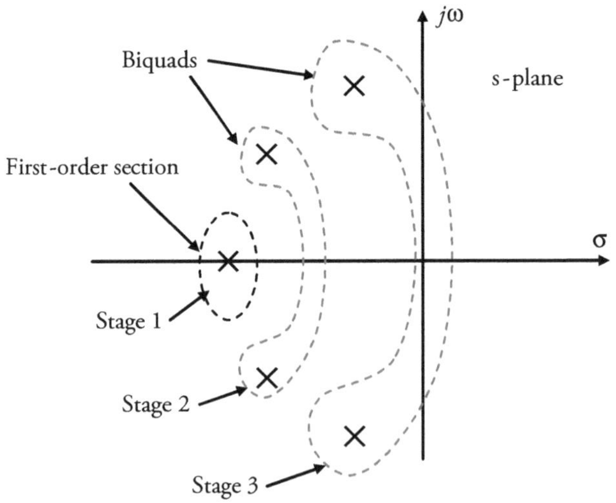

FIGURE 11.13 Partition of the transfer function into low-order sections based on poles/zeros.

Major advantages of the cascade-based filter design methods are as follows: (1) it may enable higher reusability since lower-order filter sections (typically the first- and second-order filter sections) can take very standard forms, (2) it may provide optimal filter circuit implementation with reduced active devices through the use of well-developed single-amplifier biquad (SAB)–based circuit structures, such as Sallen-Key circuits, (3) it may provide simplified design and design verification since each filter stage has dedicated response specifications, and (4) it can be verified separately.

11.2.1 First-Order Active-RC Filter Sections

A general first-order filter section can be expressed by an s-domain transfer function as

$$H(s) = \frac{Y(s)}{X(s)} = \frac{a_1 s + a_0}{b_1 s + b_0} = \frac{a_1 + a_0 s^{-1}}{b_1 + b_0 s^{-1}} \tag{11.16}$$

Tables 11.1 and 11.2 provide some commonly used first-order active-RC filter sections.

11.2.2 Second-Order Active-RC Filter Sections

A general normalized second-order filter (or biquad) section can be expressed in the s-domain transfer function as follows:

$$H(s) = \frac{Y(s)}{X(s)} = \frac{a_2 s^2 + a_1 s + a_0}{s^2 + \frac{1}{Q}s + 1} \tag{11.17}$$

The above transfer function can be expressed using lossless integrators suitable for SFG-based implementation as

$$Y(s) = a_2 X + s^{-1}\left[\left(a_1 X - \frac{1}{Q}Y\right) + s^{-1}(a_0 X - Y)\right] \tag{11.18}$$

or

$$\begin{cases} Y(s) = (a_2 + s^{-1}a_1 + s^{-2}a_0)U(s) \quad (A) \\ U(s) = X(s) - \left(s^{-1}\frac{1}{Q} + s^{-2}\right)U(s) \quad (B) \end{cases} \tag{11.19}$$

The two SFG representations of the above transfer function are shown in Fig. 11.14.

11.2.3 Low-Pass Filter Sections

The low-pass prototype-filter section can be expressed using the s-domain transfer function as follows:

$$H(s) = \frac{Y(s)}{X(s)} = \frac{1}{s^2 + \frac{1}{Q}s + 1} \tag{11.20}$$

An SFG representation of such filter sections is shown in Fig. 11.15.

This SFG can be directly implemented into the VLSI active-RC circuit as shown in Fig. 11.16.

$H(s) = \dfrac{Y(s)}{X(s)} = -\dfrac{K}{s}$	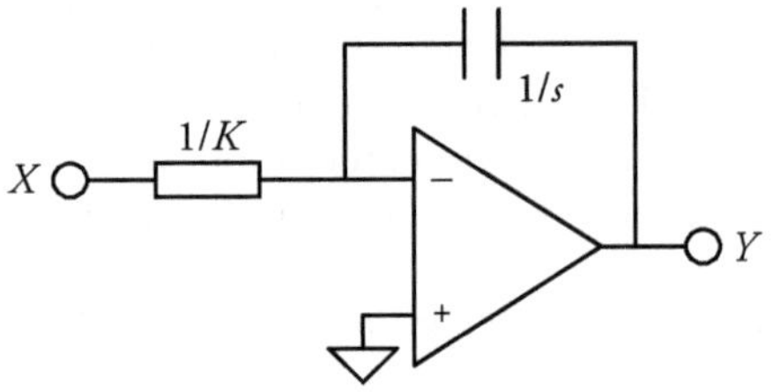
$H(s) = \dfrac{Y(s)}{X(s)} = -\dfrac{K}{s+p}$	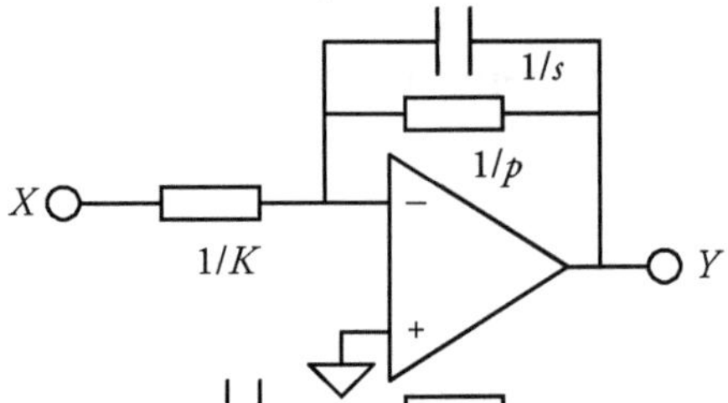
$H(s) = \dfrac{Y(s)}{X(s)} = -K(s+z)$	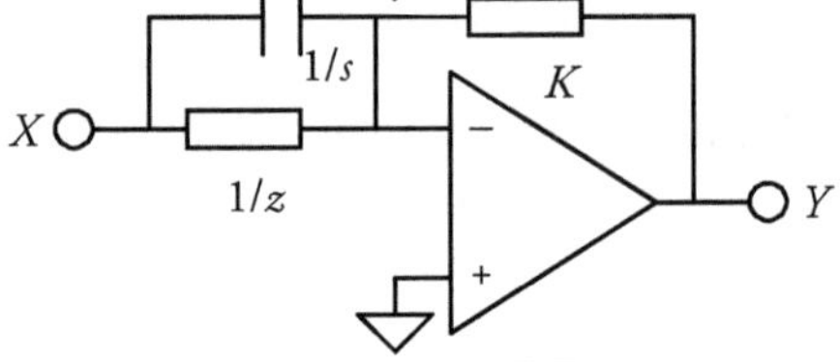
$H(s) = \dfrac{Y(s)}{X(s)} = \dfrac{-Ks}{s+p}$	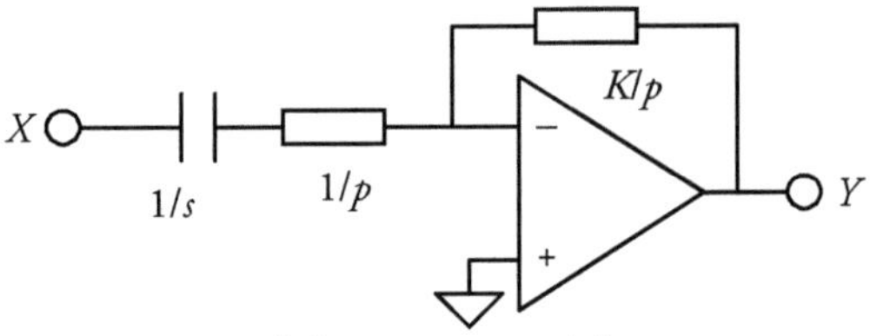
$H(s) = \dfrac{Y(s)}{X(s)} = -K\,\dfrac{s+z}{s+p}$	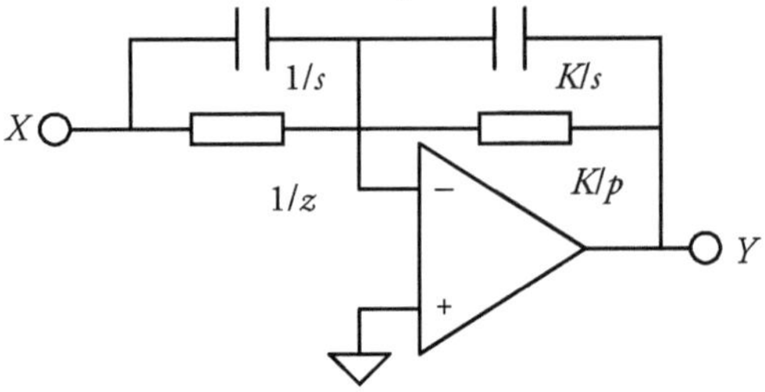
$H(s) = \dfrac{Y(s)}{X(s)} = -K\,\dfrac{s+z}{s+p}$	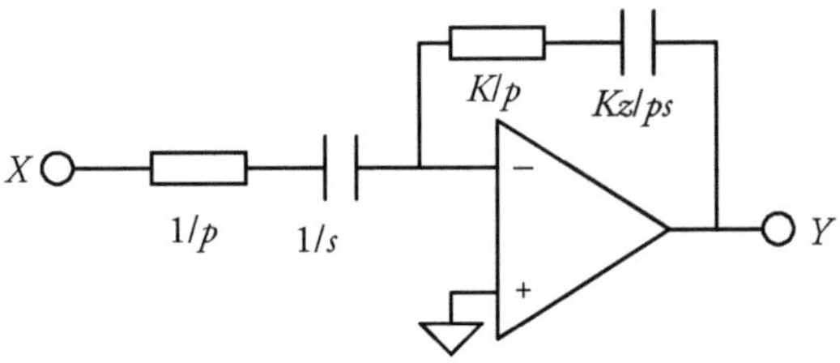

TABLE 11.1 First-Order Inverted Active-RC Filter Sections

$$H(s) = \frac{Y(s)}{X(s)} = \frac{s+z}{s+p} \qquad z > p$$

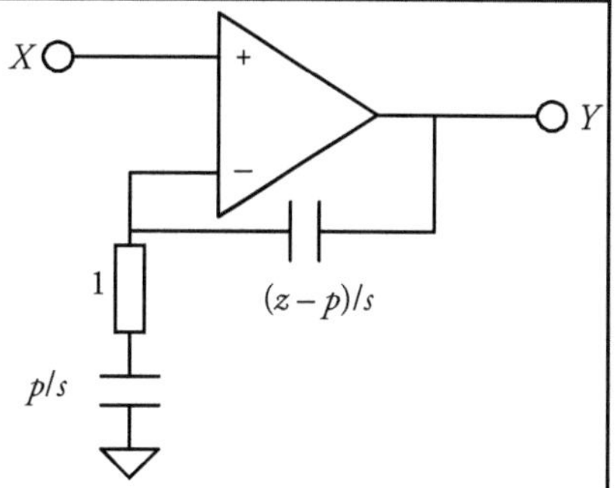

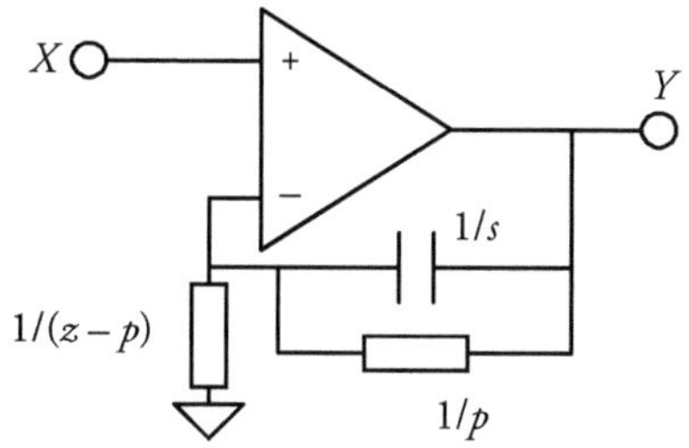

$$H(s) = \frac{Y(s)}{X(s)} = \frac{s+z}{s+p} \qquad z > p$$

$$H(s) = \frac{Y(s)}{X(s)} = \frac{p}{z}\,\frac{s+z}{s+p} \qquad p > z$$

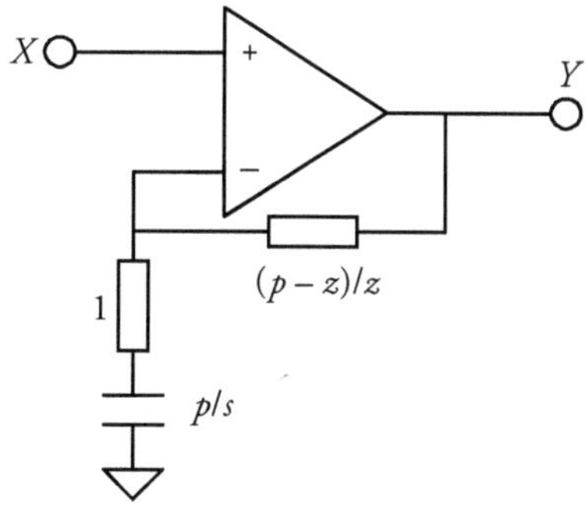

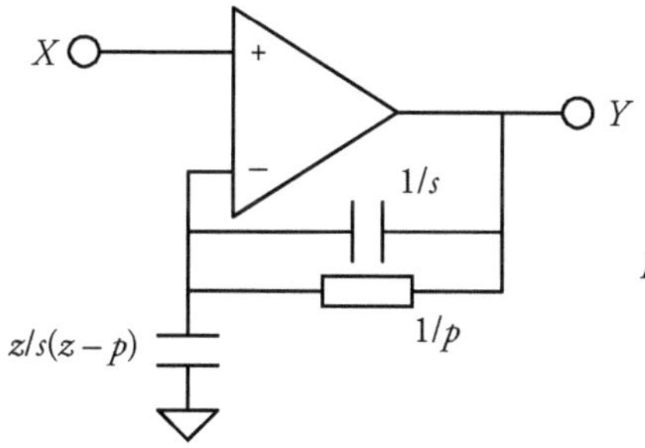

$$H(s) = \frac{Y(s)}{X(s)} = \frac{p}{z}\,\frac{s+z}{s+p} \qquad p > z$$

$$H(s) = \frac{Y(s)}{X(s)} = \frac{1}{z}\,(s+z)$$

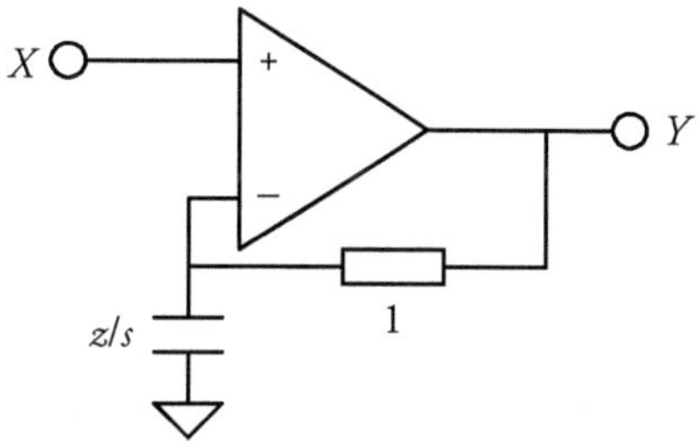

Table 11.2 First-Order Noninversion Active-RC Filter Sections

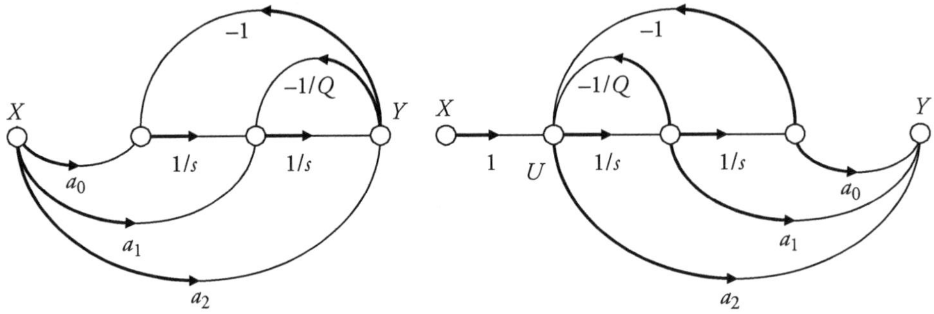

(a) Option A (b) Option B

FIGURE 11.14 SFG representations of the general second-order filter section.

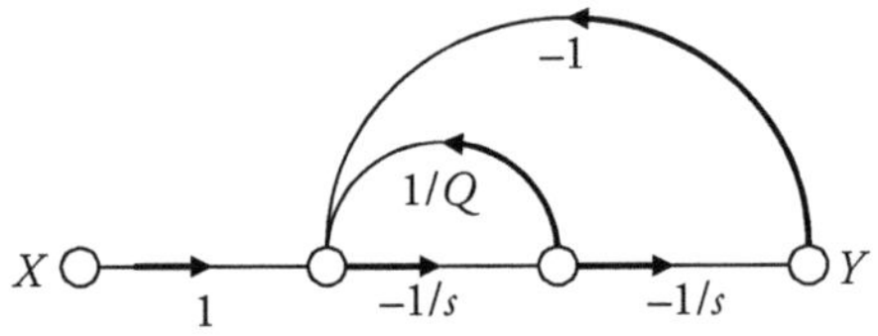

FIGURE 11.15 SFG of second-order low-pass filter section.

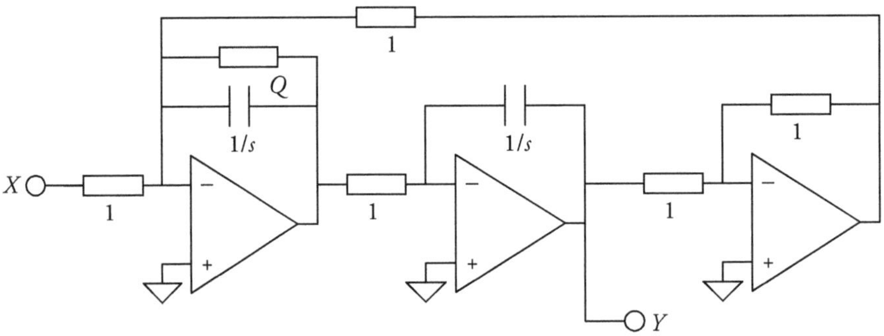

FIGURE 11.16 VLSI second-order active-RC low-pass filter circuit section.

11.2.4 Band-Pass Filter Sections

The band-pass prototype-filter section can be expressed using the s-domain transfer function as follows:

$$H(s) = \frac{Y(s)}{X(s)} = \frac{s}{s^2 + \frac{1}{Q}s + 1} \tag{11.21}$$

An SFG representation of the filter section is shown in Fig. 11.17.

The SFG can be directly implemented into the VLSI active-RC circuit as shown in Fig. 11.18.

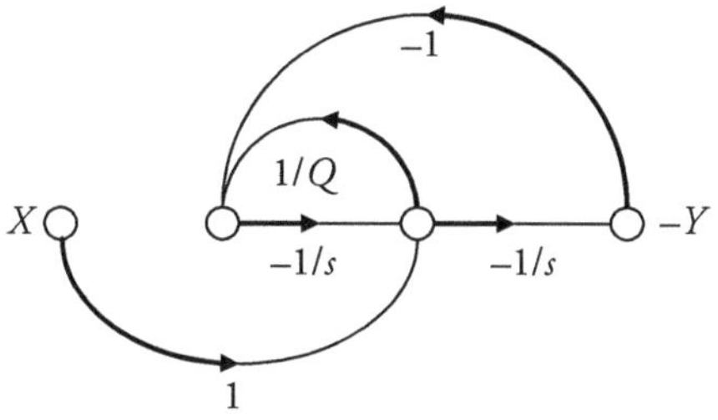

FIGURE 11.17 SFG of second-order band-pass filter sections.

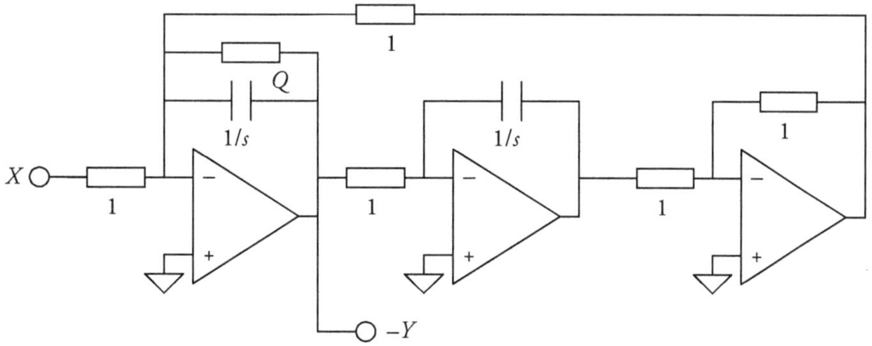

FIGURE 11.18 Second-order VLSI active-RC band-pass filter circuit section.

11.2.5 High-Pass Filter Sections

The high-pass prototype-filter section can be expressed using the following s-domain transfer function:

$$H(s)=\frac{Y(s)}{X(s)}=\frac{s^2}{s^2+\frac{1}{Q}s+1}=1-\frac{\frac{1}{Q}s+1}{s^2+\frac{1}{Q}s+1} \qquad (11.22)$$

An SFG representation of the filter section is shown in Fig. 11.19.

The SFG can be implemented into the VLSI active-RC circuit as shown in Fig. 11.20.

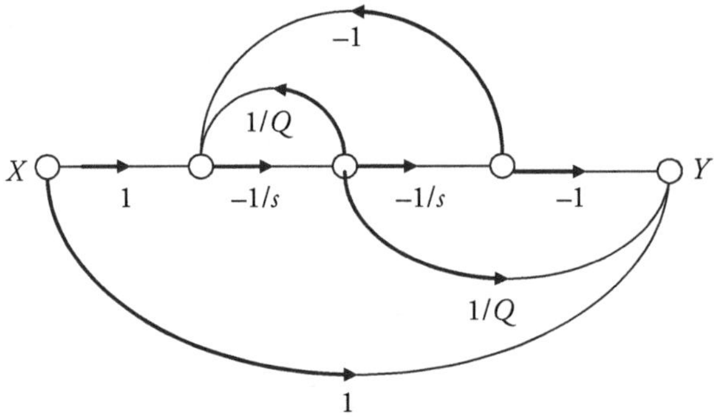

FIGURE 11.19 SFG of second-order high-pass filter sections.

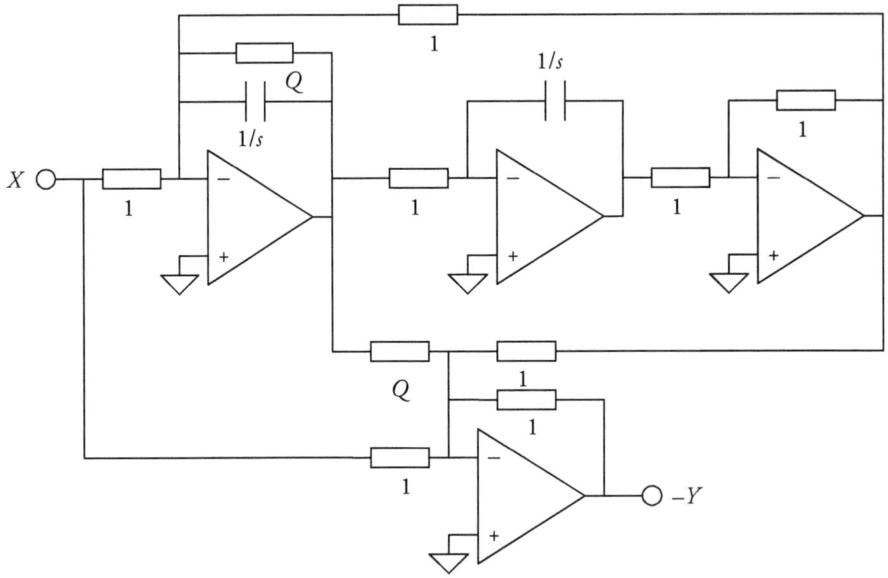

FIGURE 11.20 Second-order VLSI active-RC high-pass filter circuit section.

11.3 Sallen-Key Circuits

The Sallen-Key circuits are a class of active-RC circuits that realized second-order filter sections using only one OPAMP circuit. The general structure of the Sallen-Key circuit is shown in Fig. 11.21, where K is the voltage gain of the buffer circuit, and $\{G_i i = 1, 2, 3, 4, 5\}$ are the general conductance of the passive components. The s-domain transfer function of the general Sallen-Key circuit structure is given as follows:

$$\frac{V_o}{V_i} = \frac{KG_1G_2}{(G_2 + G_4)(G_1 + G_2 + G_3 + G_5) - KG_2G_3 - G_2G_2} \tag{11.23}$$

11.3.1 Low-Pass Sallen-Key Circuits

The low-pass prototype biquad can be realized by selecting the following device components:

$$\begin{cases} G_1 = G_2 = 1 \\ G_3 = G_4 = s \\ G_5 = 0 \\ Q \equiv 1/(3 - K) \end{cases} \tag{11.24}$$

We then have the following:

$$\Rightarrow \frac{V_o}{V_i} = \frac{K}{s^2 + \frac{1}{Q}s + 1} \tag{11.25}$$

The corresponding Sallen-Key circuit structure is shown in Fig. 11.22.

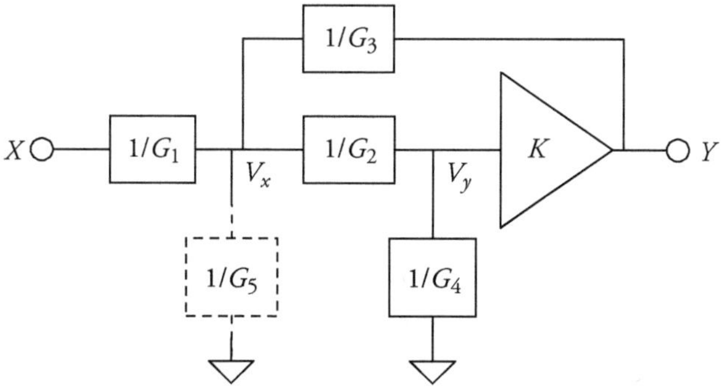

FIGURE 11.21 General Sallen-Key circuit structure.

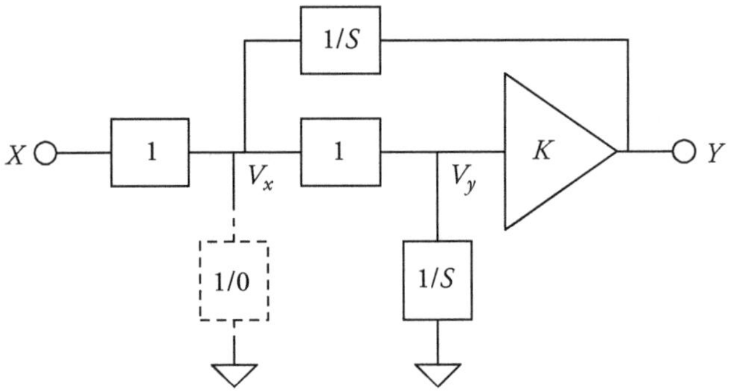

FIGURE 11.22 The low-pass prototype Sallen-Key circuit structure.

11.3.2 High-Pass Sallen-Key Circuits

The high-pass prototype biquad can be realized by selecting device parameters as follows:

$$\begin{cases} G_1 = G_2 = s \\ G_3 = G_4 = 1 \\ G_5 = 0 \\ Q \equiv 1/(3-K) \end{cases} \tag{11.26}$$

We then have the following equation:

$$\Rightarrow \frac{V_o}{V_i} = \frac{Ks^2}{s^2 + \frac{1}{Q}s + 1} \tag{11.27}$$

The corresponding high-pass Sallen-Key circuit is shown in Fig. 11.23.

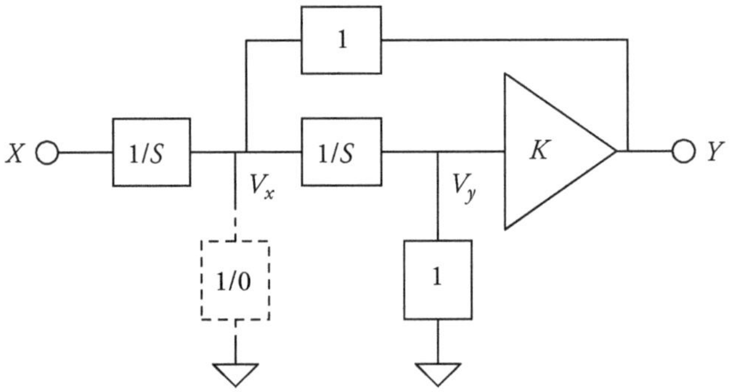

FIGURE 11.23 The high-pass prototype Sallen-Key circuit structure.

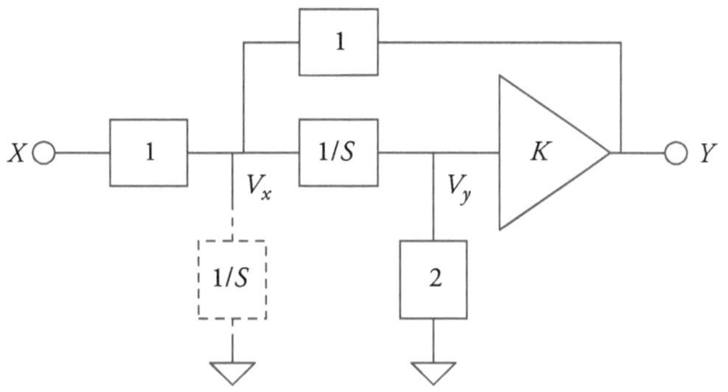

FIGURE 11.24 The band-pass prototype Sallen-Key circuit structure.

11.3.3 Band-Pass Sallen-Key Circuits

The band-pass prototype biquad can be realized by selecting device parameters as follows:

$$\begin{cases} G_1 = G_3 = 1 \\ G_2 = G_5 = s \\ G_5 = 1/2 \\ Q \equiv 1/(3-K) \end{cases} \tag{11.28}$$

We then have the following:

$$\Rightarrow \frac{V_o}{V_i} = \frac{Ks}{s^2 + \frac{1}{Q}s + 1} \tag{11.29}$$

The corresponding band-pass Sallen-Key circuit is shown in Fig. 11.24.

11.4 Delyiannis-Friend Circuits

Similar to the Sallen-Key circuits, the Delyiannis-Friend circuits are a different group of VLSI active-RC circuits, which can be used to realize second-order band-pass filter sections based on only one OPAMP circuit. These circuits are based on the bridge-T RC circuits with the OPAMP to provide negative feedback. The general structure of the Delyiannis-Friend circuit, which realizes the following transfer function, is shown in Fig. 11.25.

$$\frac{V_o}{V_i} = \frac{-2Qs}{s^2 + \frac{1}{Q}s + 1} \tag{11.30}$$

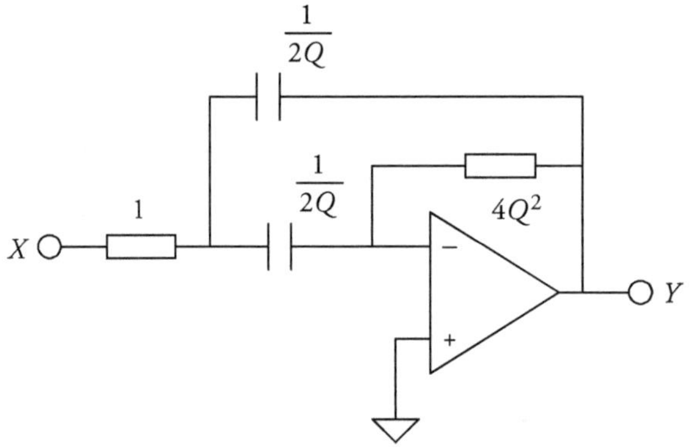

FIGURE 11.25 General Delyiannis-Friend circuit structure.

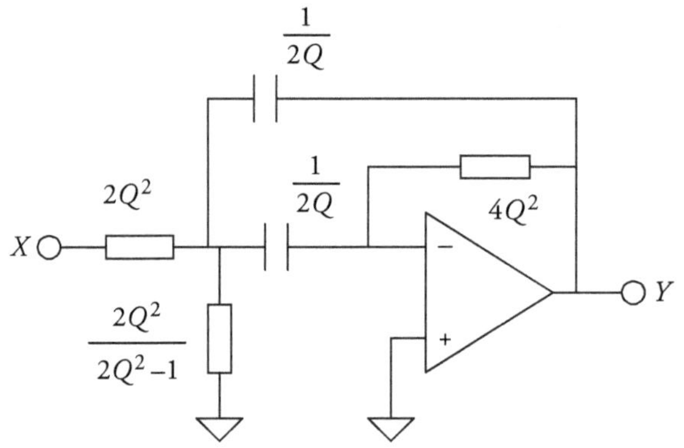

FIGURE 11.26 General Delyiannis-Friend circuit structure.

The second-order band-pass filter section, as described in Fig. 11.25, can be converted into the standard form by using the Delyiannis-Friend circuit shown in Fig. 11.26.

$$\frac{V_o}{V_i} = \frac{-\frac{1}{Q}s}{s^2 + \frac{1}{Q}s + 1} \tag{11.31}$$

11.5 Design Example

The fifth-order Butterworth prototype low-pass filter shown by the s-domain has the following transfer function (see Fig. 11.27):

$$H(s) = \frac{1}{s^5 + 3.236s^4 + 5.236s^3 + 5.236s^2 + 3.236s + 1} \tag{11.32}$$

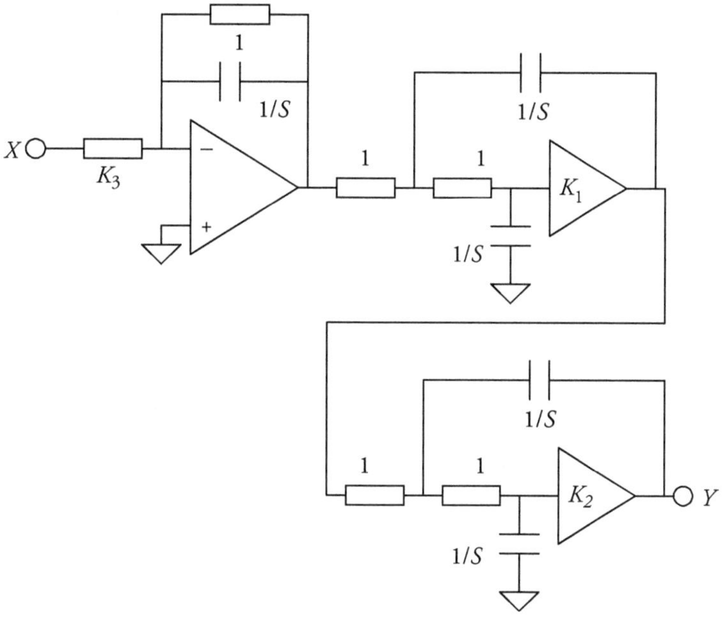

FIGURE 11.27 Active-RC realization of fifth-order Butterworth prototype filter.

This transfer function can be rewritten into a cascaded form using three sections as follows:

$$H(s) = \frac{1}{1+s} \cdot \frac{1}{s^2 + \frac{1}{0.62}s + 1} \cdot \frac{1}{s^2 + \frac{1}{1.62}s + 1} \tag{11.33}$$

Such a transfer function can be realized by using the first- and second-order sections as shown in Fig. 11.27, where

$$\begin{cases} K_1 = 3 - 1/0.62 \\ K_2 = 3 - 1/1.62 \\ K_3 = K_1 \cdot K_2 \end{cases} \tag{11.34}$$

By using the frequency-scaling technique, this circuit can further be scaled to a given frequency response.

References

1. R. Schaumann, "The Design of Continuous-Time Fully Integrated Filters: A Review," *IEE Proceedings, Part G*, vol. 136, pp. 184–190, August 1989.
2. A. M. Durham, W. Redman-White, J. B. Hughes, "Low Distortion VLSI Compatible Self-Tuned Continuous-Time Monolithic Filters," *IEEE ISCAS'91*, pp. 1448–1451, June 1991.

3. T. Hollman, S. Lindfors, M. Lansirinne, K. Jussila, A. I. Halonen, "A 2.7-V CMOS Dual-Mode Baseband Filter for PDC and WCDMA," *IEEE J. Solid-State Circuits*, vol. 36, no. 7, pp. 1148–1153, July 2001.
4. R. Hogervorst, J. P. Tero, R. G. H. Eschauzier, J. H. Huijsing, "A Compact Power Efficient Rail-To-Rail Input/Output Operational Amplifier for VLSI Cell Libraries," *IEEE J. Solid-State Circuits*, vol. SC-29, pp. 1505–1512, December 1994.
5. R. G. H. Eschauzier, J. H. Huijsing, "A Programmable 1.5V CMOS Class-AB Operational Amplifier with Hybrid Nested Miller Compensation for 120dB Gain and 6 MHz UGF," *IEEE J. Solid-State Circuits*, vol. SC-29, no. 12, pp. 1497–1504, December 1994.
6. P. V. A. Mohan, "Noise and Sensitivity Optimization of Actively Compensated Active RC Filters Based on Friend-Deliyannis SAB," *ISCAS'88*, vol. 3, pp. 2673–2676, June 1988.
7. M. J. Paschal, L. P. Huelsman, "Compensation of the Infinite-Gain Multiple-Feedback Low-Pass Active RC Filter," *Proceedings, the 37th Midwest Symposium on Circuits and Systems*, vol. 2, pp. 772–775, August 1994.
8. M. S. Zarnik, F. Novak, S. Macek, "Oscillation Test Structures of Active RC Filters," *IEE Proceedings, Circuits, Devices and Systems*, vol. 147, no. 5, October 2000.
9. Huanzhang Huang, E. K. F. Lee, "Low Voltage Technique for Active RC Filter," *Electronics Letters*, vol. 34. no. 15, July 1998.
10. C. Acar, H. Kuntman, "Limitations on Input Signal Level in Current-mode Active-RC Filters using CCIIs," *Electronics Letters*, vol. 32, no. 16, pp. 1461–1462, August 1996.
11. M. Santo Zarnik, F. Novak, S. Macek, "Design of Oscillation-Based Test Structures for Active RC filters," *Proceedings, ED&TC*, p. 618, March 1997.
12. S. Signell, K. Mossberg, "Design of High-Frequency Active RC-Filters using Passive Compensation Methods," *IEEE ISCAS'88*, vol. 3, pp. 2669–2672, June 1988.
13. A. Borys, N. Fliege, "Complementary Transformation in Mildly Nonlinear Feedback Systems," *European Conference on Circuit Theory and Design*, pp. 445–447, September 1989.
14. M. Soderstrand, S. Mitra, "Design of Active RC Filters with Zero Gain-Sensitivity Product," *IEEE Trans. on Circuits and Systems*, vol. 20, no. 4. pp. 441–445, July 1973.
15. Kong-Pang Pun, Chiu-Sing Choy, Cheong-Fat Chan, J. E. da Franca, "Digital Frequency Tuning Technique Based on Current Division for Integrated Active RC Filters," *Electronics Letters*, vol. 39, no. 19, pp. 1366–1367, September 2003.
16. M. Banu, Y. Tsividis, "Fully Integrated Active RC Filters in MOS Technology," *IEEE ISSCC*, vol. XXVI, pp. 244–245, February 1983.
17. D. Jurisic, G. S. Moschytz, N. Mijat, "Low-Noise Low-Power All pole Active-RC Filters Minimizing Resistor level," *Proceedings, ISCAS'03*, vol. 1, pp. I-469–I-472, May 2003.
18. J. Jussila, K. Halonen, "A 1.5V Active-RC Filter for WCDMA Applications," *Proceedings, ICECS*, vol. 1, pp. 489–492, September 1999.
19. N. G. Maratos, G. T. Ioannidis, "A Method for Dynamic Range Maximization for a Class of Integrated Active-RC Filters," *ISCAS'94*, vol. 5, pp. 763–766, May 1994.
20. S. Hirano, E. Hayahara, "A Compensation Technique for High Frequency Responses of Active RC Filter using Functional Trimming," *Proceedings, the 36th Midwest Symposium on Circuits and Systems*, vol. 2, pp. 1229–1232, August 1993.
21. H. Tang, R. J. Mack, "A Dictionary Approach to Fault Location in Linear Circuits," *IEEE ISCAS'91*, vol. 4, pp. 2064–2067, June 1991.

22. G. T. Volpe, L. M. Freeman, "Design of a Hybrid Integrated Circuit Elliptic Active RC Filter," *IEEE J. Solid-State Circuits*, vol. 9, no. 2, pp. 76–79, April 1974.
23. W. Khalil, Tsung-Yuan Chang, X. Jiang, S. R. Naqvi, B. Nikjou, J. Tseng, "A Highly Integrated Analog Front-End for 3G," *IEEE J. Solid-State Circuits*, vol. 38, no. 5, pp. 774–781, May 2003.
24. A. M. Durham, W. Redman-White, J. B. Hughes, "High-Linearity Continuous-Time Filter in 5V VLSI CMOS," *IEEE J. Solid-State Circuits*, vol. 27, no. 9, pp. 1270–1276, September 1992.
25. Y. Ishibashi, "Active RC Filter Based on Simulation of Dissipative LC Filters," *IEE Proceedings, Circuits, Devices and Systems*, vol. 141, no. 2, pp. 101–104, April 1994.
26. D. Jurisic, N. Mijat, "Impedance Tapering Effects on "Low-Q" SAB Band-Pass Active-RC Filter," *Proceedings, the First International Workshop on Image and Signal Processing and Analysis*, pp. 229–234, June 2000.
27. C. Frost, G. Levy, B. Allison, "A CMOS 2 MHz Self-Calibrating Bandpass Filter for Personal Area Networks," *ISCAS'03*, vol. 1, pp. I-485–I-488, May 2003.
28. J. K. Pyykonen, "A Low Distortion Wideband Active-RC Filter for a Multicarrier Base Station Transmitter," *ISCAS'01*, vol. 1, pp. 244–247, May 2001.
29. C. S. Wong, "A 3V GSM Baseband Transmitter," *Digest of Technical Papers, 1998 Symposium on VLSI Circuits*, pp. 120–123, June 1998.
30. O. Vainio, S. J. Ovaska, "A Class of Active-RC Filters with Predictive Characteristics," *Proceedings, IEEE, IMTC*, p. 358, April 1995.
31. J. Padalino, A. Tuszynski, "Note on the Analysis and Synthesis of Active RC Filters," *IEEE Trans. on Circuits and Systems*, vol. 15, no. 4, pp. 503–505, December 1968.
32. M. Soderstrand, S. Mitra, "Gain and Sensitivity limitations of Active RC Filters," *IEEE Trans. on Circuits and Systems*, vol. 18, no. 6, pp. 600–609, November 1971.
33. W. Holmes, "On Canonical Second-Order Active RC Filters," *IEEE Trans. on Circuits and Systems*, vol. 18, no. 6, pp. 721–723, November 1971.
34. D. Akerberg, K. Mossberg, "A Versatile Active RC Building Block with Inherent Compensation for the Finite Bandwidth of the Amplifier," *IEEE Trans. on Circuits and Systems*, vol. 21, no. 1, pp. 75–78, January 1974.
35. R. Srinivasagopalan, G. Martens, "Comparison of a Class of Active Filters with respect to the Operational-Amplifier Gain-Bandwidth Product," *IEEE Trans. on Circuits and Systems*, vol. 21, no. 3, pp. 377–381, May 1974.
36. K. Laker, M. Ghausi, "Synthesis of a Low-Sensitivity Multiloop Feedback Active RC Filter," *IEEE Trans. on Circuits and Systems*, vol. 21, no. 2, pp. 252–259, March 1974.
37. P. A. BrackettSedra, "Active Compensation for High-frequency Effects in OPAMP Circuits with Applications to Active RC Filters," *IEEE Trans. on Circuits and Systems*, vol. 23, no. 2, pp. 68–72, February 1976.
38. M. Reddy, "An Insensitive Active-RC Filter for High Q and High Frequencies," *IEEE Trans. on Circuits and Systems*, vol. 23, no. 7, pp. 429–433, July 1976.
39. M, Reddy, "Operational-Amplifier Circuits with Variable Phase Shift and Their Application to High-Q Active RC-filters and RC-oscillators," *IEEE Trans. on Circuits and Systems*, vol. 23, no. 6, pp. 384–389, June 1976.
40. S. Mitra, P. Vaidyanathan, B. Anderson, "A General Theory and Synthesis Procedure for Low-Sensitivity Active RC Filters," *IEEE Trans. on Circuits and Systems*, vol. 32, no. 7, pp. 687–699, July 1985.

41. M. Ishikawa, T. Kimura, N. Tamaki, "A CMOS Automatic Line Equalizer LSI Chip using Active-RC Filtering," *IEEE J. Solid-State Circuits*, vol. 19, no. 4, pp. 506–509, August 1984.
42. C. S. Wong, "A 3-V GSM Baseband Transmitter," *IEEE J. Solid-State Circuits*, vol. 34, no. 5, pp. 725–730, May 1999.
43. D. Jurisic, G. S. Moschytz, N. Mijat, "Low-sensitivity Active-RC High- and Band-Pass Second-Order Sallen & Key Allpole Filters," *ISCAS'02*, vol. 4, pp. IV-241–IV-244, May 2002.
44. S. Natarajan, "A Comparison of Two Actively Compensated Active-RC Filters," *IEEE Trans. on Circuits and Systems*, vol. 37, no. 2, pp. 296–299, February 1990.
45. D. A. Johns, W. M. Snelgrove, A. S. Sedra, "Orthonormal Ladder Filters," *IEEE Trans. on Circuits and Systems*, vol. 36, no. 3, pp. 337–343, March 1989.
46. D. Jurisic, G. S. Moschytz, N. Mijat, "Low Active-Sensitivity All Pole Active-RC Filters using Impedance Tapering," *ISCAS'04*, vol. 1, pp. 89–92, May 2004.
47. C. Durdodt, A. Hanke, S. Heinen, U. Langmann, "Comparison of an Inductorless Low-IF and Zero-IF Receiver for Bluetooth," *The 2002 45th Midwest Symposium on Circuits and Systems,* vol. 1, pp. I563–I566, August 2002.
48. P. Aronhime, M. Desai, "Offset Independent Active RC Filters," *Proceedings of the 35th Midwest Symposium on Circuits and Systems,* vol. 2, pp. 1460–1463, August 1992.
49. D. Jurisic, G. S. Moschytz, "Low-Noise Active-RC Low-, High- and Band-pass All Pole Filters using Impedance Tapering," *MELECON*, vol. 2, pp. 591–594, 2000.
50. L. Maurer, W. Schelmbauer, H. Pretl, B. Adler, A. Springer, R. Weigel, "On the Design of a Continuous-Time Channel Select Filter for a Zero-IF UMTS Receiver," *Proceedings, Vehicular Technology Conference,* vol. 1, pp. 650–654, May 2000.

CHAPTER 12

Design of VLSI MOS-C Filters

VLSI analog filters based on the MOS-C circuit techniques employing the MOS VCR offer the continuous resistance value tuning capability. Such a circuit feature combined with various filter-tuning-control circuit techniques provides effective circuit solutions to compensate for the time constant or characteristic frequency variations for the VLSI circuit implementation under the PVT corner and the application condition variations.

MOS-C–based VLSI analog filter circuits can, in general, be designed based on similar approaches as the VLSI active-RC filter circuits, including the SFG-based, the cascade-based, the component simulation–based, and the ladder simulation–based design solutions.

12.1 SFG-Based VLSI MOS-C Filter Design Methods

The SFG-based filter design methods described in VLSI active-RC filter designs can also be used for the VLSI MOS-C filter design. Based on the VLSI MOS-C circuit techniques discussed in Chap. 4, MOS-C integration, scaling, and addition functional circuits can also be developed for VLSI signal processing operations.

12.1.1 MOS-C Circuit Realization of Integration

One of the most commonly used basic building blocks for the MOS-C filter is the integrator. The MOS-C integrator with the sum of weighted inputs is shown in Fig. 12.1.

$$Y = \frac{-1}{s}(\alpha_1 X_1 + \alpha_2 X_2 + \cdots + \alpha_n X_n) \tag{12.1}$$

Note that in practical cases, a single VCR control signal is preferred to simultaneous control of all VCR elements. Such a design approach can yield a simple and reliable design (see Fig. 12.2). In such a case, the weight of the inputs can be configured through the sizes of the devices:

$$\alpha_i = \beta_i (V_C - V_T) \qquad i = 1, 2, \ldots, n \tag{12.2}$$

Alternatively, the two-transistor differential VCR or four-transistor differential VCR structures can be used to realize the basic building circuit element as shown in Figs. 12.3 and 12.4.

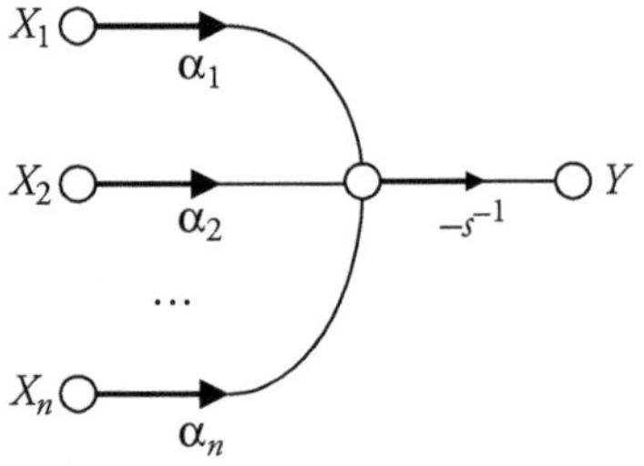

FIGURE 12.1 SFG of integration with weighted inputs.

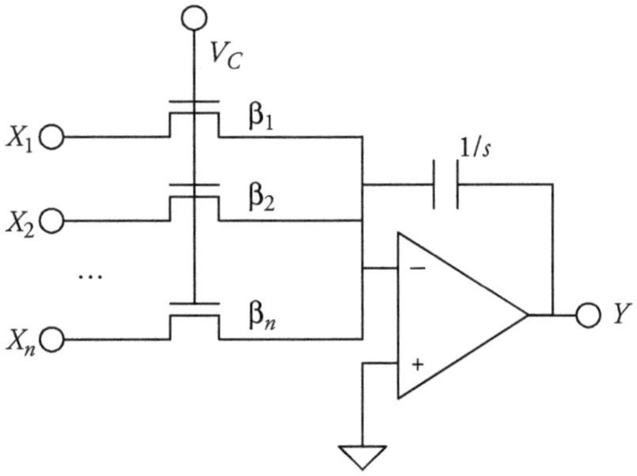

FIGURE 12.2 Simple MOS VCR integration with weighted inputs.

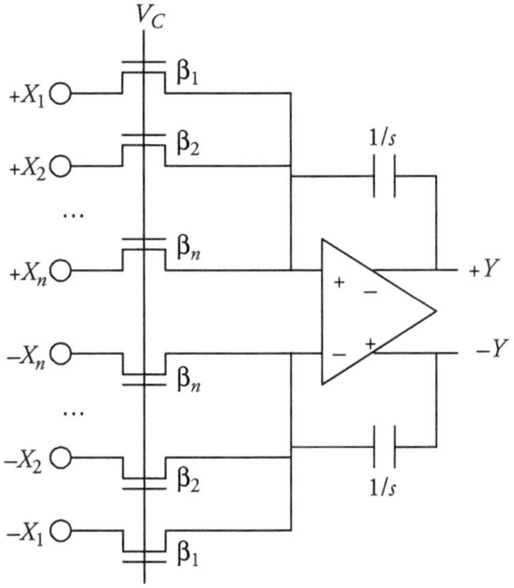

FIGURE 12.3 VLSI two-transistor differential MOS VCR integration with weighted inputs.

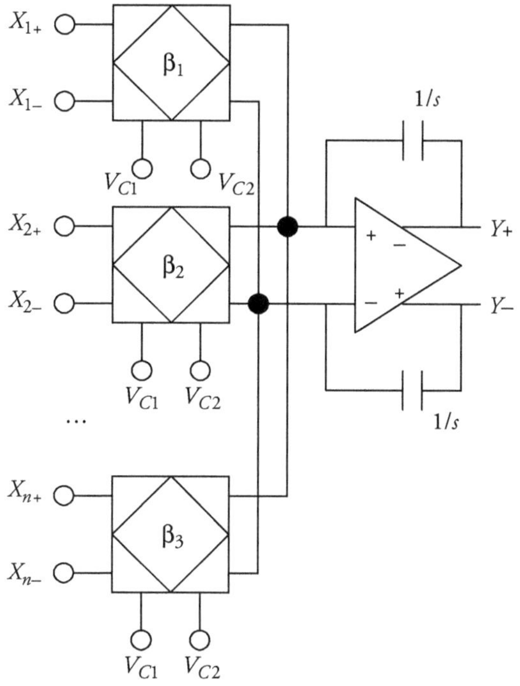

Figure 12.4 VLSI four-transistor differential MOS VCR integration with weighted inputs.

Such four-transistor MOS VCR–based MOS-C integrator circuit can be used to offer better circuit performance with the penalty of increased device count.

12.1.2 MOS-C Circuit Realization of Addition

The transfer function and SFG of the basic addition building blocks are shown in Fig. 12.5. The MOS-C circuit implementations using VLSI simple VCR, the two-transistor and four-transistor differential VCR circuit structures, are shown in Figs. 12.6 to 12.8.

$$Y = -1(\alpha_1 X_1 + \alpha_2 X_2 + \cdots + \alpha_n X_n) \tag{12.3}$$

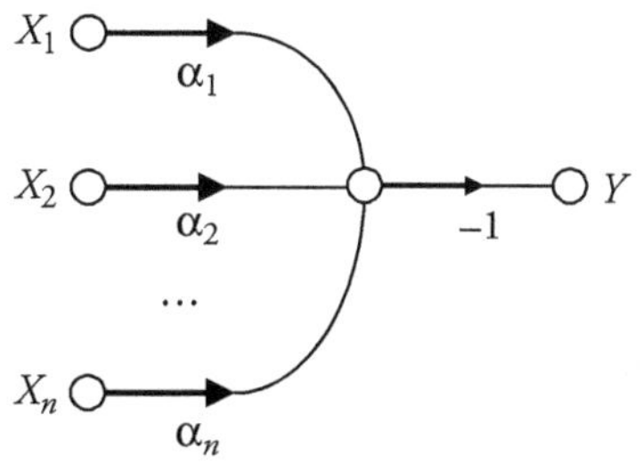

Figure 12.5 SFG of weighted addition.

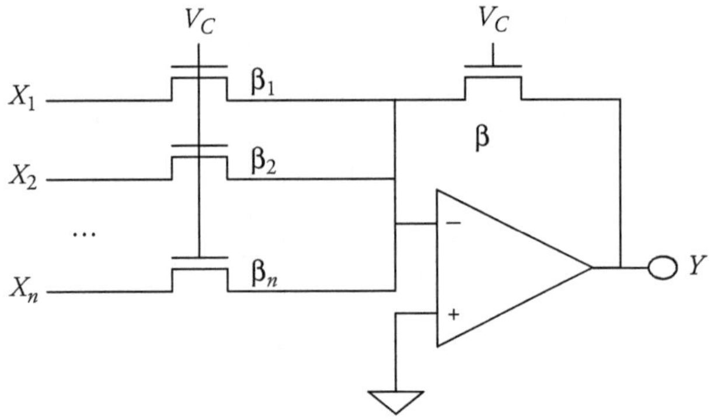

FIGURE 12.6 MOS VCR integration with weighted inputs.

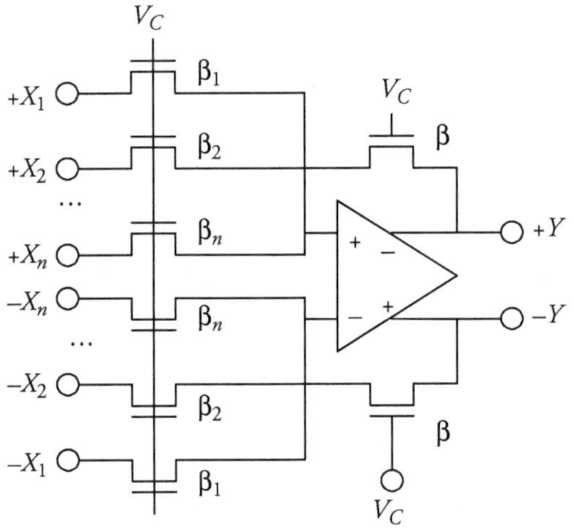

FIGURE 12.7 Two-transistor differential MOS VCR integration with weighted inputs.

where the coefficients are given as

$$\alpha_i = \beta_i / \beta \qquad i = 1, 2, \ldots, n \tag{12.4}$$

12.1.3 MOS-C Circuit Realization of Scaling

In Fig. 12.9, note that the building blocks readily offer the scaling function, which is provided by the weights of the inputs. We may simply use one path of the circuits to realize the scaling operation when a separate scaling function is required. However, practically in every VLSI filter design, it is desired to implement the scaling function along with other processing functions to minimize the implementation cost.

$$Y = -\alpha_1 X \tag{12.5}$$

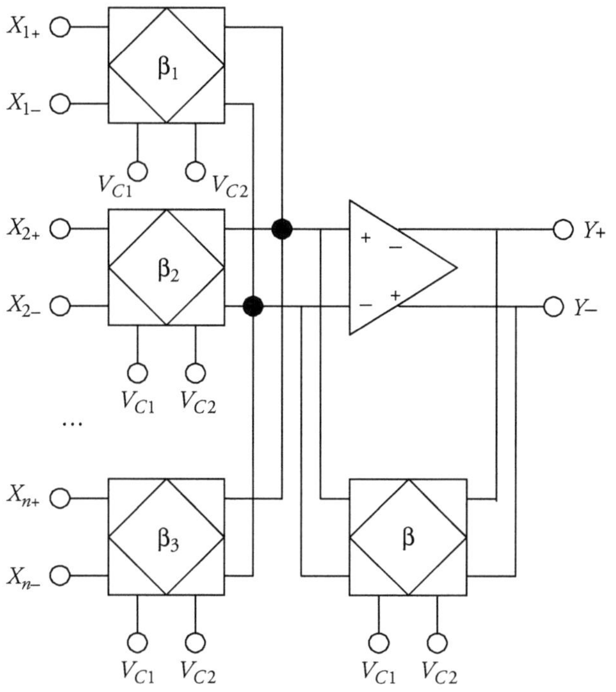

FIGURE 12.8 Four-transistor differential MOS VCR integration with weighted inputs.

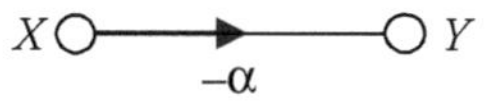

FIGURE 12.9 Transfer function and SFG of scaling operation.

With all the above basic filter building blocks, we may now implement a given filter transfer function following the filter-realization method by mapping SFG to MOS-C circuits.

12.1.4 Design Example

For the filter transfer function given (where $\omega_o = 100$ kHz and $Q = 10$) as

$$\frac{Y}{X} = \frac{1}{\left(\frac{s}{\omega_o}\right)^2 + \frac{1}{Q}\left(\frac{s}{\omega_o}\right) + 1} \tag{12.6}$$

we can obtain the normalized (or prototype) filter SFG similar to the example in the active-RC filter design chapter as shown in Fig. 12.10.

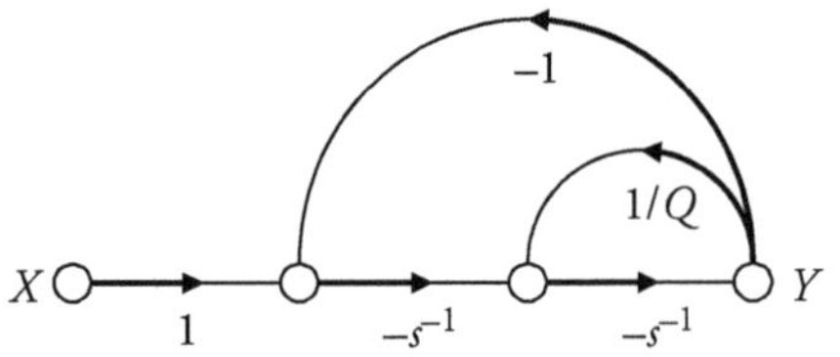

FIGURE 12.10 SFG of the given prototype filter.

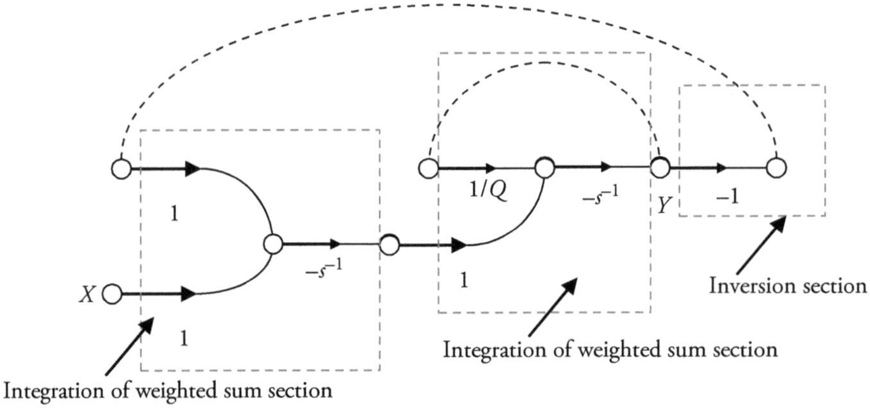

FIGURE 12.11 Normalized SFG suitable for VLSI MOS-C realization.

Similarly, the SFG in Fig. 12.10 can be partitioned into basic filter building blocks as shown in Fig. 12.11. Such an SFG representation is suitable for the VLSI MOS-C circuit implementation.

We may now directly map the normalized SFG into the MOS-C circuit as shown in Fig. 12.12.

The filter components can be further scaled to the value suitable for VLSI implementation as

$$\frac{\beta(V_{C1} - V_{C2})}{C} = 2\pi \cdot 100 \cdot 10^3 \tag{12.7}$$

By selecting C = 10 pF, we have the following:

$$\beta(V_{C1} - V_{C2}) = 2\pi \cdot 10^{-5} (\mathrm{A/V}) \tag{12.8}$$

Finally, the MOC-C building circuit, which satisfies the above conditions, can be designed.

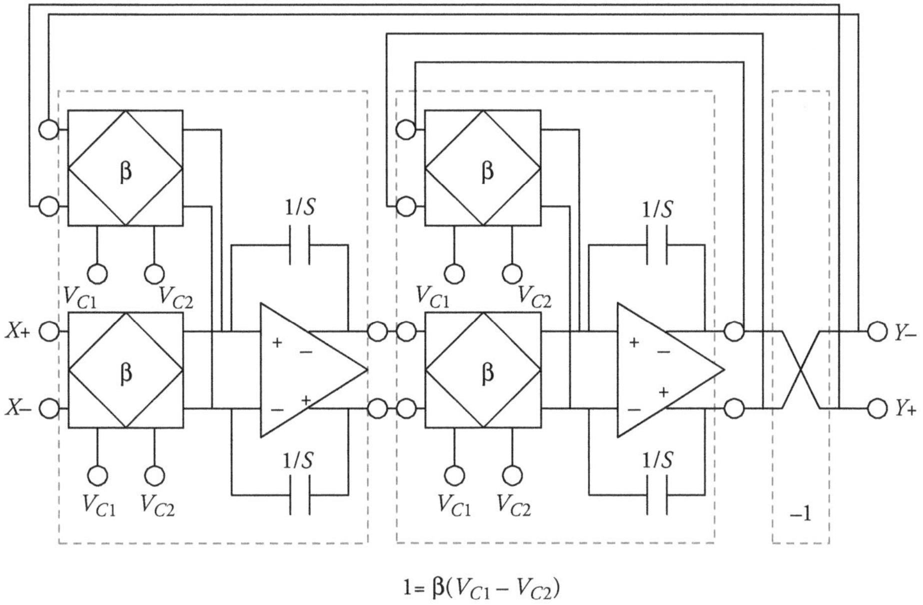

FIGURE 12.12 Four-transistor differential MOS-C realization of the normalized filter.

12.2 Cascade-Based VLSI MOS-C Filter Design Methods

The cascade-based VLSI MOS-C filter design approach is exactly the same as the active-RC cascade-based filter design, which constructs the filter through a cascade of low-order filter sections. Due to the similarity of the active-RC and MOS-C circuit structures, the first- and second-order MOS-C filter sections can be directly constructed from their active-RC counterparts with the technique discussed in the previous sections.

References

1. Y. Tsividis, M. Banu, J. Knoury, "Continuous-Time MOSFET-C Filters in VLSI," *IEEE J. Solid-State Circuits*, vol. SC-21, no. 1, pp. 15–30, February 1986.
2. M. Banu, Y. Tsividis, "Fully Integrated Active-RC Filters in MOS Technology," *IEEE J. Solid-State Circuits*, vol. SC-18, no. 6, pp. 644–651, December 1983.
3. M. Banu, Y. Tsividis, "An Elliptic Continuous-Time CMOS Filter with On-Chip Automatic Tuning," *IEEE J. Solid-State Circuits*, vol. SC-20, no. 6, pp. 1114–1121, December 1985.
4. M. Banu, Y. Tsividis, "Detailed Analysis of Nonidealities in MOS Fully Integrated Active-RC Filters Based on Balanced Networks," *IEE Proceedings*, vol. *131, Pt. G*, no. 5, pp. 190–196, October 1984.
5. J. M. Khoury, Y. Tsividis, "Analysis and Compensation of High-Frequency Effects in Integrated MOSFET-C Continuous-Time Filters," *IEEE Trans. Circuits Syst.*, vol. CAS-34, no. 8, pp. 862–875, August 1987.

6. A. Van, Bezooijen, N. Ramalho, J. O. Voorman, "Balanced Integrator Filters at Video Frequencies," *Digest ESSSCIRC'91*, pp. 1–4, 1991.
7. Z. Czarnul, "Modification of Banu-Tsivids Continuous-Time Integrator Structure," *IEEE Trans. Circuits Syst.*, vol. CAS-33, no. 7, pp. 714–716, July 1986.
8. Z. Czarnul, "Novel MOS Resistive Circuit for Synthesis of Fully Integrated Continuous-Time Filters," *IEEE Trans. Circuits Syst.*, vol. CAS-33, no. 7, pp. 718–721, July 1986.
9. M. Ismail, S. V. Smith, R. G. Beale, "A New MOSFET-C Universal Filter Structure for VLSI," *IEEE J. Solid-State Circuits*, vol. SC-22, no. 1, pp. 183–194, February 1988.
10. A. Yoshizawa, Y. P. Tsividis, "Anti-Blocker Design Techniques for MOSFET-C Filters for Direct Conversion Receivers," *IEEE J. Solid-State Circuits*, vol. 37, no. 3, pp. 757–760, March 2002.
11. S. P. Singh, J. V. Hansom, J. Vlach, "A New Floating Resistor for CMOS Technology," *IEEE Trans. on Circuits and Systems*, vol. 36, no. 9, pp. 1217–1220, September 1989.
12. H. P. Schmid, G. S. Moschytz, "Active-MOSFET-C Single-Amplifier Biquadratic Filters for Video Frequencies," *IEE Proceedings, Circuits, Devices and Systems*, vol. 147, no. 1, pp. 35–41, February 2000.
13. M. W. L. Cunha, S. N. Filho, M. C. Schneider, A. L. Dalcastagne, "Automatic Tuning of MOSFET-C Filters using Digitally Programmable Current Attenuators," *ISCAS'97*, vol. 1, pp. 329–332, June 1997.
14. J. I. Osa, A. Carlosena, "MOSFET-C Sinusoidal Oscillator with Variable Frequency and Amplitude," *ISCAS'00*, vol. 2, pp. 725–728, May 2000.
15. B. K. Ahuja, "Implementation of Active Distributed RC Anti-Aliasing/Smoothing Filters," *IEEE J. Solid-State Circuits*, vol. SC-17, no. 6, pp. 1076–1080, December 1982.
16. S. Ramet, "A Low-Distortion Anti-Aliasing/Smoothing Filter for Sampled Data Integrated Circuits," *IEEE J. Solid-State Circuits*, vol. SC-23, no. 5, pp. 1267–1270, October 1988.
17. Y. P. Tsividis, "Minimal Transistor-Only Micropower Integrated VHF Active Filter," *El. Letters*, vol. 23, no. 15, pp. 777–778, July 1987.
18. L. J. Pu, Y. P. Tsividis, "Transistor-Only Frequency-Selective Circuits," *IEEE J. Solid-State Circuits*, vol. SC-25, no. 3, pp. 821–832, October 1990.

CHAPTER 13

Design of VLSI Gm-C Filters

VLSI Gm-C filter circuits based on Gm-C circuit techniques offer attractive features like wideband operation and inherent tuning capability that are benefits of the VLSI Gm circuit structures. Such type of circuits also offer analog signal processing capabilities in various communication applications ranging from low-frequency (audio) to very high-frequency (RF) bands.

General design approaches to VLSI Gm-C filters are similar to other VLSI CT circuit types such as the active-RC or MOS-C filter circuits, including the SFG-based, cascade-based, component simulation–based, and ladder simulation–based approaches.

Although the basic building blocks of the Gm-C filter can be selected to be the same as the active-RC and MOS-C filters, the current domain behavior of the Gm circuit output offers the flexibilities of the VLSI circuit implementations in Gm-C filters, such as the circuit component count minimization employing the wired-addition function of the Gm-C outputs.

13.1 SFG-Based VLSI Gm-C Filter Design Methods

In principle, the VLSI Gm-C lossless integration, addition, and scaling circuits provide a complete set of building circuit blocks for the VLSI Gm-C filter circuit implementations. Practical VLSI circuit designs based on the SFG design approach usually employ the combined functional circuit blocks, including integration with weighted inputs, weighted addition, and scaling elements.

13.1.1 Gm-C Circuit Realization of Integration

For the inverted integration with weighted inputs, the SFG representation of building elements and the Gm-C circuit realization is shown in Fig. 13.1. The inverted integration is usually selected based on the fact that there is usually an inherent inversion for the simplest VLSI Gm-C circuit implementation. Such a selection method can usually lead to more efficient VLSI Gm-C circuit implementation.

$$Y = -\frac{1}{s}(\alpha_1 X_1 + \alpha_2 X_2 + \cdots + \alpha_n X_n) \tag{13.1}$$

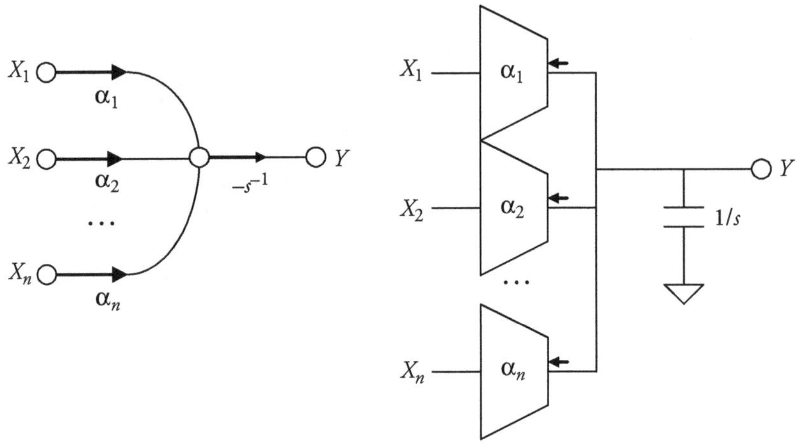

FIGURE 13.1 The SFG and Gm-C circuit of integration with weighted inputs.

13.1.2 Gm-C Circuit Realization of Addition

The SFG and Gm-C circuit implementation of the inverted weighted addition is shown in Fig. 13.2.

$$Y = -(\alpha_1 X_1 + \alpha_2 X_2 + \cdots + \alpha_n X_n) \tag{13.2}$$

13.1.3 Gm-C Circuit Realization of Scaling

For the scaling operation shown in Eq. (13.3), the SFG and Gm-C circuit implementation is shown in Fig. 13.3.

$$Y = -\alpha X \tag{13.3}$$

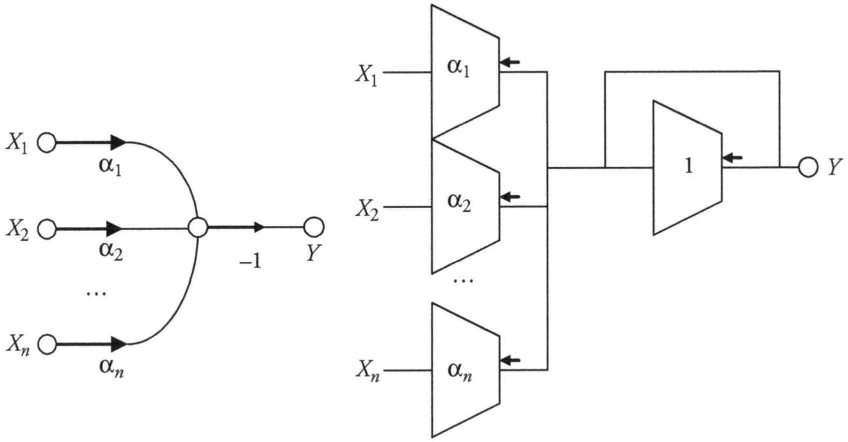

FIGURE 13.2 SFG and Gm-C circuit realization for weighted addition.

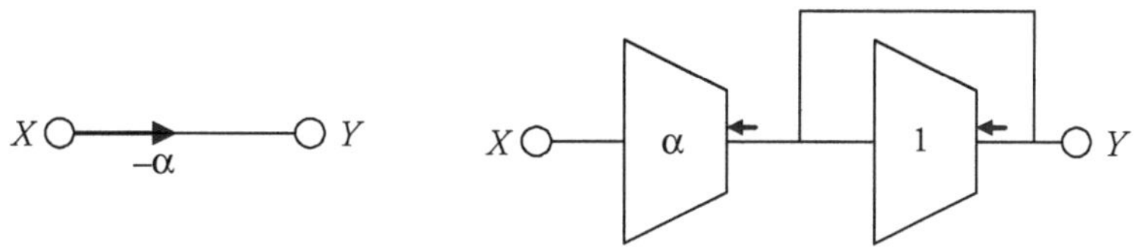

Figure 13.3 SFG and Gm-C circuit of inverted scalar.

13.1.4 Design Example

The design example in Fig. 13.4 can be used to demonstrate the SFG-based VLSI Gm-C filter circuit design procedure. Assume the filter transfer function is given below with $f_o = 100$ kHz and $Q = 10$:

$$\frac{Y}{X} = \frac{1}{\left(\frac{s}{\omega_o}\right)^2 + \frac{1}{Q}\left(\frac{s}{\omega_o}\right) + 1} \tag{13.4}$$

The SFG of the prototype transfer function can be derived by using the same way as shown in Sec. 12.1.4.

This SFG can be directly mapped to a prototype VLSI Gm-C circuit using the basic building blocks (see Fig. 13.5).

Applying the frequency-scaling technique to the prototype filter in Fig. 13.5 and selecting $R_s = 100$ kΩ, we can obtain the device parameters for Gm-C implementation of the targeted filter as follows:

$$\begin{aligned} R_s &= 100\ \text{k}\Omega \quad \Rightarrow \omega_s = 2\pi \cdot 100\ \text{k}\Omega \\ g_m &= (g_m)_n \cdot \frac{1}{R_s} = 10 \cdot (g_m)_n (\mu\text{S}) \\ C &= \frac{C_n}{R_s \omega_s} = 15.9 \cdot C_n (\text{pF}) \end{aligned} \tag{13.5}$$

Finally, the VLSI Gm-C circuit implementation of the specified filter is given in Fig. 13.6.

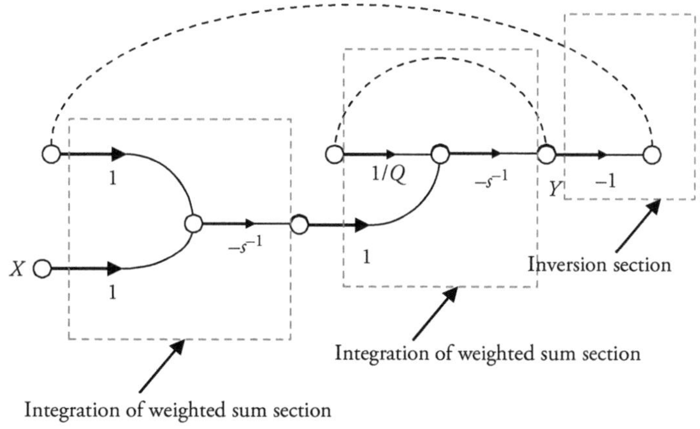

Figure 13.4 SFG of the specified filter.

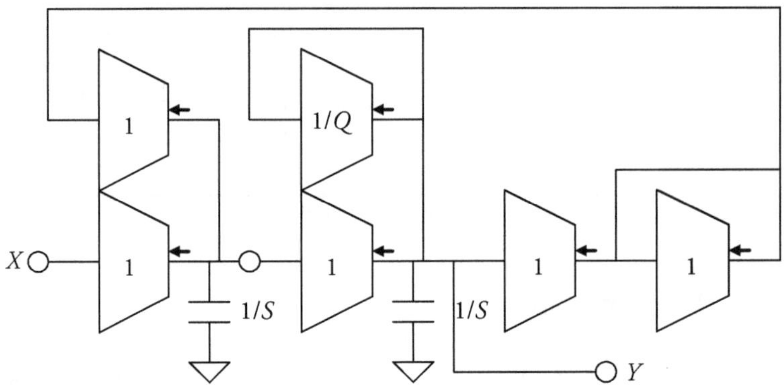

Figure 13.5 VLSI Gm-C circuit implementation of the prototype filter.

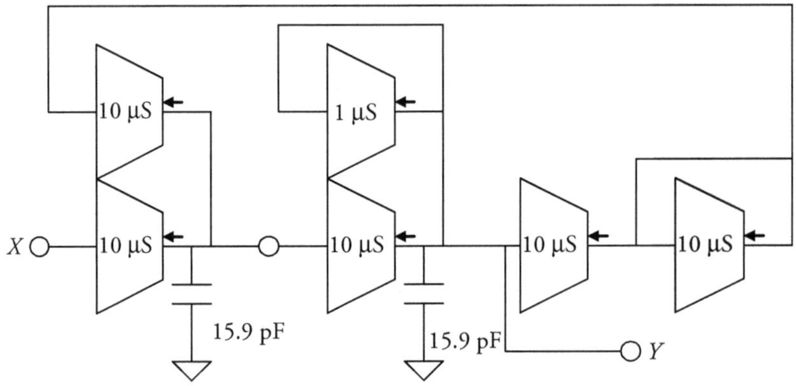

Figure 13.6 VLSI Gm-C filter circuit implementation for the given transfer function.

13.2 Cascade-Based VLSI Gm-C Filter Design Methods

Similar to the cascade-based filter design based on other VLSI circuit techniques, the cascaded Gm-C filter design method is based on the partition and implementation of the given filter transfer function using the first- and second-order sections. Consequently, one of the main tasks of VLSI Gm-C filter circuit design is to develop common first- and second-order building blocks that can be easily configured for various Gm-C filter design applications.

13.2.1 First-Order Gm-C Filter Sections

A general-purpose first-order VLSI Gm-C filter circuit section is shown in Fig. 13.7.

It can be seen that such a circuit realizes a general first-order s-domain transfer function as follows:

$$H(s) = \frac{\alpha s + \gamma}{s + \beta} \tag{13.6}$$

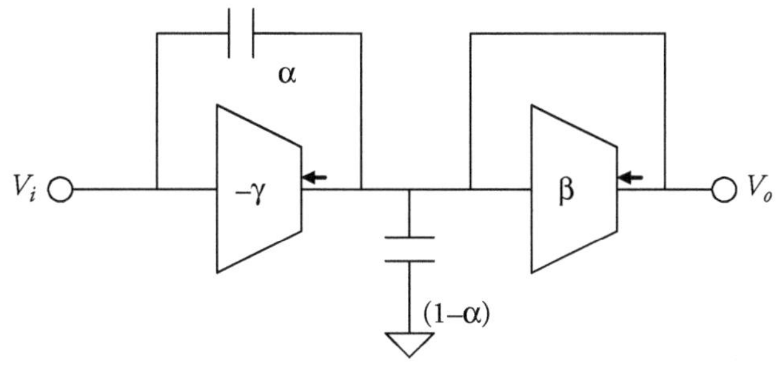

FIGURE 13.7 The first-order VLSI Gm-C filter section.

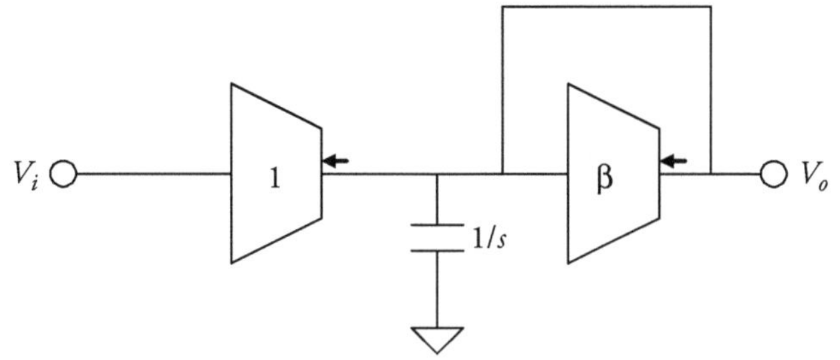

FIGURE 13.8 VLSI Gm-C lossy integrator.

Theoretically speaking, the circuit in Fig. 13.7 can be used to realize any given first-order filter transfer function for proper programming of the parameters of the circuit. For most all-pole filter applications, a commonly used first-filter section is the lossy integrator (see Fig. 13.8), which can be realized from the above VLSI Gm-C circuit structure as

$$H(s) = -\frac{1}{s+\beta} \tag{13.7}$$

13.2.2 Second-Order Gm-C Filter Sections

Various second-order Gm-C filter circuit sections can be realized through the SFG-based Gm-C filter-realization technique. Figure 13.9 and Eq. (13.8) are the SFG and the s-domain transfer function of a low-pass prototype-filter section, respectively:

$$H(s) = \frac{1}{s^2 + \frac{1}{Q}s + 1} \tag{13.8}$$

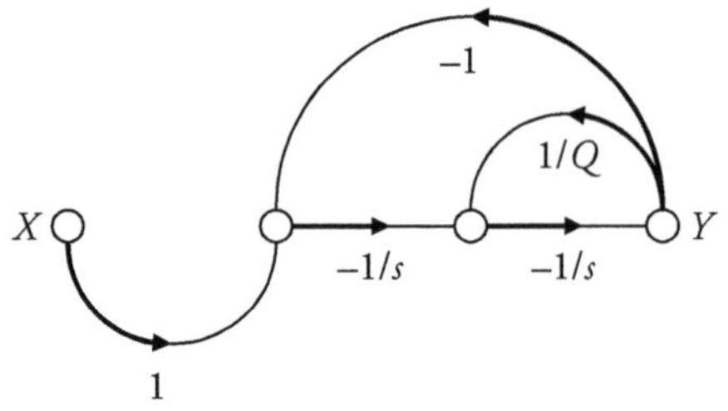

Figure 13.9 SFG and transfer function of second-order low-pass prototype-filter section.

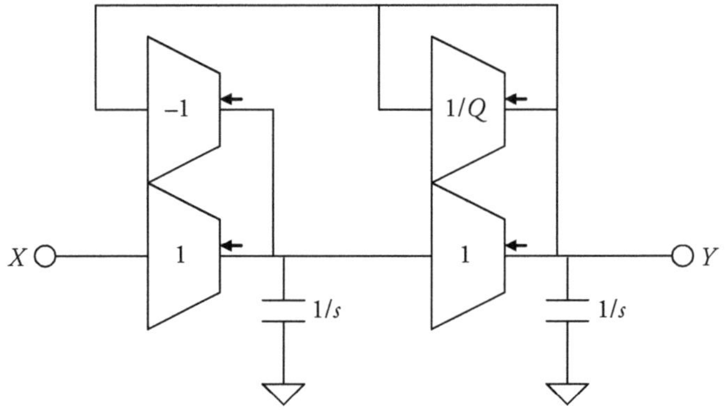

Figure 13.10 VLSI Gm-C second-order low-pass prototype circuit section.

The Gm-C circuit implementation of the filter section is shown in Fig. 13.10. Similarly, the SFG and Gm-C circuit implementations of the second-order high-pass and band-pass prototype sections are shown in Figs. 13.11 to 13.14.

$$H(s) = \frac{s}{s^2 + \frac{1}{Q}s + 1} \tag{13.9}$$

$$H(s) = \frac{s^2}{s^2 + \frac{1}{Q}s + 1} \tag{13.10}$$

Figure 13.15 shows an alternative general second-order section that can be used for low-pass, high-pass, or band-stop sections.

$$Y = \frac{X_1 + s^2 X_2}{s^2 + \frac{1}{Q}s + 1} \tag{13.11}$$

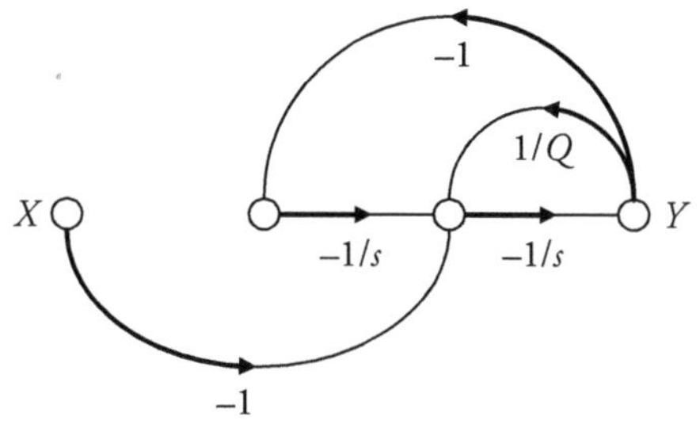

Figure 13.11 SFG and transfer function of second-order band-pass section.

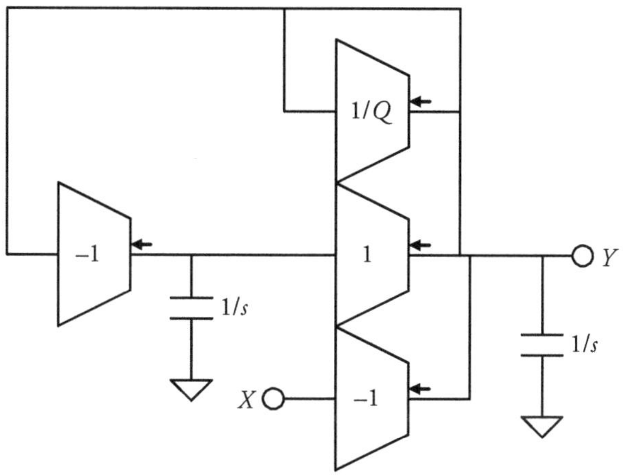

Figure 13.12 VLSI Gm-C second-order band-pass prototype circuit section.

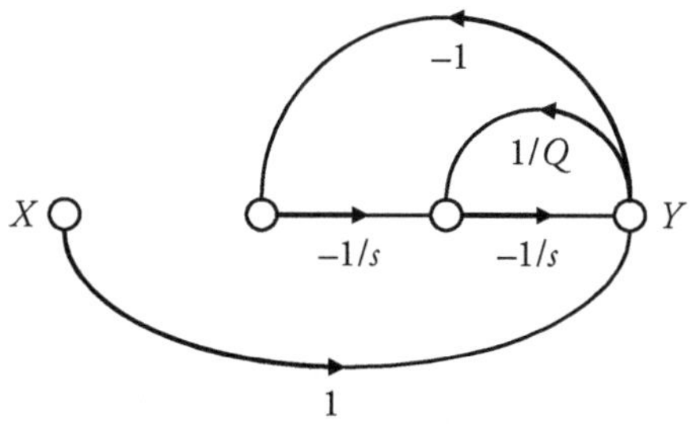

Figure 13.13 SFG and transfer function of second-order high-pass section.

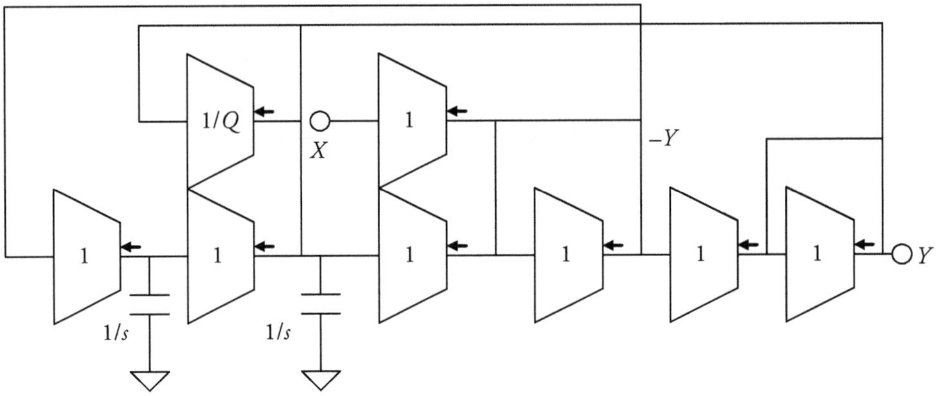

Figure 13.14 VLSI Gm-C second-order high-pass prototype circuit section.

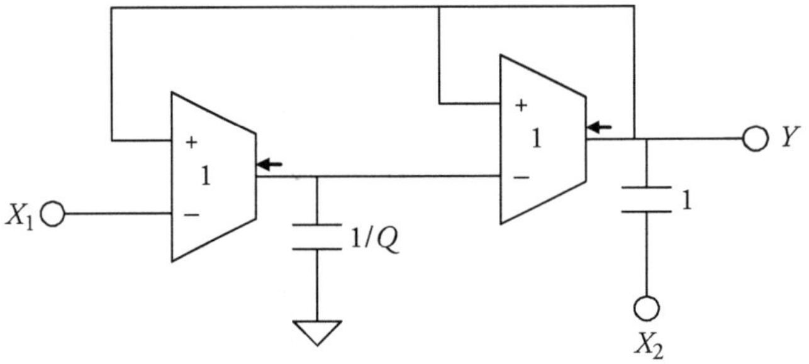

Figure 13.15 General-purpose second-order Gm-C section.

References

1. A. P. Nedungadi, R. L Geiger, "High-Frequency Voltage-Controlled Continuous-Time Lowpass Filter Using Linearised CMOS Integrators," *El. Letters*, vol. 22. no. 14, pp. 729–731, July 3, 1986.
2. C. S. Park, R. Schaumann, "Design of a 4-MHz Analog Integrated CMOS Transconductance-C Bandpass Filter," *IEEE J. Solid-State Circuits*, vol. SC-23, no. 4, pp. 897–996, August 1988.
3. H. Khorramabadi, P. R. Gray, "High-Frequency CMOS Continuous-Time Filters," *IEEE J. Solid-State Circuits*, vol. SC-19, no. 6, pp. 939–948, December 1984.
4. J. Pennock, P. Frith, R. G. Barker, "CMOS Triode Transconductor Continuous-Time Filters," *IEEE CICC*, pp. 378–381, 1986.
5. A. Kaiser, "A Micropower CMOS Continuous-Time Low-Pass Filter," *IEEE J. Solid-State Circuits*, vol. SC-24, no. 3, pp. 736–743, June 1989.
6. V. Gopinathan, Y. P. Tsividis, K.-S. Tan, R. K. Hester, "Design Considerations for High-Frequency Continuous-Time Filters and Implementation of an Antialiasing Filter for Digital Video," *IEEE J. Solid-State Circuits*, vol. SC-25, no. 6, pp. 1368–1378, December 1990.

7. Y.-T. Wang, A. A. Abidi, "CMOS Active Filter Design at Very High Frequencies," *IEEE J. Solid-State Circuits*, vol. SC-25, no. 6, pp. 1562–1574, December 1990.
8. F. Krummenacher, N. Joehl, "A 4-MHz CMOS Continuous-Time Filter with On-Chip Automatic Tuning," *IEEE J. Solid-State Circuits*, vol. SC-23, no. 3, pp. 750–758, June 1988.
9. F. Krummenacher, "Design Considerations in High-Frequency CMOS Gm Amplifier Capacitor (TAC) Filters," *Proc. IEEE ISCAS*, pp. 100–105, 1989.
10. M. Steyaert, J. Silva-Martinez, W. Sansen, "High Performance OTA-R-C Continuous-Time Filters with Full CMOS Low Distortion Floating Resistors," *Proc. ESSCIRC'91*, pp. 5–8, 1991.
11. F. Krummenacher, G. Van Ruymbeke, "Integrated Selectivity for Narrow-Band FM IF Systems," *IEEE J. Solid-State Circuits*, vol. SC-25, no. 3, pp. 757–760, June 1990.
12. J.-Y. Lee, C.-C. Tu, W.-H. Chen, "A 3V Linear Input Range Tunable CMOS Transconductor and Its Application to a 3.3V 1.1MHz Chebyshev Low-Pass Gm-C Filter for ADSL," Proceedings, *IEEE Custom Integrated Circuit Conference*, pp. 387–390, May, 2000.
13. A. Tajalli, M. Atarod, A. Adibi, "A 1.5-V Supply, Video Range Frequency, Gm-C Filter," *2000 IEEE International Symposium on Circuits and Systems*, vol. 2, pp. 148–151, 2000.
14. C. Durdodt, A. Hanke, S. Heinen, U. Langmann, "Comparison of an Inductorless Low-IF and Zero-IF Receiver for Bluetooth," *MWSCAS-2002*, vol. 1, pp. I563–566, August 2002.
15. E. A. M. Klumperink, B. Nauta, "Systematic Comparison of HF CMOS Transconductors," *IEEE Trans. on Circuits and Systems II: Analog and Digital Signal Processing*, vol. 50, no. 10, pp. 728–741, October 2003.
16. C.-C. Huang, Kari A. I. Halonen, M. Ismail, V. Porra, A. Hyogo, "A Low-Voltage, Low-Power CMOS Fifth-Order Elliptic Gm-C Filter for Baseband Mobil, Wireless Communication," *IEEE Trans. on Circuits and Systems for Video Technology*, vol. 7, no. 4, pp. 584–593, August 1997.
17. S. Szczepanski, J. Jakusz, R. Schaumann, "A Linear Fully Balanced CMOS OTA for VHF Filtering Applications," *IEEE Trans. on Circuits and Systems-II: Analog And Digital Signal Processing*, vol. 44, no. 3, pp. 174–187, March 1997.
18. P. Bowron, K. A. Mezher, A. A. Muhieddine, "Dynamic-Range Comparison of Second-Order Active-RC and OTA-C Filters," *Proceedings, the 37th Midwest Symposium on Circuits and Systems*, vol. 2, pp. 776–779, August 1994.
19. S. D'Amico, A. Baschirotto, "An Active-Gm-RC Continuous-Time Biquad Low-Pass Cell," *ICECS*, vol. 2, pp. 551–554, December 2003.
20. P. V. Ananda Mohan, "Novel OTA-C Filter Structures Using Grounded Capacitors," *ISCAS'91*, vol. 3, pp. 1347–1350, June 1991.
21. P. Bowron, H. M. Dahir, "Generalization and Evaluation of Second-Order Analogue OTA-C Filters," *IEE Colloquium on Electronic Filters*, pp. 4/1–4/4, June 1989.
22. Z. Ni, D. Gu, "Multi-loop OTA-C Filter Realization with Chain-like Feedback Structure," *Proceedings, 1991 International Conference on Circuits and Systems*, vol. 2, pp. 684–687, June 1991.
23. C.-C. Hsu, W.-S. Feng, "Testable Design of Multiple-Stage OTA-C Filters," *IEEE Transactions on Instrumentation and Measurement*, vol. 49, no. 5, pp. 929–934, October 2000.

24. A. C. M. de Queiroz, L. P. Caloba, "OTA-C Filters Derived from Unbalanced Lattice Passive Structures," *ISCAS'93*, vol. 4, pp. 2256–2256, May 1993.
25. Z. Shujiang, "Designing OTA-C Filters by SFG Approach," *ISCAS'93*, vol. 4, pp. 2248–2251, May 1993.
26. C.-M. Chang, B. M. Al-Hashimi, "Analytical Synthesis of Current-Mode High-Order OTA-C Filters," *IEEE Trans. on Circuits and Systems I: Fundamental Theory and Applications*, vol. 50, no. 9, pp. 1188–1192, September 2003.
27. F. Galvez-Durand, L. Caloba, A. C. M. de Queiroz, "Stability of High-Pass and Band-Reject Analog Continuous-Time OTA-C Filters," *Proceedings of the 40th Midwest Symposium on Circuits and Systems*, vol. 1, pp. 64–67, August 1997.
28. H. Nevarez-Lozano, J. A. Hill, E. Sanchez-Sinencio, "Frequency Limitations of Continuous-Time OTA-C Filters," *IEEE International Symposium on Circuits and Systems*, vol. 3, pp. 2169–2172, June 1988.
29. P. V. A. Mohan, "Generation of OTA-C Filter Structures from Active RC Filter Structures," *IEEE Transactions on Circuits and Systems*, vol. 37, no. 5, pp. 656–660, May 1990.
30. K.-J. Lee, K.-S. Huang, W.-C. Wang, "Concurrent Test Method for OTA-C Filters," *Electronics Letters*, vol. 33, no. 1, January 1997.
31. A. Wyszynski, R. Schaumann, "Using Multiple-Input Transconductors to Reduce Number of Components in OTA-C Filter Design," *Electronics Letters*, vol. 28, no. 3, January 1992.
32. H. W. Su, Y. Sun, "Performance Comparison of Continuous-Time IFLF and LF Low-Pass All-Pole OTA-C Filters," *IEEE APCCAS'00*, pp. 271–274, 2000.
33. R. Wilcock, B. M. Al-Hashimi, "Application of Group Delay equalization in Testing Fully-Balanced OTA-C Filters," *ISCAS'02*, vol. 4, pp. IV-643–IV-646, May 2002.
34. K. A. Mezher, P. Bowron, "Noise-Flow-Graph Analysis of OTA-C Filters," *ISCAS'01*, vol. 1, pp. 695–698, May 2001.
35. R. Lojacono, M. Salerno, F. Sargeni, "A Novel Architecture of Digitally Programmable Continuous-Time OTA-C Filters," *Proceedings, 37th Midwest Symposium on Circuits and Systems* vol. 2, pp. 1103–1106, August 1994.
36. J. E. Cousseau, L. P. Caloba, "Frequency Bounds on OTA-C Filters," *ISCAS'93*, vol. 4, pp. 2252–2255, May 1993.
37. Q. Q. Han, "Fully Integrated Continuous Time OTA-C Filter with Leapfrog Structure," *Proceedings, 1991 International Conference on Circuits and Systems*, vol. 2, pp. 696–698, June 1991.
38. A. C. M. de Queiroz, L. P. Caloba, "Some Practical Problems in OTA-C Filters Related with Parasitic Capacitances," *ISCAS'90*, vol. 3, pp. 2279–2282, May 1990.
39. Y. Sun, J. K. Fidler, "General Multiple Loop Feedback OTA-C Filter Architectures," *IEE Third One-day Colloquium on Analog Signal Processing*, pp. 3/1–3/6, November 1996.
40. Y. Sun, J. K. Fidler, "Some Design Methods of OTA-C and CCII-RC Filters," *IEE Colloquium on Digital and Analogue Filters and Filtering Systems*, pp. 7/1–7/8, November 1993.
41. A. C. M. de Queiroz, L. P. Caloba, "Reactive Loop Elimination and the Use of Special Approximations in the Design of OTA-C Filters," *Proceedings of the 38th Midwest Symposium on Circuits and Systems*, vol. 2, pp. 1341–1344, August 1995.
42. C. A. Barbargires, "Explicit Design of General High-Order FLF OTA-C Filters," *Electronics Letters*, vol. 35, no. 16, pp. 1289–1290, August 1999.

43. Y. Sun, "Second-Order OTA-C Filters Derived from Nawrocki-Klein Biquad," *Electronics Letters*, vol. 34, no. 15, pp. 1449–1450, July 1998.
44. D. Chen, H. Chen, F. Qi, Q. Gao, S. Qin, "A 5th Order LP Butterworth OTA-C Filter with Automatic Tuning System," *Proceedings, 5th International Conference on ASIC*, vol. 1, pp. 518–521, October 2003.
45. L. Caloba, F. Galvez-Durandde, A. C. M. Queiroz, "OTA-C Filters with Finite Zeros and without Floating Capacitors," *IEEE 39th Midwest Symposium on Circuits and Systems*, vol. 2, pp. 925–928, August 1996.
46. B. Pankiewicz, M. Wojcikowski, S. Szczepanski, Y. Sun, "A CMOS Field Programmable Analog Array and Its Application in Continuous-Time OTA-C Filter Design," *ISCAS'01*, vol. 1, pp. 5–8, May 2001.
47. J. Silva-Martinez, "Design Issues for UHF OTA-C Filter Realizations," *2001 Southwest Symposium on Mixed-Signal Design*, pp. 93–98, February 25–27, 2001.
48. A. B. Bhattacharyya, S. Dey, "Sub-Circuit Analysis for Power Supply Rejection Ratio in Regulated Cascode Operational Gm Amplifiers and Filters," *Proceedings, Twelfth International Conference on VLSI Design*, pp. 164–168, January 1999.
49. Y. Sun, K. Fidler, "Fully-Balanced Structures of Continuous-Time MLF OTA-C Filters," *1998 IEEE International Conference on Electronics, Circuits and Systems*, vol. 1, pp. 157–160, September 1998.
50. B. Pankiewicz, M. Wojcikowski, S. Szczepanski, Y. Sun, "A Field Programmable Analog Array for CMOS Continuous-Time OTA-C Filter Applications," *IEEE J. Solid-State Circuits*, vol. 37, no. 2, pp. 125–136, February 2002.
51. P. Bowron, K. A. Mezher, A. A. Muhieddine, "Dynamic-Range Comparison of Second-Order Active-RC and OTA-C Filters," *Proceedings of the 37th Midwest Symposium on Circuits and Systems*, vol. 2, pp. 776–779, August 1994.
52. F. Rezzi, V. Pisati, R. Castello, R. Alini, "Novel linearization Circuit for BiCMOS Transconductors Used in High Frequency OTA-C Filters," *ISCAS'93*, vol. 2, pp. 1132–1135, May 1993.
53. L. P. Caloba, F. Galvez-Durand, A. C. M. de Queiroz, "Parasitic Effects in OTA-C Filters with Finite Zeros," *Proceedings of the 34th Midwest Symposium on Circuits and Systems*, vol. 2, pp. 1049–1052, May 1991.
54. Y. Sun, J. K. Fidler, "Performance Analysis of Multiple Loop Feedback OTA-C Filters," *IEE Colloquium on Digital and Analogue Filters and Filtering Systems*, pp. 9/1–9/7, November 1994.
55. K.-J. Lee, W.-C. Wang, K.-S. Huang, "A Current-mode Testable Design of Operational Gm Amplifier-Capacitor Filters," *IEEE Trans. on Circuits and Systems II: Analog and Digital Signal Processing*, vol. 46, no. 4, pp. 401–413, April 1999.
56. Y. Sun, J. K. Fidler, "OTA-C Realization of General High-Order Transfer Functions," *Electronics Letters*, vol. 29, no. 12, pp. 1057–1058, June 1993.
57. T. Inoue, F. Ueno, Y. Aramaki, O. Matsumoto, M. Suefuji, "A Design of CMOS OTA's Using Simple Linearizing Techniques and Their Application to High-Frequency Continuous-Time Filters," *IEEE International Symposium on Circuits and Systems*, vol. 3, pp. 1741–1744, June 1991.
58. P. M. Van Peteghem, B. J. Haby, H. M. Fossati, G. L. Rice, "A Very-Linear CMOS Gm Stage for OTA-C Filters," *Proceedings of the IEEE 1989 Custom Integrated Circuits Conference*, pp. 25.1.1/1–25.1.1/4, May 1989.

59. Y. Sun, J. K. Fidler, "High-Order Current-Mode Continuous-Time Multiple Output OTA Capacitor Filters," *IEE 15th SARAGA Colloquium on Digital and Analogue Filters and Filtering Systems*, pp. 8/1–8/6, November 1995.
60. E. Hayahara, M. Enomoto, "Deriving Technique of Equivalent Forms from General Filter Circuits and Its Application," *IEEE Trans. on Circuits and Systems I: Fundamental Theory and Applications,* vol. 46, no. 9, pp. 1037–1041, September 1999.
61. Y. Sun, J. K. Fidler, "Structure Generation and Design of Multiple loop Feedback OTA-Grounded Capacitor Filters," *IEEE Trans. on Circuits and Systems I: Fundamental Theory and Applications,* vol. 44, no. 1, pp. 1–11, January 1997.
62. A. Wyszynski, R. Schaumann, S. Szczepanski, P. Van Halen, "Design of a 2.7-GHz Linear OTA and a 250-MHz Elliptic Filter in Bipolar Transistor-Array Technology," *IEEE Transactions on Circuits and Systems II: Analog and Digital Signal Processing,* vol. 40, no. 1, pp. 19–31, January 1993.
63. J. Ramirez-Angulo, E. Sanchez-Sinencio, "Active Compensation of Operational Gm Amplifier Filters Using Partial Positive Feedback," *IEEE Journal of Solid-State Circuits,* vol. 25, no. 4, pp. 1024–1028, August 1990.
64. E. Sanchez-Sinencio, J. Silva-Martinez, "CMOS Gm Amplifiers, Architectures and Active Filters: A Tutorial," *IEE Proceedings Circuits, Devices and Systems,* vol. 147, no. 1, pp. 3–12, February 2000.
65. Y. Sun, "OTA-C Filter Design Using Inductor Substitution and Bruton Transformation Methods," *Electronics Letters*, vol. 34, no. 22, pp. 2082–2083, October 1998.
66. Y. Sun, J. K. Fidler, "Synthesis and Performance Analysis of Universal Minimum Component Integrator-Based IFLF OTA-Grounded Capacitor Filter," *IEE Proceedings Circuits, Devices and Systems,* vol. 143, no. 2, pp. 107–114, April 1996.
67. J. A. De Lima, F. M. Alcaide, "On Designing Linearly Tunable High-Q OTA-C Filters with Low Sensitivity," *ISCAS'04*, vol. 1, pp. 21–24, May 2004.
68. M. Jeong, S. Takagi, Z. Czarnul, N. Fujii, "Design of a Novel Linear 3-Input CMOS OTA and Its Application to Filter Realization," *IEEE Asia Pacific Conference on Circuits and Systems,* pp. 121–124, November 1996.
69. A. Lewinski, J. Silva-Martinez, "OTA linearity Enhancement Technique for High Frequency Applications with IM3 Below -65dB," *Proceedings of the IEEE 2003 Custom Integrated Circuits Conference*, pp. 9–12, September 2003.
70. B. A. Minch, "Synthesis of Multiple-Input Translinear Element Log-Domain Filters," *ISCAS'99*, vol. 2, pp. 697–700, June 1999.
71. S. Solis-Bustos, J. Silva-Martinez, "A 4 Hz Low-Pass Continuous-Time Filter," *1998 IEEE International Conference on Electronics, Circuits and Systems,* vol. 1, pp. 169–172, September 1998.
72. J. E. Cousseau, L. P. Caloba, "Frequency Bounds on OTA-C Filters," *ISCAS'93*, pp. 2252–2255, May 1993.
73. A. C. M. de Queiroz, L. P. Caloba, "OTA-C Filters Derived from Unbalanced Lattice Passive Structures," *ISCAS'93*, pp. 2256–2256C, May 1993.
74. Shujing Zhao, "Designing OTA-C Filters by SFG Approach," *ISCAS'93*, pp. 2248–2251, May 1993.
75. F. Rezzi, V. Pisati, R. Castello, R. Mini, "Novel Linearization Circuit for BiCMOS Transconductors Used in High Frequency," *ISCAS'93*, pp. 1132–1135, May 1993.

76. L. P. Caloba, A. C. M. de Queiroz, "Constant-Resistance All-Pass OTA-C Filters," *Proceedings of the 33rd Midwest Symposium on Circuits and Systems*, pp. 488–491, August 1990.
77. L. P. Caloba, A. C. M. de Queiroz, E. Sanchez-Sinencio, "Signal-Flow Graph OTA-C Band-Pass and Band-Reject Integrated Filters," *IEEE International Symposium on Circuits and Systems*, vol. 3, pp. 1624–1627, May 1989.
78. F. Baruqui, L. P. Caloba, "Stability and Active Compensation of OTA-C Continuous Time High Frequency Filters," *Proceedings of the 32nd Midwest Symposium on Circuits and Systems*, vol. 1, pp. 169–172, August 1989.
79. Y.-S. Hwang, S.-I. Liu, D. S. Wu, Y.-P. Wu, "Table-Based Linear Transformation Filters Using OTA-C Techniques," *Electronics Letters*, vol. 30, no. 24, pp. 2021–2022, November 1994.
80. B. Pankiewicz, M. Solecki, S. Szczepanski, "A Fully Differential CMOS OTA for Continuous-Time Filter Applications," *Proceedings, ICCSC'02*, pp. 42–45, June 2002.
81. E. Ibaragi, S. Nishioka, A. Hyogo, K. Sekine, "A CMOS OTA Free from Second Order Effects with a High Input Resistance Gm Control Terminal," *ISCAS'01*, vol. 1, pp. 300–303, May 2001.
82. J. Silva-Martinez, J. S. Suner, "Very Low Frequency IC Filters," *Proceedings of the 38th Midwest Symposium on Circuits and Systems*, vol. 2, pp. 1325–1328, August 1995.
83. D. Y. Kim, S. W. Choi, J. C. N. AhnFujii, "The Design and Comparison of Elliptic Filters with an OTA-C Structure," *Proceedings of the 33rd Midwest Symposium on Circuits and Systems*, vol. 1, pp. 484–487, August 1990.
84. J. Ramirez-Angulo, E. Sanchez-Sinencio, "Active Compensation of Operational Gm Amplifier Filters Using Partial Positive Feedback," *Proceedings of the 32nd Midwest Symposium on Circuits and Systems*, vol. 2, pp. 1206–1210, August 1989.
85. C.-C. Hung, K. Halonen, M. Ismail, V. Porra, "Micropower CMOS GM-C Filters for Speech Signal Processing," *ISCAS'97*, vol. 3, pp. 1972–1975, June 1997.
86. C.-C. Hung, K. A. I. Halonen, M. Ismail, V. Porra, A. Hyogo, "A Low-Voltage, Low-Power CMOS Fifth-Order Elliptic GM-C Filter for Baseband Mobile, Wireless Communication," *IEEE Transactions on Circuits and Systems for Video Technology*, vol. 7, no. 4, pp. 584–593, August 1997.
87. C-C Hung, K. Halonen, V. Porra, M. Ismail, "Low-Voltage, Micropower Weak-Inversion CMOS GM-C Filter," *ICECS'96*, vol. 2, pp. 1178–1181, October 1996.
88. J. Shin, S. Min, S. Kim, J. Choi, S. Lee, H. Park, J. Kim, "3V Baseband Gm-C Filters for Wireless Transceiver Applications," *ISCAS'03*, vol. 1, pp. I-457–I-460, May 2003.
89. T. Itakura, T. Ueno, H. Tanimoto, A. Yasuda, R. Fujimoto, T. Arai, H. Kokatsu, "A 2.7V, 200-kHz, 49-dBm, Stopband-IIP3, Low-Noise, Fully Balanced Gm-C Filter IC," *IEEE Journal of Solid-State Circuits*, vol. 34, no. 8, pp. 1155–1159, August 1999.
90. C.-C. Hung, K. Halonen, V. Porra, M. Ismail, "Low-Voltage CMOS GM-C Filter with Rail-to-Rail Common-Mode Voltage," *IEEE 39th Midwest Symposium on Circuits and Systems*, vol. 2, pp. 921–924, August 1996.
91. C. Yoo, W. Kim, "A ±1.5 V, 4 MHz Low-Pass Gm-C Filter in CMOS," *Proceedings of the ASP-DAC'98*, pp. 341–342, February 1998.
92. J. Mahattanakul, C. Toumazou, "Modular Log-Domain Filters Based upon Linear Gm-C Filter Synthesis," *IEEE Transactions on Circuits and Systems I: Fundamental Theory and Applications*, vol. 46, no. 12, pp. 1421–1430, December 1999.
93. L. Ramezani, "An Adjustable Bandwidth Analog CMOS Gm-C Filter," *ICECS'03*, vol. 2, pp. 420–422, December 2003.

94. S. Mehrmanesh, H. A. Aslanzadeh, M. B. Vahidfar, M. Atarodi, "A 1.8V High Dynamic Range CMOS Gm-c Filter for Portable video Systems," *The 14th International Conference on Microelectronics*, pp. 38–41, December 2002.
95. A. Tajalli, M. Atarodi, A. Adibi, "A 1.5-V Supply, Video Range Frequency, Gm-C Filter," *ISCAS'00*, vol. 2, pp. 148–151, May 2000.
96. J.-Y. Lee, C.-C. Tu, W.-H. Chen, "A 3 V Linear Input Range Tunable CMOS Transconductor and Its Application to a 3.3 V 1.1 MHz Chebyshev Low-Pass Gm-C Filter for ADSL," *CICC'00*, pp. 387–390, May 2000.
97. C. Yoo, K. Jung, J.-W. Lee, W. Kim, "A 15 MHz, 2.6 mW, Sixth-Order Bandpass Gm-C Filter in CMOS," *Proceedings of the 40th Midwest Symposium on Circuits and Systems*, vol. 1, pp. 316–319, August 1997.
98. S. Mehrmanesh, M. B. Vahidfar, H. A. Aslanzadeh, M. Atarodi, "An Ultra Low-Voltage Gm-C Filter for Video Applications," *ISCAS'03*, vol. 1, pp. I-561–I-564, May 2003.
99. J. Lee, K. Hayatleh, F. J. Lidgey, J. Drew, "Tunable Linear Gm Cell for Gm-C Filter Applications," *ISCAS'02*, vol. 4, pp. IV-253–IV-256, May 2002.
100. A. Hassan, K. Sharaf, H. El-Ghitani, H. F. Ragaie, "A New Low Power Transconductor for Gm-C Filters," *ISCAS'02*, vol. 4, pp. IV-249–IV-252, May 2002.
101. A. Gharbiya, M. Syrzycki, "Highly Linear, Tunable, Pseudo Differential Transconductor Circuit for the Design of Gm-C Filters," *IEEE CCECE 2002*, vol. 1, pp. 521–526, May 2002.
102. O. Naess, Y. Berg, "Tunable Floating-Gate Low-Voltage Transconductor," *ISCAS'02*, vol. 4, pp. IV-663–IV-666, May 2002.
103. A. C. Carusone, D. A. Johns, "A 5th Order Gm-C Filter in 0.25um CMOS with Digitally Programmable Poles and Zeroes," *ISCAS'02*, vol. 4, pp. IV-635–IV-638, May 2002.
104. E. Rodriguez-Villegas, A. Yufera, A. Rueda, "A 1.25-V Micropower Gm-C Filter Based on FGMOS Transistors Operating in Weak Inversion," *IEEE J. Solid-State Circuits*, vol. 39, no. 1, pp. 100–111, January 2004.
105. Y. Palaskas, Y. Tsividis, "Dynamic Range Optimization of Weakly nonlinear, Fully Balanced, Gm-C Filters with Power Dissipation Constraints," *IEEE Trans. on Circuits and Systems II: Analog and Digital Signal Processing*, vol. 50, no. 10, pp. 714–727, October 2003.
106. F. Munoz, A. Torralba, R. G. Carvajal, J. Ramirez-Angulo, "Two New VHF Tunable CMOS Low-Voltage Linear Transconductors and Their Application to HF GM-C Filter Design," *ISCAS'00*, vol. 5, pp. 173–176, May 2000.
107. A. Tajalli, S. Mojtaba Atarodi, "Design Considerations for a 1.5-V, 10.7-MHz Bandpass Gm-C Filter in a 0.6um Standard CMOS Technology," *ISCAS'03*, vol. 1, pp. I-521–I-524, May 2003.
108. A. Hassan, K. Sharaf, H. El-Ghitani, H. F. Ragai, "The Design and Implementation of a Bandpass GM-C Filter for Bluetooth," *45th Midwest Symposium on Circuits and Systems*, vol. 2, pp. II-629–II-632, August 2002.
109. C. Yoo, S.-W. Lee, W. Kim, "A ±1.5-V, 4-MHz CMOS Continuous-Time Filter with a Single-Integrator Based Tuning," *IEEE J. Solid-State Circuits*, vol. 33, no. 1, pp. 18–27, January 1998.
110. F. Rezzi, I. Bietti, M. Cazzaniga, R. Castello, "A 70-mW Seventh-Order Filter with 7-50 MHz Cutoff Frequency and Programmable Boost and Group Delay Equalization," vol. 32, no. 12, pp. 1987–1999, December 1997.

111. S. D'Amico, A. Baschirotto, "Low-Power Compact Gm-C Filters Structures," *ISCAS'04*, vol. 1, pp. 873–876, May 2004.
112. O. Naess, Y. Berg, "Tunable Ultra-Low Voltage Gm Amplifier and GmC Filter," *ISCAS'00*, vol. 2, pp. 709–712, May 2000.
113. J. van der Tang, D. Kasperkovitz, A. Bretveld, "A 65 mW, 0.4-2.3 GHz Bandpass Filter for Satellite Receivers," *CICC'00*, pp. 383–386, May 2000.
114. G. J. Allbutt, D. G. Nairn, "Tunable Capacitors for Digital CMOS Compatible High Frequency Analog Filters," *ISCAS'96*, vol. 1, pp. 89–92, May 1996.
115. M. Ugajin, A. Yamagishi, J. Kodate, M. Harada, T. Tsukahara, "A 1-V CMOS/SOI Bluetooth RF Transceiver for Compact Mobile Applications," *2003 Symposium on VLSI Circuits 2003*, pp. 123–126, June 2003.
116. W. Jendernalik, S. Szczepanski, "A CMOS OTA-C Channel-Select Filter for Mobile Receiver," *Proceedings, ICCSC'02*, pp. 50–53, June 2002.
117. S. Mehrmanesh, M. Atarodi, "A High Dynamic Range CMOS Variable Gain Filter for ADSL," *ISCAS'02*, vol. 4, pp. IV-257–IV-260, May 2002.
118. E. Raisanen-Ruotsalainen, K. Lasanen, M. Sijander, J. Kostamovaara, "A Low-Power 5.1.2 kHz CMOS Gm-C Bandpass Filter with On-chip Center Frequency Tuning," *ISCAS'02*, vol. 4, pp. IV-651–IV-654, May 2002.
119. C. Dualibe, P. Jespers, M. Verleysen, "Embedded Fuzzy Control for Automatic Channel Equalization after Digital Transmissions," *ISCAS'01*, vol. 3, pp. 173–176, May 2001.

CHAPTER 14

Design of VLSI Continuous-Time Current-Mode Filters

Similar to other VLSI CT filter circuits, VLSI CTI filter circuits can also be implemented using the SFG-based, the cascade-based, the component simulation–based, and the ladder simulation–based circuit techniques. Due to the current-mode nature of the signals in the VLSI CTI circuit implementations, CTI-based VLSI filter circuits offer some attractive properties compared with other voltage-mode VLSI CT filter circuit implementations such as the ability of using the wired addition in CTI circuit design implementations. Such a property can be used to significantly simplify the circuit implementation of the filter circuits. On the other hand, the basic circuit building components in VLSI CTI circuits are slightly different from other VLSI voltage circuit techniques.

14.1 SFG-Based CTI Filter Design Methods

The CTI lossy integrator, rather than the lossless integrator, actually provides the filter circuit implementation with minimum device number. In order to fully utilize the benefits of the wired addition operation offered by the VLSI current-mode circuit technique, weights of the input signals are usually limited to one. The outputs of signal are much flexible and can be regenerated with desired weights. As a result, an optimal selection of the basic VLSI CTI building blocks can usually be given as lossy integration with weighted outputs, wired addition, and scaling with weighted outputs.

14.1.1 CTI Circuit Realization of Integration

The SFG and CTI circuit implementation of lossy integration with the weighted outputs represented in Eq. (14.1)

$$Y_i = -\frac{1}{s+1}(\alpha_i X) \qquad i = 1, 2, 3, \ldots, n \tag{14.1}$$

is shown in Fig. 14.1, where the bias current and the NMOS devices are sized to be $\{\alpha_i, i = 1, 2, \ldots, n\}$ times the input device. In the circuit, an explicit resistor (or VCR) and a capacitor are added to degenerate the circuit operation to lower frequency so that the

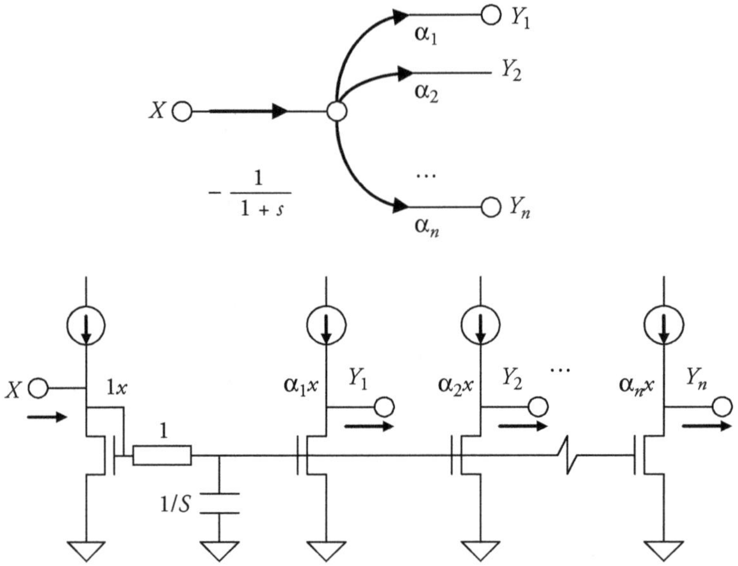

FIGURE 14.1 CTI realization of integration with weighted outputs.

parasitic resistance and capacitance effects can be ignored. Such a reduced-bandwidth design method significantly simplifies the CTI circuit design procedure.

14.1.2 CTI Circuit Realization of Addition

The SFG and CTI circuit of the wired addition based on Eq. (14.2) is shown in Fig. 14.2.

$$Y = X_1 + X_2 + \cdots + X_n \tag{14.2}$$

Such an operation is common to all current-mode signal processing circuits, and it offers a very effective way for analog signal processing operation.

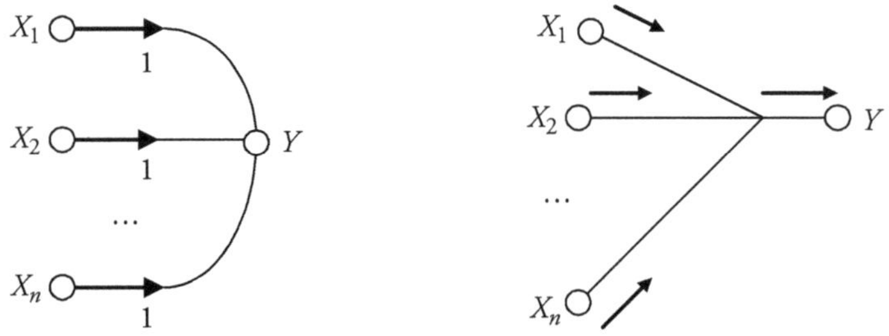

FIGURE 14.2 CTI circuit realization of addition.

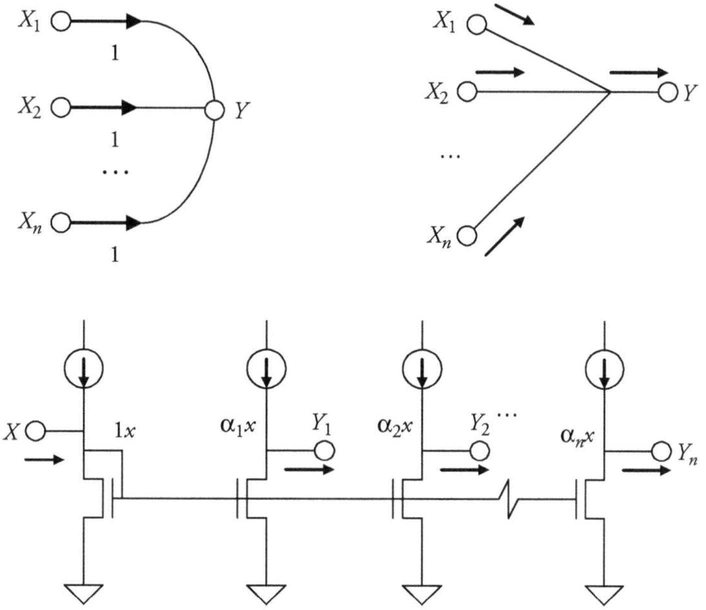

FIGURE 14.3 CTI realization of scaling with weighted branch output.

14.1.3 CTI Circuit Realization of Scaling

The scaling function of the SFG and the CTI circuit realization is shown in Fig. 14.3.

$$Y_i = -(\alpha_i X) \qquad i = 1, 2, 3, \ldots, n \tag{14.3}$$

14.1.4 Design Example

In this design example, we will demonstrate the procedure of designing a lossless CTI integrator based on the lossy CTI integrator, adder, and scaler. The s-domain transfer function of a normalized lossless integration is given as follows:

$$\frac{Y}{X} = -\frac{1}{s} \tag{14.4}$$

By introducing the normalized lossy integration as

$$G \equiv -\frac{1}{s+1} \tag{14.5}$$

the transfer function of the lossless integrator based on the variable G can be expressed as

$$\Rightarrow \frac{Y}{X} = -\frac{1}{s} = \frac{G}{G+1} \tag{14.6}$$

or

$$\Rightarrow Y = G(X - Y) \tag{14.7}$$

An SFG based on the CTI lossy integration can be easily constructed as shown in Fig. 14.4.

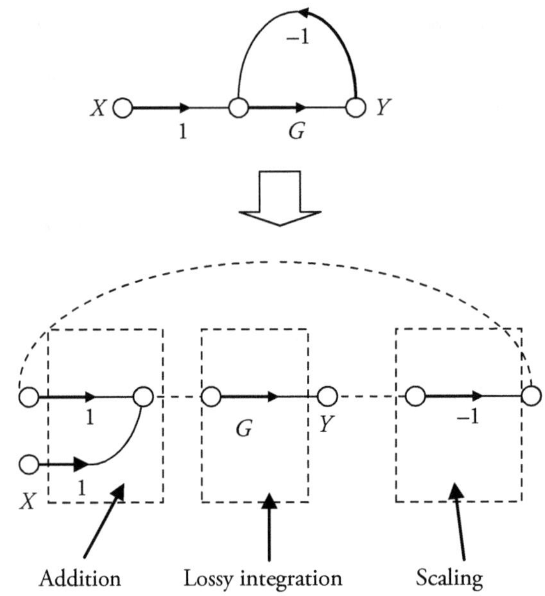

Figure 14.4 SFG of the lossless integrator based on the lossy integrator.

This SFG can be directly mapped to the VLSI CTI circuit using the basic building blocks as shown in Fig. 14.5.

Note that the lossless integration is typically not the simplest integrator form in practical VLSI CTI filter circuit implementations. However, for some filter circuit design applications where the filter circuits are implemented through the transformation from other types of VLSI filter circuit, the lossless integrator–based CTI filter design approaches offer advantages regarding design reuse since there is a direct mapping

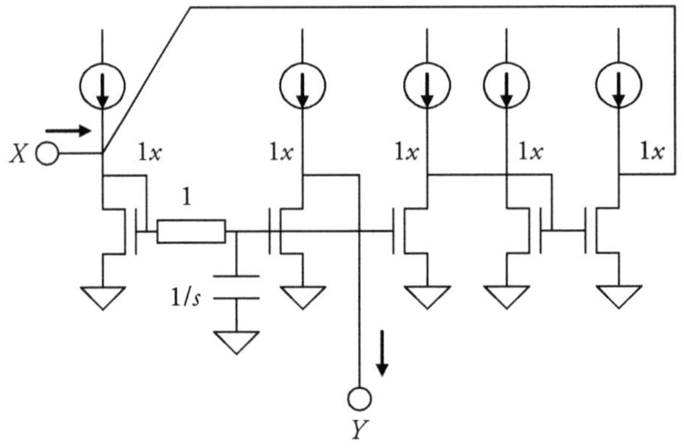

Figure 14.5 CTI lossless integrator circuit.

between these filter circuit types. Of course, such a design is usually not optimized with respect to the device count in the filter implementation.

In the second design example, we will design a second-order low-pass filter with the transfer function given as

$$\frac{Y}{X}=\frac{1}{\left(\frac{s}{\omega_o}\right)^2+\frac{1}{Q}\left(\frac{s}{\omega_o}\right)+1} \tag{14.8}$$

with $\omega_o = 2\pi \times 100$ kHz and $Q = 10$.

We can start the design of the prototype filter of the above transfer function:

$$\frac{Y}{X}=\frac{1}{s^2+\frac{1}{Q}s+1} \tag{14.9}$$

Let

$$G \equiv -\frac{1}{s+1} \tag{14.10}$$

We have that

$$\Rightarrow \frac{Y}{X}=\frac{1}{\left(\frac{G+1}{G}\right)^2-\frac{1}{Q}\left(\frac{G+1}{G}\right)+1}=\frac{G^2}{G^2\left(2-\frac{1}{Q}\right)+G\left(2-\frac{1}{Q}\right)+1} \tag{14.11}$$

or

$$\Rightarrow Y = G^2X-\left(2-\frac{1}{Q}\right)G^2Y-\left(2-\frac{1}{Q}\right)GY \tag{14.12}$$

The SFG based on the CTI lossy integrator is shown in Fig. 14.6.

This SFG can be modified to a form suitable for the VLSI CTI realization as shown in Fig. 14.7.

Finally, the SFG can be implemented into the VLSI CTI circuit as shown in Fig. 14.8:

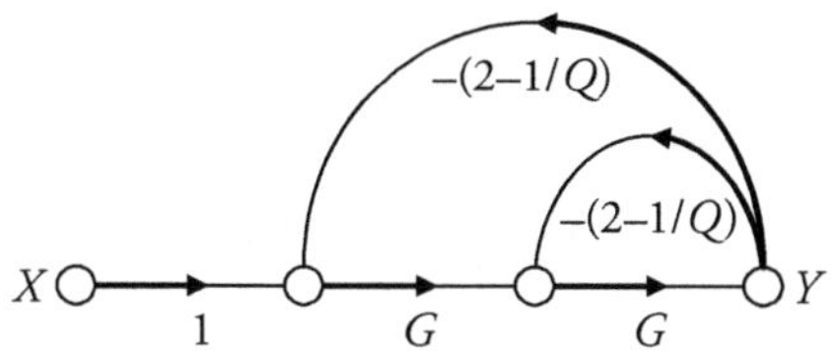

FIGURE 14.6 SFG of the prototype filter.

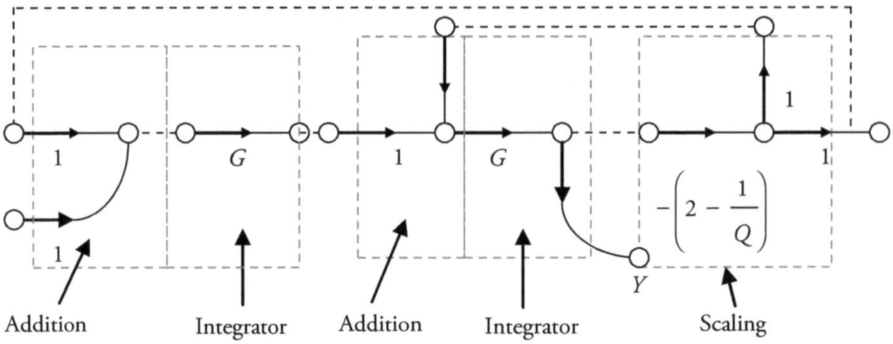

FIGURE 14.7 Modified SFG of the prototype filter.

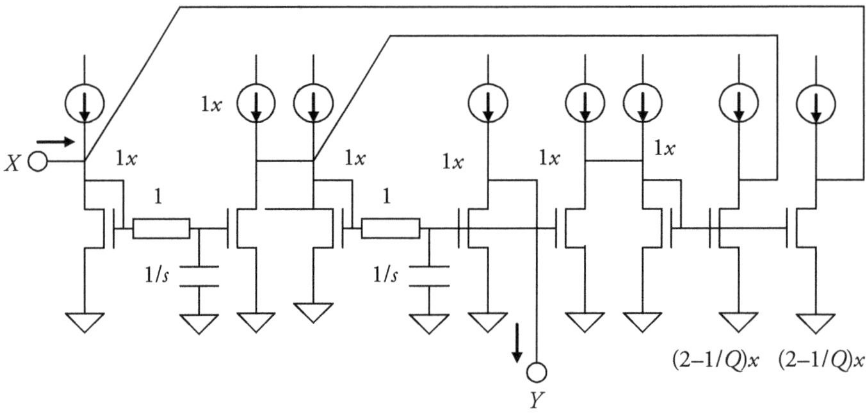

FIGURE 14.8 CTI circuit implementation of prototype filter.

Applying the frequency-scaling technique to the prototype filter in Fig. 14.8 and letting $R_s = 100$ kΩ, we have the following:

$$R_s = 100\ \text{k}\Omega$$

$$\Rightarrow \omega_s = 2\pi \cdot 100\ \text{k}\Omega$$

$$g_m = (g_m)_n \cdot \frac{1}{R_s} = 10 \cdot (g_m)_n\ (\mu\text{S}) \tag{14.13}$$

$$C = \frac{C_n}{R_s \omega_s} = 15.9 \cdot C_n\ (\text{pF})$$

14.2 Cascade-Based CTI Filter Design Methods

Cascade-based CTI filter design is very similar to cascade filter approaches in other VLSI technologies where the entire filter transfer function is usually partitioned into products of the first- and second-order filter sections. As a result, the key step for cascade CTI filter design is to find the common first- and second-order building blocks, which can be easily configured for different applications.

References

1. S.-S. Lee, R. H. Zele, D. J. Allstot, G. Liang, "CMOS Continuous-Time Current-Mode Filters for High-Frequency Applications," *IEEE J. Solid-State Circuits*, vol. 28, no. 3, pp. 323–329, June 1993.
2. V. C. Vincence, C. G. Montoror, M. C. Schneider, "A High-Swing MOS Cascode Bias Circuit for Operation at Any Current Level," *IEEE ISCAS'00*, May 28, 2000.
3. C. Thamviriyakul, A. Worapishet, C. Tourmazou, "A Low-Voltage Linear CMOS First-Order Two-Pair for Integrated CMA Filters," *IEEE ISCAS'97*, June 1997.
4. J. Tingleff, C. Toumazou, "Integrated Current Mode Wave Active Filters Based on Lossy Integrators [CMOS Chip]," *IEEE Trans. on Circuits and Systems I: Fundamental Theory and Applications*, vol. 42, no. 5, pp. 237–244, May 1995.
5. H. F. A. Hamed, "A Low Voltage Digitally Programmable Current-Mode Filter," *Proceedings of the 15th International Conference Microelectronics, 2003*, pp. 413–416, December 2003.

CHAPTER 15

Design of VLSI Switched-Capacitor Filters

VLSI switched-capacitor (SC) filter circuits belong to a large class of VLSI analog discrete-time signal processing systems, which occupy the position between the analog and the digital signal processing systems where the signals are represented as electric charges that are processed by discrete-time controlled charge transfers among capacitors within the SC circuits.

There are several commonly used methods to implement the VLSI SC filter circuits based on the given transfer-function specifications, including (1) the SFG-based circuit partitioning and implementation using basic VLSI SC circuit building blocks, (2) the s-to-z transformation–based implementation employing the related s-domain transfer function and mapping between the s- and z-domain circuit operations, (3) the direct z-domain transfer-function realization, (4) the resistor simulation–based methods with the realization of the resistors in corresponding VLSI active-RC filter circuit using VLSI switched capacitors, and (5) the ladder simulation–based implementation from the passive LC ladder filter.

15.1 SFG-Based VLSI Oversampling SC Filter Design Methods

Similar to the VLSI continuous-time signal processing circuits, VLSI discrete-time signal processing circuits can be realized from their transfer functions based on the selected basic building blocks of signal processing circuits. For SC filter circuits, the basic building components can typically be selected as the integration with weighted inputs, the addition with weighted inputs, and the scaling elements. Mathematically, it is always possible to partition a given z-domain transfer function based on defined building blocks such that the constructed SFG from the transfer function can be directly mapped to the VLSI SC circuit form.

Shown in Fig. 15.1 is a forward Euler integration–based SC circuit-implementation method for the general z-domain transfer function given below as

$$\frac{Y}{X} = \frac{b_m z^{-m} + b_{m-1} z^{-(m-1)} + \cdots + b_1 z^{-1} + b_0}{a_m z^{-m} + a_{m-1} z^{-(m-1)} + \cdots + a_1 z^{-1} + 1} \tag{15.1}$$

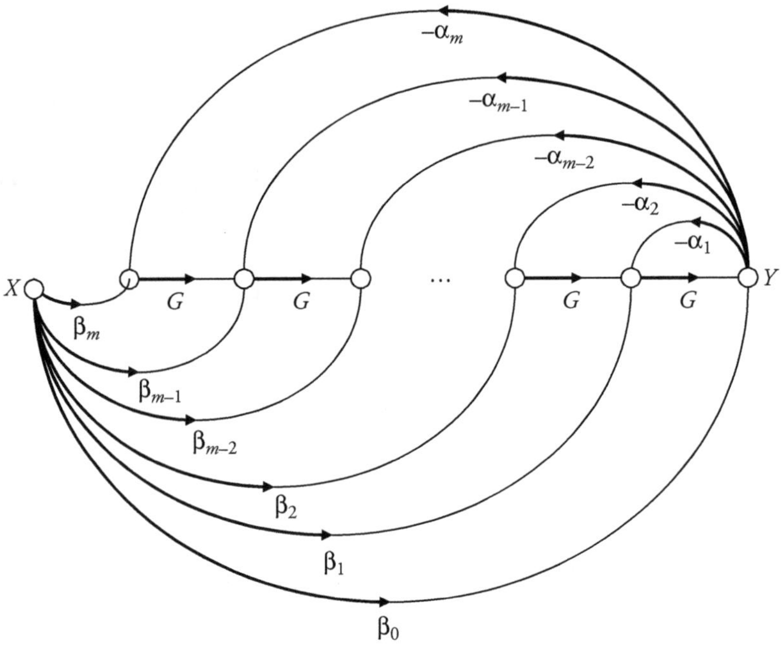

FIGURE 15.1 SFG expression of general z-domain transfer function based on SC integration.

We can substitute the delay operation $1/z$ in the transfer function using the forward Euler integration block G defined as

$$G \equiv \frac{1}{z-1} \quad \text{or } z = G^{-1} + 1 \tag{15.2}$$

The targeted z-domain transfer function can now be expressed in terms of the introduced variable G and a set of constants $\{\alpha_i, \beta_i, i = 0, 1, m\}$ as

$$\Rightarrow \frac{Y}{X}\Big|_{z=G^{-1}+1} = \frac{\beta_m G^m + \beta_{m-1} G^{(m-1)} + \cdots + \beta_1 G + \beta_0}{\alpha_m G^m + \alpha_{m-1} G^{(m-1)} + \cdots + \alpha_1 G + 1} \tag{15.3}$$

We can get the expression of the output Y as the function of the input X and the feedback of Y as follows:

$$\begin{aligned} \Rightarrow Y = {} & (\beta_m X - \alpha_m Y)G^m + (\beta_{m-1} X - \alpha_{m-1} Y)G^{(m-1)} \\ & + \cdots + (\beta_1 X - \alpha_1 Y)G + \beta_0 X \end{aligned} \tag{15.4}$$

The corresponding SFGs based on the two implementation approaches are shown in Fig. 15.1.

Alternatively, we also can obtain a different transfer-function expression by introducing a state variable U as follows:

$$\Rightarrow \frac{Y}{X}\Big|_{z=D+1} = \frac{\beta_m G^m + \beta_{m-1} G^{(m-1)} + \cdots + \beta_1 G + \beta_0}{\alpha_m G^m + \alpha_{m-1} G^{(m-1)} + \cdots + \alpha_1 G + 1} \cdot \frac{U}{U} \tag{15.5}$$

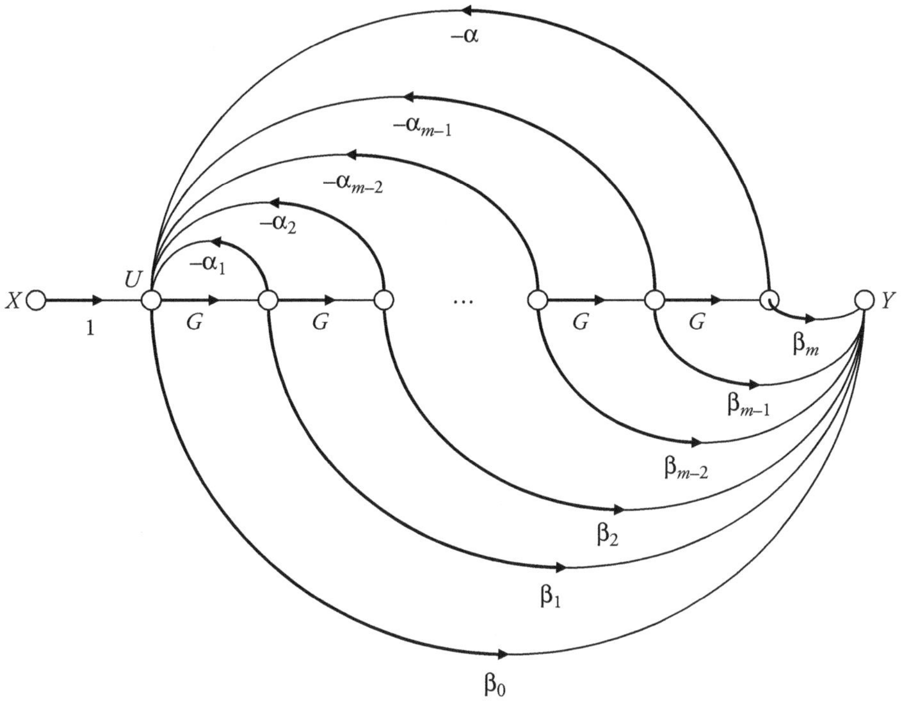

FIGURE 15.2 Alternative SFG expression of general z-domain transfer function.

$$\Rightarrow \begin{cases} U = X - (\alpha_m G^m + \alpha_{m-1} G^{(m-1)} + \cdots + \alpha_1 G)U \\ Y = (\beta_m G^m + \beta_{m-1} G^{(m-1)} + \cdots + \beta_1 G + \beta_0)U \end{cases} \tag{15.6}$$

The corresponding SFGs based on the two implementation approaches are shown in Fig. 15.2.

15.1.1 SC Circuit Realization of Integration

For the forward Euler integration operation with weighted inputs, the SFG element and SC circuit realization can be expressed as shown in Figs. 15.3 and 15.4, respectively:

$$Y = \frac{1}{z-1}(\alpha_1 X_1 + \alpha_2 X_2 + \cdots \alpha_n X_n). \tag{15.7}$$

The above implementation provides integrations with weighted input of the same polarity. For this type of approach, the inversion can be introduced directly by swapping the differential signals or by adding signal inverter (negative scaling). Alternatively, negative weight of the addition in the SC circuit can be easily realized by modifying the switch phases. The integration operations with mixed input weight polarities are shown in Figs. 15.5 and 15.6. Note that although such a clock phase modification provides an easy way for input polarity inversion, it also leads to the modification of the transfer function because of the clock phase difference among different input paths of different

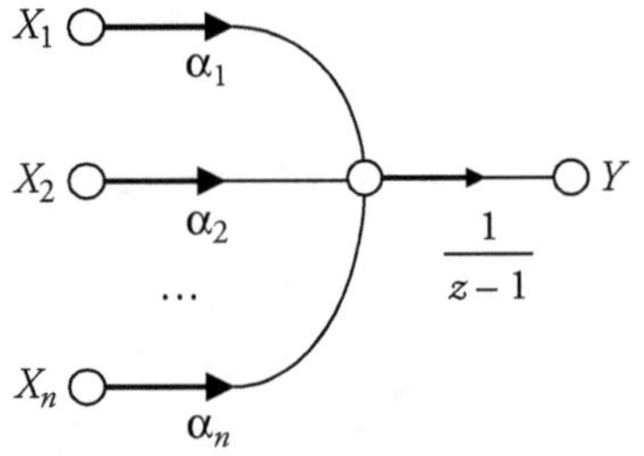

Figure 15.3 SFG expression of integration with weighted inputs.

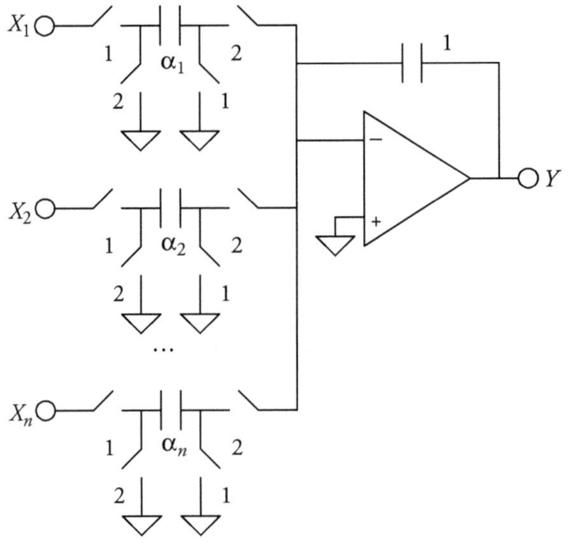

Figure 15.4 SC circuit realization of integration with weighted inputs.

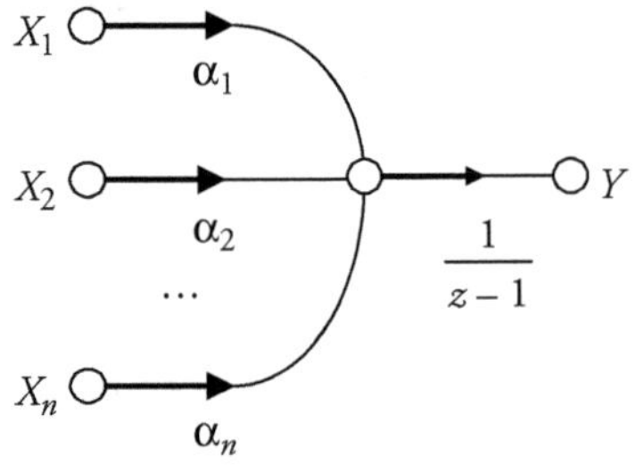

Figure 15.5 SFG expression of integration with weighted inputs with mixed polarities.

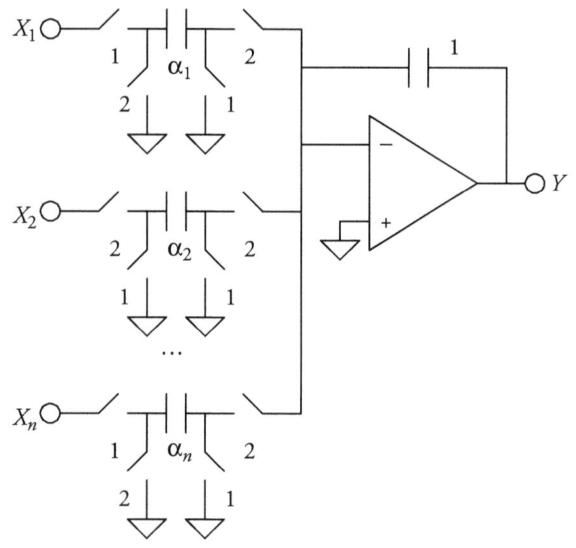

Figure 15.6 SC circuit realization of integration with weighted inputs with mixed polarities.

polarities. It can be seen that the paths with negative weights actually represent the backward Euler integrations while the positive weight paths represent the forward Euler integrations. This will be an issue if the SC circuit is required to exactly represent the given transfer function. However, in most SC filter applications, this difference may be ignored under high-oversampling-ratio (OSR >> 1) operation conditions.

$$Y \approx \frac{1}{z-1}(\alpha_1 X_1 - \alpha_2 X_2 + \cdots \alpha_n X_n) \tag{15.8}$$

15.1.2 SC Circuit Realization of Addition

The SFG and transfer function of the addition operation with weighted inputs of same polarity are shown in Fig. 15.7. The VLSI SC circuit implementation of such an operation is shown in Fig. 15.8. It is important to note that the output of the circuit is only valid in the ϕ_2 phase. A direct negative feedback of the OPAMP circuit is introduced in the ϕ_1 phase to ensure the proper operation of the OPAMP circuit. In practical application,

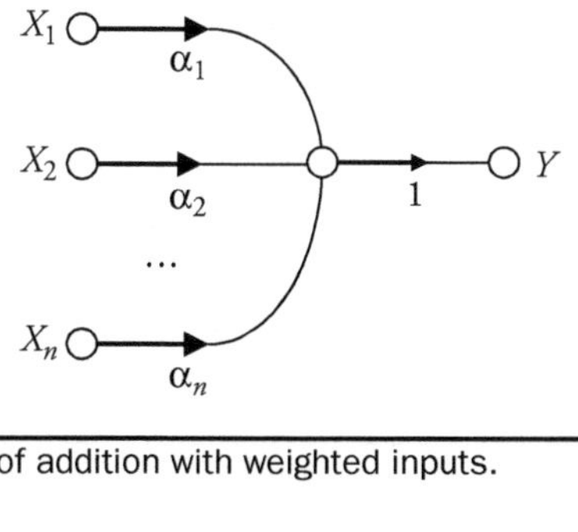

Figure 15.7 SFG expression of addition with weighted inputs.

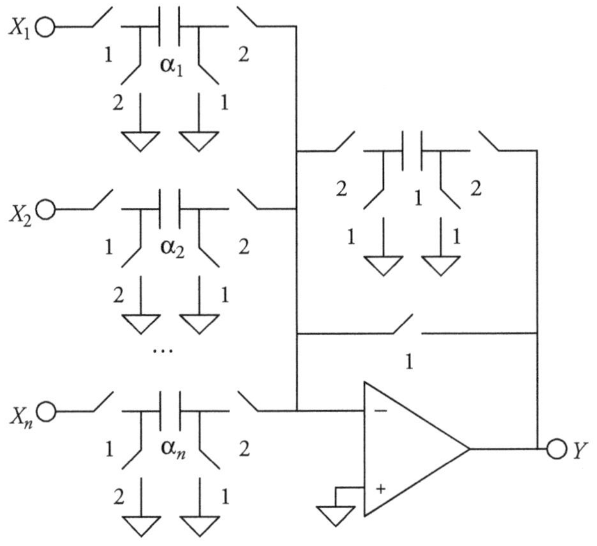

FIGURE 15.8 SC circuit realization of addition with weighted inputs.

addition operations are preferred to be realized together with integration operation for better implementation efficiency.

$$Y = (\alpha_1 X_1 + \alpha_2 X_2 + \cdots \alpha_n X_n) \tag{15.9}$$

Similarly, an addition with mixed input signal polarities may also be realized through modifying the phases of switching clocks as shown in Figs. 15.9 and 15.10.

$$Y = (\alpha_1 X_1 - \alpha_2 X_2 + \cdots + \alpha_n X_n) \tag{15.10}$$

15.1.3 SC Circuit Realization of Scaling

The SFG and SC circuit realizations of scaling operations are shown in Figs. 15.11 and 15.12, respectively.

$$Y = \alpha X \tag{15.11}$$

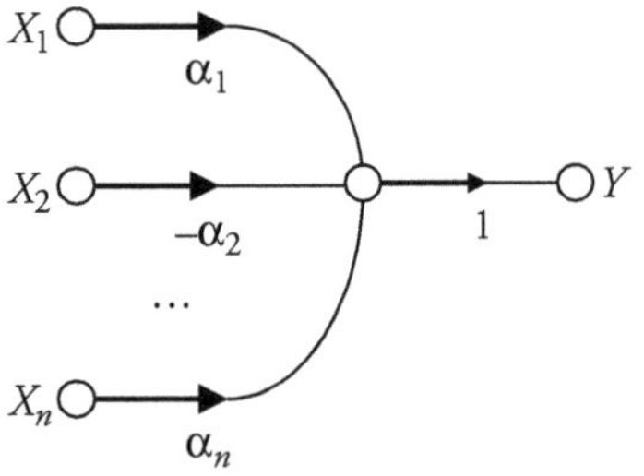

FIGURE 15.9 SFG expression of addition with mixed input polarities.

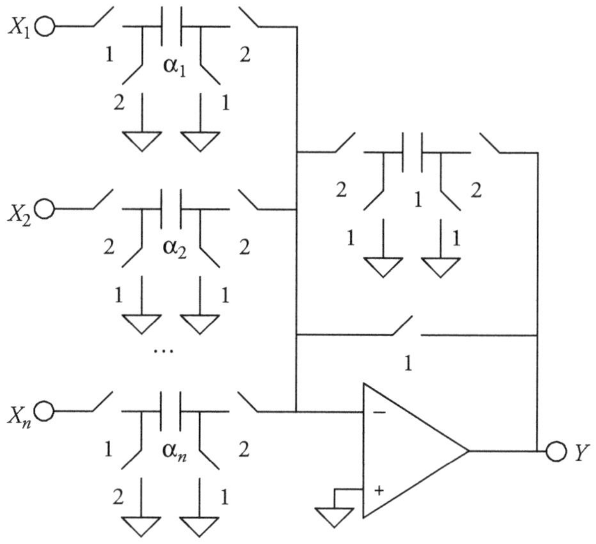

FIGURE 15.10 SC circuit realization of addition with mixed input polarities.

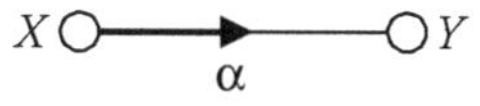

FIGURE 15.11 SFG expression and transfer function of scaling.

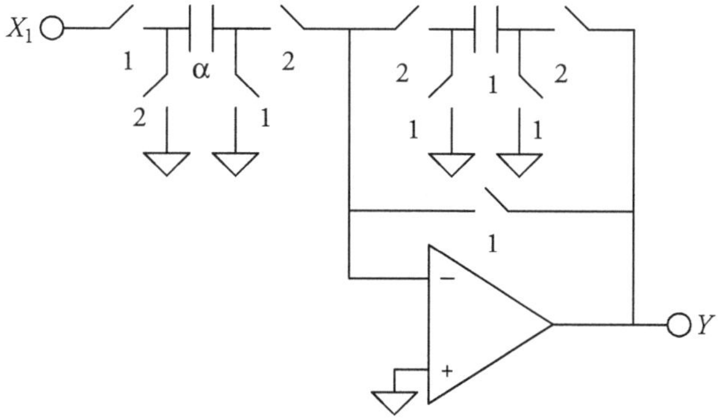

FIGURE 15.12 SC circuit expression of the scaling.

Alternatively, the SFG and SC circuit implementations of scaling with negative polarity are shown in Figs. 15.13 and 15.14, respectively.

$$Y = -\alpha X \tag{15.12}$$

It is important to point out that the dedicated scaling operation should be avoided in practical design application since a scaling operation can usually be realized with either the integration or the addition of circuit implementations.

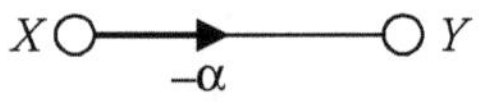

FIGURE 15.13 SFG expression of negative scaling.

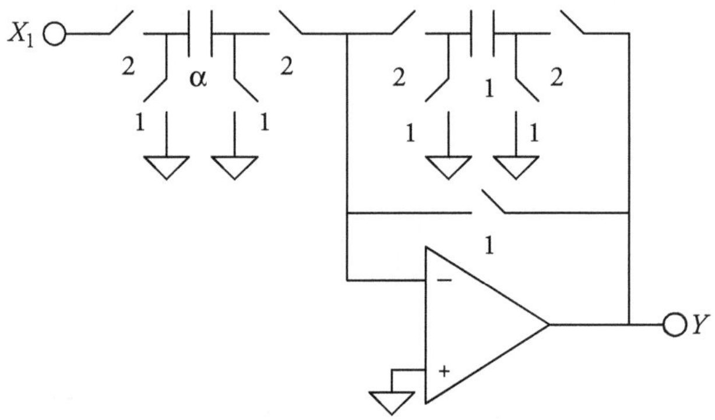

FIGURE 15.14 SC circuit implementation of negative scaling.

15.1.4 Design Example

To realize the following z-domain transfer function, we assume that the clock frequency is much higher than the frequency band of the signal (i.e., OSR >> 1).

$$H(z) = \frac{Y}{X} = \frac{z+1}{z-0.5} \tag{15.13}$$

We can introduce a variable G as

$$G \equiv \frac{1}{z-1} \quad \text{or } z = G^{-1} + 1 \tag{15.14}$$

Substituting the variable z in the transfer function by variable G, we have that

$$\Rightarrow H(z) = \frac{Y}{X} = \frac{G^{-1}+2}{G^{-1}+0.5} = \frac{1+2G}{1+0.5G} \tag{15.15}$$

or

$$\Rightarrow Y = X + G(2X - 0.5Y) \tag{15.16}$$

An SFG based on the circuit block G can be constructed as shown in Fig. 15.15.

The SFG can be directly mapped to the SC circuit using the building blocks introduced previously as in Fig. 15.16.

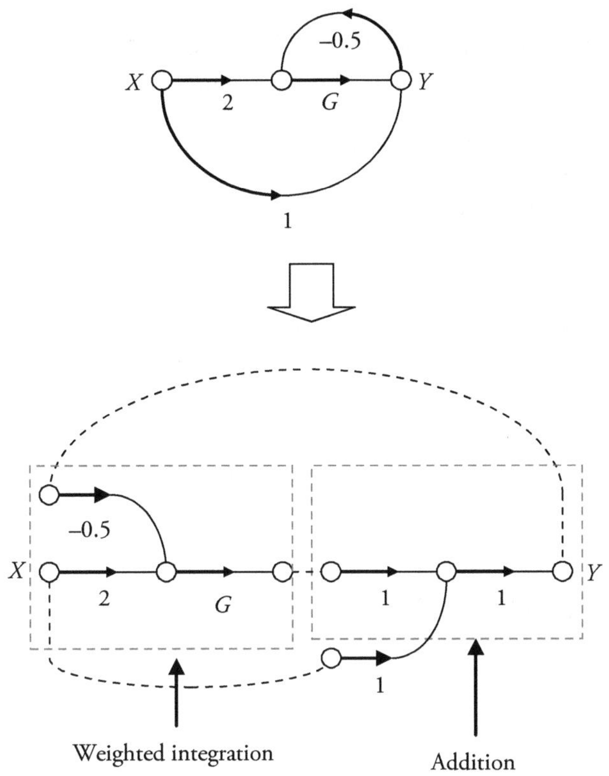

Figure 15.15 SFG expression of the design example.

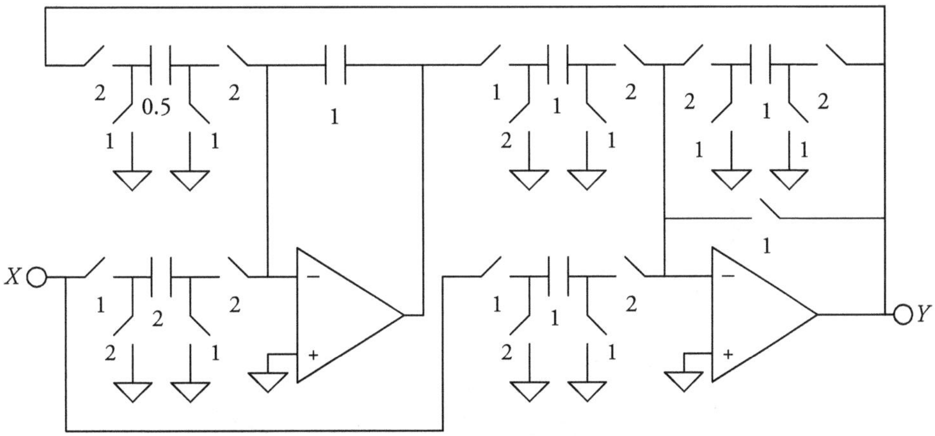

Figure 15.16 SC circuit realization of the given transfer function.

15.2 S-to-Z Transform–Based SC Filter Design Methods

Based on the s-to-z transform for the oversampling signal processing operation, an SC filter can be directly designed based on the s-domain transfer function through the s-to-z transformation. Similar to other continuous-time VLSI filter circuits, the VLSI SC filter circuit implementation may be started from a known s-domain transfer function. The realization of the SFG for a general s-domain transfer function based on the lossless integration operation is shown in Fig. 15.17.

Next, instead of mapping the SFG into a CT filter structure, the s-domain SFG is mapped into the z-domain through the s-to-z transformation by transforming the s-domain integration into its z-domain approximation. To demonstrate this procedure, we look at the forward Euler transformation given as

$$\frac{1}{s} \Rightarrow \frac{T}{z-1} \equiv G \tag{15.17}$$

Such a transformation directly maps the s-domain SFG into a z-domain SFG as shown in Fig. 15.18. Finally, the z-domain SFG can be mapped into the SC circuits as shown in Figs. 15.19 and 15.20.

15.2.1 Design Example

In the following design example, a second-order low-pass SC filter circuit will be designed from the given s-domain transfer function. In this design, we assume that the

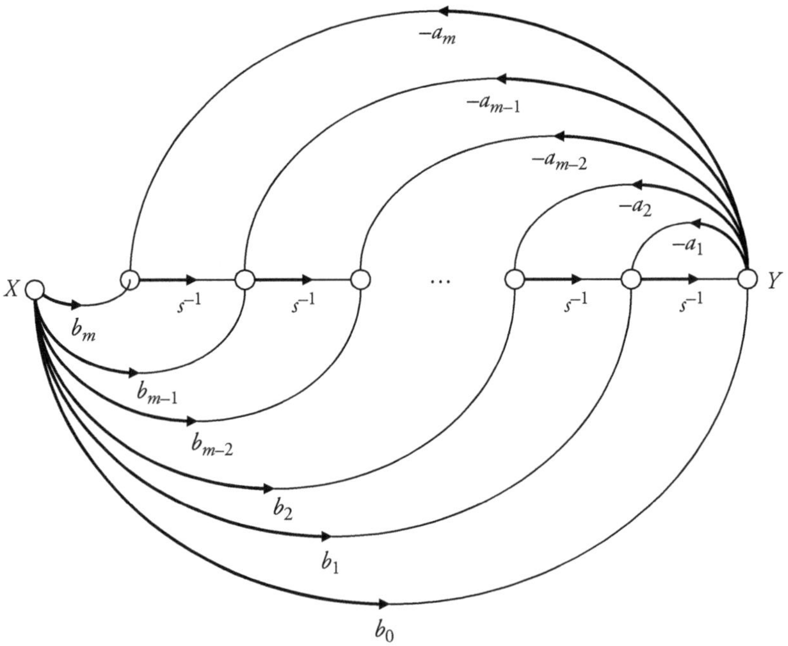

FIGURE 15.17 The targeted s-domain SFG expression for SC filter circuit implementation.

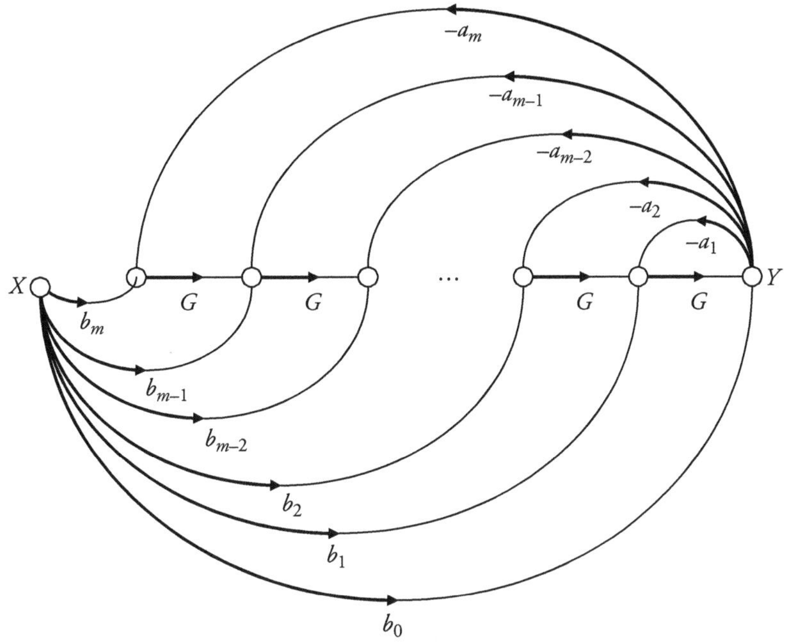

Figure 15.18 Z-domain SFG expression constructed based on s-to-z transformation.

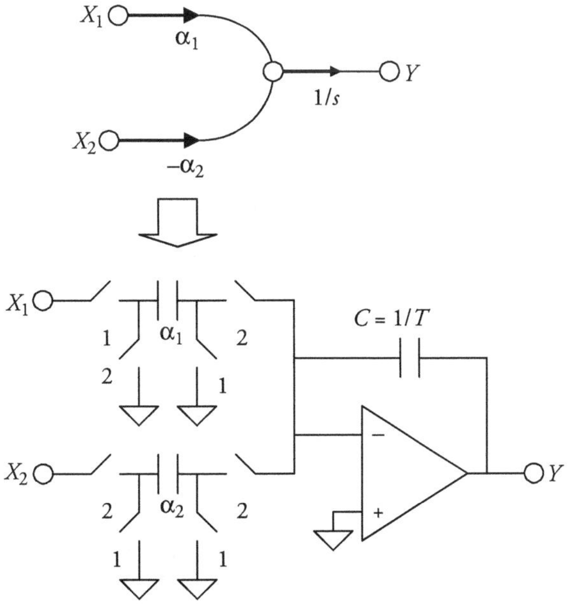

Figure 15.19 Mapping of s-domain integrator SFG to VLSI SC circuits.

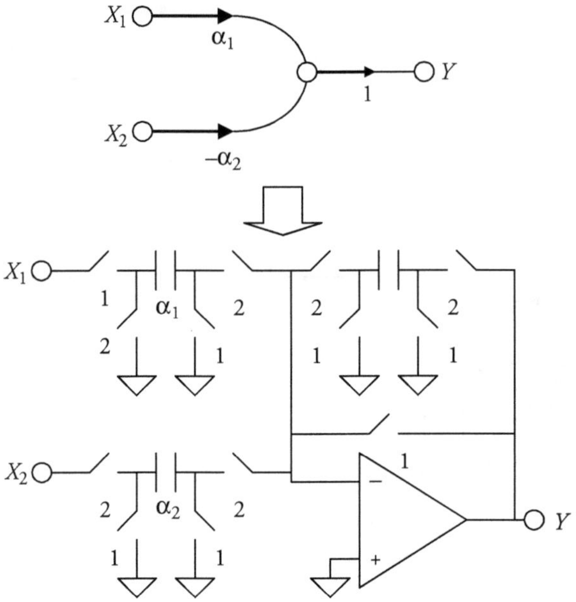

Figure 15.20 Mapping of SFG to VLSI SC circuit.

sample clock frequency ($f = 1/T$) is much higher than the frequency band of the signal to be processed.

$$\frac{Y}{X} = \frac{1}{\left(\frac{s}{\omega_o}\right)^2 + \frac{1}{Q}\left(\frac{s}{\omega_o}\right) + 1} \tag{15.18}$$

The s-domain SFG based on the above transfer function can be generated as shown in Fig. 15.21:

The s-domain SFG can be converted into the z-domain through the forward Euler s-to-z transformation as follows (see Fig. 15.22):

$$\frac{1}{s} = \frac{T}{z-1} \tag{15.19}$$

Finally, the z-domain SFG of the targeted filter can be mapped to the VLSI SC circuit as shown in Fig. 15.23.

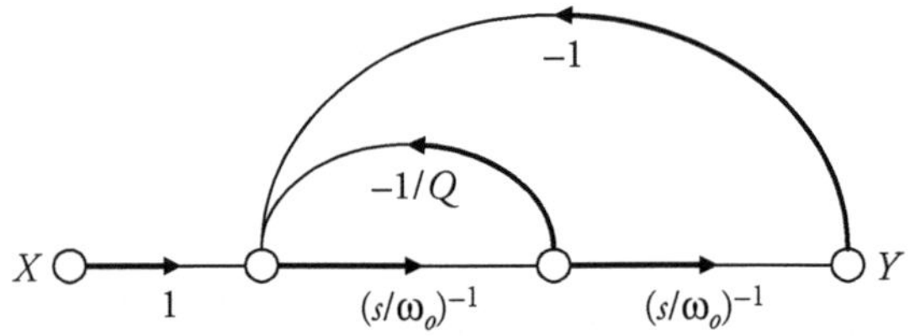

Figure 15.21 SFG of the s-domain transfer function.

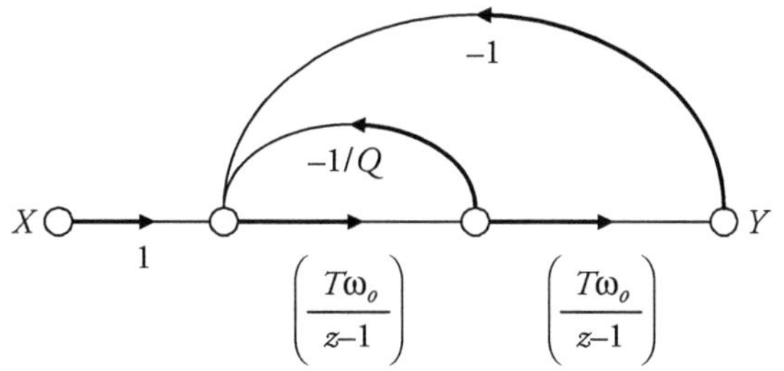

FIGURE 15.22 The z-domain SFG representation of the filter.

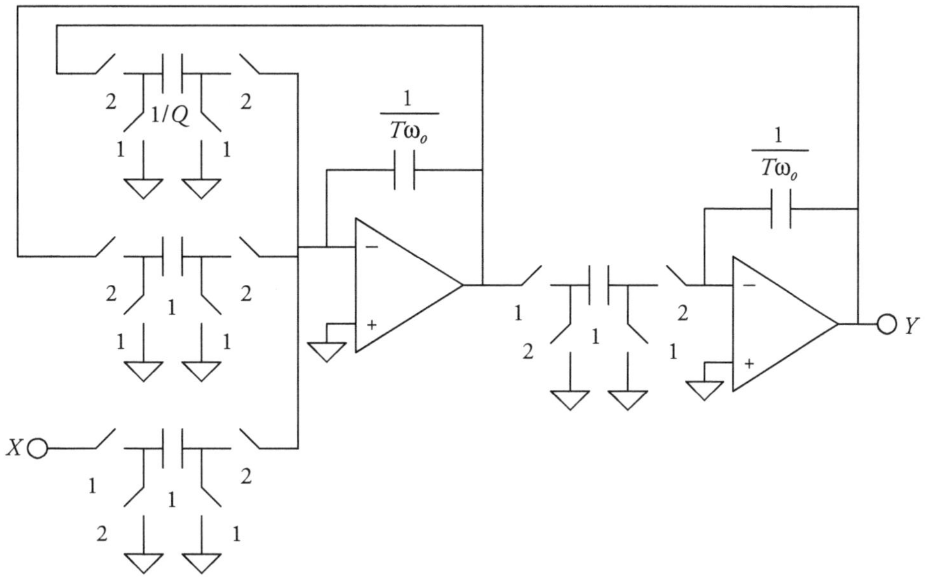

FIGURE 15.23 VLSI SC circuit realization of filter.

15.3 SC Filter Design from Active-RC Circuit Structures

An SC filter may also be designed from an active-RC filter circuit structure. In this design approach, an SC filter can be realized by simply replacing the resistors in the active-RC circuit structure using the SC circuit structure as shown in Fig. 15.24.

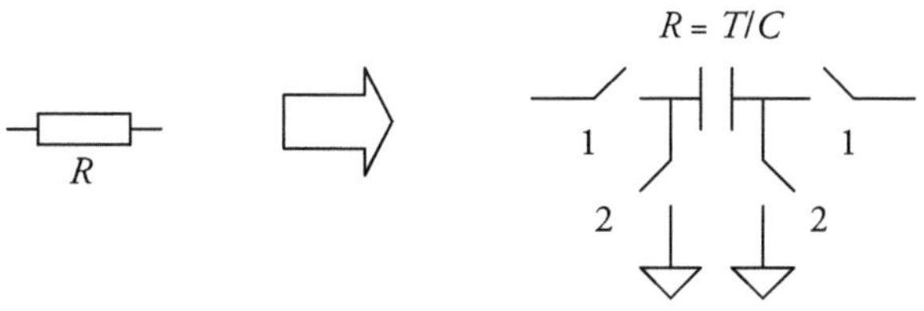

FIGURE 15.24 SC circuit simulation of resistor.

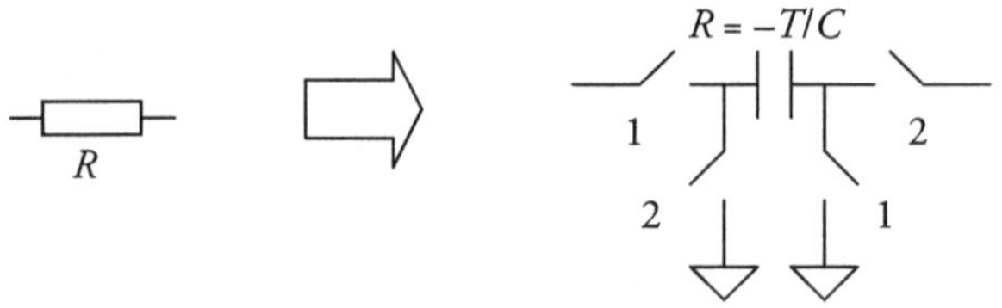

Figure 15.25 SC circuit simulation of negative resistance.

It is important to note that in the active-RC–based SC filter design approach, an equivalent negative resistor can also be implemented in the SC circuit by modifying the control clock phases of the SC circuit as shown in Fig. 15.25.

15.3.1 Design Example

In the following design example, we demonstrate the realization of an SC circuit based on an active-RC structure. In this example, we assume that the sample clock frequency ($f = 1/T$) is much higher than the frequency band of the signal to be processed.

$$\frac{Y}{X} = \frac{1}{\left(\frac{s}{\omega_o}\right)^2 + \frac{1}{Q}\left(\frac{s}{\omega_o}\right) + 1} \tag{15.20}$$

The s-domain SFG representation of the given transfer function can be constructed as shown in Fig. 15.26.

This SFG can be directly mapped into an active-RC filter shown in Fig. 15.27 based on the method previously discussed in the active-RC filter design chapter. It is important

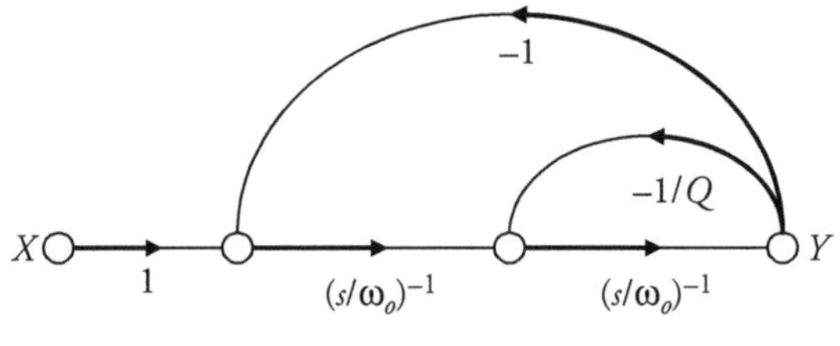

Figure 15.26 SFG of the s-domain transfer function.

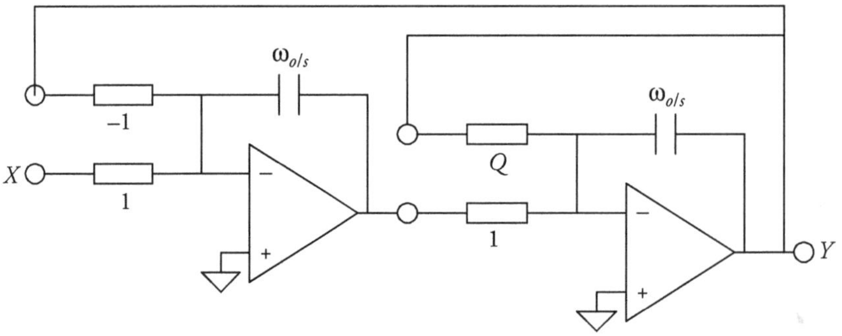

Figure 15.27 Active-RC circuits of the prototype filter.

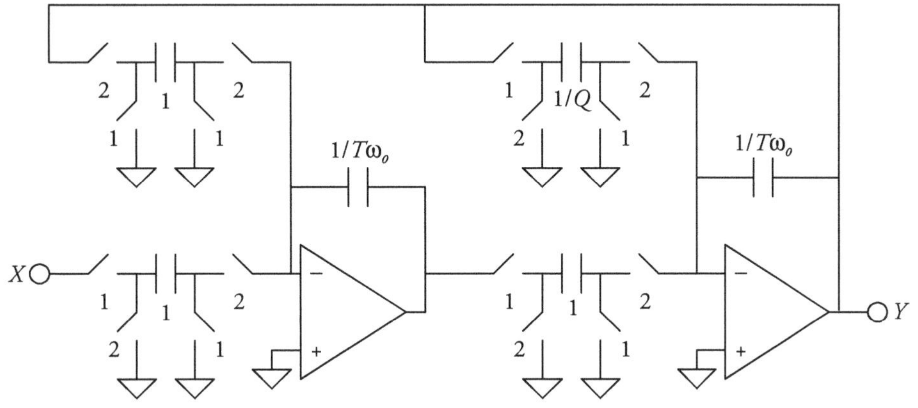

FIGURE 15.28 SC circuit realization of the design example.

to note that the negative resistances are used in this case to simplify the design so they can be realized in the SC circuit structure.

By replacing the resistors in the active-RC circuit structure with the SC resistor structures, the active-RC circuit can be converted into the VLSI SC circuit structure shown in Fig. 15.28.

15.4 Exact SFG-Based SC Filter Design Methods

The SC filter design approaches discussed above are based on the assumption of oversampling operation that allows the use of various numerical integration approximations—such as the forward or backward Euler integrations, the bilinear integration, the LDI integration, and the multistep integration—in the SC filter circuit implementations.

A more accurate SC filter design method that provides an exact z-domain transfer-function operation valid for all sampling frequencies down to the Nyquist rate (OSR = 1) is introduced in this section.

15.4.1 Switching Sequence Effect in SC Circuits

For the noninverted SC integration circuit shown in Fig. 15.29, the driver circuit and the input circuit of the SC circuit under study are clocked at the ϕ_2 phase and the ϕ_1 phase, respectively.

Such a circuit can be analyzed based on the charge redistribution method as follows:

- At $t = (n + 1/2)T$, the sampling capacitor C_R and the storage capacitor C are charged to voltage values as follows:

$$\begin{cases} V_i(nT + T/2) = V_i(nT) \\ V_o(nT + T/2) = V_o(nT) \end{cases} \tag{15.21}$$

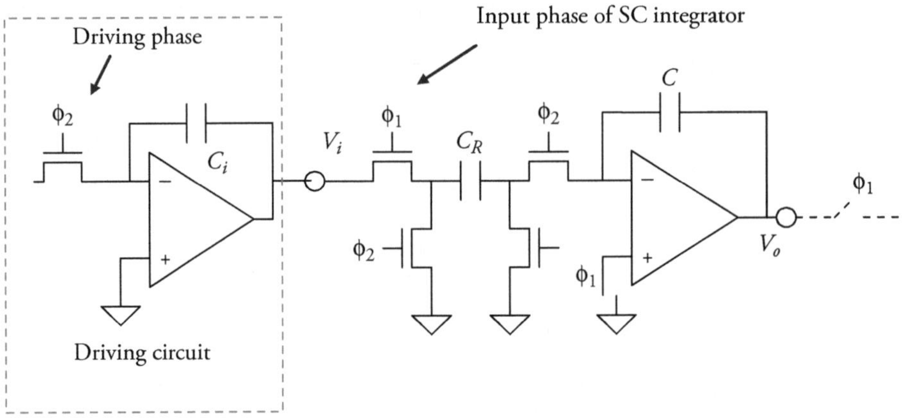

FIGURE 15.29 VLSI SC integrator circuit with ϕ_2 phase driver.

- At $t = (n+1)T$, the charge on the sampling capacitor is transferred to the storage capacitor. The output voltage at end of this phase is given as

$$V_o(nT+T) = V_o(nT+T/2) + \frac{C_R}{C} V_i(nT+T/2) \tag{15.22}$$

or

$$V_o(nT+T) = V_o(nT) + \frac{C_R}{C} V_i(nT) \tag{15.23}$$

- The z-domain transfer function of the SC circuit can be derived as

$$\Rightarrow \frac{V_o(z)}{V_i(z)} = \frac{C_R}{C} \frac{z^{-1}}{1-z^{-1}} \tag{15.24}$$

It can be seen that this SC circuit represents a forward Euler integration operation. Shown in Fig. 15.30 are the control clocks of the SC integrator in Fig. 15.29.

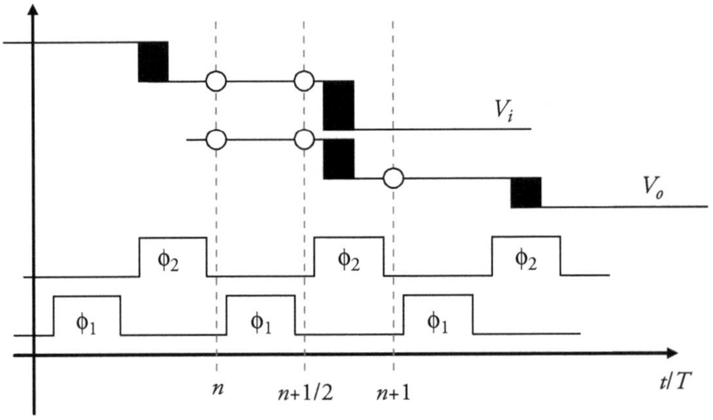

FIGURE 15.30 Control clocks for SC integrator in Fig. 15.29.

In the second circuit configuration, the driver circuit and the input stage are controlled by the ϕ_1 phase as shown in Fig. 15.31.

In this case, we have that

$$\begin{cases} V_i(nT + T/2) = V_i(nT + T) \\ V_o(nT + T/2) = V_o(nT) \end{cases} \tag{15.25}$$

and

$$V_o(nT + T) = V_o(nT) + \frac{C_R}{C} V_i(nT + T) \tag{15.26}$$

The z-domain transfer function of the SC circuit is given as

$$\Rightarrow \frac{V_o(z)}{V_i(z)} = \frac{C_R}{C} \frac{1}{1 - z^{-1}} \tag{15.27}$$

It can be seen that it represents a backward Euler integration operation. Shown in Fig. 15.32 are the control clocks of the SC integrator in Fig. 15.31.

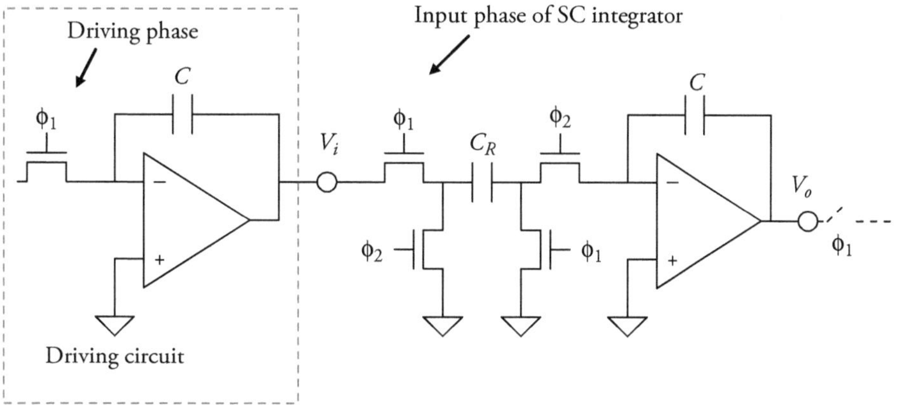

FIGURE 15.31 SC integrator circuit with ϕ_1 phase driver.

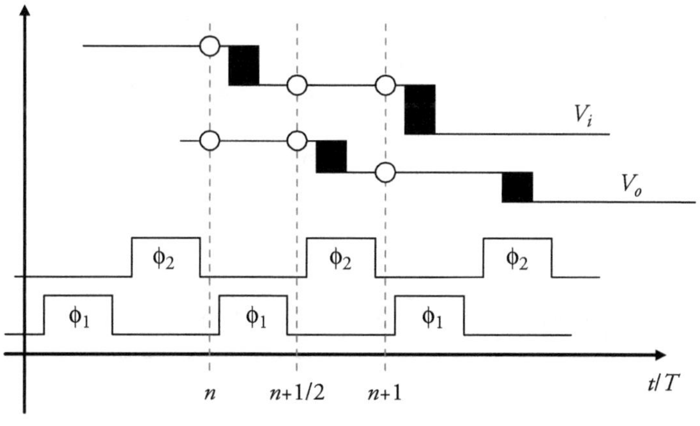

FIGURE 15.32 Control clocks for SC integrator in Fig. 15.31.

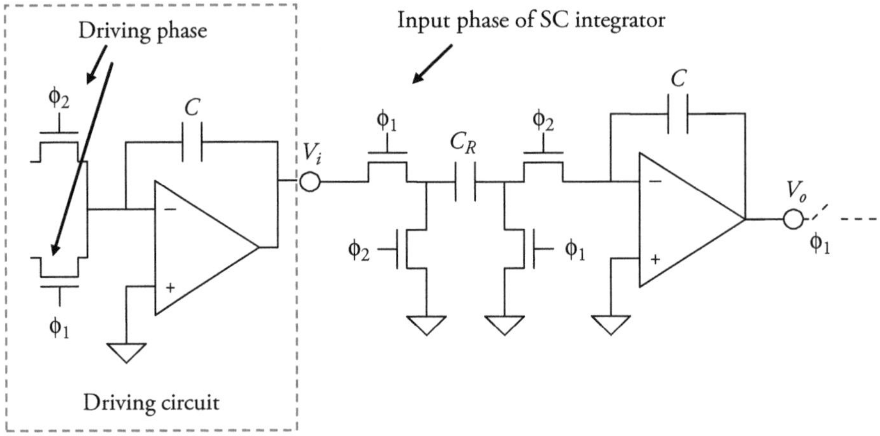

FIGURE 15.33 SC integrator with ϕ_1 and ϕ_2 phase inputs.

In the third case, the driver is controlled in the double sampled clock as shown in Fig. 15.33.

We have that at $t = (n + 1/2)T$,

$$\begin{cases} V_i(nT + T/2) \neq V_i(nT) \\ V_i(nT + T/2) \neq V_i(nT + T) \\ V_o(nT + T/2) = V_o(nT) \end{cases} \tag{15.28}$$

At $t = (n + 1)T$,

$$V_o(nT + T) = V_o(nT) + \frac{C_R}{C} V_i(nT + T/2) \tag{15.29}$$

Shown in Fig. 15.34 are the control clocks of the SC integrator in Fig. 15.33.

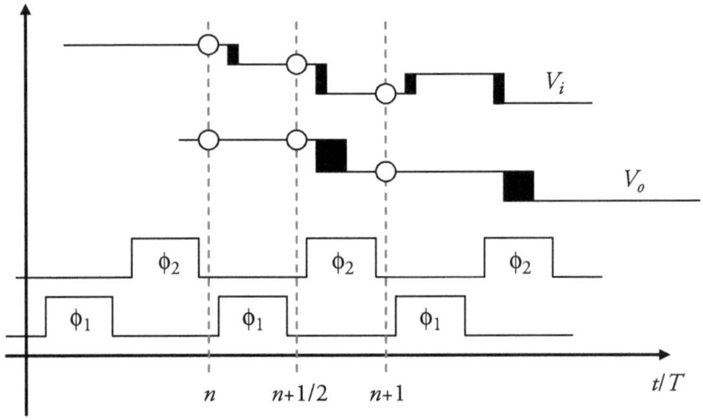

FIGURE 15.34 Control clocks for SC integrator in Fig. 15.33.

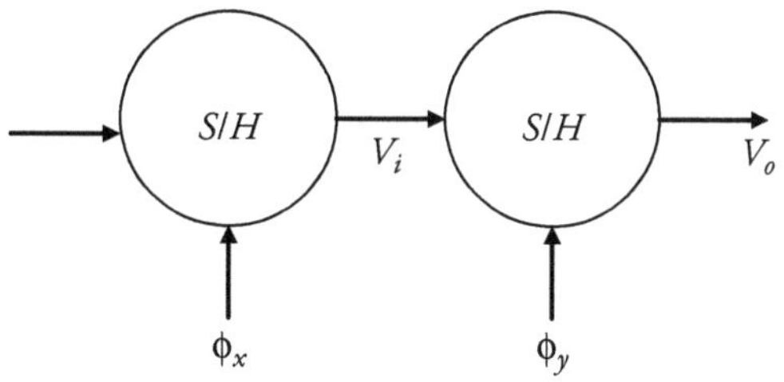

FIGURE 15.35 Cascade of sampled-data circuit.

The z-domain transfer function of the SC circuit is given as

$$\Rightarrow \frac{V_o(z)}{V_i(z)} = \frac{C_R}{C}\frac{z^{-1/2}}{1-z^{-1}} \tag{15.30}$$

It can be seen that it represents an LDI or middle-point integration function. Note that for oversampling SC operation, all the above SC circuit structures provide different approximation methods to the same s-domain transfer function. However, they are not equivalent to each other under the Nyquist rate operation. Consider a general case with two SC S/H circuit stages connected together and controlled by the ϕ_x and ϕ_y phases, respectively, as shown in Fig. 15.35.

We can see that the output of the S/H circuit is a delayed (by Δt) version of the input signal, that is,

$$V_o(t+\Delta t) = V_i(t) \tag{15.31}$$

or

$$\Rightarrow \frac{V_o}{V_i} = z^{-\frac{\Delta t}{T}} \tag{15.32}$$

Shown in Fig. 15.36 are the control clocks of the SC integrator in Fig. 15.35.

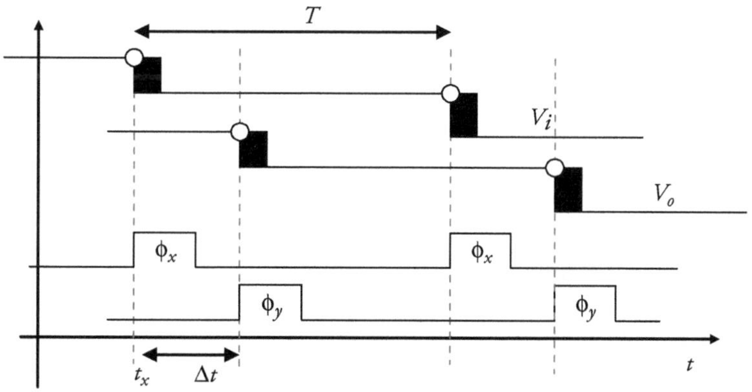

FIGURE 15.36 Control clocks for circuit in Fig. 15.35.

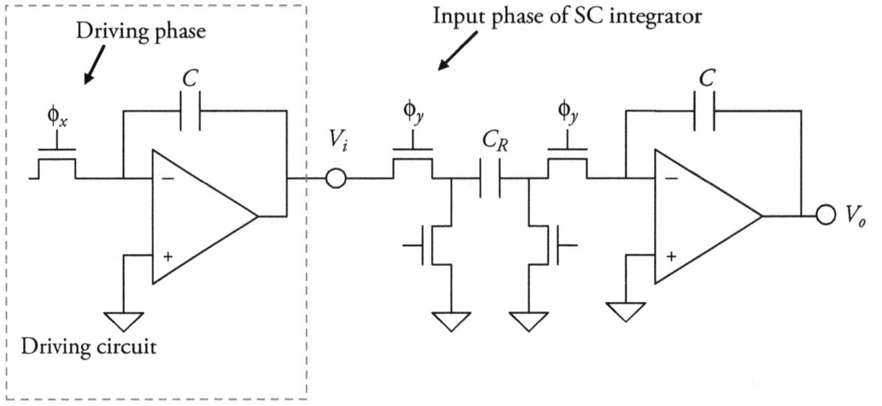

FIGURE 15.37 Cascade of SC circuit blocks.

Similarly, for the integrator circuit shown in Fig. 15.37, we have the following:

$$V_o(t+\Delta t) = V_o(T+\Delta T - t) - \frac{C}{C_R} V_i(t) \tag{15.33}$$

$$\frac{V_o}{V_i} = -\frac{C}{C_R}\frac{z^{-\frac{\Delta t}{T}}}{1-z^{-1}} \tag{15.34}$$

For the following special cases, we have the following:

- For $\phi_x = \phi_y = \phi_2$, $\Delta t = 0$:

$$\Rightarrow \frac{V_o}{V_i} = -\frac{C}{C_R}\frac{1}{1-z^{-1}} \tag{15.35}$$

- For $\phi_x = \phi_1$ and $\phi_y = \phi_2$:

$$\Rightarrow \frac{V_o}{V_i} = -\frac{C}{C_R}\frac{z^{-1/2}}{1-z^{-1}} \tag{15.36}$$

For the noninverted integrator shown in Fig. 15.38, which does not have transparent path from the input to the output, we know that $\Delta t = T$. The transfer function is given as follows:

$$\frac{V_o}{V_i} = \frac{C}{C_R}\frac{z^{-1}}{1-z^{-1}} \tag{15.37}$$

15.4.2 SC Building Blocks for Exact SC Filter Design

The forward Euler integration with weighted inputs, because of the nontransparent property, provides a robust building component for implementing the integration in the exact SC realization (see Figs. 15.39 and 15.40). Note that by this method, only positive weights can be directly realized.

$$Y = \frac{1}{z-1}(\alpha_1 X_1 + \alpha_2 X_2 + \cdots + \alpha_n X_n) \tag{15.38}$$

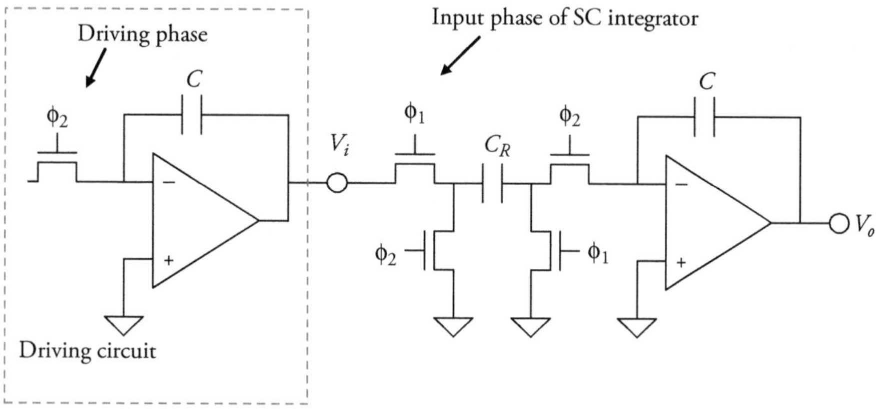

FIGURE 15.38 Cascade of SC circuit blocks without transparent path.

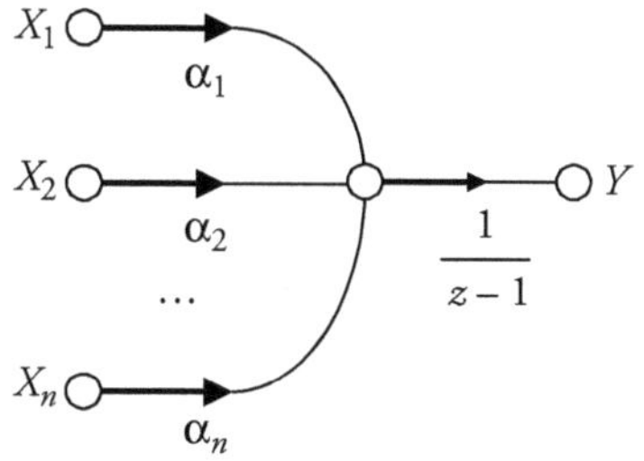

FIGURE 15.39 Integration with positive weighted inputs for exact SC filter realization.

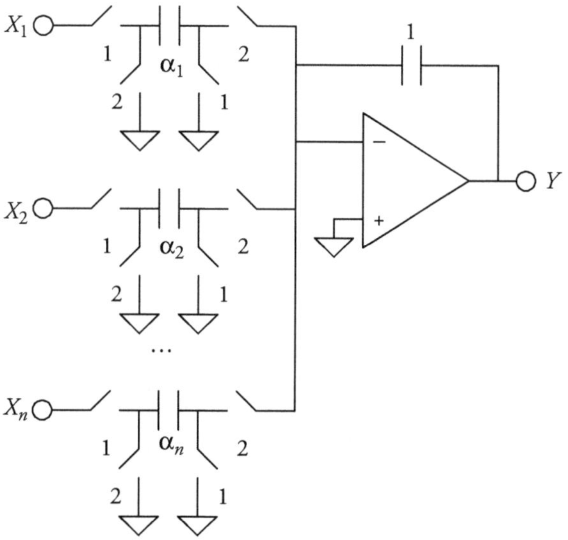

FIGURE 15.40 SC circuit realization of integration with weighted inputs.

The addition and scaling building blocks for exact SC filter realization are shown in Figs. 15.41 and 15.42, respectively.

$$Y = -(\alpha_1 X_1 + \alpha_2 X_2 + \cdots \alpha_n X_n) \quad (15.39)$$

An important property of the two circuits is that they do not introduce operation extra delay; therefore, only the transparent circuit type can be used. In practical filter design application, it also desired that these functions are merged with the integration operation whenever possible to minimize the realization circuits.

The SC scaling circuit block can be viewed as a special case of addition with a weighted input where only one input signal path is used.

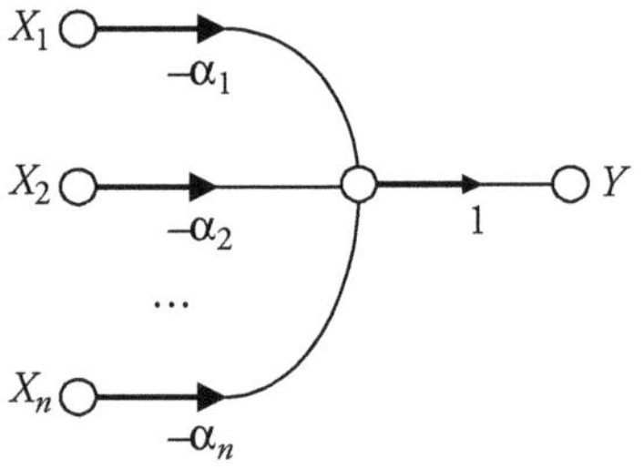

Figure 15.41 Exact SC circuit realization of addition with weighted inputs.

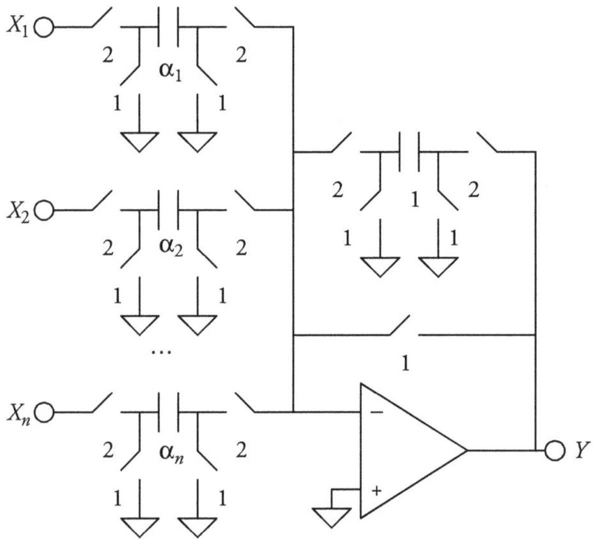

Figure 15.42 Exact SC circuit realization of addition with weighted inputs.

15.4.3 Design of Exact SC Filters

In exact SC filter design method, the design procedure starts from a given z-domain transferred directly as

$$\frac{Y}{X}=\frac{b_m z^{-m}+b_{m-1}z^{-(m-1)}+\cdots+b_1 z^{-1}+b_0}{a_m z^{-m}+a_{m-1}z^{-(m-1)}+\cdots+a_1 z^{-1}+1} \tag{15.40}$$

The transfer function can be expressed by a different variable G through the following substitution as

$$G\equiv\frac{1}{z-1} \quad \text{or } z=G^{-1}+1 \tag{15.41}$$

$$\Rightarrow \frac{Y}{X}\bigg|_{z=G^{-1}+1}=\frac{\beta_m G^m+\beta_{m-1}G^{(m-1)}+\cdots+\beta_1 G+\beta_0}{\alpha_m G^m+\alpha_{m-1}G^{(m-1)}+\cdots+\alpha_1 G+1} \tag{15.42}$$

The modified transfer function can then be rewritten into two standard forms for constructing the SFGs as follows:

$$\begin{aligned}\Rightarrow Y=&(\beta_m X-\alpha_m Y)G^m+(\beta_{m-1}X-\alpha_{m-1}Y)G^{(m-1)}\\&+\cdots+(\beta_1 X-\alpha_1 Y)G+\beta_0 X\end{aligned} \tag{15.43}$$

$$\Rightarrow \begin{cases} U=X-(\alpha_m G^m+\alpha_{m-1}G^{(m-1)}+\cdots+\alpha_1 G)U \\ Y=(\beta_m G^m+\beta_{m-1}G^{(m-1)}+\cdots+\beta_1 G+\beta_0)U \end{cases} \tag{15.44}$$

The two SFG expressions constructed according to the above transfer functions are shown in Figs. 15.43 and 15.44, respectively.

Finally, the SFGs can be directly mapped to the SC circuits using the basic building blocks.

15.4.4 Design Example

Considering the following transfer function:

$$H(z)=\frac{Y}{X}=\frac{z+1}{z-0.5} \tag{15.45}$$

the given z-domain transfer function can be expressed in terms of G as

$$\Rightarrow H(z)\big|_{G\equiv\frac{1}{z-1}}=\frac{G^{-1}+2}{G^{-1}+0.5}=\frac{1+2G}{1+0.5G} \tag{15.46}$$

or

$$\Rightarrow Y=X+G(2X-0.5Y) \tag{15.47}$$

An SFG can be constructed based on the G components as shown in Figs. 15.45 and 15.46.

The SFG can now be directly mapped to the SC circuit using the building blocks introduced in the section as shown in Fig. 15.47.

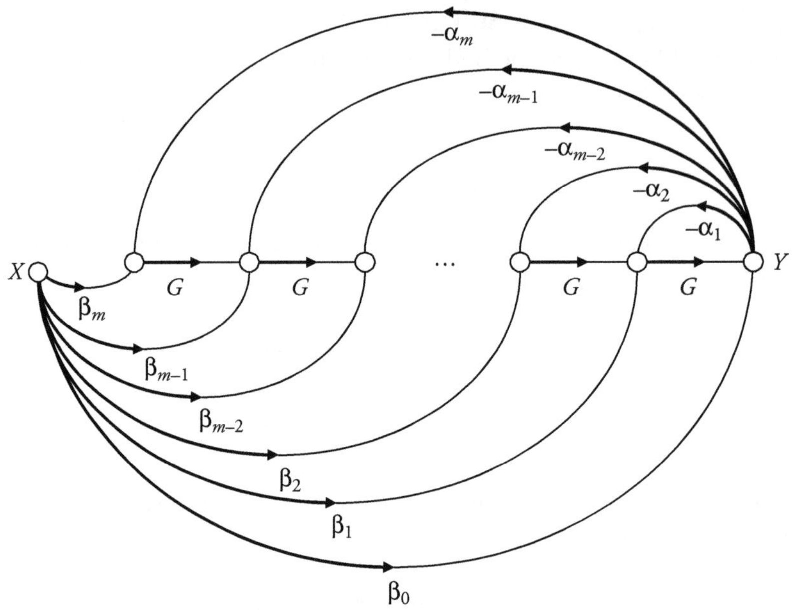

Figure 15.43 SFG expression based on SC integration.

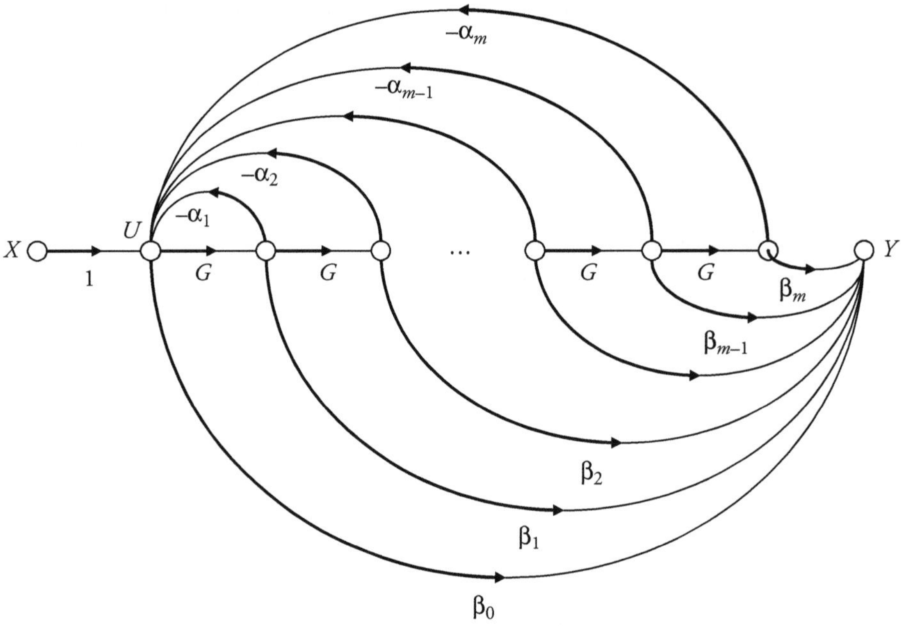

Figure 15.44 Alternative SFG expression based on SC integration.

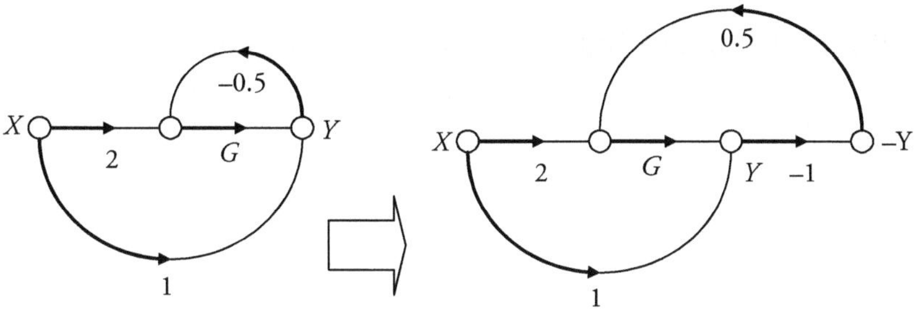

FIGURE 15.45 SFG expression of the design example.

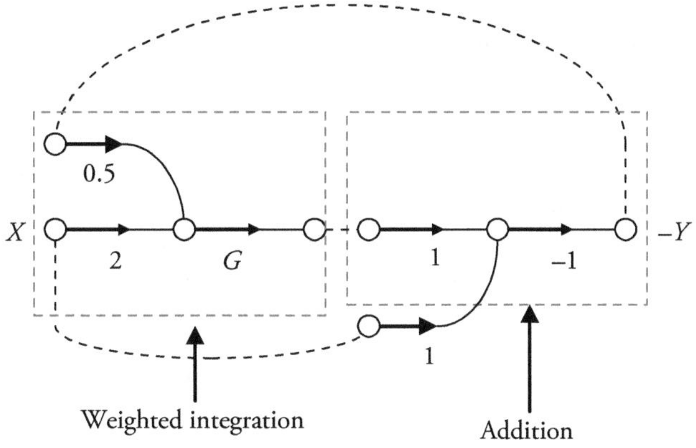

FIGURE 15.46 Modified SFG of the given transfer function.

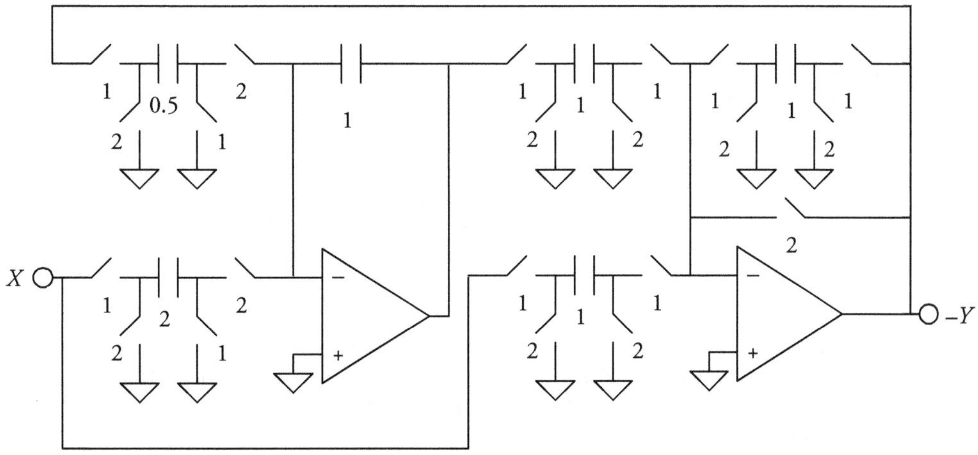

FIGURE 15.47 Exact SC circuit realization of the given transfer function.

15.5 Cascade-Based SC Filter Design Methods

Using the SC filter design introduced in the previous sections, it is possible to construct simple filter sections. Complicated SC filters can then be constructed by the cascading of lower order sections.

References

1. G. M. Jacobs, D. J. Allstot, R. W. Brodersen, P. Gary, "Design Techniques for MOS Switched Capacitor Ladder Filters," *IEEE Trans. on Circuits and Systems*, vol. CAS-25, no. 12, pp. 1014-1021, December 1978.
2. K. Martin, A. S. Sedra, "Exact Design of Switched-Capacitor Bandpass Filter Using Coupled-Biquad Structures," *IEEE Trans. on Circuits and Systems*, vol. CAS-27, no. 6, pp. 469–475, December 1980.
3. M. S. Lee, G. C. Tempes, C. Chang, M. B. Ghaderi, "Bilinear Switched-Capacitor Ladder Filters," *IEEE Trans. on Circuits and Systems*, vol. CAS-28, no. 8, pp. 881–821, December 1981.
4. F. Krummenacher, "Micropower Switched-Capacitor Biquadratic Cell," *IEEE J. Solid-State Circuits*, vol. SC-17, no. 3, pp. 507–512, June 1982.
5. R. Gregorian, K. W. Martin, G. C. Tempes, "Switched-Capacitor Circuit Design," *Proc. of the IEEE*, vol. 71, no. 8, pp. 941–966, August 1983.
6. R. B. Datar, A. S. Sedra, "Exact Design of Stray-Insensitive Switched-Capacitor Ladder Filters," *IEEE Trans. on Circuits and Systems*, vol. CAS-30, no. 12, pp. 888–898, December 1983.
7. R. K. Henderson, J. I. Sewell, "Matrix Methods for Switched-Capacitor Filter Design," *ISCAS'88*, vol. 2, pp. 1021–1024, June 1988.
8. K. Chen, S. Eriksson, "Z-domain Synthesis of Switched-Capacitor LDI Ladder Filters," *ISCAS'88*, vol. 2, pp. 1013–1016, June 1988.
9. U.-K. Moon, "CMOS High-Frequency Switched-Capacitor Filters for Telecommunication Applications," *IEEE J. Solid-State Circuits*, vol. 35, no. 2, pp. 212–220, February 2000.
10. E. M. Saad, "Switched Capacitor Circuits and Their Applications, an Overview," *Proceedings, Radio Science Conference2002*, pp. 51–59, March 2002.
11. A. S. Korotkov, "Synthesis of Narrow-Band Coupled Resonator SC-Filters," *MELECON 98*, vol. 1, pp. 554–557, May 1998.
12. W. Aloisi, G. Giustolisi, G. Palumbo, "Exploiting the High-Frequency Performance of Low-Voltage Low-Power SC Filters," *IEEE Trans. on Circuits and Systems II: Express Briefs*, vol. 55, no. 2, pp. 77–84, February 2004.
13. L. Yue, J. I. Sewell, "A Comparison Study of SC Biquads in the Realization of SC Filters," *IEE Colloquium on Digital and Analogue Filters and Filtering Systems*, pp. 101/1–101/9, November 1993.
14. J. S. Pereira, A. Petraglia, "Low-Sensitivity Direct-Form IIR SC Filters with Improved Phase Linearity," *ISCAS'00*, vol. 3, pp. 169–172, May 2000.
15. K. Nakayama, M. Zhiqiang, G. Yamamoto, "A Method to Minimize Total Capacitance in Cascade Realization of SC Filters," *Proceedings of the 32nd Midwest Symposium on Circuits and Systems*, vol. 2, pp. 968–972, August 1989.
16. A. Handkiewicz, "Two-Dimensional SC Filters with Circularly Symmetric Response," *European Conference on Circuit Theory and Design*, pp. 252–255, September 1989.

17. M. D. Lutovac, D. Tosic, D. Novakovic, "Programmable Low-Pass/High-Pass SC-Filters," *MELECON 98*, vol. 1, May 1998.
18. Y. Lu, J. I. Sewell, "A Comparison Study of SC Biquads in the Realization of SC Filters," *ISCAS'94*, vol. 5, pp. 711–714, June 1994.
19. M. Biey, F. Montecchi, A. Premoli, "Performances of Bandpass Cascaded SC Filters with Reduced Pole-Qs," *IEEE International Symposium on Circuits and Systems*, vol. 2, pp. 1253–1256, June 1988.
20. A. T. Younis, R. E. Massara, "On the Design of Optimal Switched Capacitor Filters Based on the Use of Lossy Frequency-Dependent Negative Resistance (FDNR) Prototype Structures," *IEE 1988 Saraga Colloquium on Electronic Filters*, pp. 2/1–2/6, May 1988.
21. E. Sanchez-Sinencio, R. Geiger, J. Silva-Martinez, "Tradeoffs Between Passive Sensitivity, Output Voltage Swing, and Total Capacitance in Biquadratic SC Filters," *IEEE Trans. on Circuits and Systems*, vol. 31, no. 11, pp. 984–987, November 1984.
22. A. S. Korotkov, H. Hauer, R. Unbehauen, "Practical Design of Element Simulation Type SC Filters," *IEEE Trans. on Circuits and Systems I: Fundamental Theory and Applications*, vol. 47, no. 6, pp. 934–940, June 2000.
23. H. Baher, "Microelectronic Switched Capacitor Filters," *IEEE Circuits and Devices Magazine*, vol. 7, no. 1, pp. 33–36, January 1991.
24. J. J. F. Rijns, H. Wallinga, "Spectral Analysis of Double-Sampling Switched-Capacitor Filters," *IEEE Trans. Circuits and Systems*, vol. 38, no. 11, pp. 1269–1279, November 1991.
25. G. W. Roberts, A. S. Sedra, "Synthesizing Switched-Current Filters by Transposing the SFG of Switched-Capacitor Filter Circuits," *IEEE Trans. on Circuits and Systems*, vol. 38, no. 3, pp. 337–340, March 1991.
26. J. Silva-Martinez, E. Sanchez-Sinencio, "Strategic SC Filter Design Based on a Comparative Study of Various S-to-z Mappings," *IEEE Trans. on Circuits and Systems*, vol. 36, no. 11, pp. 1465–1472, November 1989.
27. M. D. Lutovac, D. Novakovic, I. Markoski, "Selective SC-Filters with Low Passive Sensitivity," Electronics Letters, vol. 33, no. 8, pp. 674–675, April 1997.
28. B. Nowrouzian, "Theory and Design of LDI Lattice Digital and Switched-Capacitor Filters," *IEE Proceedings Circuits, Devices and Systems*, vol. 139, no. 4, pp. 517–526, August 1992.
29. A. Baschirotto, D. Bijno, R. Castello, F. Montecchi, "A 1 V 1.2 μW 4th Order Bandpass Switched-OPAMP SC Filter for a cardiac Pacer Sensing Stage," *ISCAS'00 Geneva*, vol. 3, pp. 173–176, May 2000.
30. J. S. Pereira, A. Petraglia, M. F. Quelhas, "Approximating Linear Phase with IIR SC Filters," *ISCAS'02*, vol. 4, pp. IV-623–IV-626, May 2002.
31. J. L. Ausin, G. Torelli, J. F. Duque-Carrillo, J. M. Carrillo, P. Merchan, "Programmable Time-Multiplexed SC Filters without Dynamic Range Degradation," *IEEE International Conference on Electronics, Circuits and Systems*, vol. 1, pp. 373–376, September 1998.
32. P. Filoramo, G. Giustolisi, G. Palmisano, G. Palumbo, "An Approach to the Design of Low-voltage SC Filters," *ISCAS'98*, vol. 1, pp. 265–268, June 1998.
33. J. Goette, A. Kaelin, W. Guggenbuhl, "Criteria and Methods for the Selection of Optimized capacitor values in SC-filters," *ISCAS'89*, vol. 3, pp. 1696–1700, May 1989.
34. J. Silva-Martinez, E. Sanchez-Sinencio, "A Sparing SC Filter Design Approach with Reduced Transmission zeros," *ISCAS'88*, vol. 2, pp. 1721–1724, June 1988.

35. B. G. Lofmark, "SC-Filter Structures with Reduced Requirements on Op-amp Bandwidth and Settling Time," *ISCAS'88*, vol. 2, pp. 1473–1477, June 1988.
36. E. S. Sinencio, J. S. Martinez, R. Geiger, "Biquadratic SC Filters with Small GB Effects," *IEEE Trans. on Circuits and Systems*, vol. 31, no. 10, pp. 876–884, October 1984.
37. P. Van Peteghem, W. Sansen, "Power Consumption Versus Filter Topology in SC Filters," *IEEE Trans. on Circuits and Systems*, vol. 33, no. 2, pp. 150–157, February 1986.
38. A. S. Korotkov, K. H. Feistel, R. Unbehauen, "Synthesis of Switched-Capacitor Parametric Bandpass Filters," *IEEE Trans. on Circuits and Systems I: Fundamental Theory and Applications*, vol. 46, no. 4, pp. 484–490, April 1999.
39. D. Vazquez, A. Rueda, J. L. Huertas, E. Peralias, "A High-Q Bandpass Fully Differential SC Filter with Enhanced Testability," *IEEE J. Solid-State Circuits*, vol. 33, no. 7, pp. 976–986, July 1998.
40. A. Baschirotto, R. Castello, "A 1-V 1.8-MHz CMOS Switched-OPAMP SC Filter with Rail-to-Rail Output Swing," *IEEE J. Solid-State Circuits*, vol. 32, no. 12, pp. 1979–1986, December 1997.
41. B. Maundy, E. I. El-Masry, "A Switched Capacitor Bidirectional Associative Memory," *IEEE Trans. on Circuits and Systems*, vol. 37, no. 12, pp. 1568–1572, December 1990.
42. R. Castello, A. G. Grassi, S. Donati, "A 500-nA Sixth-Order Bandpass SC Filter," *IEEE J. Solid-State Circuits*, vol. 25, no. 3, pp. 669–676, June 1990.
43. P. M. Van Peteghem, "On the Relationship between PSRR and Clock Feed Through in SC Filters," *IEEE J. Solid-State Circuits*, vol. 23, no. 4, pp. 997–1004, August 1988.
44. P. Filoramo, G. Giustolisi, G. Palmisano, G. Palumbo, "Approach to the Design of Low-Voltage SC Filters," IEE Proceedings, Circuits, Devices and Systems, vol. 147, no. 3, pp. 196–200, June 2000.
45. M. C. Schneider, C. Galup-Montoro, J. C. M. Bermudez, "Explicit Formula for Harmonic Distortion in SC Filters with Weakly Nonlinear Capacitors," *IEE Proceedings Circuits, Devices and Systems*, vol. 141, no. 6, pp. 505–509, December 1994.
46. A. Dabrowski, U. Menzi, G. S. Moschytz, "Design of Switched-Capacitor FIR Filters with Application to a Low-Power MFSK Receiver," *IEE Proceedings, Circuits, Devices and Systems*, vol. 139, no. 4, pp. 450–466, August 1992.
47. M. F. Quelhas, A. Petraglia, A. P. Baruqui, "Power Consumption Estimation in SC Filters," *IEEE International Symposium on Industrial Electronics*, vol. 1, pp. 298–300, June 2003.
48. A. S. Korotkov, "Predistortion Technique for Element Simulation Type SC-Filters," *MELECON '96*, vol. 3, pp. 1259–1262, May 1996.
49. J. L. Ausin, R. Perez-Aloe, J. F. Duque-Carrillo, G. Torelli, E. Sanchez-Sinencio, "High-Selectivity SC Filters with Continuous Digital Q-factor Programmability," *ISCAS'02*, vol. 4, pp. IV-631–IV-634, May 2002.
50. R. Harjani, B. Vinnakota, "Digital Detection of Analog Parametric Faults in SC Filters," *Proceedings, 36th Design Automation Conference*, pp. 772–777, June 1999.
51. U. Kleine, T. Pasch, "Low-Voltage Medium Q-SC Filters for High Frequency Communication Applications," *1998 IEEE International Conference on Electronics, Circuits and Systems*, vol. 2, pp. 39–42, September 1998.
52. U. Marschner, W.-J. Fischer, E.-G. Kranz, "Low Expense Architectures for a Dynamic Spectrum Analyzer based on SC-filters," *ISCAS'98*, vol. 2, pp. 276–279, June 1998.

53. A. Baschirotto, F. Montecchi, R. Castello, "A 150 Msample/s 20 mW BiCMOS Switched-Capacitor Biquad Using Precise Gain Op Amps," *42nd ISSCC*, pp. 212–213, February 1995.
54. G. H. Wang, K. Watanabe, "Current-Mode Switched-Capacitor Circuit Synthesis by C-invariant Dual Transformation," *ISCAS'91*, vol. 3, pp. 1553–1556, June 1991.
55. X. F. Wania, D. A. Johns, A. S. Sedra, "Programmable Multiplexed Switched-Capacitor Filters," *Proceedings of the 33rd Midwest Symposium on Circuits and Systems*, vol. 2, pp. 973–976, August 1990.
56. A Baschirotto, R Castello, F. Montecchi, "Finite Gain Compensation Techniques for High-Q Bandpass SC Filters," *ISCAS'90*, vol. 4, pp. 23813–2816, May 1990.
57. R. Geiger, E. Sanchez-Sinencio, "Operational Amplifier Gain-Bandwidth Product Effects on the Performance of Switched-Capacitor Networks," *IEEE Trans. on Circuits and Systems*, vol. 29, no. 2, pp. 96–106, February 1982.
58. R. Castello, P. Gray, "Performance Limitations in Switched-Capacitor Filters," *IEEE Trans. on Circuits and Systems*, vol. 32, no. 9, pp. 865–876, September 1985.
59. D. Bruckmann, U. Kleine, "Novel Voltage Inverter Switches with Minimum Sensitivity Properties," *IEEE Trans. on Circuits and Systems*, vol. 32, no. 7, pp. 723–726, July 1985.
60. P. Li, J. I. Sewell, "The LUD Approach to Switched-Capacitor Filter Design," *IEEE Trans. on Circuits and Systems*, vol. 34, no. 12, pp. 1611–1614, December 1987.
61. P. M. Van Peterghem, W. M. C. Sansen, "Power Consumption versus Filter Topology in SC Filters," *IEEE J. Solid-State Circuits*, vol. 21, no. 1, pp. 40–47, February 1986.
62. P. Landau, D. Michel, D. Melnik, "A Reduced Capacitor Spread Algorithm for Elliptic Bandpass SC Filters" *IEEE J. Solid-State Circuits*, vol. 22, no. 4, pp. 624–626, August 1987.
63. W. M. C. Sansen, H. Qiuting, K. A. I. Halonen, "Transient Analysis of Charge-Transfer in SC Filters-Gain Error and Distortion," *IEEE J. Solid-State Circuits*, vol. 22, no. 2, pp. 268–276, April 1987.
64. J. L. Ausin, J. F. Duque-Carrillo, G. Torelli, E. Sanchez-Sinencio, "Switched-Capacitor Circuits with Periodical Nonuniform Individual Sampling," *IEEE Trans. on Circuits and Systems II: Analog and Digital Signal Processing*, vol. 50, no. 8, pp. 404–414, August 2003.
65. L. Lentola, G. M. Corelazzo, E. Malavasi, A. Baschirotto, "Design of SC Filters for Video Applications," *IEEE Trans. on Circuits and Systems for Video Technology*, vol. 10, no. 1, pp. 14–22, February 2000.
66. B. Vinnakota, R. Harjani, "DFT for Digital Detection of Analog Parametric Faults in SC Filters," *IEEE Trans. on Computer-Aided Design of Integrated Circuits and System*, vol. 19, no. 7, pp. 789–798, July 2000.
67. R. P. Martins, J. E. Franca, "Cascade Switched-Capacitor IIR Decimating Filters," *IEEE Trans. on Circuits and Systems I: Fundamental Theory and Applications*. vol. 42, no. 7, pp. 367–376, July 1995.
68. R. P. Martins, J. E. Franca, F. Maloberti, "An Optimum CMOS Switched-Capacitor Anti-aliasing Decimating Filter," *IEEE J. Solid-State Circuits*, vol. 28, no. 9, pp. 962–970, September 1993.
69. J. L. Huertas, A. Rueda, D. Vazquez, "Testable Switched-Capacitor Filters," *IEEE J. Solid-State Circuits*, vol. 28, no. 7, pp. 719–724, July 1993.

70. A. Baschirotto, R. Castello, F. Montecchi, "IIR Double-Sampled Switched-Capacitor Decimators for High-Frequency Applications," *IEEE Trans. on Circuits and Systems I: Fundamental Theory and Applications*, vol. 39, no. 4, pp. 300–304, April 1992.
71. P. Mehta, M. Darwish, T. Thomson, "Switched-Capacitor Filters," *IEEE Trans. on Power Electronics*, vol. 5, no. 3, pp. 331–336, July 1990.
72. C-Y Wu, Tsai-Chung Yu, Shin-Shi Chang, "New Monolithic Switched-Capacitor Differentiators with Good Noise Rejection," *IEEE J. Solid-State Circuits*, vol. 24, no. 1, pp. 177–180, February 1989.
73. J. C. M. Bermudez, M. C. Schneider, C. G. Montoro, "Compatibility of Switched Capacitor Filters with VLSI Processes," *IEE Proceedings G, Circuits, Devices and Systems*, vol. 139, no. 4, pp. 413–418, August 1992.
74. E. I. El-Masry, E. G. Hix, "Novel Switched-Capacitor Integrator for High-Frequency Applications," *IEE Proceedings G Circuits, Devices and Systems*, vol. 136, no. 5, pp. 263–267, October 1989.
75. D. Novakovic, B. Loncar, P. Osmokrovic, "Efficient High-Pass Low-Sensitive Selective SC-Filters," *22nd International Conference on Microelectronics*, 2000, vol. 2, pp. 739–742, May 2000.
76. I. Seo, R. M. Fox, "Low-Power Switched-Capacitor Filters Using Charge Transfer Integrators," *ISCAS'03*, vol. 1, pp. I-609–I-612, May 2003.
77. K. Hajek, J. Sedlacek, "General Multiple LC Prototype Filter Solutions and Optimization," *9th International Conference on Electronics, Circuits and Systems*, vol. 1, pp. 165–168, September 2002.
78. J. M. Canive, A. Petraglia, "On the Testability of SC Filters based on Allpass Sections," *ISCAS'01*, vol. 1, pp. 65–68, May 2001.
79. A. Baschirotto, "IIR Double-Sampled Switched-Capacitor Building Blocks for High-Frequency Decimators," *ISCAS'91*, vol. 3, pp. 1673–1676, June 1991.
80. B. Nowrouzian, "A Novel Approach to the Exact Design of LDI Symmetrical Digital and Switched-Capacitor Filters," *Proceedings of the 33rd Midwest Symposium on Circuits and Systems*, vol. 2, pp. 967–972, August 1990.
81. R. E. Massara, A. T. Younis, "An Efficient Design Method for Optimal MOS Integrated Circuit Switched-Capacitor LDI Ladder Filters," *Proceedings of the 33rd Midwest Symposium on Circuits and Systems*, 1990, vol. 2, pp. 956–959, August 1990.
82. R. Raut, B. B. Bhattacharrya, S. M. Faruque, "An Application of Systolic Array Design Architecture to Switched Capacitor Filter Circuits," *ICASSP-89*, vol. 4, pp. 2401–2404, May 1989.
83. M. Nalecz, J. J. Mulawka, "Parasitic-Compensated Building Blocks for Active Switched-Capacitor Filters," *ISCAS'88*, vol. 2, pp. 1483–1486, June 1988.
84. A. M. Davis, H. P. Nguyen, "Exact Synthesis of SC Filters Using Forward Difference Integrators," *ISCAS'88*, vol. 2, pp. 1009–1012, June 1988.
85. S. Setty, C. Toumazou, K. Manetakis, P. Cheung, R. Spence, "Ideas and Concepts for the Automation of Analogue Filters Using Qualitative Reasoning," *IEE 15th SARAGA Colloquium on Digital and Analogue Filters and Filtering Systems*, pp. 7/1–7/9, November 1995.
86. H. Shafeeu, A. Betts, J. Taylor, "Approaches to Ultra-Narrow-Band Analogue IC Filter Design Using Switched-Capacitors," *IEE Colloquium on Advances in Analogue VLSI*, pp. 2/1–2/9, May 1991.

87. G. Temes, H. Orchard, M. ahanbegloo, "Switched-Capacitor Filter Design Using the Bilinear z-Transform," *IEEE Trans on Circuits and Systems,* vol. 25, no. 12, pp. 1039–1044, December 1978.
88. M. Lee, C. Chang, "Switched-Capacitor Filters Using the LDI and Bilinear Transformations," *IEEE Trans. on Circuits and Systems*, vol. 28, no. 4, pp. 265–270, April 1981.
89. E. El-Masry, "Strays-Insensitive State-Space Switched-Capacitor Filters," *IEEE Trans. on Circuits and Systems*, vol. 30, no. 7, pp. 474–488, July 1983.
90. J. Garcia-Vazquez, E. Sanchez-Sinencio, "Finite Gain-Bandwidth Product Effects on a Pair of Pseudo-N-Path SC Filters," *IEEE Trans. on Circuits and Systems*, vol. 31, no. 6, pp. 583–584, June 1984.
91. C. Campbell, K. Reineck, "A Pole/Zero Prewarping Procedure in SC Filter Design," *IEEE Trans. on Circuits and Systems*, vol. 31, no. 9, pp. 821–825, September 1984.
92. G. Fischer, G. Moschytz, "SC Filters for High-Frequencies with Compensation for Finite-Gain Amplifiers," *IEEE Trans. on Circuits and Systems*, vol. 32, no. 10, pp. 1050–1056, October 1985.
93. T. Inoue, F. Ueno, "On Robustness of the Stability of Multiphase Switched-Capacitor Two-Ports," *IEEE Trans. on Circuits and Systems*, vol. 32, no. 6, pp. 522–529, June 1985.
94. J. Da Franca, "A Single-path Frequency-translated Switched- Capacitor Bandpass Filter System," *IEEE Trans. on Circuits and Systems*, vol. 32, no. 9, pp. 938–944, September 1985.
95. A. Davis, "Flow Graph Synthesis of Darlington – Cauer Switched- Capacitor Filters," *IEEE Trans. on Circuits and Systems*, vol. 32, no. 7, pp. 727–732, July 1985.
96. M. H. Fino, J. E. Franca, A. Steiger-Garcao, "Automatic Symbolic Analysis of Switched-Capacitor Filtering Networks Using Signal Flow Graphs," *IEEE Trans. on Computer-Aided Design of Integrated Circuits and Systems*, vol. 14, no. 7, pp. 858–867, July 1995.
97. J. Crols, M. Steyaert, "Switched-OPAMP: An Approach to Realize Full CMOS Switched-Capacitor Circuits at Very Low Power Supply Voltages" *IEEE J. Solid-State Circuits*, vol. 29, no. 8, pp. 936–942, August 1994.
98. C.-Y. Wu, P.-H. Lu, M.-K. Tsai, "Design Techniques for High-frequency CMOS Switched-Capacitor Filters Using Non-OPAMP-Based Unity-Gain Amplifiers" *IEEE J. Solid-State Circuits*, vol. 26, no. 10, pp. 1460–1466, October 1991.
99. A. Handkiewicz, "Two-dimensional Switched Capacitor Filter Design System for Real-time Image Processing," *IEEE Trans. on Circuits and Systems for Video Technology*, vol. 1, no. 3, pp. 241–246, September 1991.
100. K. Nakayama, Y. Takahashi, Y. Sato, Y. Nukada, "An Adaptive SC Line Equalizer System for Four-Wire Full-Duplex and Multirate Digital Transmission," *IEEE Trans. on Circuits and Systems*, vol. 35, Issue: 9, pp. 1073–1081, September 1988.
101. A. E Salama, "Factored State-space Stray-Insensitive Switched-Capacitor Filters," *IEE Proceedings Circuits, Devices and Systems*, vol. 143, no. 6, pp. 325–330, December 1996.
102. T. F. Yong, M. J. Long, "Systematic Method for Design of Switched Capacitor Filters," Electronics Letters, vol. 28, no. 3, pp. 309–310, January 1992.
103. L. Qiang, H. Yifeng, X. Ke, M. Hao, "A Novel Offset Compensation Biquad Switched-Capacitor Filter Design," *5th International Conference on ASIC*, vol. 1, 21–24 pp. 643–646, October 2003.

104. J. L. Ausin, J. F. Duque-Carillo, G. Torelli, E. Sanchez-Sinencio, F. Maloberti, "Periodical Nonuniform Individually Sampled Switched-Capacitor Circuits," *ISCAS'00*, vol. 5, pp. 449 – 452, May 2000.
105. H. Matsumoto, Z. Tang, O. Ishizuka, "A Buffer-Based Switched-Capacitor Integrator with Reduced Capacitance Ratio," *Proceedings of the 33rd Midwest Symposium on Circuits and Systems*, vol. 2, pp. 719–720, August 1990.
106. J. A. Hegt, "Signal Flow Graph Based Synthesis of Strays-Insensitive Switched-Capacitor Filters," *ISCAS'90*, vol. 3, pp. 2177–2180, May 1990.
107. I. M. Sahawaneh, A. A. Sakla, E. I. El-Masry, "Comparison between Cascade and Parallel Switched Capacitor Structures," *Proceedings of the 32nd Midwest Symposium on Circuits and Systems*, vol. 2, pp. 973–976, August 1989.
108. A. K. Betts, J. T. Taylor, D. G. Haigh, "Synthesis Method for Switched-Capacitor FIR Decimators and Interpolators," *ISCAS'88*, vol. 3, pp. 2463–2466, June 1988.
109. Q. Huang, "A Novel Technique for the Reduction of Capacitance Spread in High Q SC Circuits" *ISCAS'88*, vol. 2, pp. 1249–1252, June 1988.
110. C. Kurth, G. Moschytz, "Two-port Analysis of Switched-Capacitor Networks Using Four-Port Equivalent Circuits in the z-Domain," *IEEE Trans. on Circuits and Systems*, vol. 26, no. 3, pp. 166–180, March 1979.
111. U. Brugger, D. von Grunigen, G. Moschytz, "A Comprehensive Procedure for the Design of Cascaded Switched-Capacitor Filters," *IEEE Trans. on Circuits and Systems*, vol. 28, no. 8, pp. 803–810, August 1981.
112. K. Martin, A. Sedra, "Effects of the Op Amp Finite Gain and Bandwidth on the Performance of Switched-Capacitor Filters," *IEEE Trans. on Circuits and Systems*, vol. 28, no. 8, pp. 822–829, August 1981.
113. S. Wong, C. Salama, "A Switched Differential Op-amp with Low Offset and Reduced 1/f Noise," *IEEE Trans. on Circuits and Systems*, vol. 33, no. 11, pp. 1119–1127, November 1986.
114. F. A. P. Baruqui, A. Petraglia, J. E. Franca, S. K. Mitra, "Efficient Design of Integrated Switched-Capacitor Decimation Filters," *IEEE Trans. on Circuits and Systems II: Analog and Digital Signal Processing*, vol. 47, no. 11, pp. 1314–1318, November 2000.
115. B. Raahemi, A. Opal, "Group Delay and Group Delay Sensitivity of Periodically Switched Linear Networks," *IEEE Trans. on Circuits and Systems I: Fundamental Theory and Applications*, vol. 47, no. 1, pp. 96–104, January 2000.
116. A. Petraglia, "Random Capacitance Ratio Error Effects in the Frequency Response of Switched-Capacitor Filters," *IEEE-CAS Region 8 Workshop on Analog and Mixed IC Design*, pp. 112–116, September 1996.
117. K. van Hartingsveldt, P. Quinn, A. van Roermund, "A 10.7MHz CMOS SC Radio IF Filter with variable Gain and a Q of 55," *ISSCC'00*, pp. 152–153, 452, February 2000.
118. M. M. Amourah, R. L. Geiger, "Gain and Bandwidth Boosting Techniques for High-speed Operational Amplifiers," *ISCAS'01*, vol. 4, pp. 674–677, May 2001.
119. A. Hossein Nejad-Malayeri, T. Manku, "A 270MHz, 1.8V Fully Differential CMOS Operational Amplifier for Switched Capacitor Channel Select Filters in Wide-Band Wireless Applications," *ISCAS'01*. vol. 1, pp. 659–662, May 2001.
120. B. Manai, P. Loumeau, "A Wide-band CMOS Switched Capacitor Filter for UMTS Direct Conversion Receiver," *MWSCAS'01*, vol. 1, pp. 377–380, August 2001.
121. Z. Kaelin, G. S. Moschytz, "Exact Design of Arbitrary Parasitic-Insensitive Elliptic SC-Ladder Filters in the z-Domain," *ISCAS'88*, vol. 3, pp. 2485–2488, June 1988.

122. T. S. Fiez, B. Lee, D. J. Allstot, "CMOS Switched-Current Biquadratic Filters," *ISCAS'90*, vol. 3, pp. 2300–2303, May 1990.
123. C.-Y. Wu, T.-C. Yu, "New Forward- and Backward-Mapping SC Differentiator and Their Applications in the Design of Biquad SCF," *Proceedings of Technical Papers VLSI Technology, Systems and Applications*, pp. 169–173, May 1989.
124. M. J. Svihura, B. Nowrouzian, "A New Approach to the Design of Bilinear-LDI Switched-Capacitor Filters Having Low Passband Sensitivity," *Proceeding. IEEE Pacific Rim Conference on Communications, Computers and Signal Processing*, pp. 193–197, June 1989.
125. R. E. Vallee, E. I. El-Masry, "High Performance CMOS Operational-Amplifier," *ISCAS'89*, vol. 2, pp. 1475–1478, May 1989.
126. S. Michael, "Stray Insensitive Programmable Switched-Capacitor Filter," *Proceedings of the 32nd Midwest Symposium on Circuits and Systems*, vol. 2, pp. 1189–1192, August 1989.

CHAPTER 16

Design of VLSI Switched-Current Filters

VLSI switched-current (SI) filters operate as alternative DT analog signal processing circuits that are based on periodically switched linear current networks. SI filter circuit structures offer attractive features that are compatible to standard VLSI digital process technology, and therefore, they can be operated at a low power supply voltage. In addition, SI filter circuits can potentially be operated at a higher frequency than the SC filter circuit due to the low-impedance wide-bandwidth internal nodes of the current-mirror circuits. In the VLSI SI signal processing circuit implementations, the noncritical MOS transistor gate capacitances are used to store electric charges, which avoid the requirements of using OPAMP circuits and linear capacitors. As a result, SI circuits are fully digital CMOS process technology compatible. Fundamental limitations of SI circuits are the switch charge injection, final output impedance, and mismatch of the MOS transistors in the current-mirror circuits.

16.1 SFG-Based SI Filter Design Methods

In the VLSI SI filter circuit implementations, the unit delay SI circuit structure, with its implementation simplicity and efficiency, can be directly used as the basic building circuit component of the filter. In this approach, the basic building blocks for SI filter circuit implementation can typically be selected as the unit delay with weighted outputs, the wired addition, and the scaling element. By using the above building blocks, it is possible to convert a given z-domain transfer function to a form that can be used to construct SFG easily. Such an SFG expression then can be directly mapped into the VLSI SI circuit. Equation (16.1) is a general z-domain transfer function of a specific frequency response:

$$\frac{Y}{X} = \frac{b_n z^{-n} + b_{n-1} z^{-(n-1)} + \cdots + b_1 z^{-1} + b_0}{a_n z^{-n} + a_{n-1} z^{-(n-1)} + \cdots + a_1 z^{-1} + 1} \tag{16.1}$$

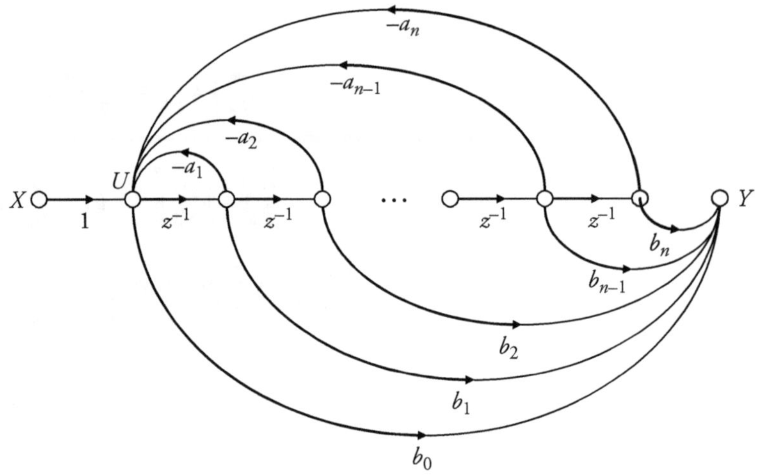

FIGURE 16.1 SFG expression based on unit delay elements for SI filter implementation.

The SFG based on the expression in Eq. (16.2) is shown in Fig. 16.1.

$$\Rightarrow Y = (b_n X - a_n Y)z^{-n} + (b_{n-1}X - a_{n-1}Y)z^{-(n-1)} + \cdots + (b_1 X - a_1 Y)z^{-1} + b_0 X \tag{16.2}$$

The alternative SFG based on an alternative transfer-function manipulation using variable U as shown in Eqs. (16.3) and (16.4) is shown in Fig. 16.2.

$$\frac{Y}{X} = \frac{b_n z^{-n} + b_{n-1}z^{-(n-1)} + \cdots + b_1 z^{-1} + b_0}{a_n z^{-n} + a_{n-1}z^{-(n-1)} + \cdots + a_1 z^{-1} + 1} \cdot \frac{U}{U} \tag{16.3}$$

$$\Rightarrow \begin{cases} U = X - (a_n z^{-n} + a_{n-1}z^{-(n-1)} + \cdots + a_1 z^{-1})U \\ Y = (b_n z^{-n} + b_{n-1}z^{-(n-1)} + \cdots + b_1 z^{-1} + b_0)U \end{cases} \tag{16.4}$$

16.1.1 SI Circuit Realization of Unit Delay

An SI delay circuit can be realized using two VLSI SI S/H circuits connected in series and controlled by two-phase nonoverlap clocks. In practical design, unit delay with weighted outputs is usually selected to minimize the implementation device count. The transfer function of the resulting unit delay building components for the SI filter is given as follows:

$$Y_i = \frac{1}{z}(\alpha_i X) \qquad i = 1, 2, 3, \ldots, n \tag{16.5}$$

The SFG expression and the SI circuit realization of the unit delay operation are shown in Figs. 16.3 and 16.4, respectively, where ϕ_1 and ϕ_2 represent the two nonoverlap control clocks.

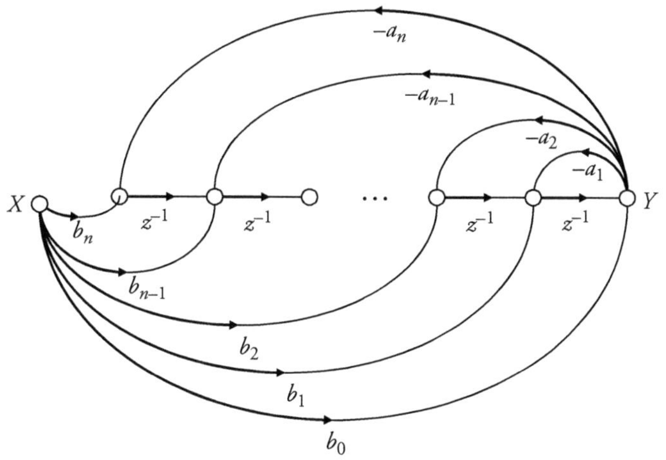

FIGURE 16.2 Alternative SFG expression of a general z-domain transfer function for SI filter design.

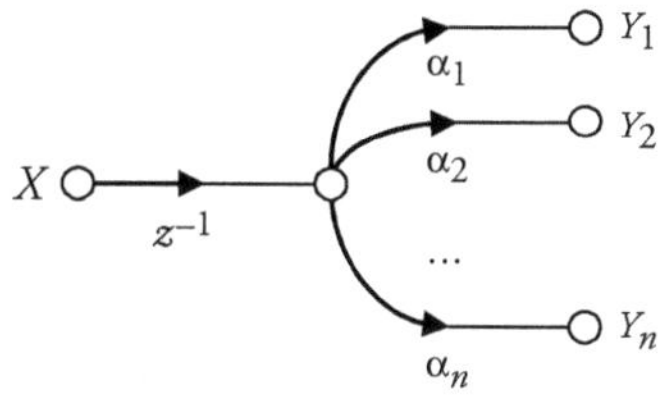

FIGURE 16.3 SFG of unit delay with weighted outputs.

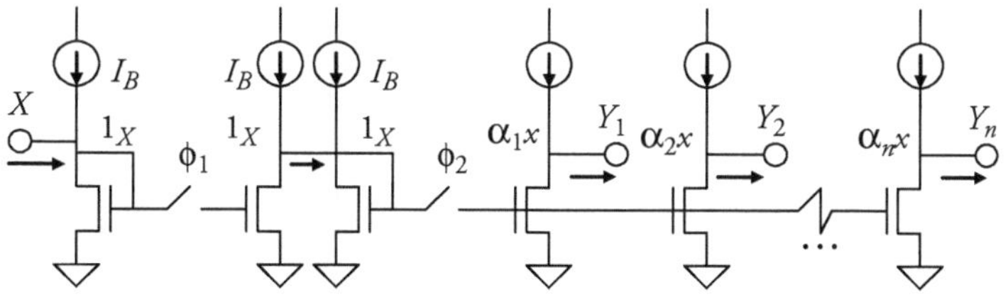

FIGURE 16.4 Delay circuit realization of delay with weighted outputs.

16.1.2 SI Circuit Realization of Addition

The current nature of the CTI signal allows the addition operation to be realized by the wired connection. For the addition signal transfer function in Eq. (16.6), the SFG expression and SI circuit realization are shown in Figs. 16.5 and 16.6, respectively.

$$Y = X_1 + X_2 + \cdots + X_n \tag{16.6}$$

Note that in order to take advantage of the wired addition operation of the current-mode signal processing circuit implementation, the weight of all the input signals for addition operation are limited to one.

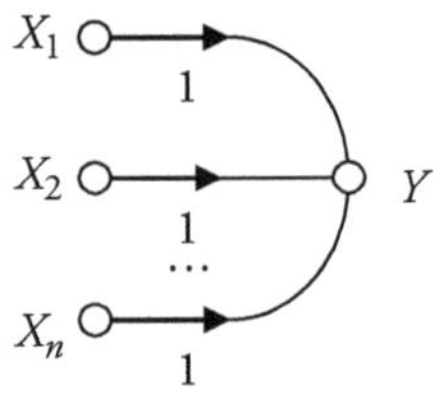

FIGURE 16.5 SFG expression for current-mode addition.

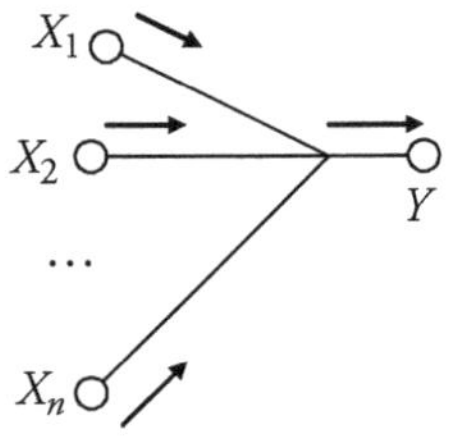

FIGURE 16.6 SI circuit realization of addition.

16.1.3 SI Circuit Realization of Scaling

Scaling operation in SI circuits can be used to generate a scaled version of multiple copies of a given current-mode signal. Such an operation is important in the VLSI current-mode circuit implementations since the current-mode addition operation cannot be used for implementing the nonunity SFG coefficients. As a result, the weighting functions in the current-mode circuit implementations must be realized outside of the addition. On the other hand, current branching functions are usually needed in the current-mode circuit for generating multiple copies of the current signals since single current signals cannot be directly sent to multiple locations.

The transfer function, the SFG structure, and the VLSI SI circuit implementation of the current signal scaling with weighted branching out are shown in Figs. 16.7 and 16.8, respectively.

$$Y_i = -(\alpha_i X) \qquad i = 1, 2, 3, \ldots, n \tag{16.7}$$

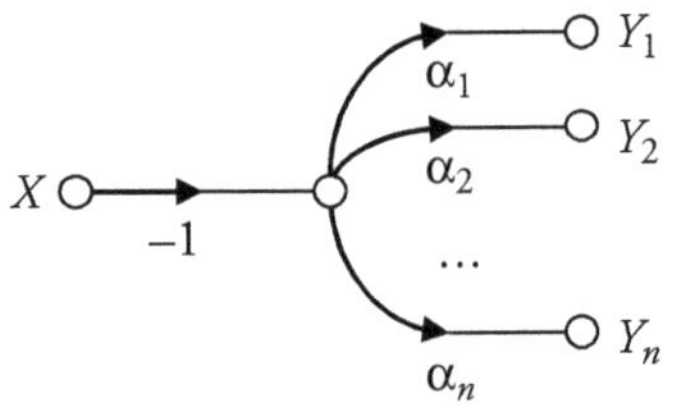

FIGURE 16.7 SFG expression of scaling with weighted multiple outputs.

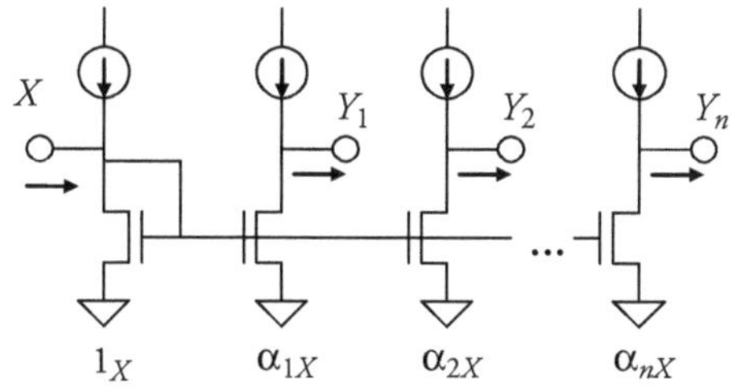

FIGURE 16.8 VLSI SI circuit implementation with weighted outputs.

16.2 S-to-Z Transform–Based SI Filter Design Methods

Although unit delay operation can naturally be selected as the basic building block, an integrator can sometimes be used as the basic building component in SI filter circuit implementation. The use of SI integrator offers the ability of designing VLSI SI filter circuits from other existing VLSI circuit approaches such as SC filter circuit structures or the CT filter circuit structures directly based on the s-to-z transformations. In these methods, the basic building components can be selected as the integrator with the weighted outputs, the wired addition, and the scaling element.

16.2.1 SI Forward Euler Integrator

The transfer function of the forward Euler integrator given in Eq. (16.8) can be realized in SFG and its SI circuit implementation as shown in Figs. 16.9 and 16.10, respectively.

$$G \equiv \frac{Y}{X} = \frac{1}{z-1} \tag{16.8}$$

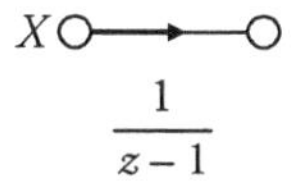

FIGURE 16.9 Forward Euler integration SFG expression.

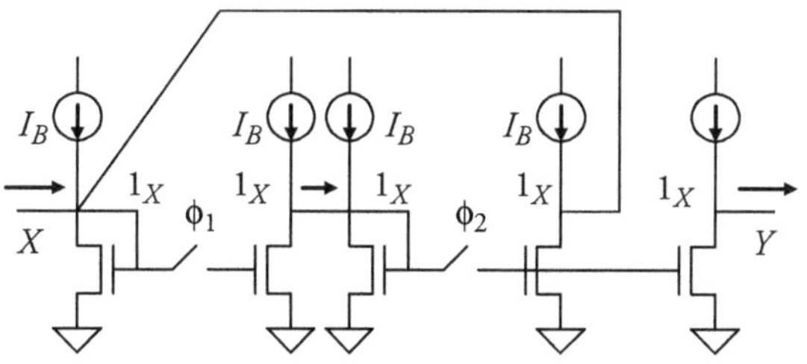

FIGURE 16.10 VLSI SI forward Euler integration implementation.

16.2.2 SI Backward Euler Integrator

Similarly, the transfer function of the backward Euler integrator is given in Eq. (16.9), and its SFG expression and SI circuit implementation are shown in Figs. 16.11 and 16.12, respectively.

$$G \equiv \frac{Y}{X} = \frac{z}{z-1} \tag{16.9}$$

In the s-to-z mapping–based SI filter design approaches, the SFG is directly constructed from its s-domain transfer function based on the lossless integrators as shown in Fig. 16.13 in a way similar to the SFG-based filter design approach of the VLSI active-RC filter design.

The SFG structure suitable for SI realization can be constructed by replacing the lossless integrators with the SI integrators directly as shown in Fig. 16.14.

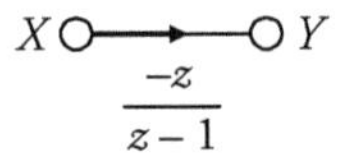

Figure 16.11 SI backward Euler integration SFG expression.

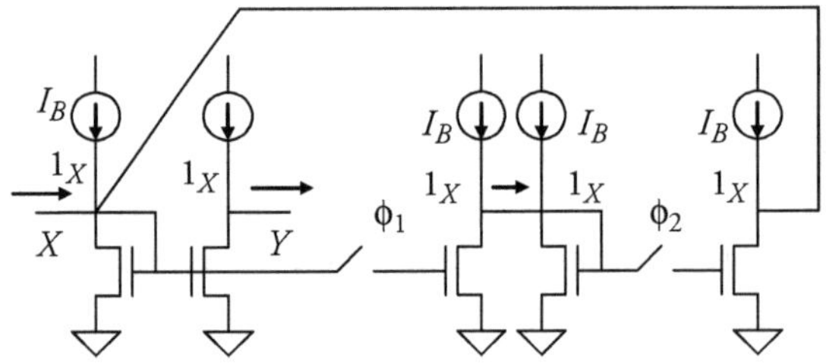

Figure 16.12 VLSI SI backward Euler integration circuit.

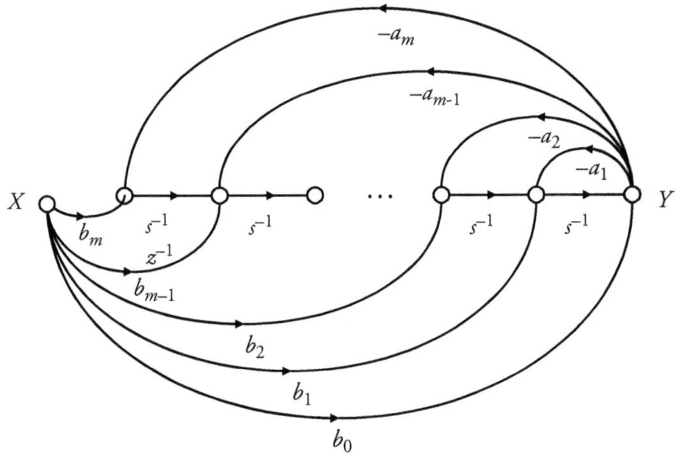

Figure 16.13 S-domain SFG expression based on lossless integration for CT filters.

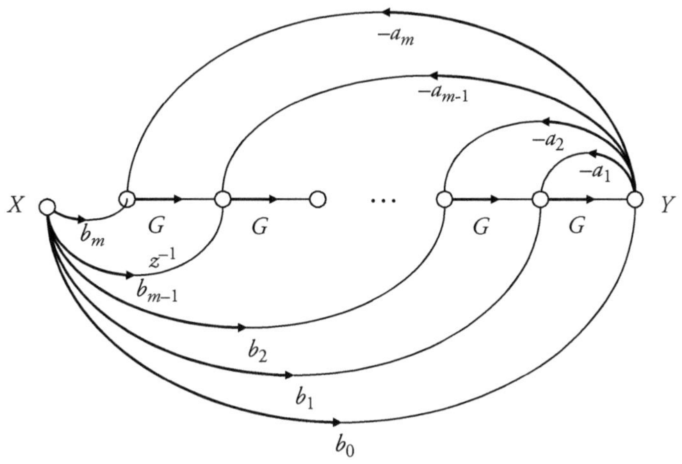

FIGURE 16.14 Z-domain SFG expression based on s-to-z transformation for DT filters.

16.3 Cascade-Based SI Filter Design Methods

Similar to the cascade-based filter design based on other VLSI circuit technologies, the cascade-based SI filter design is based on the first- and second-order CTI building blocks, which can be easily configured for different applications.

16.4 Design Examples

In the following VLSI SI filter circuit design example, we will realize an SI integrator using circuit of the transfer function given as follows:

$$\frac{Y}{X} = \frac{1}{z-1} \tag{16.10}$$

By modifying the transfer function into the form

$$Y = \left(\frac{1}{z}\right)(X+Y) \tag{16.11}$$

the SFG expressions of the lossless SI integrator are given as shown in Fig. 16.15.

Such an SFG can be directly mapped to the VLSI SI circuit using the unit delay circuit as shown in Fig. 16.16.

Note that the integrator-based SI filter circuit implementation is not the simplest form of the VLSI SI filter circuit implementation. In the second SI circuit design example, we would like to design a second-order low-pass VLSI SI filter where $\omega_o = 100$ kHz and $Q = 10$ in the following transfer function:

$$\frac{Y}{X} = \frac{1}{\left(\frac{s}{\omega_o}\right)^2 + \frac{1}{Q}\left(\frac{s}{\omega_o}\right) + 1} \tag{16.12}$$

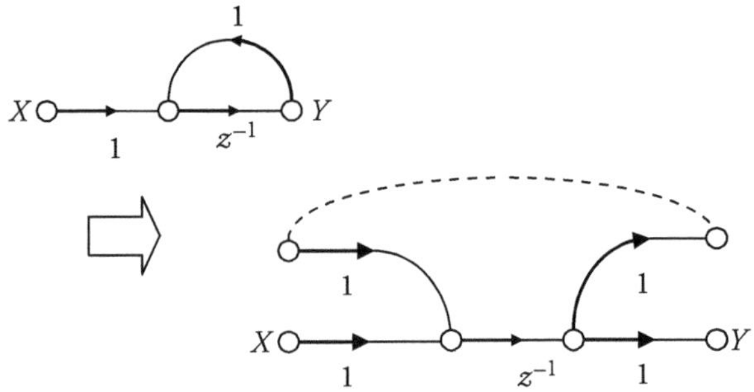

FIGURE 16.15 SFG expressions based on the unit delay elements.

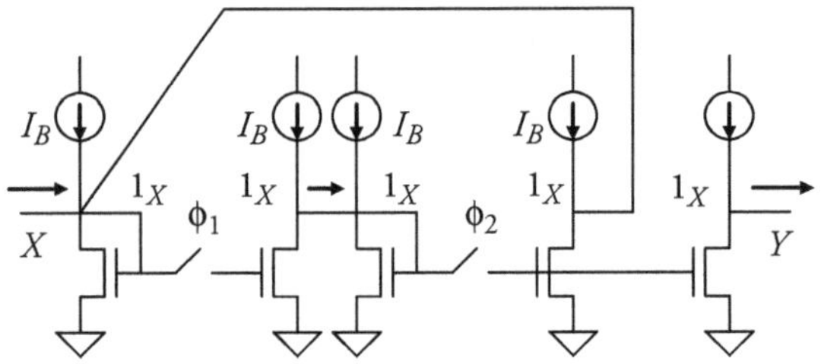

FIGURE 16.16 SI realization of forward Euler integration.

We can start the design of the prototype filter of the above transfer function as follows:

$$\frac{Y}{X} = \frac{1}{s^2 + \frac{1}{Q}s + 1} \tag{16.13}$$

The SFG expression shown in Fig. 16.17 can be developed from the above transfer function.

Mapping from the s-domain to the z-domain using the SI integrator as the building block, we have Figs. 16.18 and 16.19.

The above SFG can be directly mapped into the VLSI SI circuit using the integrator as shown in Fig. 16.20.

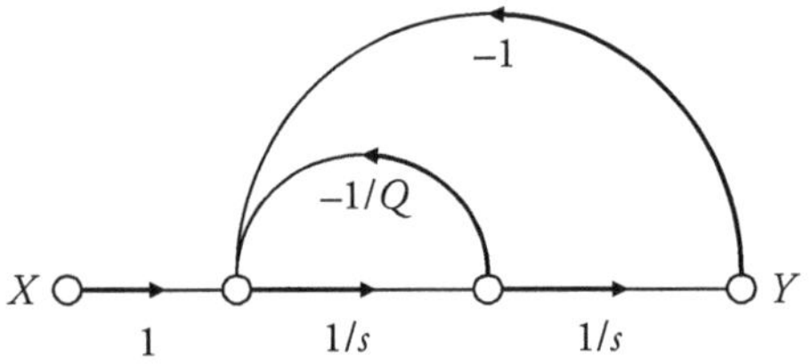

FIGURE 16.17 SFG expression of normalized second-order low-pass filter.

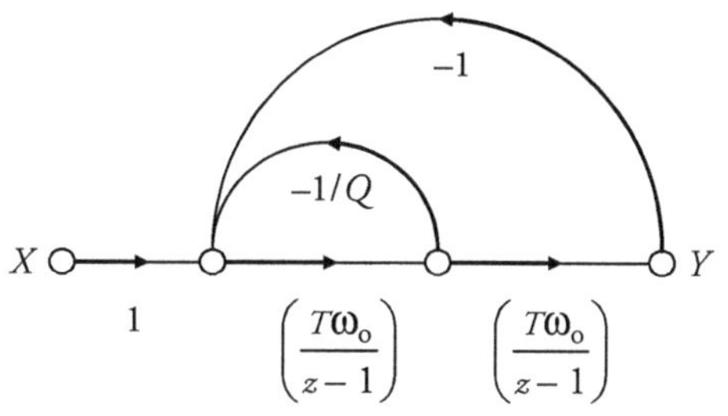

FIGURE 16.18 Z-domain filter SFG based on the s-to-z transformation.

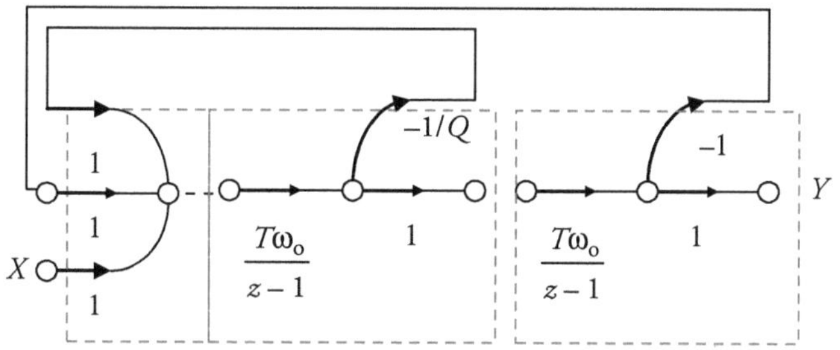

FIGURE 16.19 Partitioning of SFG for SI implementation.

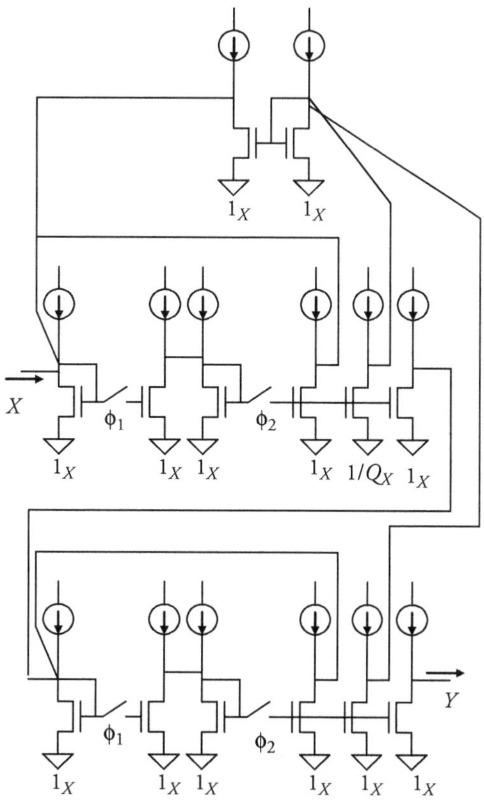

FIGURE 16.20 VLSI SI realization of the second-order filter.

References

1. A. Ng, J. I. Sewell, "Switched-Current Techniques under Pressure: High Order and Wideband Filter Design," *IEE Colloquium on Analog Signal Processing*, pp. 4/1–411, October 1998.
2. Y. Lu, J. I. Sewell, "First or Second Generation SI Cells? A Comparison of Sensitivity from an SI Filter System Viewpoint," *IEE Colloquium on Digital and Analogue Filters and Filtering Systems*, pp. 11/1–11/5, November 1994.
3. J. A. Barby, "Switched-Current Filter Models for Frequency Analysis in the Continuous-Time Domain," *ISCAS'93*, vol. 2, pp. 1427–1430, May 1993.
4. Y. Lu, J. I. Sewell, "Multirate SC and SI Filter System Design by XFILT," *ISCAS'95*, vol. 3, pp. 2257–2260, May 1995.
5. Z. Q. Shang, J. I. Sewell, "Efficient Analysis of Some Nonlinearities in SC and SI Filter Networks," *IEE Colloquium on Digital and Analogue Filters and Filtering Systems*, pp. 10/1–10/5, November 1994.
6. G. W. Roberts, A. S. Sedra, "Synthesizing Switched-Current Filters by Transposing the SFG of Switched-Capacitor Filter Circuits," *IEEE Trans. on Circuits and Systems*, vol. 38, no. 3, pp. 337–340, March 1991.
7. B. Jonsson, S. Eriksson, "A Low Voltage Wave SI Filter Implementation Using Improved Delay Elements," *ISCAS'94*, vol. 5, pp. 305–308, June 1994.
8. M. Goldenberg, R. Croman, T. S. Fiez, "Accurate SI Filters Using RGC Integrators," *IEEE J. Solid-State Circuits*, vol. 29, no. 11, pp. 1388–1395, November 1994.
9. T. S. Fiez, D. J. Allstot, "CMOS Switched-Current Ladder Filters," *IEEE J. Solid-State Circuits*, vol. 25, no. 6, pp. 1360–1367, December 1990.
10. B. Jonsson, S. Eriksson, "Current-Mode N-Port Adaptors for Wave Si Filters," *Electronics Letters*, vol. 29, no. 10, pp. 925–926, May 1993.
11. M. H. Fino, L. J. Mourao, "SymbSI-A Program for the Symbolic Signal Flow Graph Generation of Switched Current Circuits," *1998 IEEE International Conference on Electronics, Circuits and Systems*, vol. 3, pp. 211–214, September 1998.
12. M. Helfenstein, A. Muralt, G. S. Moschytz, "Direct Analysis of Multiphase Switched-Current Networks Using Signal-Flow Graphs," *ISCAS'95*, vol. 2, pp. 1476–1479, May 1995.
13. M. Helfenstein, A. Muralt, G. Fischer, P. Zbinden, D. Pfaff, F. Frey, G. S. Moschytzl, "SC and SI Filters in Baseband Applications: A Comparison," *ISCAS'97*, vol. 1, pp. 297–300, June 1997.
14. A. Handkiewicz, P. Sniatala, M. Lukowiak, "Low-Voltage High-Performance Switched Current Memory Cell," *Proceedings, Tenth Annual IEEE International ASIC Conference and Exhibit*, pp. 12–16, September 1997.
15. J. L. Chan, S. S. Chung, "Universal Switched-Current Integrator Blocks for SI Filter Design," *Proceedings of the ASP-DAC'99*, vol. 1, pp. 261–264, January 1999.
16. L. J. Mourao, M. H. Fino, "Automatic Symbolic Evaluation of Nonideal Effects in SI Circuit Behavior," *Third International Workshop on Design of Mixed-Mode Integrated Circuits and Applications*, pp. 40–43, July 1999.
17. J. Schechtman, A. C. M. De Queiroz, L. P. Caloba, "A Practical Implementation Scheme for Component Simulation SI Filters," *Proceedings., Proceedings of the 38th Midwest Circuits and Systems*, vol. 1, pp. 174–177, August 1995.
18. J. B. Hughes, K. W. Moulding, "A Switched-Current Double Sampling Bilinear z-Transform Filter Technique," *ISCAS'94*, vol. 5, pp. 293–296, June 1994.

19. N. C. Battersby, C. Toumazou, "Towards High Frequency Switched-Current Filters in CMOS and GaAs Technology," *ISCAS'93*, vol. 2, pp. 1239–1242, May 1993.
20. T. Fiez, D. Allstot, "A CMOS Switched-Current Filter Technique," *Digest of Technical Papers, 37th ISSCC*, pp. 206–207, February 1990.
21. G. K. Balachandran, P. E. Allen, "Switched-Current Circuits in Digital CMOS Technology with Low Charge-Injection Errors," *IEEE J. Solid-State Circuits*, vol. 37, no. 10, pp. 1271–1281, October 2002.
22. B. Raahemi, A. Opal, "Group Delay and Group Delay Sensitivity of Periodically Switched Linear Networks," *IEEE Trans. on Circuits and Systems I: Fundamental Theory and Applications*, vol. 47, no. 1, pp. 96–104, January 2000.
23. T.-H. Kuo, S.-Y. Lee, K.-D. Chen, "Offset Current Cancellation Based on a Multiple-Path Feedback Compensation (MPFC) Technique for Switched-current Circuits and Systems," *IEEE Trans. on Circuits and Systems II: Analog and Digital Signal Processing*, vol. 44, no. 4, pp. 299–309, April 1997.
24. A. Yufera, A. Rueda, "Studying the Effects of Mismatching and Clock-Feed Through in Switched-Current Filters Using Behavioral Simulation," *IEEE Trans. on Circuits and Systems II: Analog and Digital Signal Processing*, vol. 44, no. 12, pp. 1058–1067, December 1997.
25. A. Worapishet, R. Sitdhikorn, B. J. Hughes, "Architecture for Low-Voltage Switched-Current Complex Bandpass Filters," *Electronics Letters*, vol. 38, no. 12, pp. 535–536, June 2002.
26. A. E. J. Ng, J. I. Sewell, "Ladder Decompositions for Wideband SI Filter Applications," *IEE Proceedings, Circuits, Devices and Systems*, vol. 145, no. 5, pp. 306–313, October 1998.
27. T.-H. Kuo, S.-Y. Lee, D.-J. Lu, T. Jih, J.-J. Tsaur, "3.3 V Mixed-Mode IC Design Using Switched-Current Techniques for Speech Applications," *IEE Proceedings Circuits, Devices and Systems*, vol. 144, no. 6, pp. 367–374, December 1997.
28. K.-D. Chen, T.-H. Kuo, S.-H. Lee, "A Novel Offset Current Cancellation Technique for Switched-Current Circuits and Systems," *ISCAS'96*, vol. 1, pp. 417–420, May 1996.
29. A. Handkiewicz, M. Kropidlowski, M. Lukowiak, "Computer Tools for Switched-Current Filter Design," *MWSCAS-2002*, vol. 2, pp. II-501–II-504, August 2002.
30. Q. Gao, S. Qin, X. Jia, Y. Song, "Computer Aided Design of Switched-Current Filters," *4th International Conference on ASIC*, pp. 94–97, October 2001.
31. D. L. Shlemin, "A Design Technique for High Q-factor SI Filters," *APEIE-98*, vol. 1, pp. 142-152, September 1998.
32. I. Yusim, G. Ionis, K. Suyama, "Simulator for Switched-Current Integrated Circuits," *Proceedings of the IEEE 1998 Custom Integrated Circuits Conference*, pp. 459–462, May 1998.
33. A. Yufera, A. Rueda, J. L. Huertas, "Switched-Current Wave Analog Filters," *ISCAS'92*, vol. 2, pp. 859–862, May 1992.
34. T. S. Fiez, B. Lee, D. J. Allstot, "CMOS Switched-Current Biquadratic Filters," *IEEE International Symposium on Circuits and Systems*, vol. 3, pp. 2300–2303, May 1990.
35. J. B. Hughes, K. W. Moulding, J. Richardson, J. Bennett, W. Redman-White, M. Bracey, R. S. Soin, "Automated Design of Switched-Current Filters," *IEEE J. Solid-State Circuits*, vol. 31, no. 7, pp. 898–907, July 1996.
36. T. S. Fiez, G. Liang, D. J. Allstot, "Switched-Current Circuit Design Issues," *IEEE J. Solid-State Circuits*, vol. 26, no. 3, pp. 192–202, March 1991.

37. J. B. Hughes, "Top-Down Design of a Switched-Current Video Filter," *IEE Proceedings, Circuits, Devices and Systems*, vol. 147, no. 1, pp. 73–81, February 2000.
38. A. Handkiewicz, M. Kropidlowski, M. Lukowiak, M. Bartkowiak, "Switched-Current Filter Design for Image Processing Systems," *13th Annual IEEE International ASIC/SOC Conference*, pp. 8-12, September 2000.
39. R. T. Goncalves, S. Noceti Filho, M. C. Schneider, C. Galup-Montoro, "Digitally Programmable Switched Current Filters," *ISCAS'96*, vol. 1, pp. 258–261, May 1996.
40. A. D. Remenyuk, E. V. Astrova, L. I. Korovin, I. G. Lang, V. B. Shuman, "Diffraction Effects in Macroporous Silicon with Through Pores," *Proceedings of 2003 5th International Conference on Transparent Optical Networks*, vol. 1, pp. 273–275, July 2003.
41. X. Li, X. Zeng, D. Zhou, X. Ling, "Behavioral Modeling of Analog Circuits by Wavelet Collocation Method," *ACM IEEE International Conference on Computer Aided Design*, pp. 65–69, November 2001.
42. Y. Lu, J. I. Sewell, "A Systematic Approach for Ladder Based Switched Current Filter Design," *ISCAS'95*, vol. 3, pp. 2253–2256, May 1995.
43. R. T. Goncalves, S. N. Filho, M. C. Schneider, C. Galup-Montoro, "Programmable Switched Current Filters Using MOSFET-Only Current Dividers," *Proceedings of the 38th Midwest Circuits and Systems*, vol. 2, pp. 1046–1049, August 1995.
44. A. Yufera, A. Rueda, J. L. Huertas, "A Study of the Sensitivity of Switched-Current Wave Analog Filters to Mismatching and Clock-Feed through Errors," *ISCAS'94*, vol. 5, pp. 317–320, June 1994.

CHAPTER 17

Mixed-Mode VLSI Analog Signal Processing Circuit Configurations

In previous chapters, we have discussed the VLSI voltage-mode signal processing operations based on the VLSI active-RC, MOS-C, Gm-C, and SC filter circuits and the VLSI current-mode signal processing operations based on the VLSI CTI and SI filter circuits. These approaches represent the commonly used methods in the VLSI filter circuit implementations. Historically, the VLSI active-RC, MOS-C, Gm-C, and SC filters are typically viewed as voltage-mode circuits and the VLSI CTI and SI filters as current-mode circuits. However, as seen in this chapter, either voltage or current can be used to represent signals in all types of VLSI signal processing circuit implementations. On the other hand, in many practical system design applications of VLSI signal processing circuits, both the voltage and the current signal formats are needed in the signal processing operations.

17.1 Mixed-Mode Signal Processing Systems

Two important features of the VLSI voltage- and current-mode circuits are the wired voltage branching and the wired current addition operations. As shown in Fig. 17.1, a voltage signal can, in principle, be directly branched to multiple locations without requiring additional circuit components; and multiple current signals can be added by directly connecting together without requiring additional circuit components. These properties provide optimized VLSI circuit-implementation solutions to the VLSI signal processing system implementations.

It can be seen in Fig. 17.2 that the wired voltage signal branching and the wired current signal addition operations are directly correlated to the default branch output and the addition (merge) operations of the SFG expressions.

From the application viewpoint, either the voltage-mode or the current-mode signals may be required in the VLSI signal processing operation based on the nature and the approach of the VLSI circuit and system implementations (see Fig. 17.3).

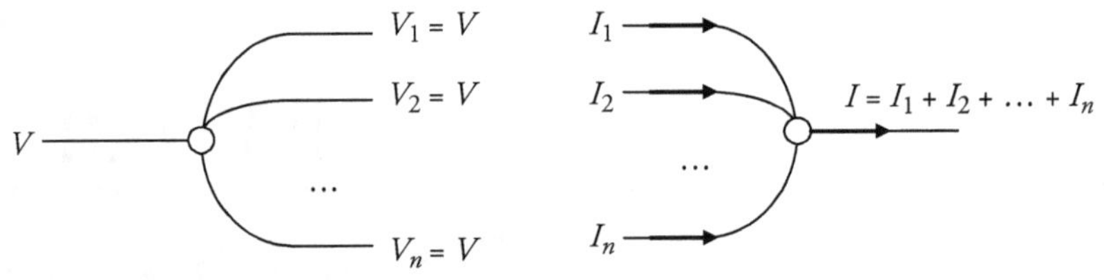

(a) The wired voltage branching operation

(b) The wired current addition operation

FIGURE 17.1 Basic VLSI wired voltage branching and current addition operations.

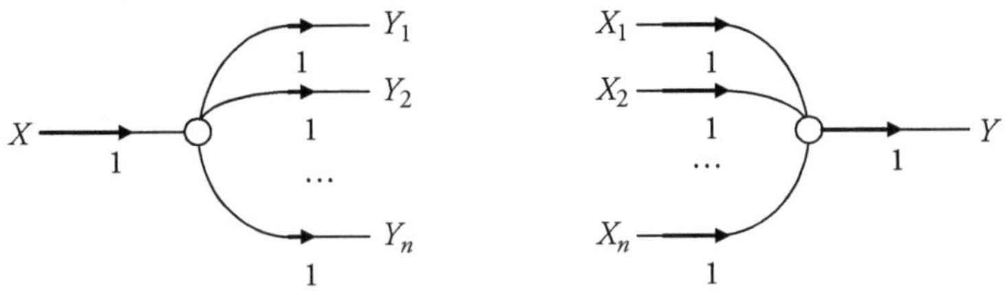

(a) The branching in SFG

(b) The addition in SFG

FIGURE 17.2 Basic signal branching and wired addition SFG configurations.

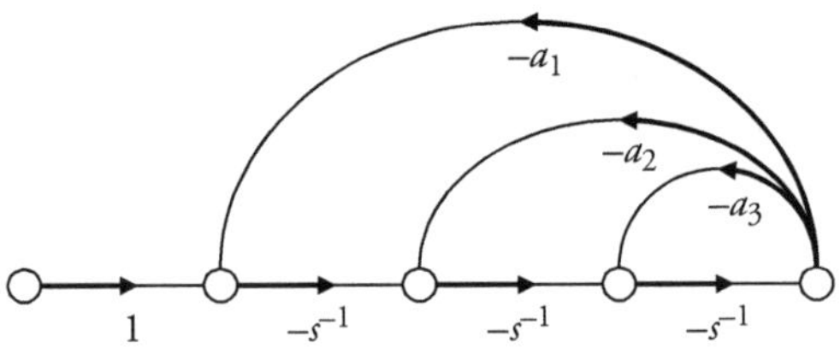

FIGURE 17.3 A typical SFG of CT signal processing system.

17.2 Basic VLSI V/I and I/V Circuit Structures

It is interesting to see that most VLSI voltage- or current-mode active signal processing circuit building blocks can be further decomposed into the more fundamental building circuit structures of the voltage-to-current (V/I) and current-to-voltage (I/V) converter. Based on the combination of these basic VLSI V/I and I/V converter circuit elements, the VLSI filter circuit in a specific VLSI circuit technology can be implemented either in the voltage mode or in the current mode.

Figure 17.4 shows the V/I and I/V decompositions of the VLSI active-RC signal processing circuit structures.

Figure 17.5 shows active-RC circuit implementations of the I/V conversion circuits.

Similarly, the V/I and I/V converter circuit structures in the VLSI Gm-C signal processing circuits are shown in Figs. 17.7 and 17.8, respectively.

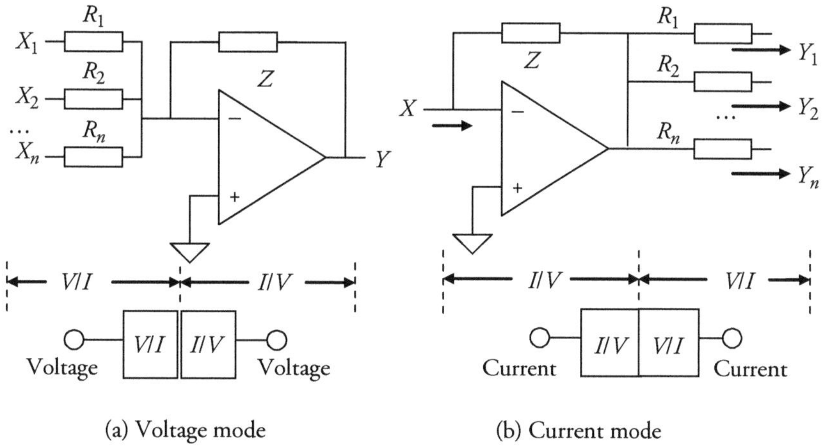

FIGURE 17.4 V/I and I/V decompositions of VLSI active-RC circuits.

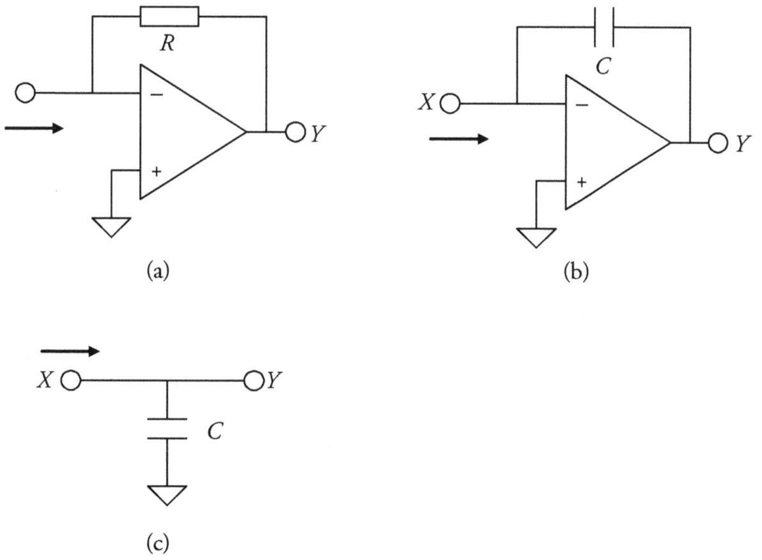

FIGURE 17.5 Basic VLSI active-RC I/V conversion circuit structures.

FIGURE 17.6 Basic VLSI active-RC V/I converter circuit structures.

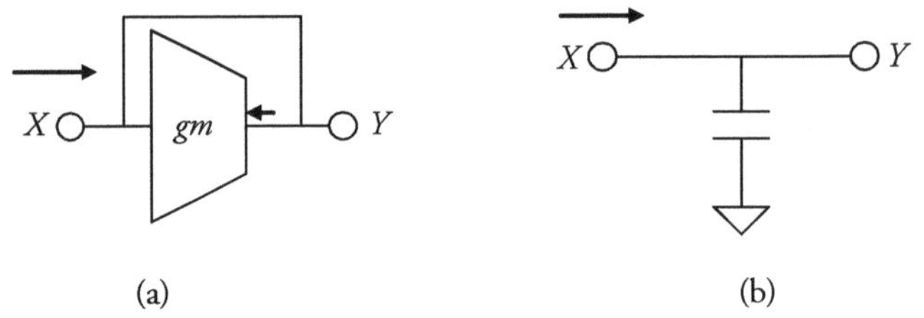

Figure 17.7 Basic VLSI Gm-C *I*/*V* converter circuit structures.

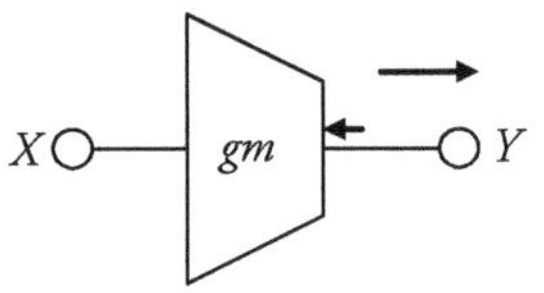

Figure 17.8 Basic VLSI Gm-C *V*/*I* converter circuit structure.

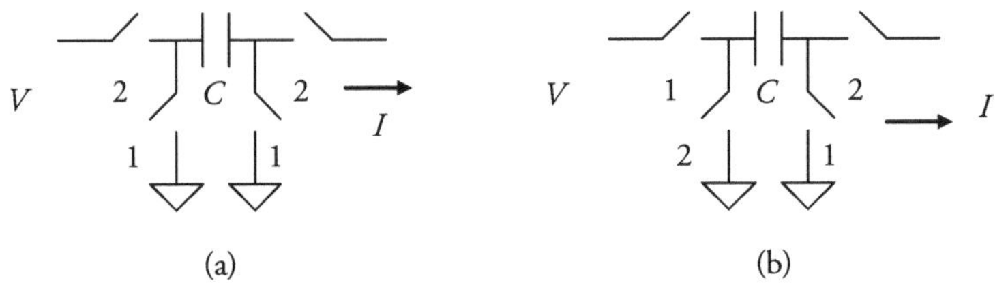

Figure 17.9 Basic VLSI SC *V*/*I* converter circuit structures.

The V/I and I/V structures for the VLSI SC signal processing circuits are shown in Figs. 17.9 and 17.10, respectively.

The I/V and V/I circuit elements for the VLSI CTI and SI signal processing circuits are shown in Figs. 17.11 and 17.12, respectively.

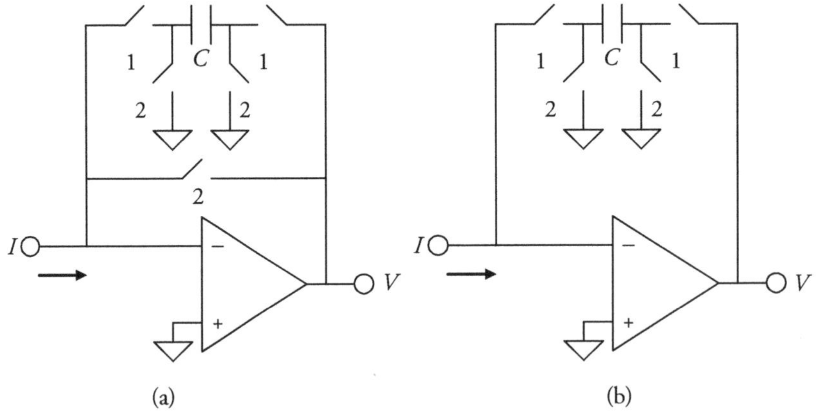

Figure 17.10 Basic VLSI SC *I*/*V* converter circuit structures.

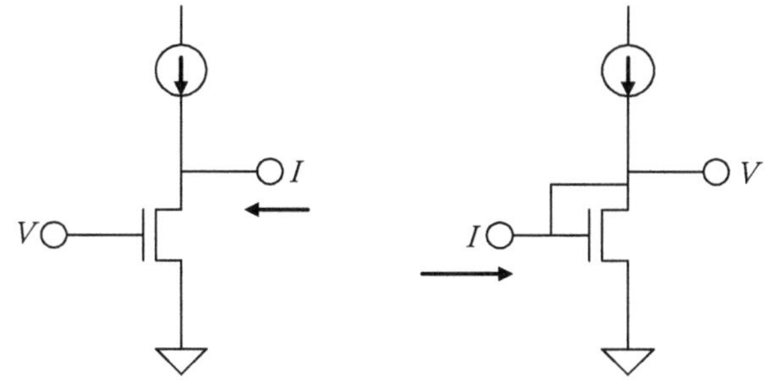

FIGURE 17.11 Basic VLSI CTI V/I and I/V converter circuit structures.

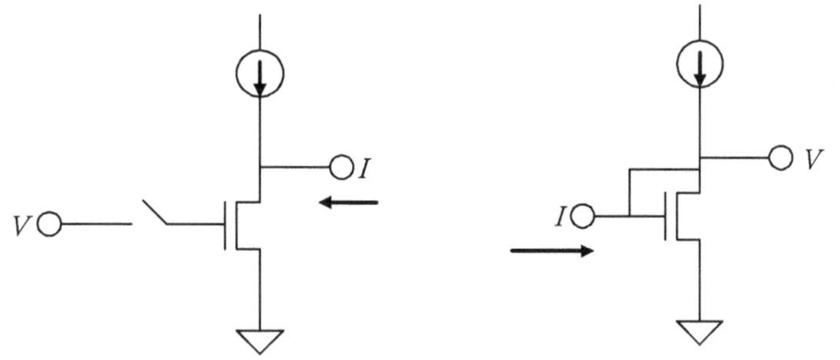

FIGURE 17.12 Basic VLSI SI V/I and I/V converter circuit structures.

With the V/I and I/V element decomposition of the VLSI circuits, it is easy to implement the signal processing circuit in either the voltage-mode or the current-mode configurations as shown in Fig. 17.13.

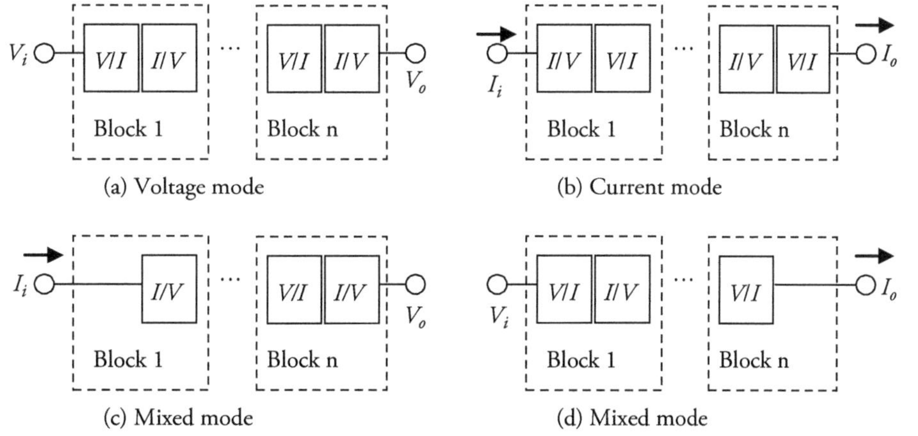

FIGURE 17.13 Mixed-mode VLSI signal processing circuit configurations.

17.3 Current-Mode-Based Design of Active-RC Filters

Typically, signals in VLSI active-RC circuits are represented in the voltage form. Such a representation provides a straightforward way for the active-RC filter circuit design and analysis. However, VLSI active-RC filter circuits can also be designed and analyzed in the current-mode form. For the VLSI active-RC filter shown in Fig. 17.14, the input and output signals of all blocks are in voltage forms. Alternatively, the same active-RC circuit can be implemented using the current-mode building block circuits shown in Fig. 17.15.

The above example demonstrates the different selections of the basic building blocks. VLSI active-RC filters can be designed and analyzed in either voltage or current modes.

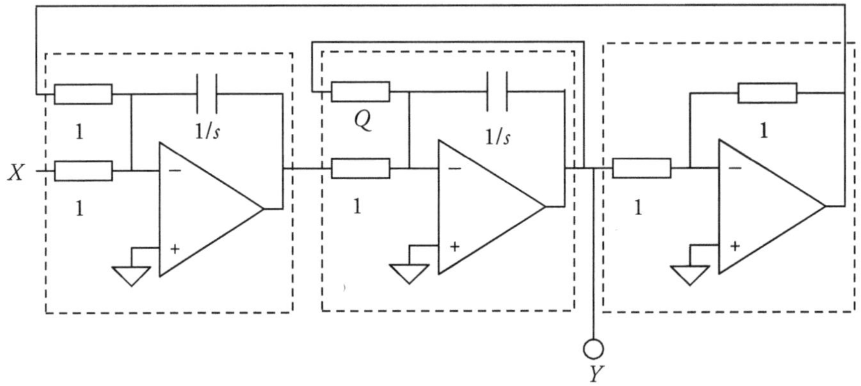

FIGURE 17.14 VLSI active-RC filter based on voltage-mode approach.

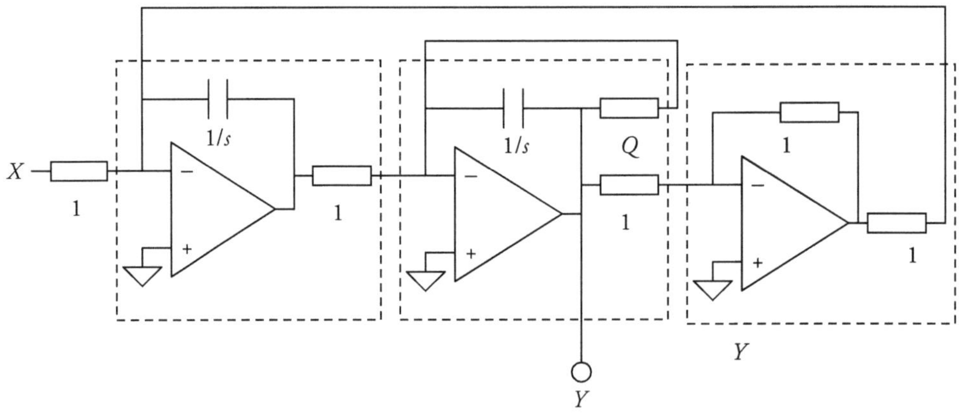

FIGURE 17.15 VLSI active-RC filter based on current-mode approach.

17.3.1 Current-Mode Active-RC Building Blocks

Figure 17.16 shows the current-mode VLSI active-RC circuit implementation and the SFG representation of the integration with weighted outputs. Note that it is assumed that the loads of the integrator circuit are also other current-mode active-RC circuits, which have low impedances. The transfer function of the building block is given as follows:

$$Y_i = -\frac{1}{R_i Cs} X_1 \ \{i = 1, 2, \ldots, n\} \tag{17.1}$$

Similarly, the VLSI current-mode active-RC scaling circuit with weighted outputs can also be constructed by the same way as shown in Fig. 17.17.

$$Y_i = -\frac{R}{R_i} X_1 \ \{i = 1, 2, \ldots, n\} \tag{17.2}$$

Similar to other current-mode circuits, the addition of a VLSI current-mode active-RC circuit can be realized through wired addition by connecting all input signals together as shown in Fig. 17.18. Note that for the wired current addition operation, all inputs can only have the unity weights; therefore, no scaling operation is allowed in such a case.

$$Y = (X_1 + X_2 + \cdots + X_n) \tag{17.3}$$

In many design applications, the signals are usually required to be converted between the voltage and current representations. The VLSI active-RC V/I converter can be realized simply using a resistor as shown in Fig. 17.19.

$$I = V/R \tag{17.4}$$

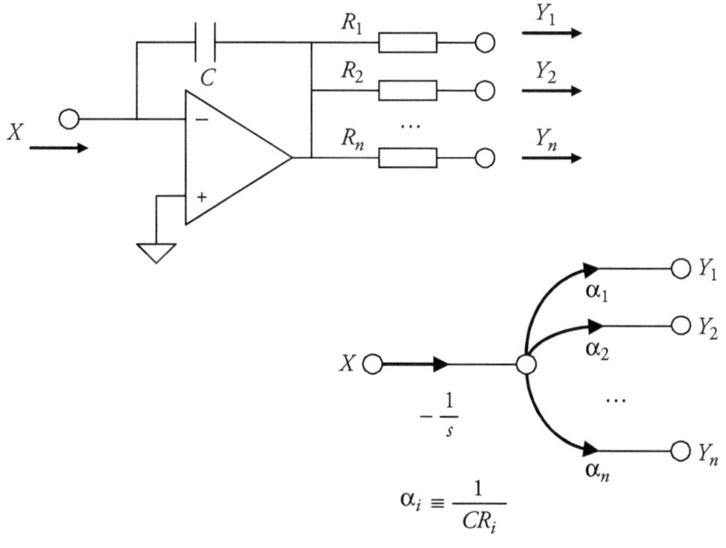

FIGURE 17.16 Current-mode VLSI active-RC integrator with weighted outputs.

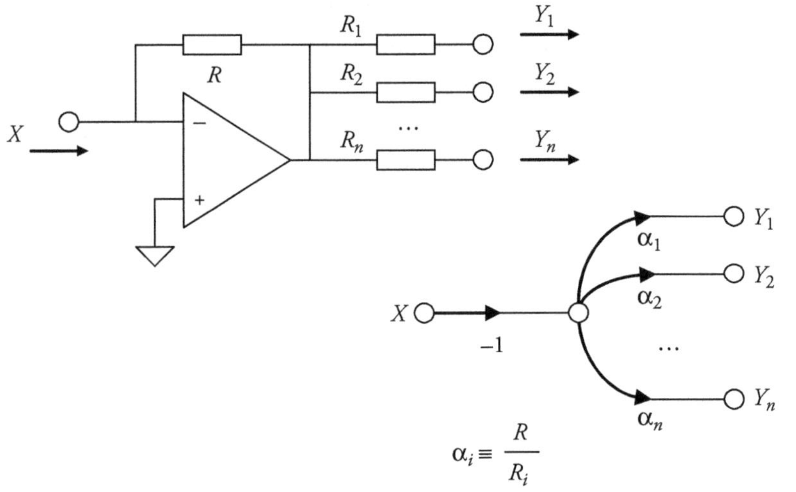

FIGURE 17.17 Current-mode VLSI active-RC implementation of scaling with weighted outputs.

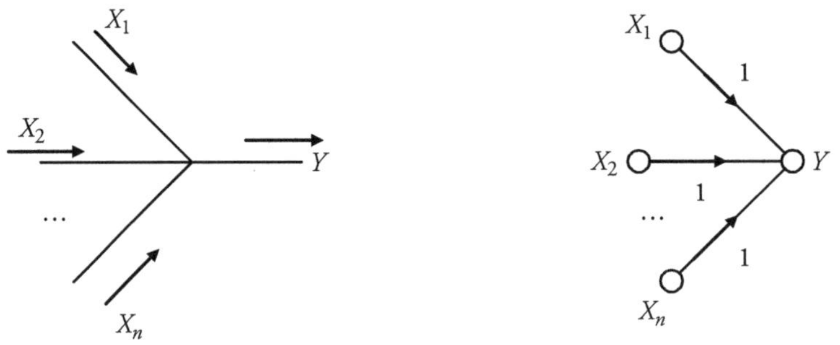

FIGURE 17.18 Current-mode addition.

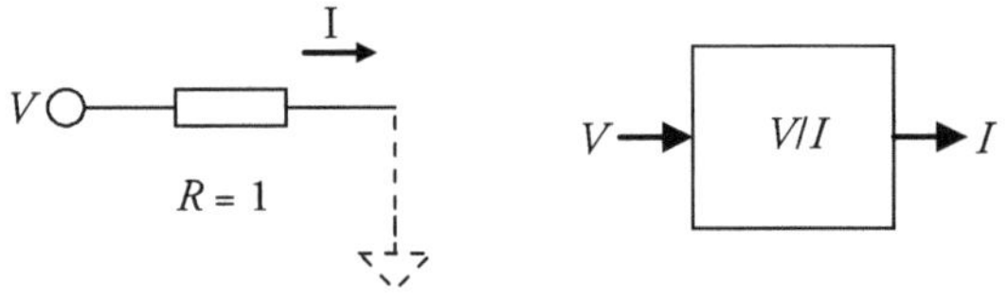

FIGURE 17.19 Active-RC V/I converter.

Converting a current signal back to a voltage in the active-RC circuit can be realized using an I/V converter as shown in Fig. 17.20.

$$V = -RI \tag{17.5}$$

For both the V/I and I/V conversion operations, the nonunity resistance R can be selected to realize both signal conversion and scaling simultaneously.

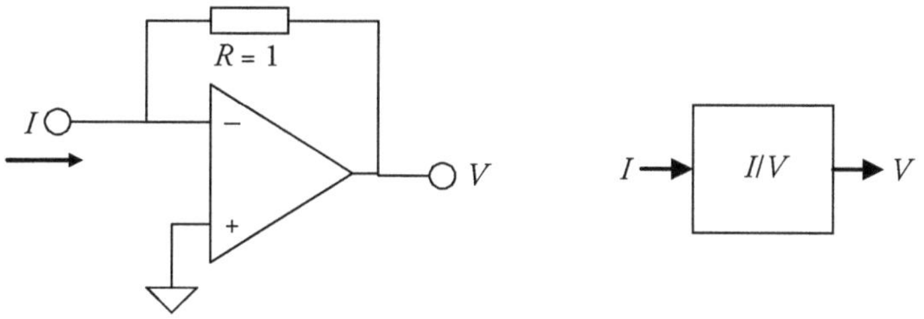

Figure 17.20 VLSI active-RC I/V converter.

17.3.2 Design Example

In this design example, we will demonstrate how to use the current-mode approach to realize the VLSI active-RC filter for the following transfer function:

$$H(s) = \frac{Y(s)}{X(s)} = \frac{s}{s^2 + \frac{1}{Q}s + 1} \tag{17.6}$$

The SFG expression for such a filter as shown in Fig. 17.21 can be easily constructed based on the method introduced in the previous chapters.

The above SFG can be further modified to the form shown in Fig. 17.22, which can be directly mapped into the VLSI current-mode active-RC filter circuit as shown in Fig. 17.23.

In the above circuit, both the input and output signals are in the current mode. By using the V/I and I/V conversion elements at the input and the output of the filter respectively, the filter circuit can be converted back to the voltage-mode circuit as shown in Fig. 17.24.

Considering the equivalent of the node Y and Y', the output branch of Y can be eliminated to save some circuits. It can be seen that the voltage-mode circuit constructed by this approach is exactly the same as the filter designed directly in the voltage mode in Chap. 11 as shown in Fig. 17.25.

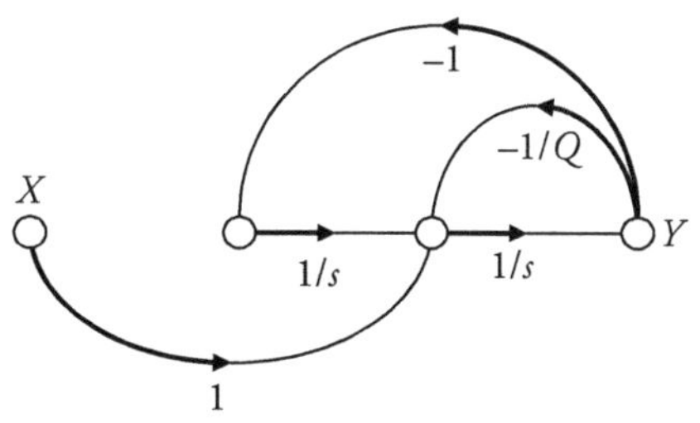

Figure 17.21 SFG of the transfer function of the design example.

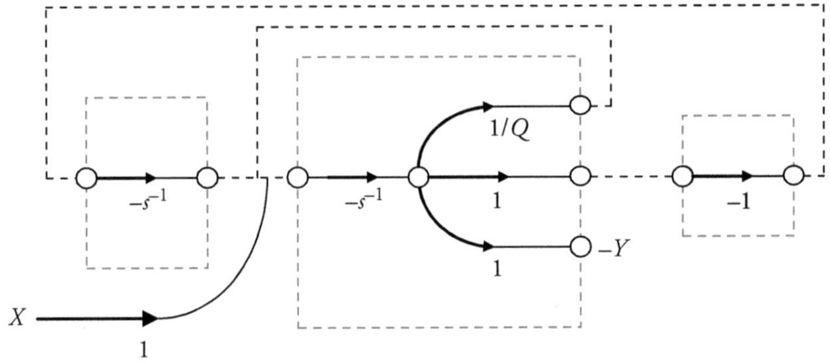

FIGURE 17.22 Modified SFG expression for current-mode active-RC realization.

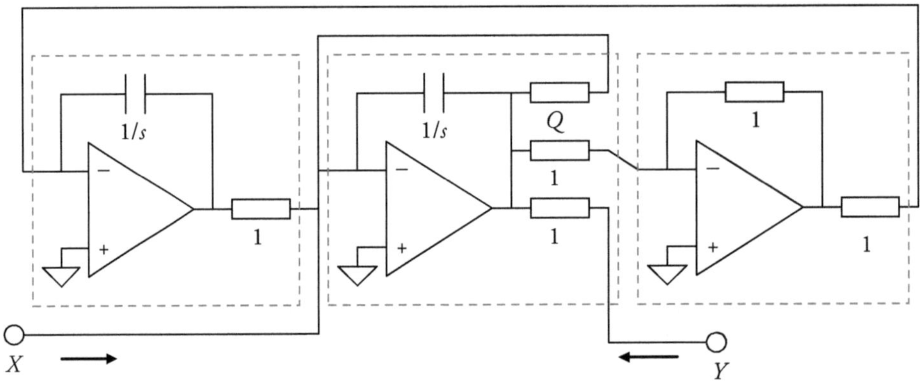

FIGURE 17.23 Current-mode realization of the given active-RC filter.

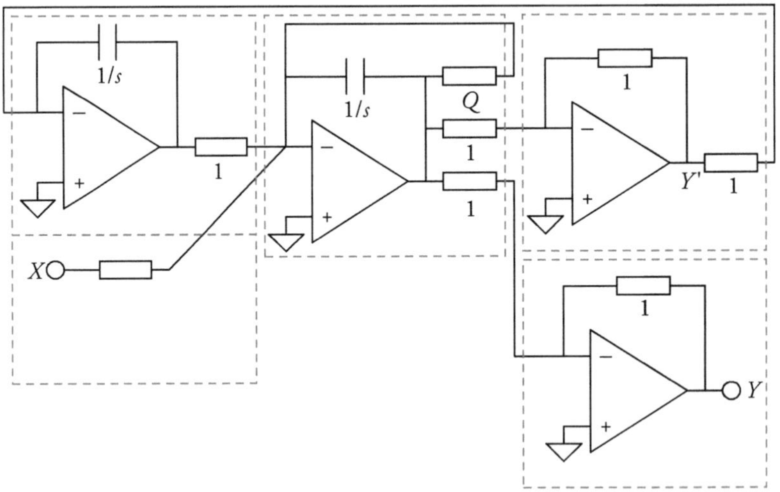

FIGURE 17.24 Conversion of current-mode active-RC filter to voltage mode.

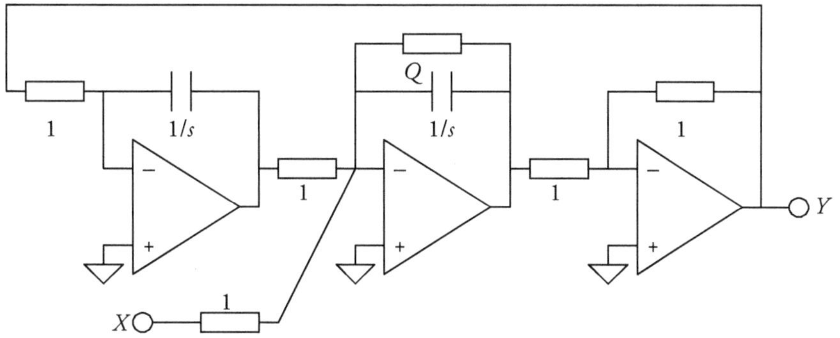

FIGURE 17.25 Voltage-mode active-RC filter converted from current mode.

17.4 Current-Mode-Based Design of Gm-C Filters

With some modifications of the basic circuit building blocks, signals in a VLSI Gm-C filter circuit can also be represented in the current form. This will convert a conventional voltage-mode Gm-C filter circuit into a current-mode Gm-C filter circuit.

17.4.1 Current-Mode Gm-C Filter Building Blocks

The Gm-C current-mode integration with weighted outputs is shown in Fig. 17.26.

$$Y_i = -\frac{1}{R_i Cs} X_1 \; \{i = 1, 2, \ldots, n\} \tag{17.7}$$

The Gm-C current-mode scaling circuit with weighted outputs is shown in Fig. 17.27.

$$Y_i = -\frac{g_{\mathrm{mi}}}{g_m} X \; \{i = 1, 2, \ldots, n\} \tag{17.8}$$

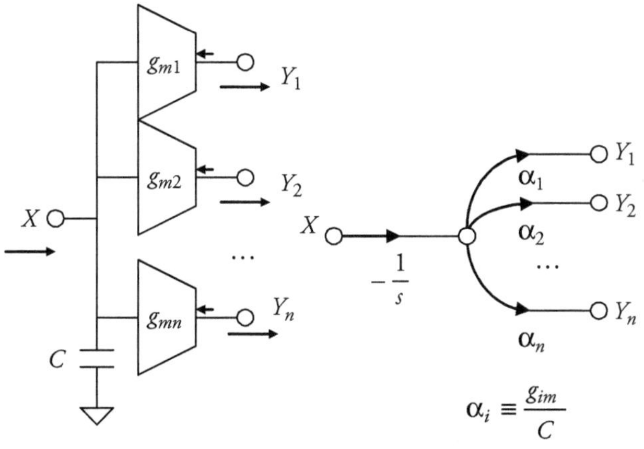

FIGURE 17.26 VLSI current-mode Gm-C integrator with weighted outputs.

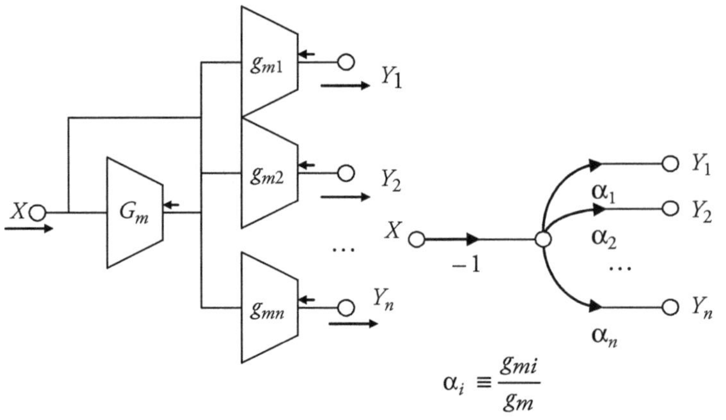

Figure 17.27 Current-mode Gm-C scaling circuit with weighted outputs.

The current-mode addition based on the Gm-C circuit can be realized exactly the same as other types of the circuit implementation through the wired addition of current-mode signals.

The Gm-C V/I and I/V elements are shown in Fig. 17.28.

$$I = -g_m \cdot V \tag{17.9}$$

The Gm-C I/V converter circuit can be realized as shown in Fig. 17.29.

$$V = \frac{1}{g_m} I \tag{17.10}$$

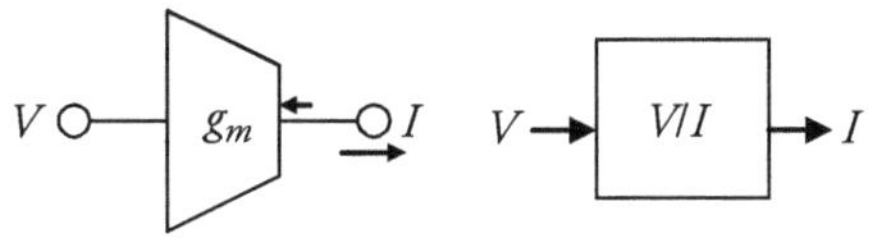

Figure 17.28 Gm-C V/I converter structures.

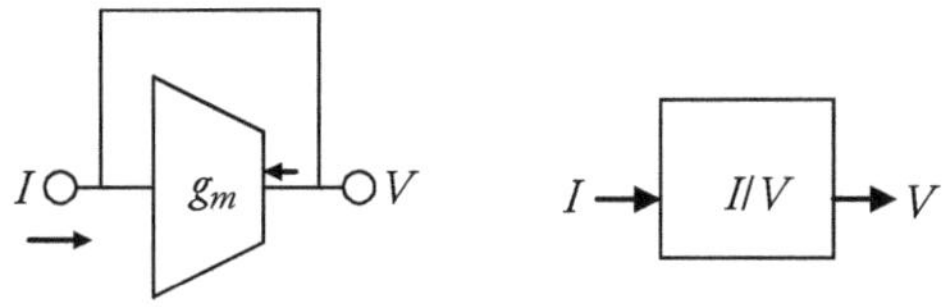

Figure 17.29 Gm-C I/V converter circuit.

17.4.2 Design Example

In this design example, we will demonstrate the use of the current-mode Gm-C circuit approach to realize the following filter transfer function:

$$H(s) = \frac{Y(s)}{X(s)} = \frac{s}{s^2 + \frac{1}{Q}s + 1} \tag{17.11}$$

The SFG expression of the filter can be constructed as shown in Fig. 17.30.

By modifying the above SFG to the form shown in Fig. 17.31, the SFG can be directly mapped into the current-mode Gm-C filter circuit as shown in Fig. 17.32.

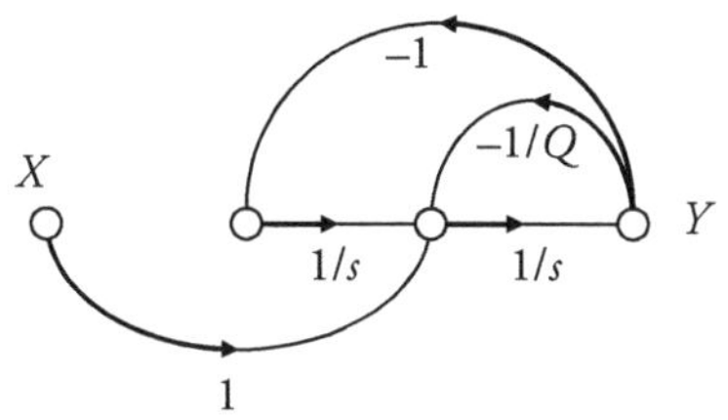

FIGURE 17.30 SFG expression of the given filter transfer function.

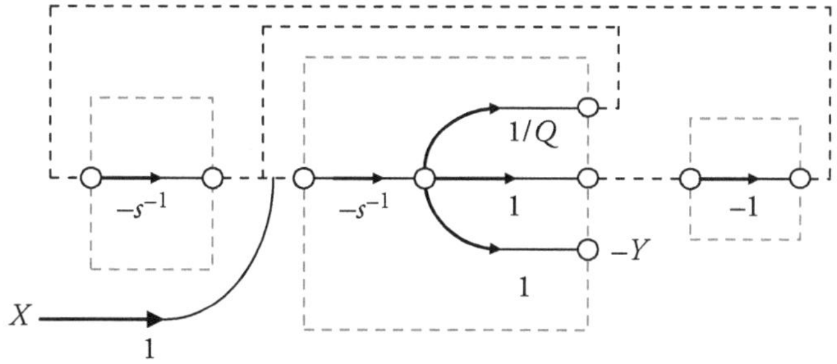

FIGURE 17.31 Alternative SFG for current-mode active-RC realization.

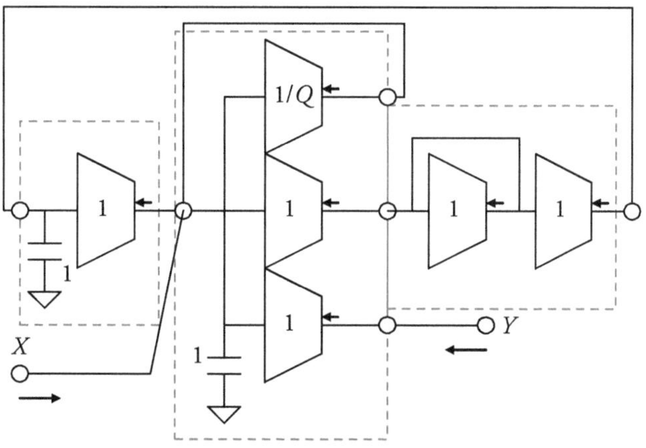

FIGURE 17.32 VLSI current-mode Gm-C circuit realization of the given filter.

17.5 Current-Mode-Based Design of SC Filters

VLSI SC filters are commonly viewed as voltage-mode circuits since the voltage-mode signals are usually used in the SC filters to represent their signal processing operations. However, similar to other VLSI filter structures, current signals can also be used to represent the signal processing operation of the SC filters.

17.5.1 Current-Mode Building Blocks of SC Filters

Figures 17.33 and 17.34 show the current-mode SC integration circuits with weighted outputs. Similar to the active-RC filters, the low impedances are assumed for the load nodes.

$$Y_i = -\frac{1}{R_i Cs} X_1 \; \{i = 1, 2, \ldots, n\} \tag{17.12}$$

$$Y_i = \frac{1}{R_i Cs} X_1 \quad \{i = 1, 2, \ldots, n\} \tag{17.13}$$

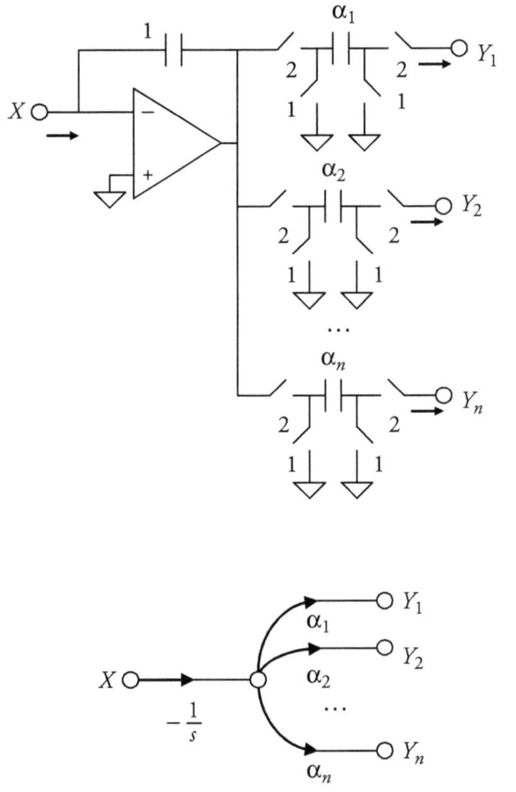

FIGURE 17.33 Current-mode SC inverted integration with weighted outputs.

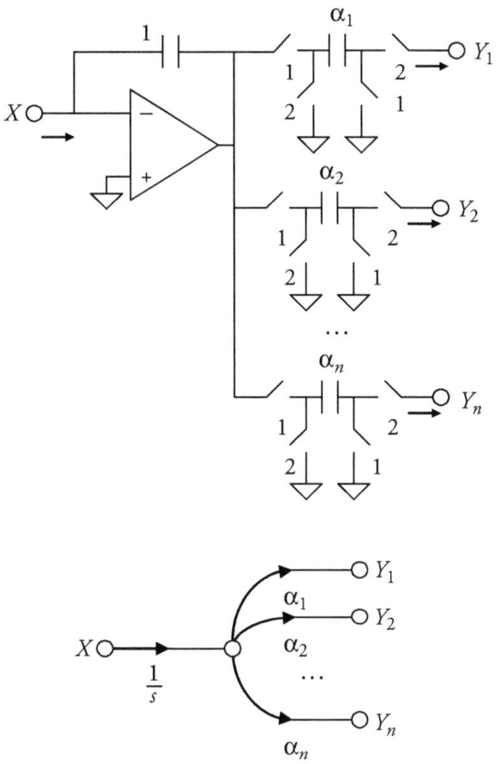

FIGURE 17.34 VLSI current-mode noninverted SC integration with weighted outputs.

Figures 17.35 and 17.36 show SC circuit implementations of current-mode scaling with weighted current branch out:

$$Y_i = \alpha_i X \quad \{i = 1, 2, \ldots, n\} \tag{17.14}$$

The VLSI SC current-mode addition operation can be realized through the wired addition operation. The SC V/I and I/V converter circuits are shown in Figs. 17.37 and 17.38, respectively.

17.5.2 Design Example

In this design example, we will demonstrate the use of the current-mode SC circuit design approach to realize the following transfer function:

$$H(s) = \frac{Y(s)}{X(s)} = \frac{s}{s^2 + \frac{1}{Q}s + 1} \tag{17.15}$$

The SFG expression of the filter can be constructed as shown in Fig. 17.39.

The above SFG expression can be modified to the form shown in Fig. 17.40, which can be directly mapped into the current-mode SC filter as shown in Fig. 17.41.

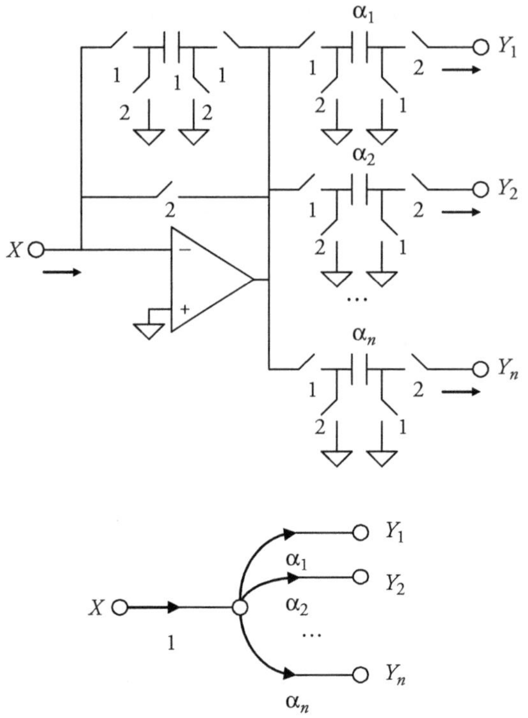

FIGURE 17.35 SC circuit realization of noninverted scaling with weighted outputs.

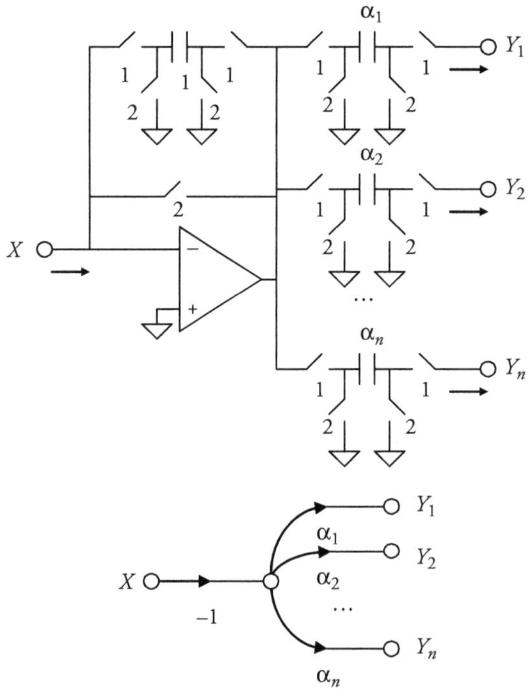

FIGURE 17.36 SC circuit realization of inverted scaling with weighted outputs.

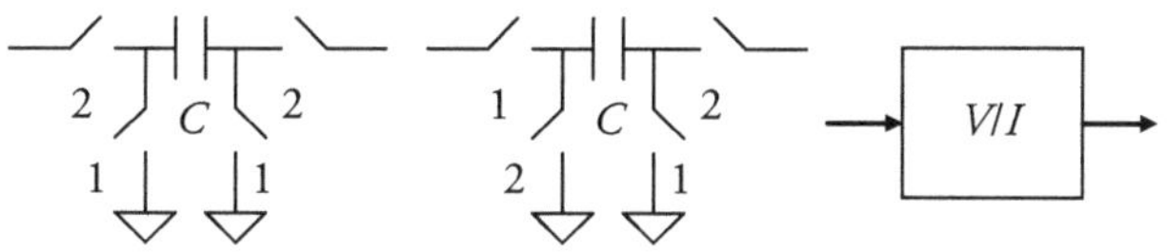

FIGURE 17.37 The VLSI SC V/I converter circuit.

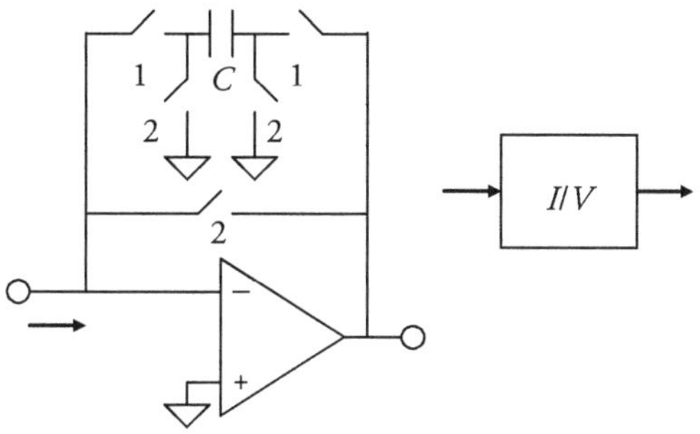

FIGURE 17.38 The VLSI SC I/V converter circuit.

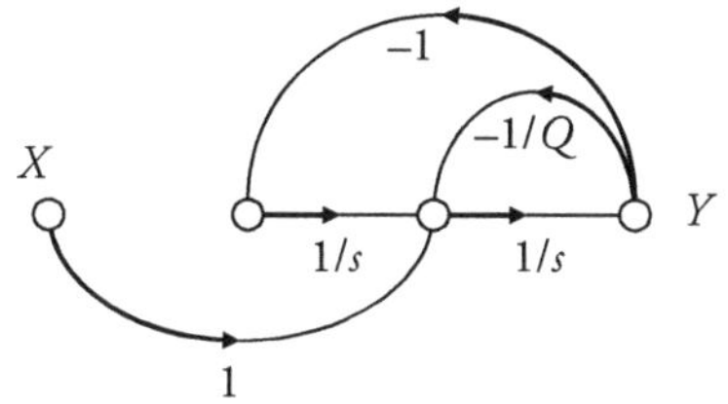

FIGURE 17.39 SFG expression of the transfer function of the design example.

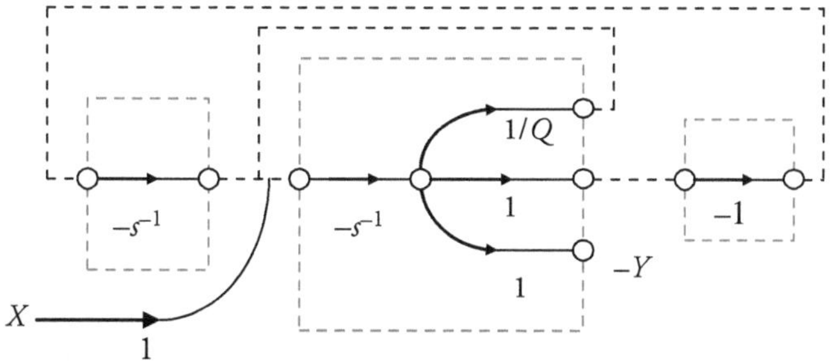

FIGURE 17.40 Modified SFG for current-mode SC circuit realization.

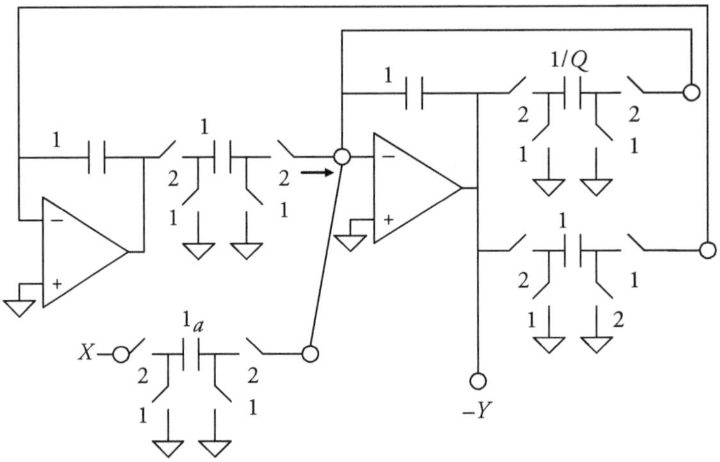

FIGURE 17.41 VLSI SC circuit implementation of given transfer function.

17.6 Voltage-Mode-Based Design of CTI Filters

VLSI CT current-mode (CTI) circuits are conventionally designed and analyzed in the current mode. However, similar to voltage-mode filters that can be designed in the current mode, CTI filters can also be designed and analyzed in the voltage mode with proper selection of the basic filter circuit building blocks.

17.6.1 Voltage-Mode CTI Filter Building Blocks

The voltage-mode CTI circuit implementation of a lossy integration with weighted inputs is shown in Fig. 17.42.

$$Y = \frac{-1}{1+s}(\alpha_1 X_1 + \alpha_2 X_2 + \cdots + \alpha_n X_n) \tag{17.16}$$

Similarly, CTI voltage-mode addition with weighted inputs can be realized as shown in Fig. 17.43.

$$Y = -(\alpha_1 X_1 + \alpha_2 X_2 + \cdots + \alpha_n X_n) \tag{17.17}$$

Scaling of the voltage can be realized using one branch of the above circuit shown in Fig. 17.44.

$$Y = -\alpha X \tag{17.18}$$

17.6.2 Design Example

In this design example, we will demonstrate the use of the voltage-mode CTI circuit design approach to realize the following second-order low-pass filter transfer function:

$$H(s) = \frac{Y(s)}{X(s)} = \frac{1}{s^2 + \frac{1}{Q}s + 1} \tag{17.19}$$

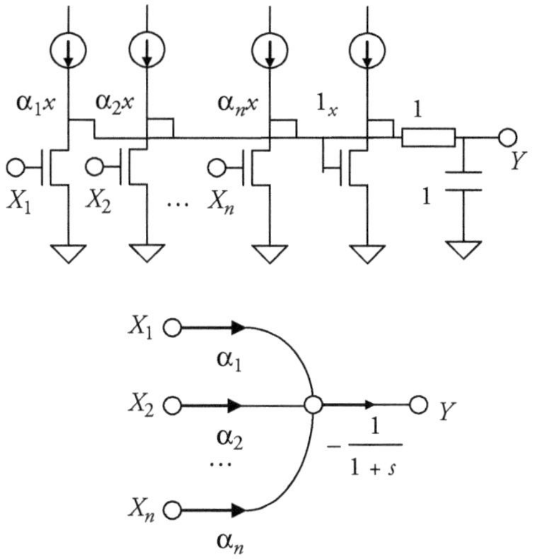

FIGURE 17.42 Voltage-mode CTI integration with weighted inputs.

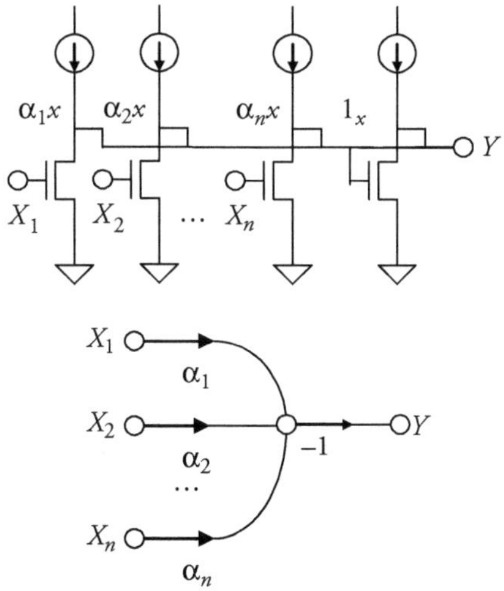

FIGURE 17.43 Voltage-mode CTI integration with weighted inputs.

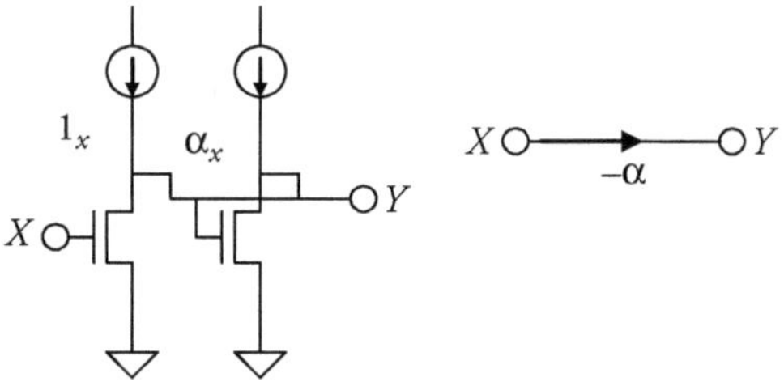

FIGURE 17.44 Voltage-mode CTI scaling circuit.

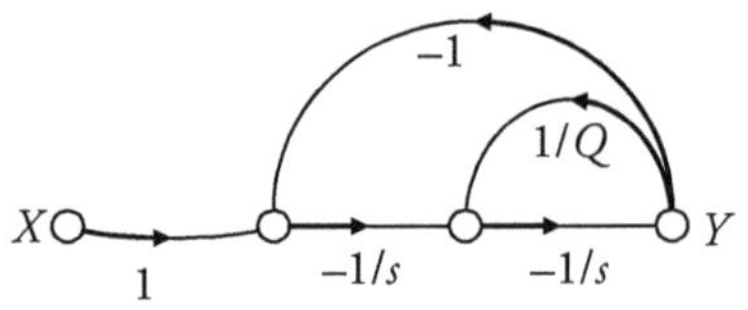

FIGURE 17.45 SFG expression of second-order low-pass filter section.

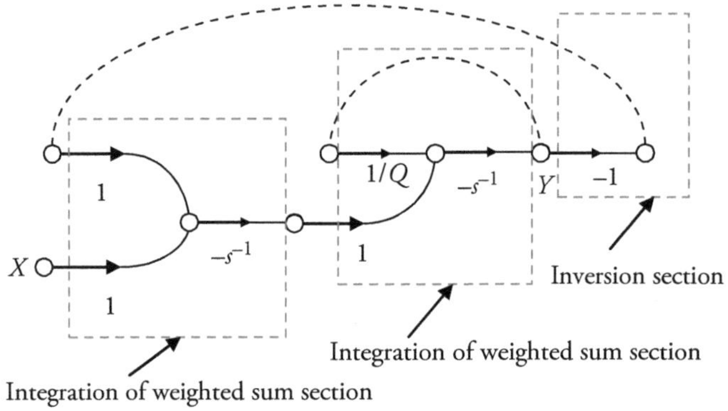

FIGURE 17.46 Normalized SFG suitable for MOS-C realization.

The SFG expression of the filter can be constructed as shown in Fig. 17.45.

The modified SFG for voltage-mode CTI filter implementation is shown in Fig. 17.46.

The SFG can be directly mapped into the CTI circuit as shown in Fig. 17.47.

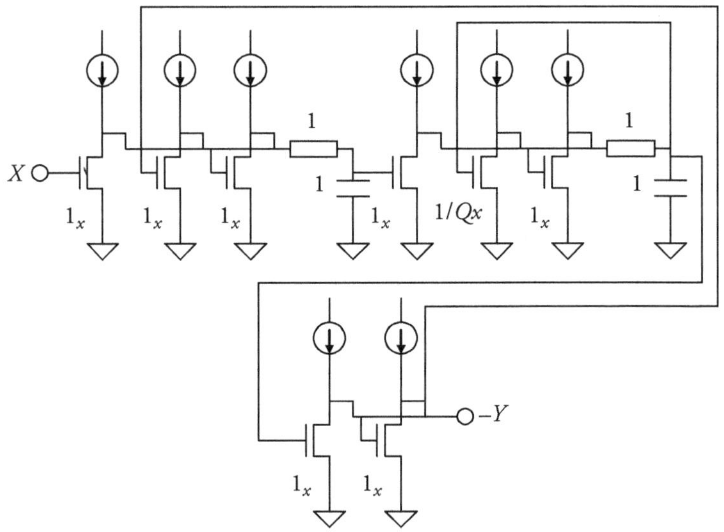

FIGURE 17.47 VLSI voltage-mode CTI filter implementation of second-order low-pass filter.

17.7 Voltage-Mode-Based Design of SI Filters

A voltage-mode SI circuit configuration is similar to the voltage-mode CTI circuit configurations except that the sampled circuit stage is used to realize the delay operation of the signal processing operation instead of the continuous-time circuit stage.

17.7.1 Voltage-Mode SI Filter Building Blocks

The voltage-mode SI unit delay with weighted inputs can be constructed as shown in Fig. 17.48.

A voltage-mode SI addition circuit with weighted inputs is exactly the same as the CTI circuit as shown in Fig. 17.49.

Scaling operation of the SI circuit can be directly realized based on the above addition with weighted inputs. In practical applications, scaling usually is realized with the delay or the addition to minimize the implementation device number.

17.7.2 Design Example

In this design example, we will demonstrate the use of the voltage-mode SI circuit design approach to realize the following filter transfer function:

$$H(z) = \frac{Y(z)}{X(z)} = \frac{1}{z^2 + az + 1} \tag{17.20}$$

The SFG expression of the filter as shown in Fig. 17.50 can be constructed based on the method introduced in the previous chapters.

The above SFG can be modified to the form shown in Fig. 17.51.

Figure 17.52 shows the voltage-mode VLSI SI circuit implementation of the transfer function given in the design example that can be obtained by direct mapping of the SFG shown in Fig. 17.51.

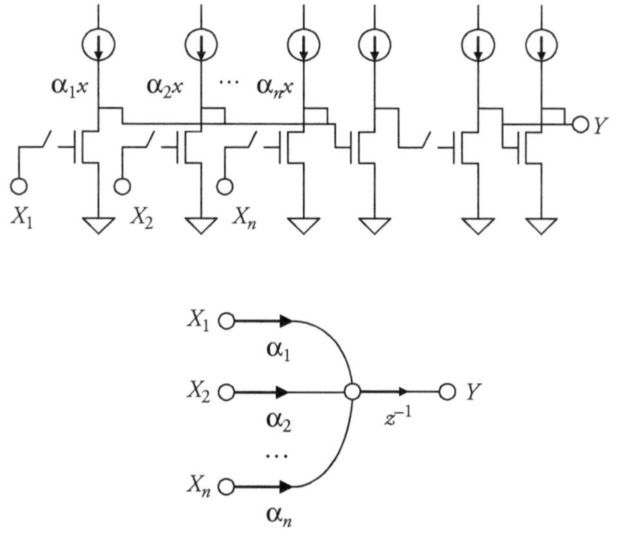

FIGURE 17.48 VLSI voltage-mode SI unit delay circuit with weighted inputs.

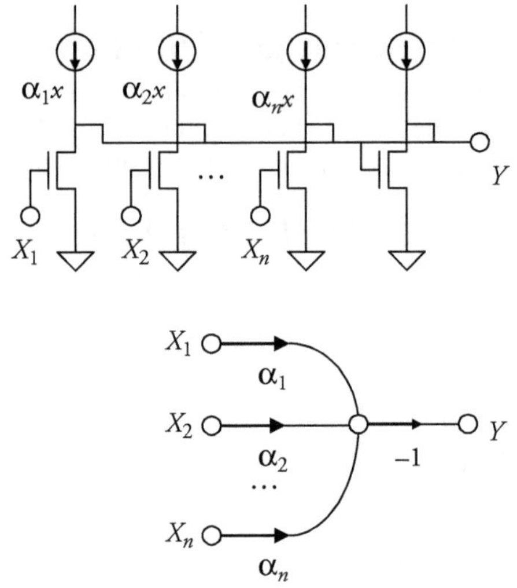

FIGURE 17.49 VLSI voltage-mode SI unit addition circuit with weighted input.

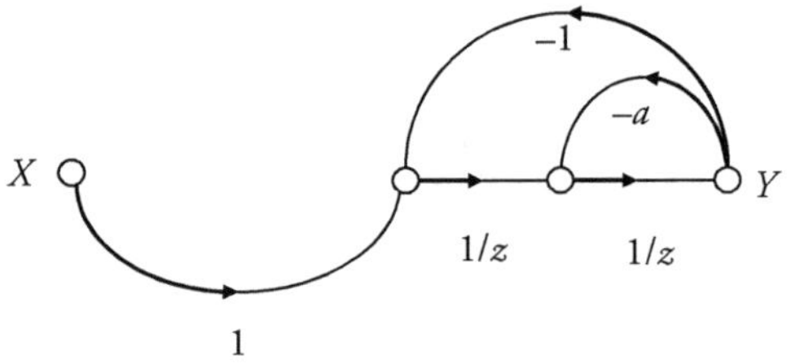

FIGURE 17.50 SFG expression of the transfer function of the design example.

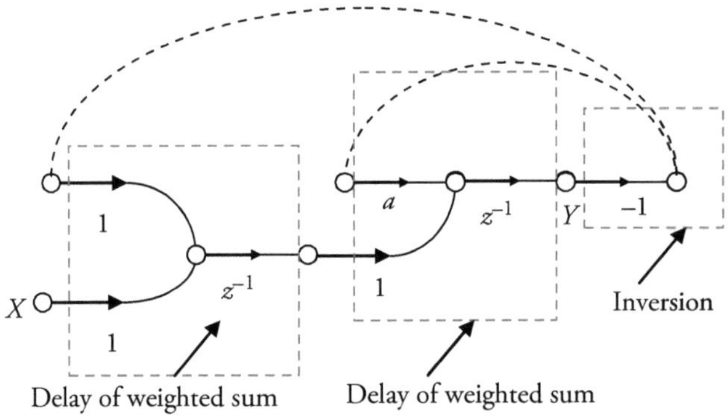

FIGURE 17.51 Alternative SFG for voltage-mode SI circuit realization.

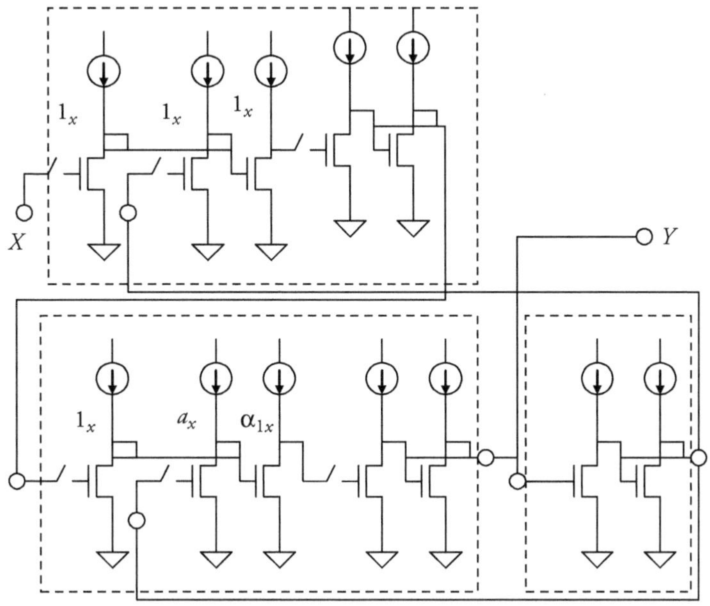

Figure 17.52 VLSI voltage-mode SI circuit implementation.

References

1. G. S. Moschytz, A. Carlosena, "A Classification of Current-Mode Single-Amplifier Biquads Based on a Voltage-to-Current Transformation," *IEEE Trans. on Circuits and Systems II: Analog and Digital Signal Processing*, vol. 41, no. 2, pp. 151–156, February 1994.
2. V. Novotny, K. Vrba, "Novel Voltage- and Current-Mode Filter Design," *2002 Asia-Pacific Conference on Circuits and Systems*, vol. 1, pp. 229–232, October 2002.
3. T.-C. Esteban, S.-R. L. Arturo, "Transforming OTA-C Filters from Voltage- to Current-Mode," *Proceedings, the Fourth IEEE International Caracas Conference on Devices, Circuits and Systems*, pp. C022-1–C022-4, April 2002.
4. Y. Sun, J. K. Fidler, "Current-Mode OTA-C Realisation of Arbitrary Filter Characteristics," *Electronics Letters*, vol. 32, no. 13, pp. 1181–1182, June 1996.
5. H. Schmid, "Why the Terms 'Current Mode' and 'Voltage Mode' Neither Divide Nor Qualify Circuits," *ISCAS'02*, vol. 2, pp. II-29–II-32, May 2002.
6. B. J. Maundy, P. B. Aronehime, E. I. El-Masry, "Switched-Capacitor Filters: The Current Mode Approach," *Proceedings of the 38th Midwest Symposium Circuits and Systems*, vol. 2, pp. 1321–1324, August 1995.
7. H. J. Song, *VLSI Analog Filter and Signal Processing Circuits*, Scholargy Customer Publishing, Tempe, AZ, 2003, ISBN 1-59247-545-0.

CHAPTER 18

Component Simulation–Based VLSI Filter Design

In the VLSI active signal processing circuit design approaches discussed in earlier chapters, selected VLSI circuit techniques are used to synthesize the specific signal processing circuit frequency responses based on their s- or z-domain transfer functions. Under such VLSI signal processing circuit design methods, the topologies of the realized VLSI circuits vary depending on the VLSI circuit technologies employed. A key advantage of such transfer-function–based VLSI active signal processing circuit design approaches is the relatively higher design flexibility of the circuit design implementation that employs the SFG- and cascade-based signal processing circuit design approaches. However, these design methods usually suffer from drawbacks of higher design sensitivities with respect to the VLSI circuit manufacture and application nonidealities such as the PVT variations and the mismatch of the VLSI circuit components. On the other hand, it has been observed that some passive filter circuit topologies, such as the ladder filter circuit structures, offer inherently lower design sensitivity to signal processing applications with respect to the circuit component nonidealities. The component simulation–based VLSI signal processing circuit design approaches are based on the fact that the original low-sensitivity passive signal processing circuit topologies can be kept unchanged through replacement or "simulation" of the physical circuit elements using selected VLSI circuit technologies. There are several key advantages of such VLSI active signal processing circuit design approaches such as (1) the potentially lower circuit sensitivities for realizing high-order signal processing circuits, (2) the lower spreads of circuit component values in the VLSI circuit implementations, and (3) the higher design reusability of most well-developed passive circuit structures such as the LC passive ladder filter circuit structures.

18.1 Active-RC General Impedance Converter Circuits

A useful VLSI active-RC circuit building block for component simulation is the so-called general impedance converter (GIC) circuit shown in Fig. 18.1.

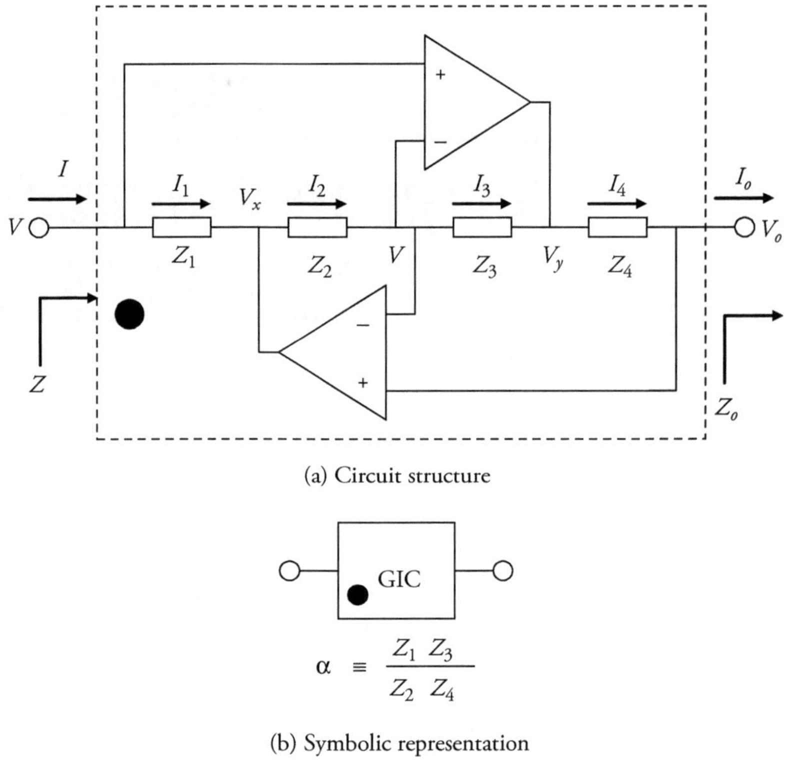

FIGURE 18.1 Active-RC general impedance converter circuit.

The active-RC GIC circuit consists of two OPAMP circuits in negative feedback connections and four passive components. The relationship of the currents and voltages in the circuit can be expressed as follows:

$$\begin{cases} I = I_1 \\ Z_1 I_1 = -Z_2 I_2 \\ I_2 = I_3 \\ Z_3 I_3 = -Z_4 I_4 \\ I_4 = I_o \end{cases} \tag{18.1}$$

Solving for the output current with respect to the input current and the impedances in the circuit, we may obtain a proportional property of the active-RC GIC circuit current.

$$I = \frac{Z_2 Z_4}{Z_1 Z_3} I_o \equiv \alpha^{-1} I_o \tag{18.2}$$

Note that the output current is proportional to the input current with a factor α determined by the impedances of the passive components.

On the other hand, as a result of the negative feedback loops offered by the two OPAMP circuits, we may obtain an identical property of the active-RC GIC circuit terminal voltage as

$$V = V_o \tag{18.3}$$

That is, the two terminal voltages of the active-RC GIC circuit are always the same.

Combining Eqs. (18.2) and (18.3), we may obtain a conversion property of the active-RC GIC circuit impedance as follows:

$$Z = \alpha Z_o \tag{18.4}$$

Note that the equivalent impedance at the input terminal is scaled by α times with respect to the load impedance at the output terminal.

18.1.1 Basic Active-RC GIC Impedance Conversion Rules

The VLSI active-RC GIC circuits offer attractive properties that can be used for the component simulation–based VLSI active-RC signal processing circuit implementation.

Considering the two circuits shown in Fig. 18.2*a* and *b*, we can see that the two circuits are equivalent to each other since both circuits provide the identical terminal voltage response with respect to the same current stimulus. Such a relation implies a fundamental impedance conversion rule offered by the active-RC GIC circuit as shown in Fig. 18.3.

Similarly, an inverse conversion rule of the active-RC GIC circuit as shown in Fig. 18.4 can be derived from the above circuit configurations.

For the active-RC GIC circuit shown in Fig. 18.5, two identical active-RC GIC circuits with mirrored directions are connected together. It can be derived that the voltage drop across two terminals, points A and B, is always zero. Since the two terminal currents are the same, such a circuit is equivalent to electrical short as shown in Fig. 18.5. Consequently, an active-RC GIC circuit with a mirrored direction is an inversion of the original GIC itself (see Fig. 18.6).

Alternatively, we may obtain the second inversion rule of the active-RC GIC circuit based on Fig. 18.7 with two GIC circuits in inversed conversion gain α and 1/α being

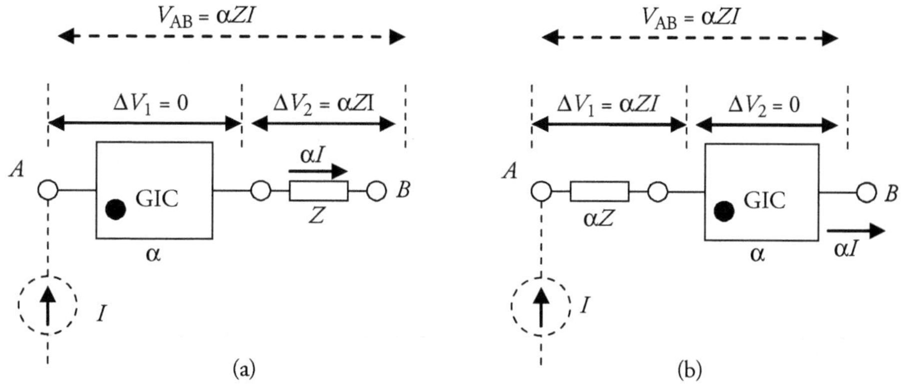

FIGURE 18.2 The equivalent active-RC GIC circuit pair.

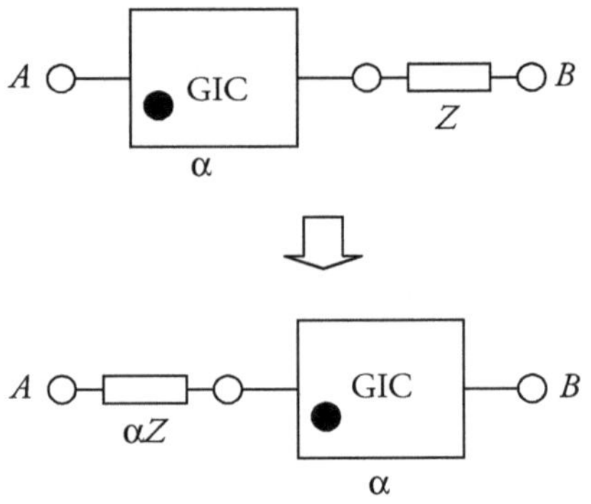

Figure 18.3 The impedance translation rule of active-RC GIC circuit.

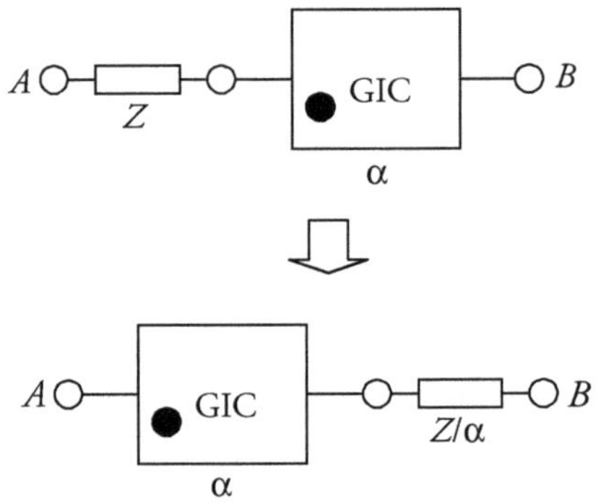

Figure 18.4 Inverse impedance conversion rule of active-RC GIC circuit.

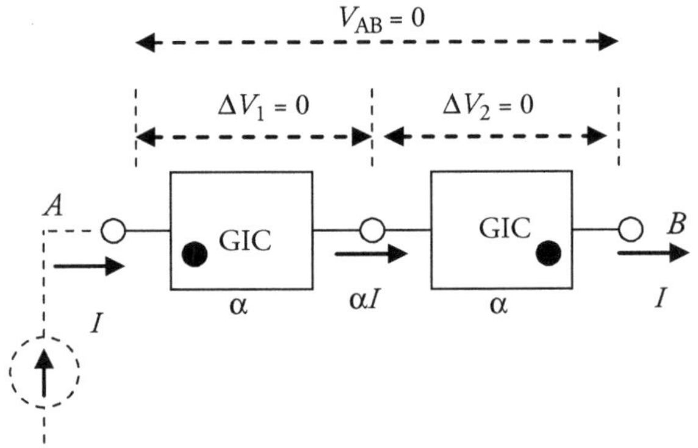

Figure 18.5 Two identical GIC connected back to back.

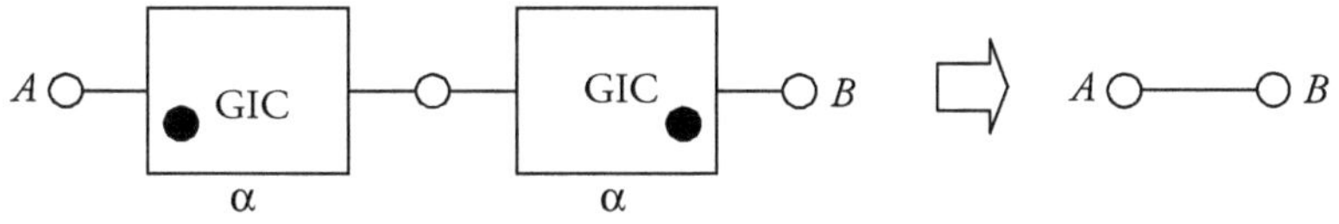

FIGURE 18.6 The first inversion rule of active-RC GIC circuit.

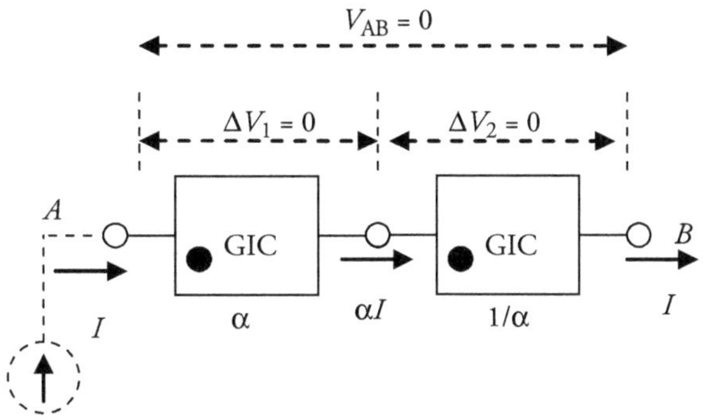

FIGURE 18.7 Two inversed gain GIC circuits in serial connection.

connected to each other. It can be seen that the voltage drop across the two terminals (A and B) is always zero with the same input and output terminal currents. As a result, flip of the active-RC GIC circuit realizes an inversion as shown in Fig. 18.8.

For an active-RC GIC circuit with one terminal short to ground as shown in Fig. 18.9, there will be no voltage drop across the GIC based on the identical terminal voltage property of the active-RC GIC circuit. As a result, the active-RC GIC circuit with

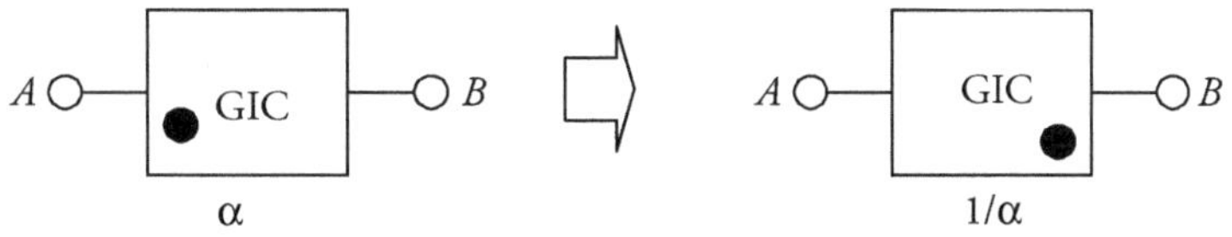

FIGURE 18.8 The second inversion rule of active-RC GIC circuit.

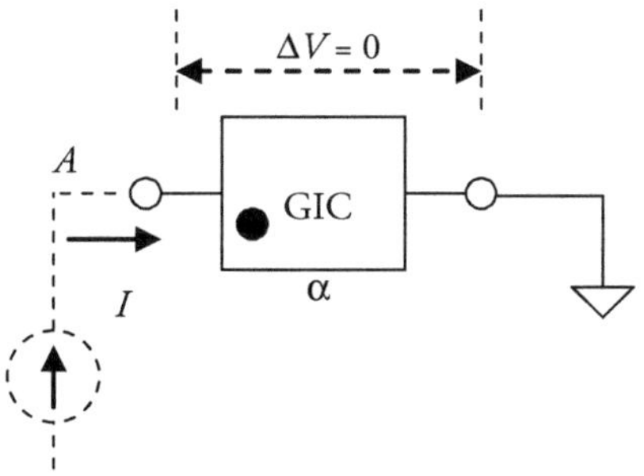

FIGURE 18.9 Active-RC GIC circuit with one terminal connected to ground.

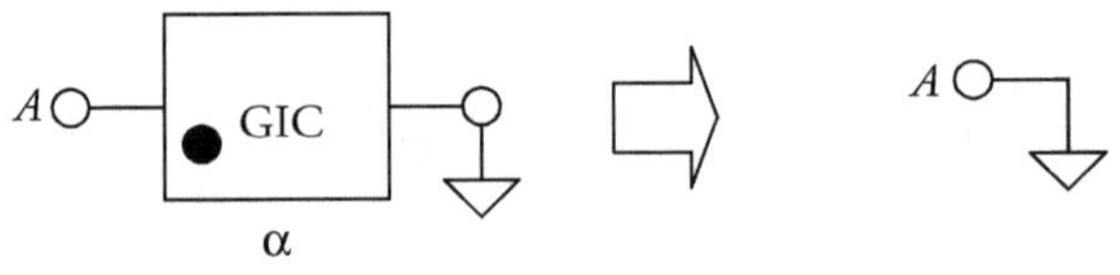

FIGURE 18.10 Basic annealing rule of active-RC GIC circuit (I).

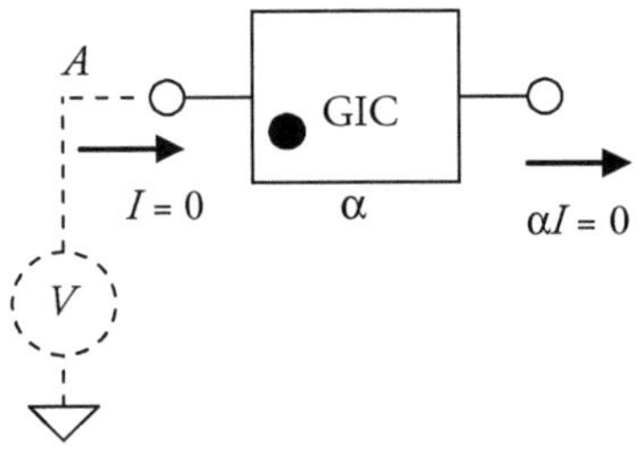

FIGURE 18.11 GIC circuit with one-terminal-open circuit.

one terminal connected ground can be eliminated from the circuit without changing circuit functionality as shown in Fig. 18.10.

Alternatively, for an active-RC GIC circuit with one terminal open as shown in Fig. 18.11, no current can flow into another terminal of the GIC circuit based on the proportional property of the input and output currents of the active-RC GIC circuit. As a result, an active-RC GIC circuit with one end open can be eliminated from the circuit without changing the circuit functionality as shown in Fig. 18.12.

It can be seen that the two active-RC GIC circuits with branches shown in Fig. 18.13*a* and *b* are equivalent to each other since they both have the same terminal voltages and currents.

We may get the basic branching rule of the active-RC GIC circuit shown in Fig. 18.14 based on the above equivalent circuit structures.

A list of useful properties of the active-RC GIC circuits is given in Table 18.1.

18.1.2 Active-RC Inductor Simulation

By selecting $Z_1 = R_1$, $Z_2 = R_2$, $Z_3 = R_3$, and $Z_4 = 1/CS$ in the active-RC GIC circuit shown in Fig. 18.15, we may realize an inductor based on the simulation.

The conversion factor α in this case is given as follows:

$$\alpha = \frac{R_1 R_3}{R_2} Cs \tag{18.5}$$

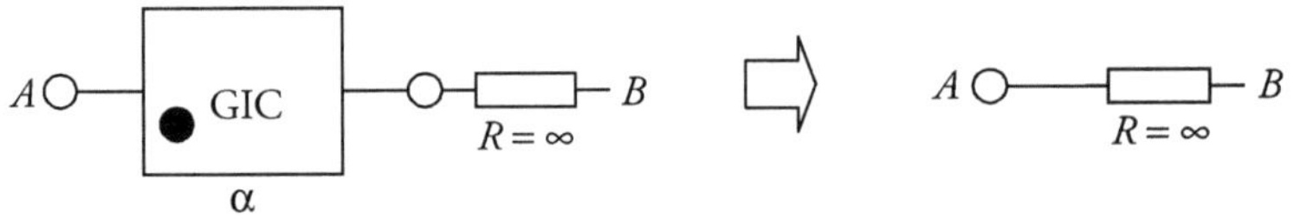

FIGURE 18.12 Basic annealing rule of active-RC GIC circuit (II).

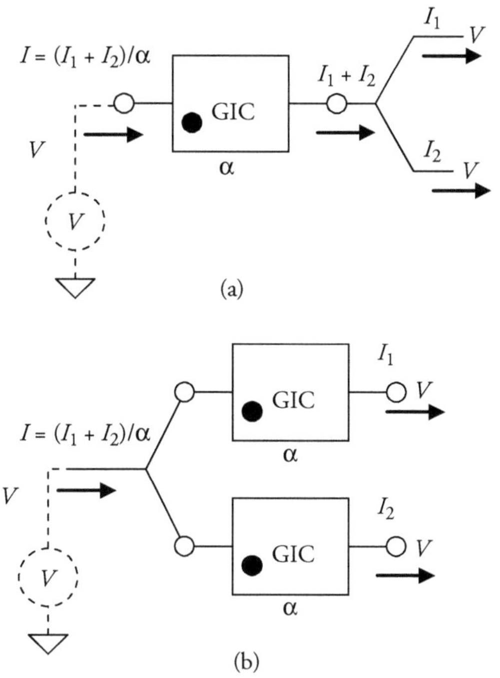

FIGURE 18.13 Active-RC GIC circuit with branches.

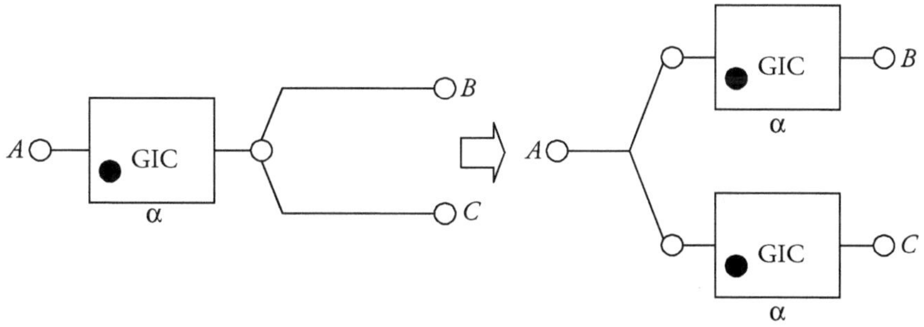

FIGURE 18.14 Basic branching rule of active-RC GIC circuits.

By applying the conversion rule 1 and rule 4 to the active-RC circuit shown in Fig. 18.16, we can see that such a circuit is equivalent to a grounded inductor.

The equivalent inductance is given as follows:

$$Z \equiv \frac{V}{I} = \frac{Z_1 Z_3}{Z_2 Z_4} Z_5 = \frac{R_1 R_3 R_5}{R_2} Cs \tag{18.6}$$

Similarly, by applying rule 1 and rule 2 to the active-RC circuit in Fig. 18.17, we may simulate the floating inductor.

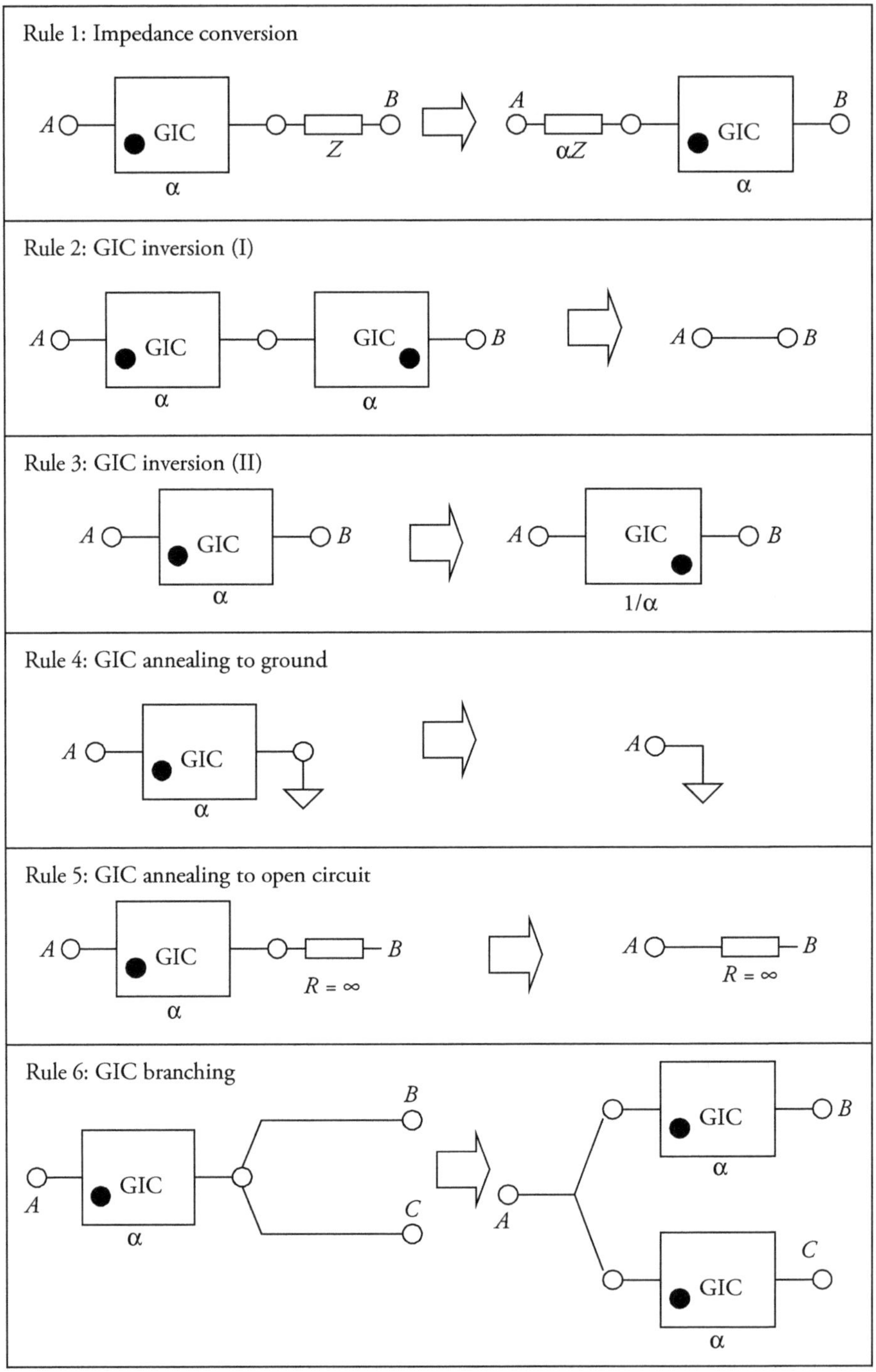

TABLE 18.1 Summary of Conversion Rules for VLSI Active-RC GIC Circuit

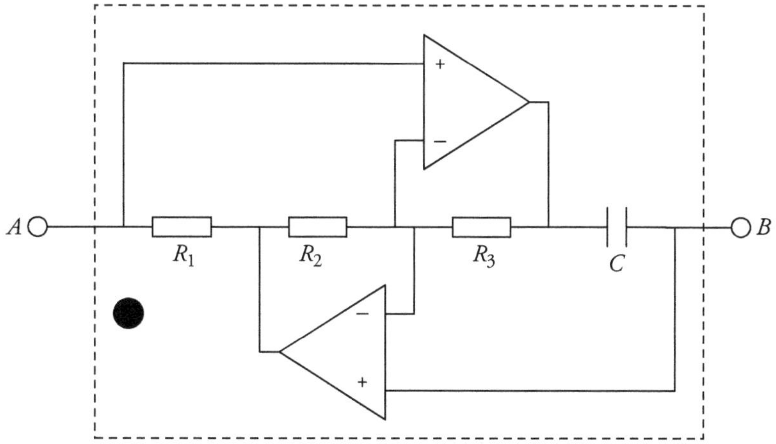

FIGURE 18.15 Active-RC GIC for inductor simulation.

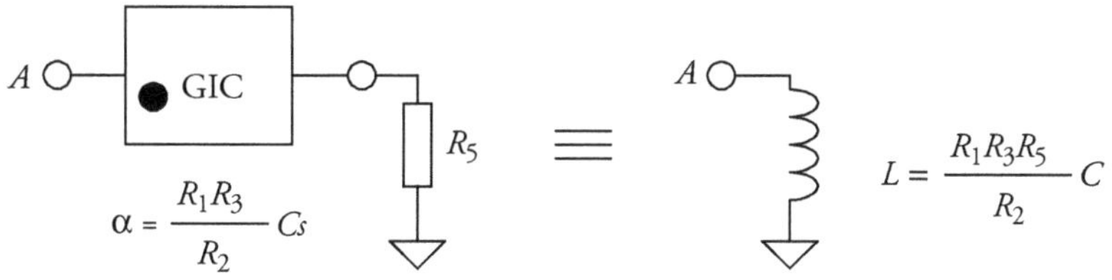

FIGURE 18.16 Active-RC simulation of grounded inductor.

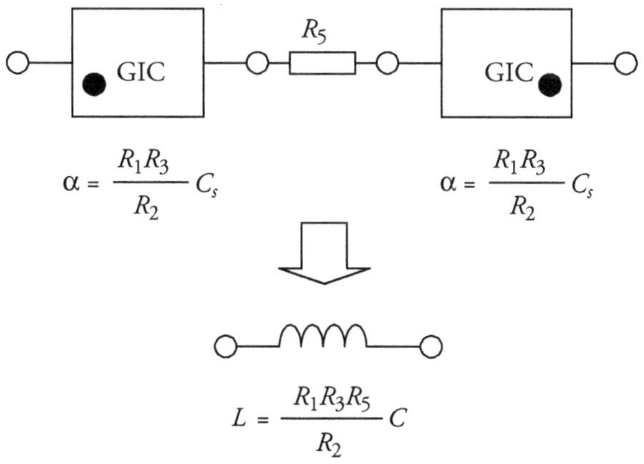

FIGURE 18.17 Active-RC simulation of floating inductor.

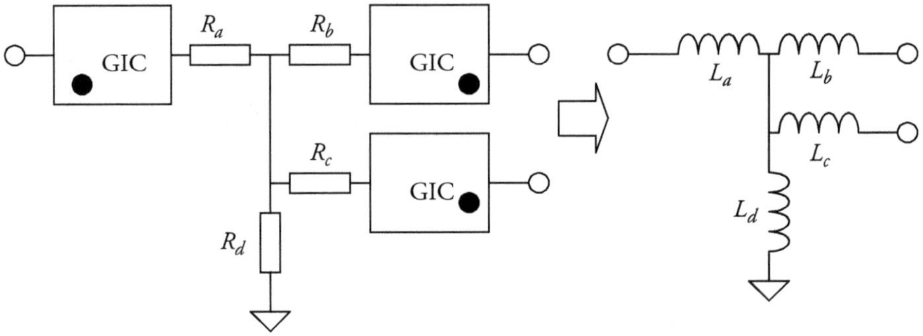

FIGURE 18.18 Active-RC simulation of inductor network.

In a more general case, by applying rule 1, rule 2, and rule 4, a resistive network can be converted into an inductive network using the active-RC GIC circuit as shown in Fig. 18.18.

18.1.3 Design Example

With the simulated inductor, the LRC filter can now be directly mapped into the active-RC circuit as shown in Fig. 18.19, where resistance R_L can be determined by the inductance L and the conversion factor α as follows:

$$R_L = \frac{R_2}{R_1 R_3 C} L \tag{18.7}$$

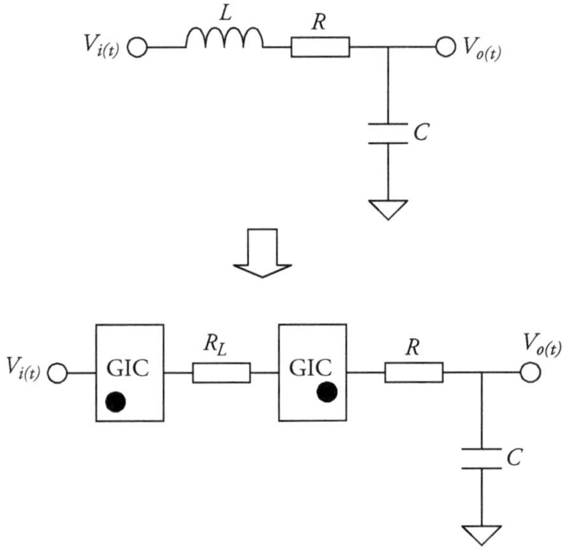

FIGURE 18.19 Realization of LRC filter using the VLSI active-RC GIC circuit.

18.2 Gm-C General Impedance Converter Circuits

A VLSI Gm-C general impedance converter (GIC) circuit consisting of two Gm cells of different polarities connected back to back in parallel is shown in Fig. 18.20.

The two terminal voltages and currents of the Gm-C circuits are given as follows:

$$\begin{cases} I_2 = -g_{m1}V_1 \\ I_1 = -g_{m2}V_2 \end{cases} \tag{18.8}$$

18.2.1 Basic Gm-C GIC Impedance Conversion Rules

The impedance with respect to the output and input terminals of the VLSI Gm-C GIC circuits shown in Fig. 18.21 are defined as follows:

$$\begin{cases} Z_o \equiv \dfrac{V_2}{I_2} \\ Z \equiv \dfrac{V_1}{I_1} \end{cases} \tag{18.9}$$

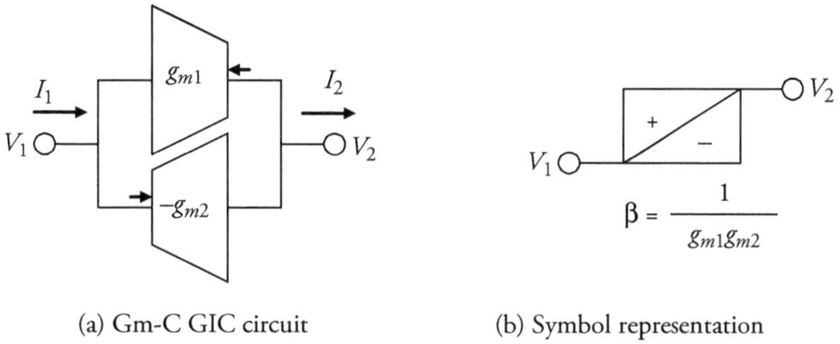

FIGURE 18.20 Gm-C general impedance converter (GIC) circuit.

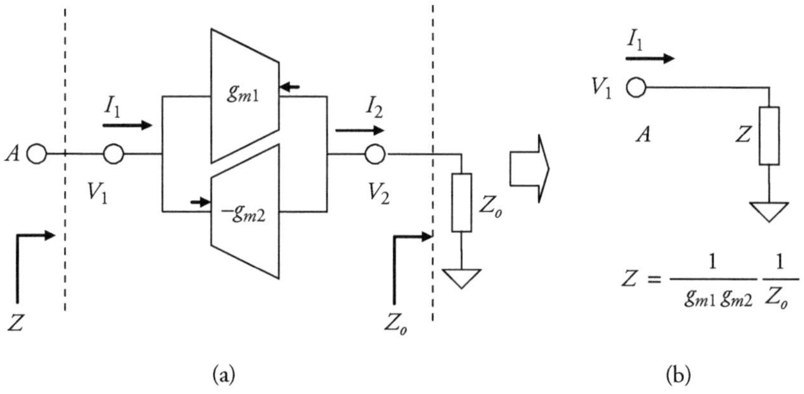

FIGURE 18.21 Gm-C GIC circuits and its equivalent circuit.

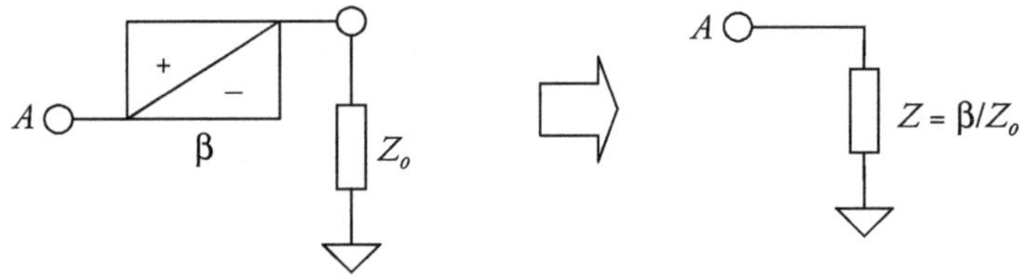

Figure 18.22 Gm-C GIC conversion of ground impedance.

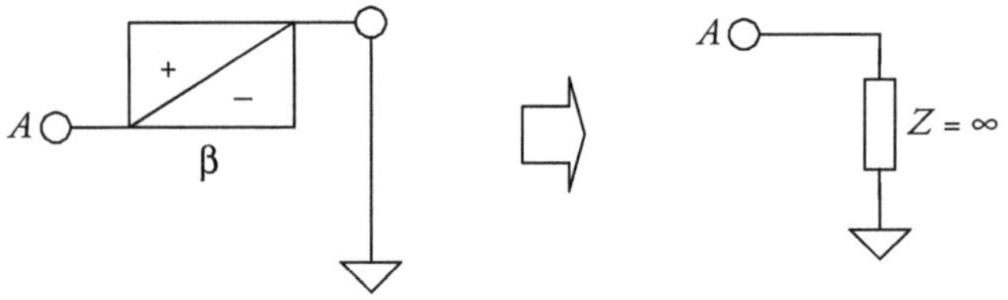

Figure 18.23 Basic annealing conversion rule of Gm-C GIC for ground circuit node.

We can obtain the basic Gm-C impedance conversion rule as follows:

$$Z = \frac{1}{g_{m1}g_{m2}}\frac{1}{Z_o} \equiv \beta\frac{1}{Z_o} \tag{18.10}$$

It can be seen that impedances at the input and output are inverted, and such an impedance conversion property is shown by the equivalent circuits in Fig. 18.21.

The above property can be expressed by the Gm-C GIC conversion rule in Fig. 18.22.

Considering the special case in the above circuit with $Z_o = 0$, we can see that a Gm-C GIC converts a short to ground into an open circuit as shown in Fig. 18.23.

On the other hand, a Gm-C GIC converts an open node into ground as shown in Fig. 18.24.

For the Gm-C GIC circuit shown in Fig. 18.25, we can see that two identical Gm-C GIC circuits connected in series canceling each other.

Such a property implies the inversion rule of the Gm-C GIC circuit shown in Fig. 18.26.

A step-by-step Gm-C GIC conversion of branch circuit is shown in Fig. 18.27.

It can be seen that the Gm-C GIC circuit converts the parallel network into a serial network as shown in Fig. 18.28.

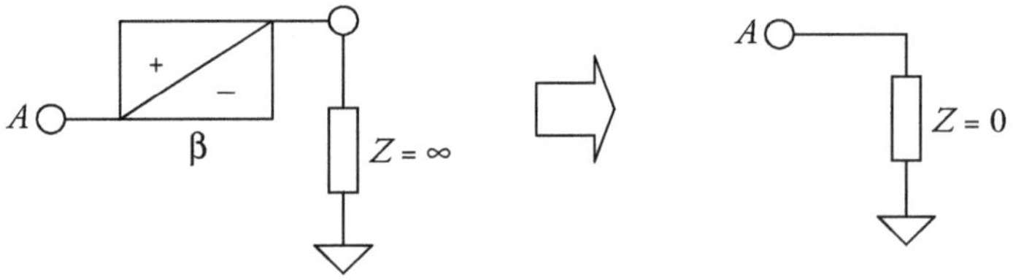

Figure 18.24 Basic annealing conversion rule of Gm-C GIC for open circuit node.

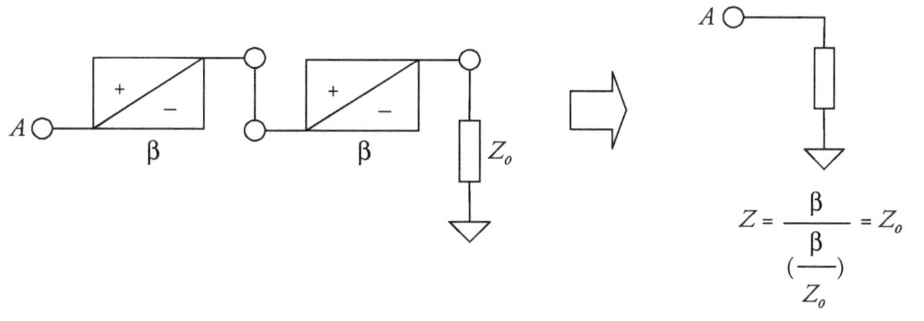

FIGURE 18.25 Serial Gm-C GIC conversion.

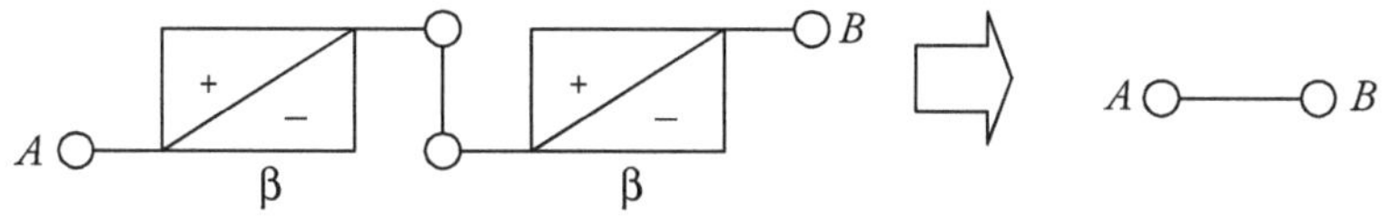

FIGURE 18.26 Basic Gm-C GIC inversion conversion rule.

On the other hand, the Gm-C GIC circuit converts a serial network to a parallel network as shown in Fig. 18.29.

The Gm-C GIC circuit can also be used to convert a grounded circuit component to a floating circuit component as shown in Fig. 18.30.

A summary of basic Gm-C GIC circuit conversion rules are shown in Table 18.2.

18.2.2 Gm-C GIC Simulation of Inductors

By using the Gm-C GIC conversion rule 1 with a capacitor load, we can simulate the inductors using the Gm-C circuit structures as shown in Fig. 18.31.

Alternatively, using conversion rule 2 with the above capacitor load, we may simulate the floating inductor as shown in Fig. 18.32.

18.2.3 Gm-C Simulation of Resistors

A grounded resistor can be simulated in Gm-C circuit technology in the way shown in Fig. 18.33.

For the Gm-C circuit shown, we have that

$$I = g_m V \tag{18.11}$$

and

$$R = \frac{V}{I} = \frac{1}{g_m} \tag{18.12}$$

It can be seen that the circuit works as a grounded resistor. Similarly, a floating resistor can be simulated in the Gm-C circuit in the way shown in Fig. 18.34, where

$$I = g_m(V_a - V_b) \tag{18.13}$$

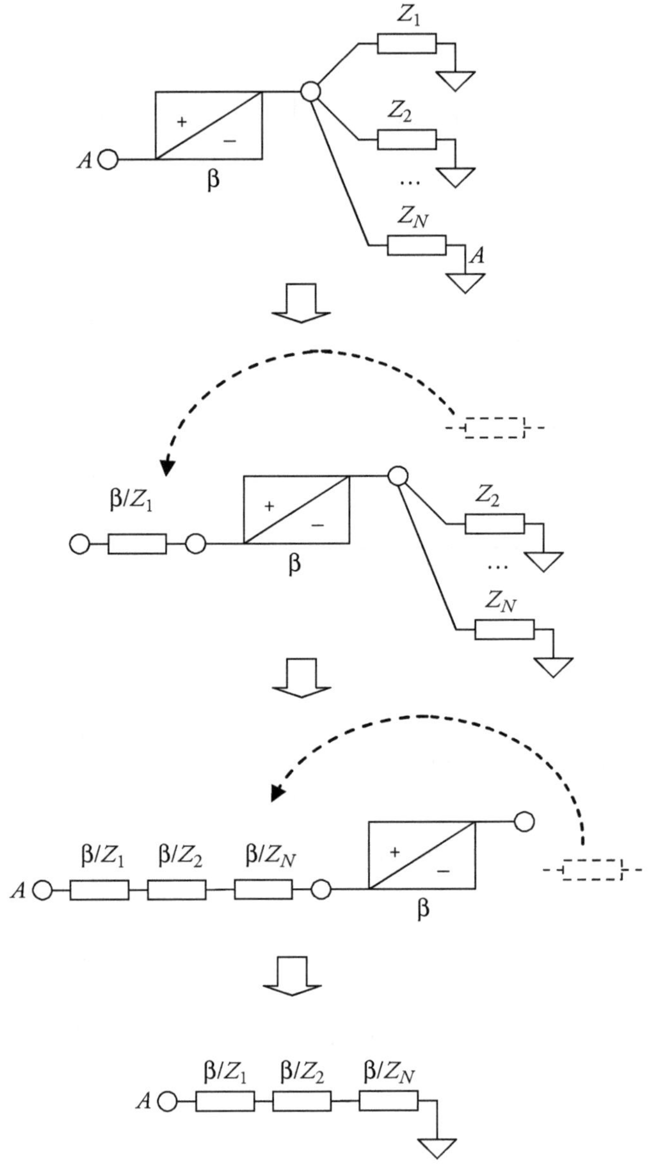

Figure 18.27 Gm-C GIC conversion of branching circuit.

and

$$R = \frac{V_a - V_b}{I} = \frac{1}{g_m} \tag{18.14}$$

By using the Gm-C circuit shown in Fig. 18.35, a buffered resistor can be simulated. The circuit can be used in practical design applications for simulation of terminate resistors in the ladder filters.

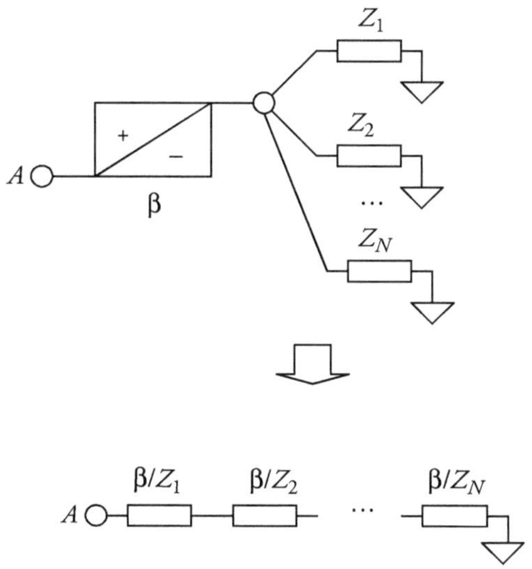

FIGURE 18.28 Basic Gm-C GIC circuit branch conversion rule.

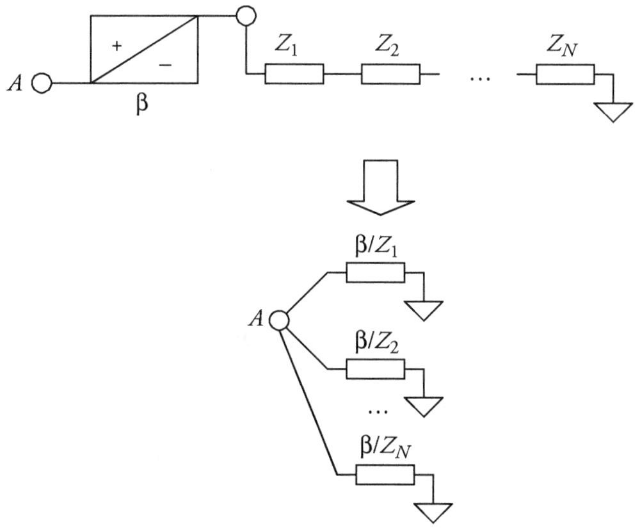

FIGURE 18.29 Basic Gm-C GIC circuit branch conversion rule.

18.2.4 Design Example

With the VLSI Gm-C GIC circuit shown in previous sections, we are ready to design the VLSI signal processing circuits through the Gm-C GIC–based component simulation. For the passive low-pass LRC ladder filter circuit shown in Fig. 18.36, we may convert this passive filter into a Gm-C circuit using the introduced conversion rules step by step as shown in Fig. 18.37.

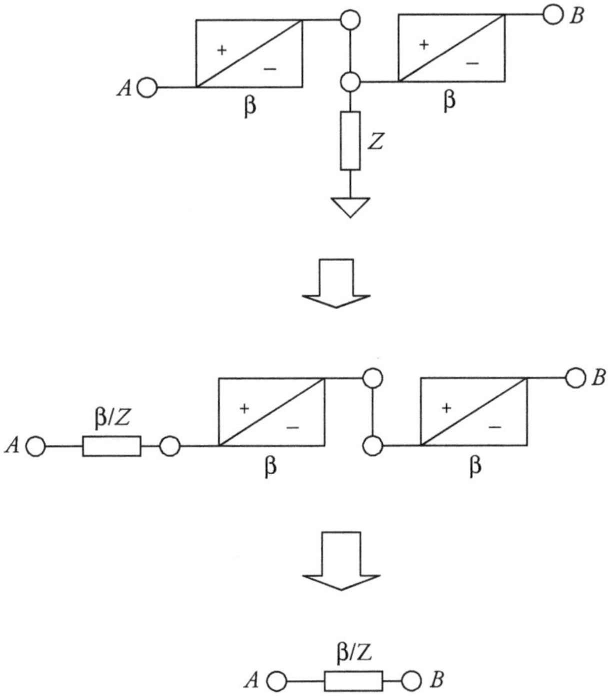

FIGURE 18.30 Basic Gm-C GIC grounded impedance to floating impedance conversion rule.

Generally speaking, by using the Gm-C GIC conversion rules, a general ladder filter structure shown in Fig. 18.38 can be directly converted into the Gm-C circuit form as shown in Fig. 18.39.

18.3 VLSI Impedance Converter Circuit Structures

By combining the impedance conversion concepts and the VLSI circuit technology available, various useful basic VLSI impedance conversion circuits can also be developed.

18.3.1 MOS Cross-Coupled Pair Impedance Converters

Figure 18.40 shows a VLSI impedance conversion circuit consisting of an NMOS transistor pair in crossed connection. In the single-ended operation, such a circuit can be used to generate negative impedance inversion as follows:

$$Z = -\frac{1}{g_{m1}g_{m2}}\frac{1}{Z_o} = -\frac{1}{2\beta I_B}\frac{1}{Z_o} \tag{18.15}$$

The above converter can also be operated in the differential mode as shown in Fig. 18.41.

The equivalent differential impedance of the circuit is given as

$$Z = -\frac{1 + Z_o g_m/2}{g_m/2} = -\frac{1 + Z_o\sqrt{\beta I_B/2}}{\sqrt{\beta I_B/2}} \tag{18.16}$$

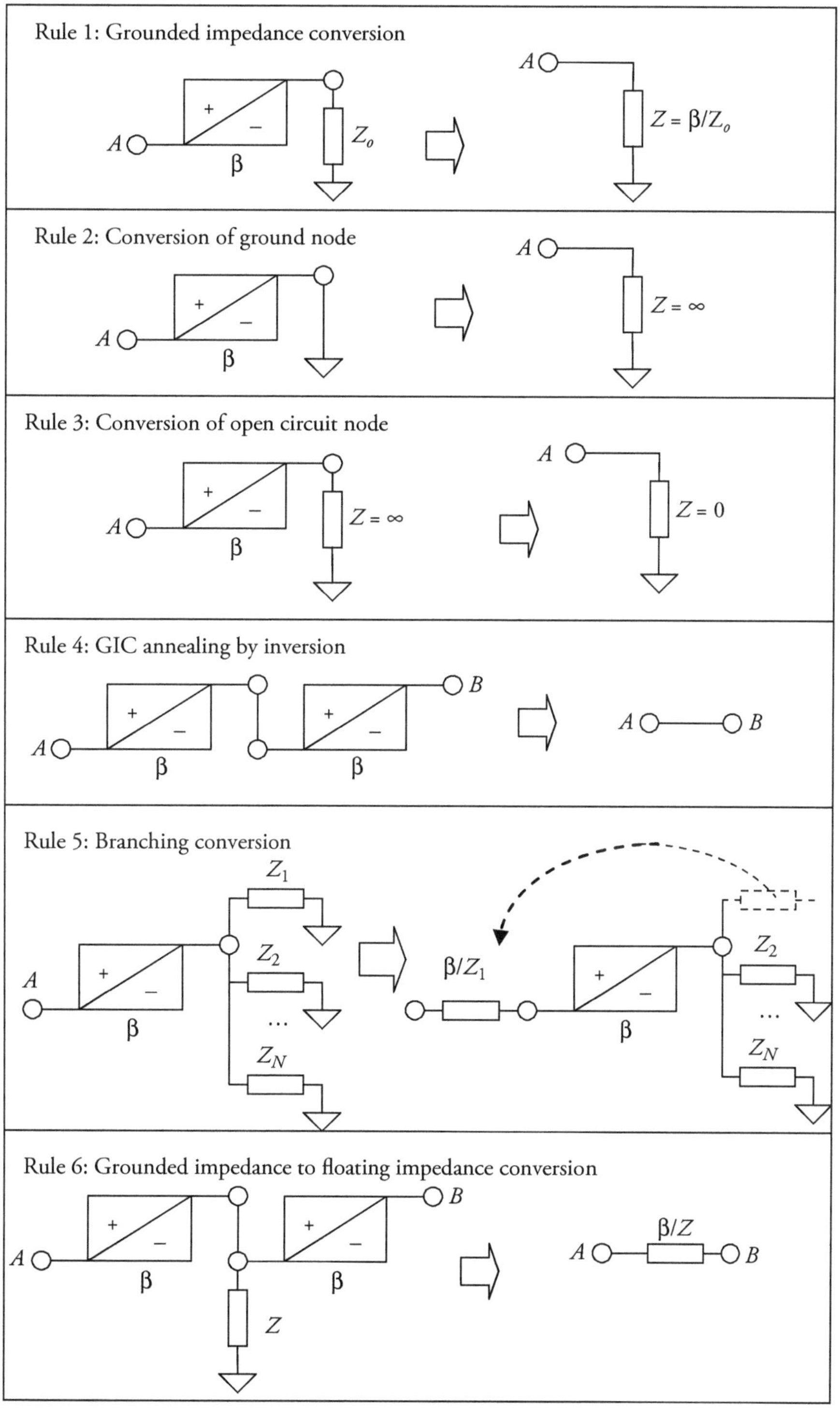

TABLE 18.2 Summary of Conversion Rules for VLSI Gm-C GIC Circuit

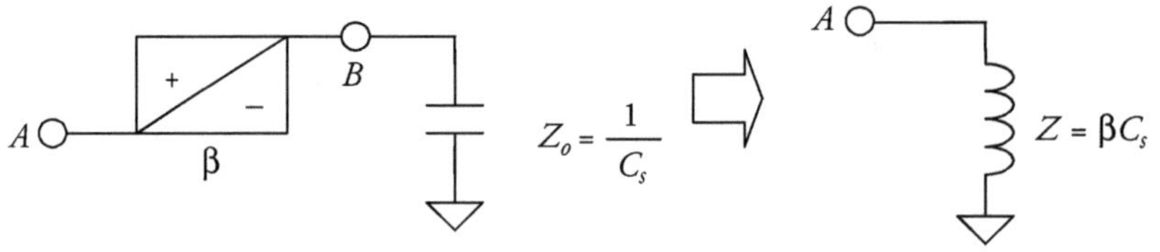

FIGURE 18.31 Gm-C GIC simulation of grounded inductor.

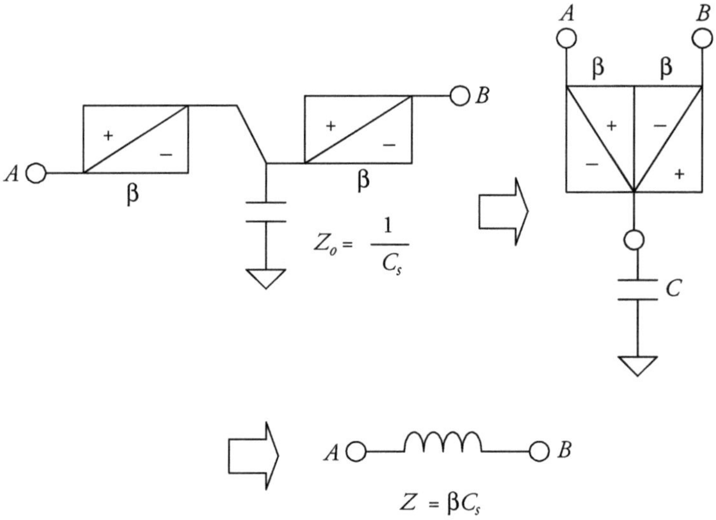

FIGURE 18.32 Gm-C GIC simulation of floating inductor.

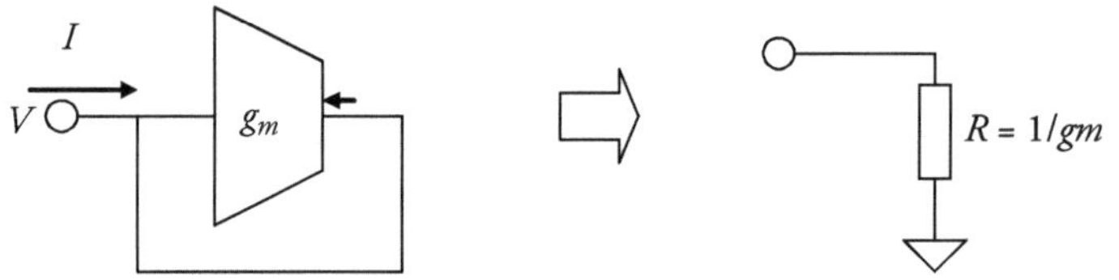

FIGURE 18.33 Gm-C GIC simulation of grounded resistor.

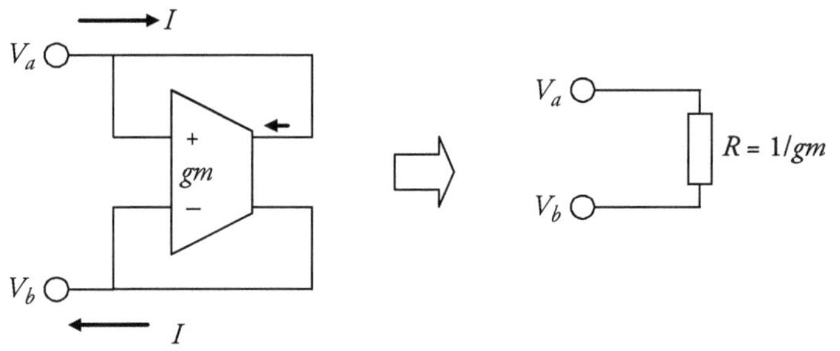

FIGURE 18.34 Gm-C simulation of floating resistor.

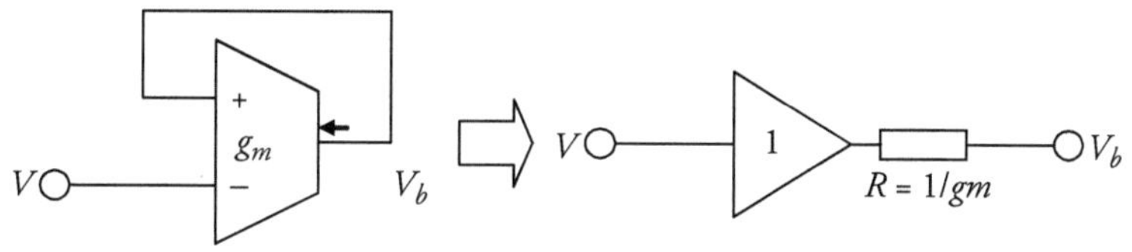

FIGURE 18.35 Gm-C simulation of buffered floating resistor.

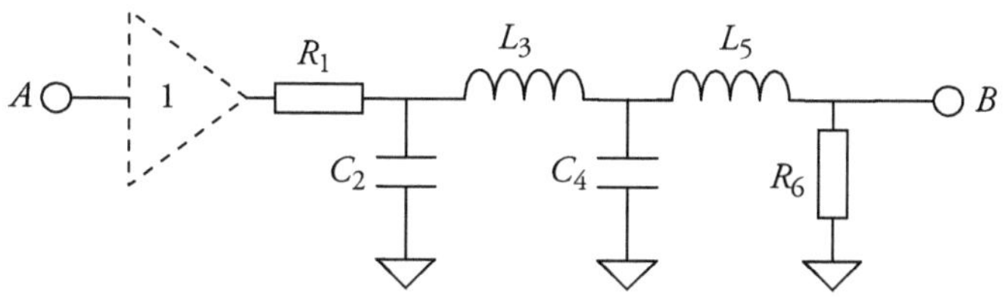

FIGURE 18.36 Passive low-pass ladder filter section.

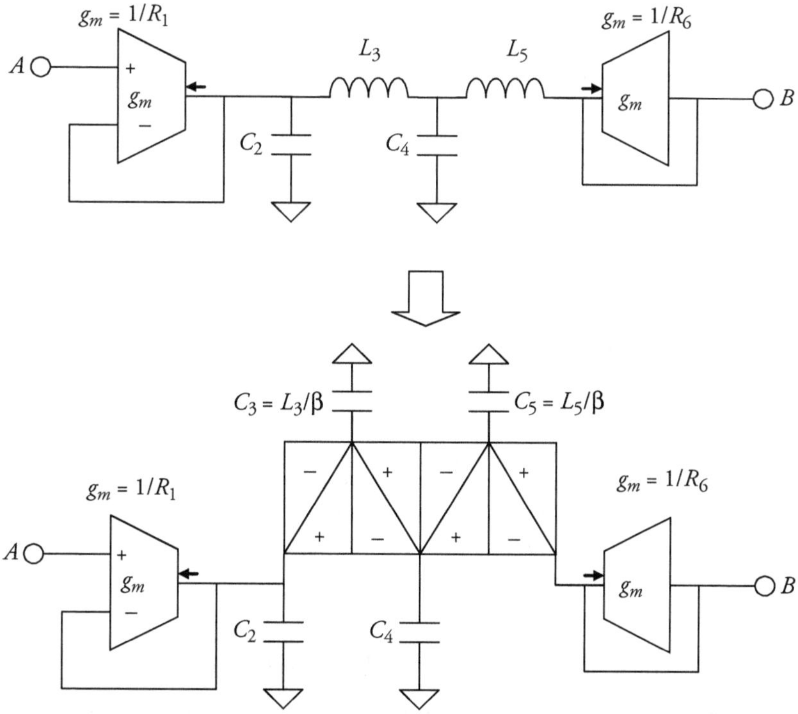

FIGURE 18.37 Simulation of passive low-pass ladder filter using VLSI Gm-C circuit.

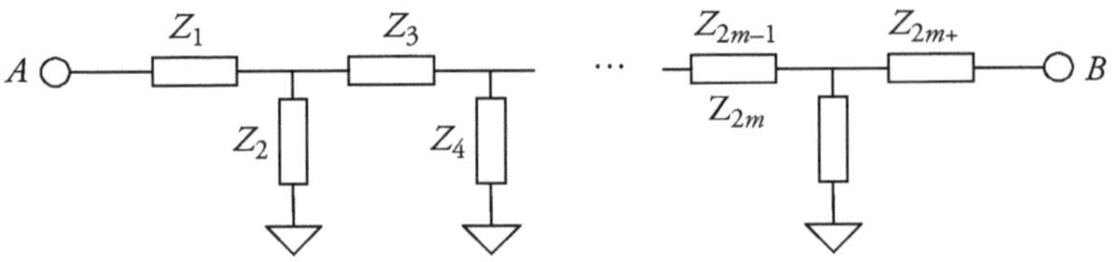

FIGURE 18.38 Passive ladder filter structure.

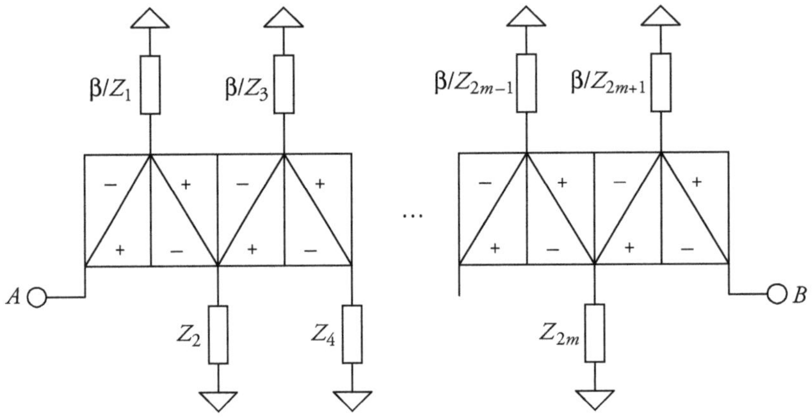

FIGURE 18.39 VLSI Gm-C component simulation–based implementation of ladder filter.

Especially for the high load impedance, the circuit can be used to generate the negative impedance:

$$Z_o >> 1/\sqrt{\beta I_B/2} \Rightarrow Z = -Z_o \tag{18.17}$$

For low load impedance, we have that

$$Z_o << 1/\sqrt{\beta I_B/2} \Rightarrow Z = -2/g_m = -1/\sqrt{\beta I_B/2} \tag{18.18}$$

The above impedance converter circuits are widely used in the VCO design applications where the negative resistance is used to compensate for the lossy resistance in the VCO to ensure its stable oscillation operation. Similarly, for the following differential converter circuit structure shown in Fig. 18.42, we can obtain the equivalent differential impedance of the circuit as

$$Z = -\frac{1 + Z_o(g_{m1} + g_{m3})/2}{(g_{m1} - g_{m3})/2} \tag{18.19}$$

In the special case where four identical transistors are used, the circuit provides very high input impedance given as

$$g_{m1} = g_{m3} \Rightarrow Z = \infty \tag{18.20}$$

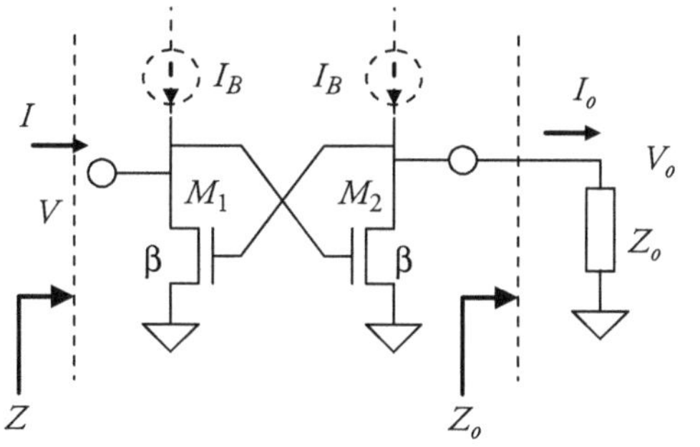

(a) VLSI impedance converter circuit

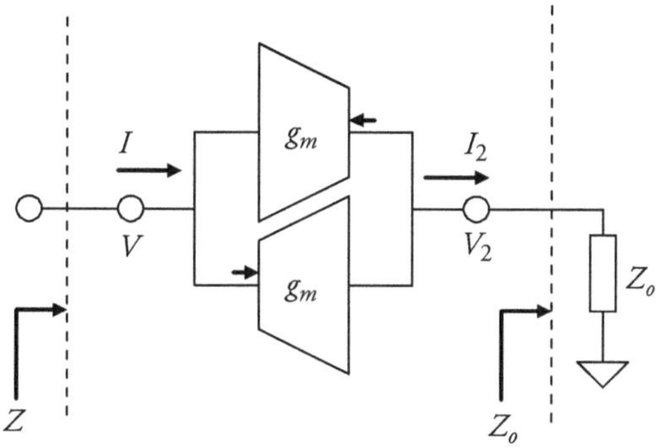

(b) Equivalent circuit

Figure 18.40 VLSI MOS pair impedance converter circuit.

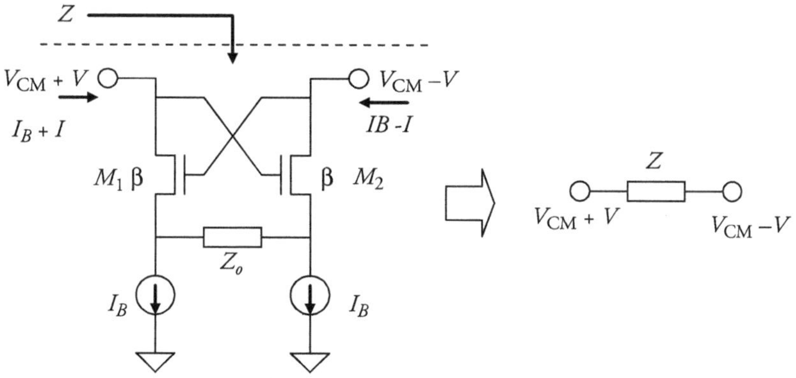

Figure 18.41 Differential VLSI impedance converter circuit.

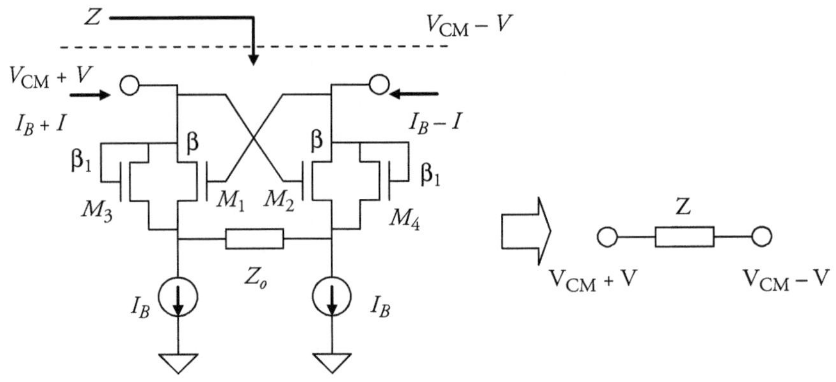

FIGURE 18.42 Differential VLSI impedance converter circuit.

18.3.2 Single MOS Impedance Converter Circuits

A commonly used VLSI impedance converter circuit in RF application (such as LNA or Low Nose Amplifier) is constructed based on a single MOS transistor as shown in Fig. 18.43.

In the s-domain, the voltage across the input gate capacitor is given as

$$V_c = I\frac{1}{C_g s} \tag{18.21}$$

The input voltage at the gate is given as follows:

$$V = V_c + V_s = I\frac{1}{C_g s} + Z_o I\left(1 + \frac{g_m}{C_g s}\right) \tag{18.22}$$

The equivalent input resistance can be derived as

$$Z = \frac{V}{I} = \frac{1}{C_g s} + Z_o\left(1 + \frac{g_m}{C_g s}\right) \tag{18.23}$$

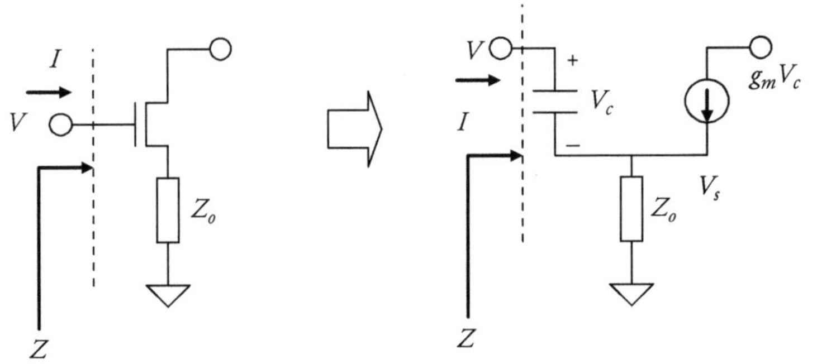

FIGURE 18.43 Single MOS impedance converter structure.

In the RF LNA application, an inductor L is usually used as the load. As a result, at the frequency of interest, the capacitive and the inductive terms in the above equation cancel each other and result into a real equivalent input impedance (typically selected to be 50 Ω for most applications):

$$LC = \left(\frac{1}{\omega_o}\right)^2 \Rightarrow Z = \frac{g_m L}{C_g} \tag{18.24}$$

References

1. A. Kaiser, "A Micro Power CMOS Continuous-Time Low-Pass Filter," *IEEE J. Solid-State Circuits*, vol. SC-24, no. 3, pp. 736–743, June 1989.
2. Y.-T. Wang, A. A. Abidi, "CMOS Active Filter Design at Very High Frequencies," *IEEE J. Solid-State Circuits*, vol. SC-25, no. 6, pp. 1562–1574, June 1990.
3. F. Krummenacher, N. Joehl, "A 4-MHz CMOS Continuous-Time Filter with On-Chip Automatic Tuning," *IEEE J. Solid-State Circuits*, vol. SC-23, no. 3, pp. 750–758, June 1988.
4. F. Krummenacher, "Design Considerations in High-Frequency CMOS Transconductance Amplifier Capacitor (TAC) Filters," *IEEE Proc. ISCAS*, pp. 100–105, 1989.
5. K. W. Moulding, G. A. Wilson, "A Fully Integrated Five-Gyrator Filter at Video Frequencies," *IEEE J. Solid-State Circuits*, vol. SC-13, no. 3, pp. 303–307, June 1978.
6. J. O. Voorman, W. H. A. Bruls, P. J. Barth, "Integration of Analog Filters in a Bipolar Process," *IEEE J. Solid-State Circuits*, vol. SC-17, no. 4, pp. 713–722, August 1982.
7. S.Takagi, H.Nitta, J.Koyama, M.Furihata, N.Fujii, M.Nagata, T.Yanagisawa, "100-MHz Monolithic Low-Pass Filters with Transmission Zeros Using NIC Integrators," *IEEE J. Solid-State Circuits*, vol. SC-26, no. 4, pp. 669–671, June 1991.
8. Y. Ishibashi, K. Tsuduki, "Synthesis and RC-Active Simulation of an LR Terminated LC Filter," *IEE Proc. Circuits Devices Syst.*, vol. 142, no. 1, pp. 49–52, February 1995.
9. H. Hagiwara, M. Kumazawa, S. Takagi, M. Furihata, M. Nagata, T. Yanagisawa, "A Monolithic Video Frequency Filter Using NIC-based Gyrators," *IEEE Journal of Solid-State Circuits*, vol. 23, no. 1, pp. 175–182, February 1988.
10. Y. Sun, "OTA-C Filter Design using Inductor Substitution and Bruton Transformation Methods," *Electronics Letters*, vol. 34, no. 22, pp. 2082–2083, October 1998.
11. P. Andreani, S. Mattisson, "On the Use of Nauta's Transconductor in Low-Frequency CMOS g_m-C Bandpass Filters," *IEEE J. Solid-State Circuits*, vol. 37, no. 2, pp. 114–124, February 2002.
12. J. Schechtman, A. C. M. de Queiroz, L. P. Caloba, "Switched-Current Filters Using Component Simulation," *ISCAS'94*, vol. 5, pp. 569–572, June 1994.
13. L. P. Caloba, A. C. M. de Queiroz, "OTA-C Simulation of Passive Filters via Embedding," *ISCAS'89*, vol. 2, pp. 1083–1086, May 1989.

CHAPTER 19

Ladder Simulation–Based VLSI Filter Design

It has been found that doubly terminated LC passive ladder networks that are designed to offer maximum power transfer from source to load over the filter passband provides very low sensitivities to variations in their circuit component values. This fact has, over the years, spurred considerable interest in finding VLSI analog filter circuit structures, which simulate the internal workings of LC ladder prototype networks.

Essentially, these filter circuit structures consist of a connection of subnetworks of first- or second-order blocks that implement the *I-V* integral relation of the *L*, *C*, and *R* passive elements in the prototype circuits.

In the VLSI signal processing circuit implementation, such a design approach provides a very effective way for implementing the ladder circuit structure using various VLSI circuit techniques that offer the VLSI process technologies.

19.1 Mixed-Mode Ladder Simulation

Figure 19.1 shows a general passive ladder structure consisting of a set of series admittance $\{Y_i; i = 1, 3, 2m - 1\}$ and shunt impedance $\{Z_j; j = 2, 4, 2m\}$.

A set of voltage and current equations that describe the output and input signal relations for each of the ladder sections can be derived using internal current and voltage node (mixed-mode) signals as follows:

$$\begin{cases} I_{2k-1} = Y_{2k-1}(V_{2k-2} - V_{2k}) \\ V_{2k} = Z_{2k}(I_{2k-1} - I_{2k+1}) \\ V_i \equiv V_0 \\ V_o \equiv V_{2m} \\ (k = 1, 2, \ldots, m) \end{cases} \tag{19.1}$$

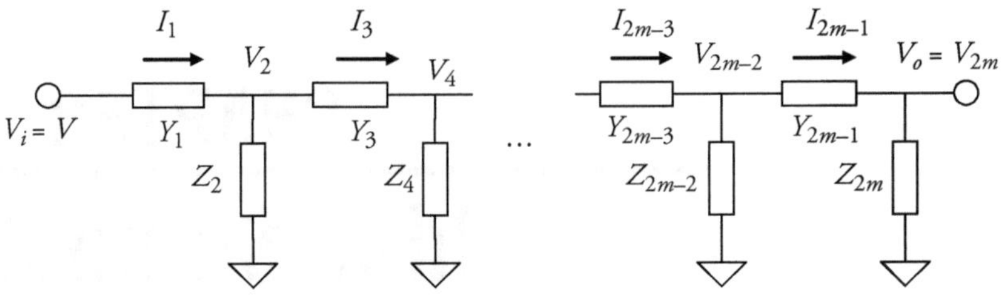

FIGURE 19.1 General passive ladder filter structure.

To simplify the analysis, we may introduce a set of variables that unify all the node voltages and branch currents in the ladder filter circuit:

$$\begin{cases} X_{2k-1} \equiv I_{2k-1} \\ X_{2k} \equiv V_{2k} \\ X_{2m} \equiv V_o \\ X_0 \equiv V_i \\ (k = 1, 2, ..., m) \end{cases} \tag{19.2}$$

Consequently, Eq. (19.1) can be rewritten based on the mixed-mode variables as follows:

$$\begin{cases} X_{2k-1} = Y_{2k-1}(X_{2k-2} - X_{2k}) \\ X_{2k} = Z_{2k}(X_{2k-1} - X_{2k+1}) \\ X_0 \equiv V_i \\ X_{2m} \equiv V_o \\ (k = 1, 2, ..., m) \end{cases} \tag{19.3}$$

From Eq. (19.3), it can be seen that a ladder filter circuit can be partitioned into the two basic series and shunt subnetworks as shown in Figs. 19.2 and 19.3.

Connecting the above subnetwork SFGs together, the SFG expression of the ladder filter for VLSI circuit implementation can be constructed as shown in Fig. 19.4.

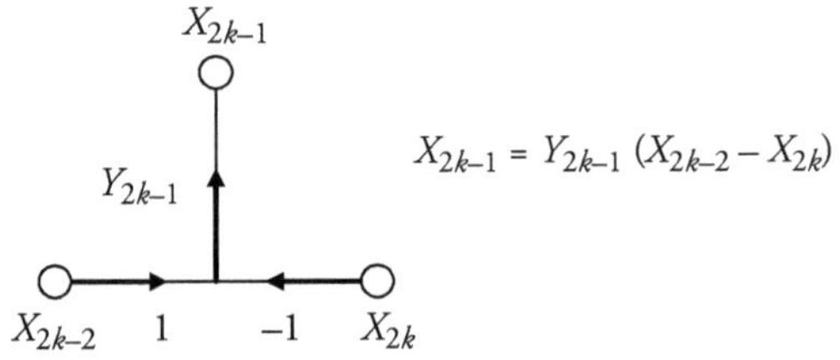

FIGURE 19.2 SFG of basic series subnetwork.

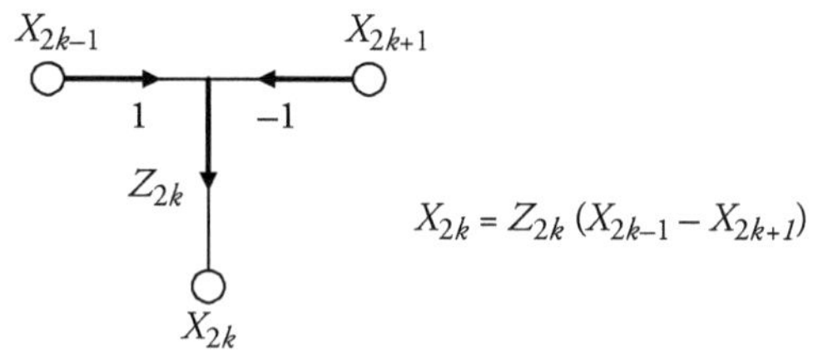

FIGURE 19.3 SFG of basic shunt subnetwork.

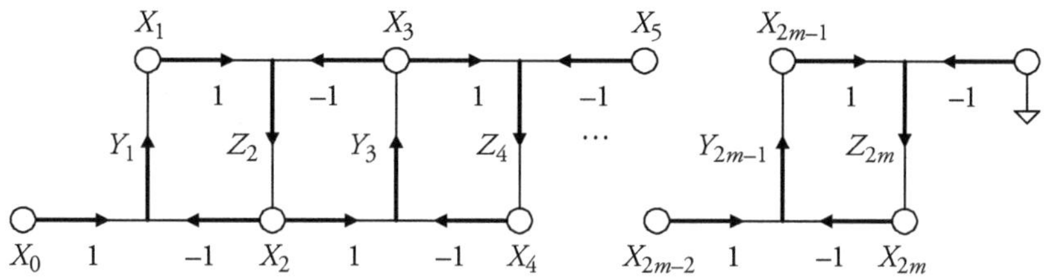

FIGURE 19.4 SFG expression of the ladder filter.

19.2 Node Voltage–Based Ladder Simulation

The above approach is based on the node voltages and branch currents in the ladder filter as the design variables. Alternatively, the ladder filter circuit can also be expressed only based on the node voltages. In this case, a set of node voltages $\{V_i;\ i = 0, 2, 2m\}$ as well as the admittances of the filter component as shown in Fig. 19.5 are used for the ladder filter circuit implementation.

From Fig. 19.6, the equations for node voltages can be expressed as follows:

$$\begin{cases} V_{2k} = W_{2k}(Y_{2k-1}V_{2k-2} + Y_{2k+1}V_{2k+2}) \\ W_{2k} \equiv \dfrac{1}{Y_{2k-1} + Y_{2k} + Y_{2k+1}} \\ V_i \equiv V_0 \\ V_o \equiv V_{2m} \\ (k = 1, 2, \ldots, m) \end{cases} \tag{19.4}$$

Note that there is only one basic general subnetwork as shown in Fig. 19.6 that is required to construct the entire ladder filter.

The node voltage–based SFG expression of the ladder filter can be constructed using the basic building components as shown in Fig. 19.7.

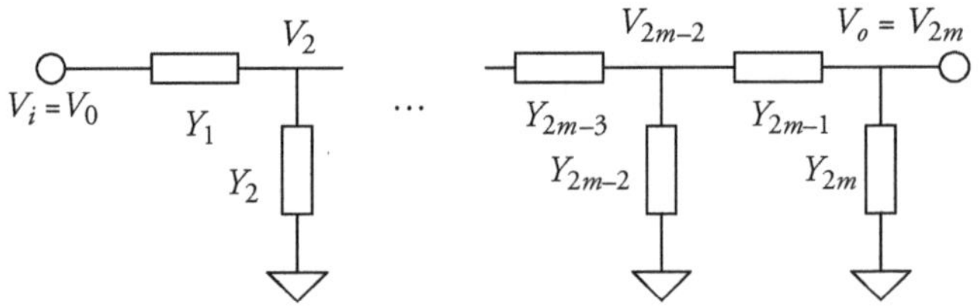

Figure 19.5 Passive ladder filter expression using only node voltages.

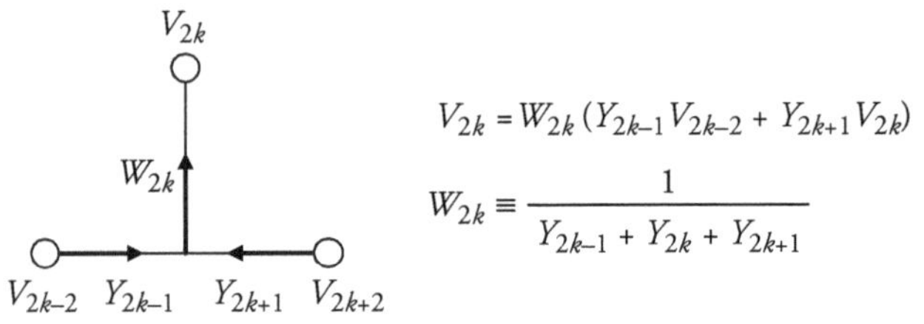

Figure 19.6 SFG of basic building components of the ladder filter.

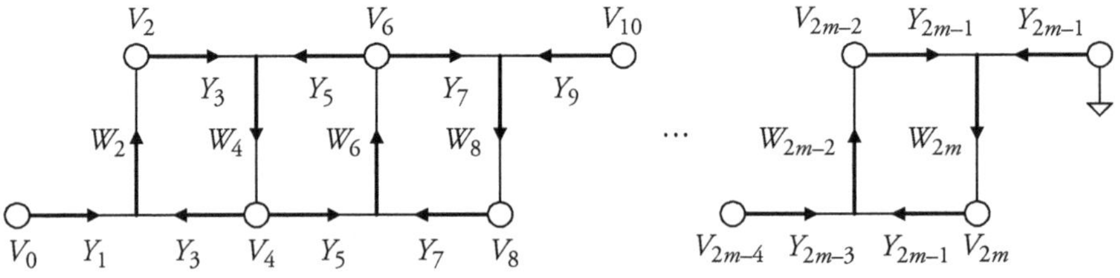

Figure 19.7 Node voltage–based SFG expression of the ladder filter.

19.3 VLSI Active Circuit Simulation of All-Pole Ladder Filters

The key for active simulation of the ladder filter is to find the basic building structures for a given VLSI technology that can be used to simulate the entire ladder filter circuit. For the double-terminated lossless (all-pole) low-pass ladder filter as shown in Fig. 19.8, all series components, except the source resistor, are inductors; and all shunt components, except the load resistor, are capacitors. The equations for mixed-mode ladder simulation can be derived as

$$\begin{cases} X_1 = (X_0 - X_2) \\ X_{2k+1} = \dfrac{R_i}{sL_{2k+1}}(X_{2k} - X_{2k+2}) \\ X_{2k} = \dfrac{1}{sR_iC_{2k}}(X_{2k-1} - X_{2k+1}) \\ X_{2m} = \dfrac{R_o}{R_i}X_{2m-1} \end{cases} \tag{19.5}$$

with $V_i = X_0$, $V_o = X_{2m}$, and $k = 1, 2, m - 1$.

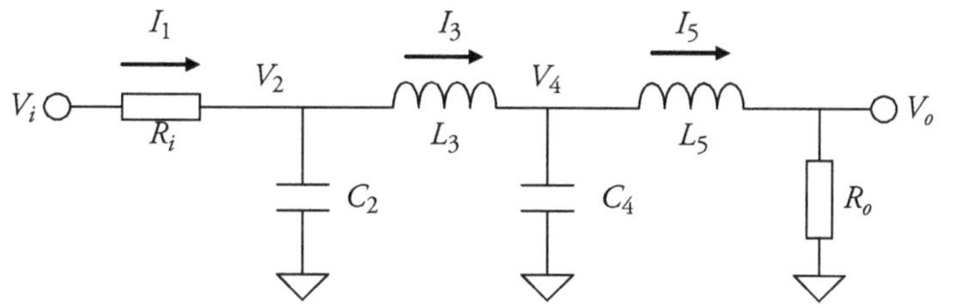

FIGURE 19.8 Example of all-pole ladder filter structure.

It can be seen that the basic building blocks for the all-pole low-pass ladder simulation is the integration with signed inputs, addition with weighted (signed) inputs, and the gain stage. In practical design, the addition and scaling equations at the input and the output terms of the ladder filter can usually be merged with other equations using the lossy integrator to simplify the circuit implementation. For various VLSI circuit techniques, such types of circuits have been developed in previous chapters that can be directly used to simulate the entire ladder filter.

19.3.1 Active-RC Circuit Simulation

The single-ended active-RC circuit implementation of integration with signed inputs is shown in Fig. 19.9.

$$X_n = \frac{1}{\tau s}(X_{n+1} - X_{n-1}) \tag{19.6}$$

Similarly, the active-RC circuit realization of addition with signed inputs is shown in Fig. 19.10.

$$X_n = (X_{n+1} - X_{n-1}) \tag{19.7}$$

Finally, the gain stage can also be easily realized using the VLSI active-RC circuit as follows (see Fig. 19.11):

$$Y = \alpha X \tag{19.8}$$

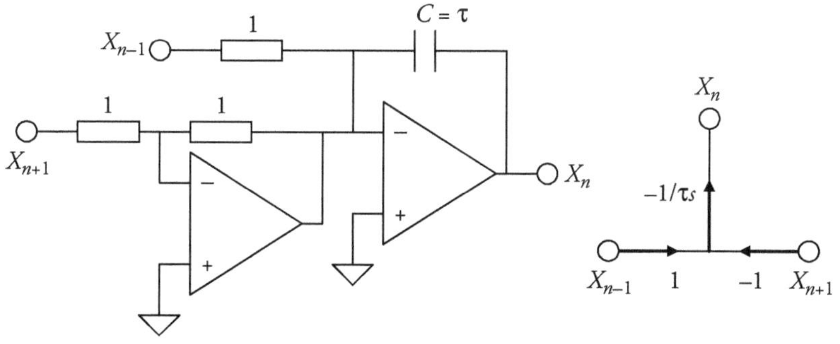

FIGURE 19.9 Active-RC circuit and SFG of integration with signed inputs.

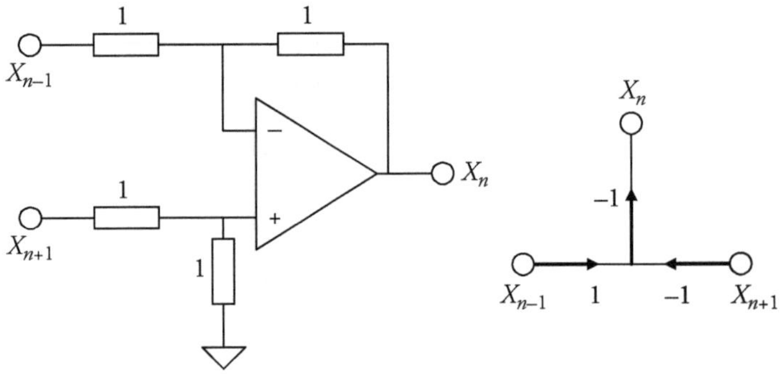

FIGURE 19.10 Active-RC circuit and SFG of addition with signed inputs.

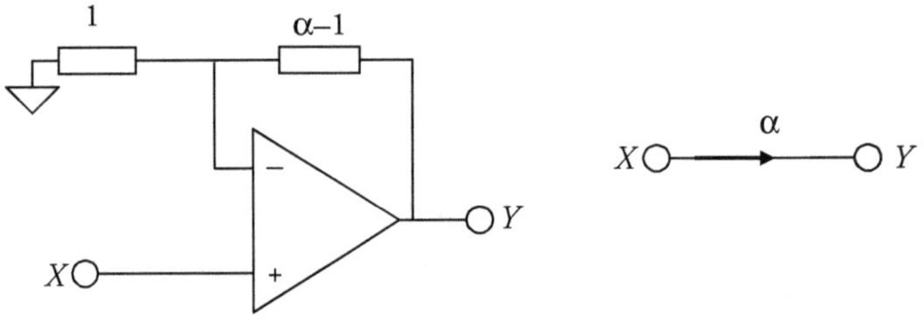

FIGURE 19.11 Active-RC circuit and SFG of gain stage.

Note that the above circuit is suitable for a circuit stage with gain. For a stage with attenuation, it may have a slightly different form. Since there is a one-to-one relationship between the active-RC and MOS-C circuits, the above building block can be directly converted into a MOS-C circuit by replacing the resistor using MOS VCRs.

19.3.2 Gm-C Circuit Simulation

The Gm-C circuit implementation of integration with signed inputs is shown in Fig. 19.12.

The Gm-C circuit realization of addition with signed inputs is shown in Fig. 19.13.

The Gm-C circuit realization of gain stage is shown in Fig. 19.14.

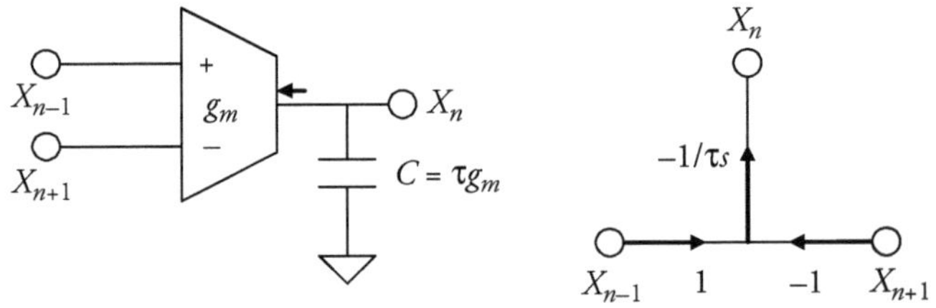

FIGURE 19.12 Gm-C circuit realization of integration with signed inputs.

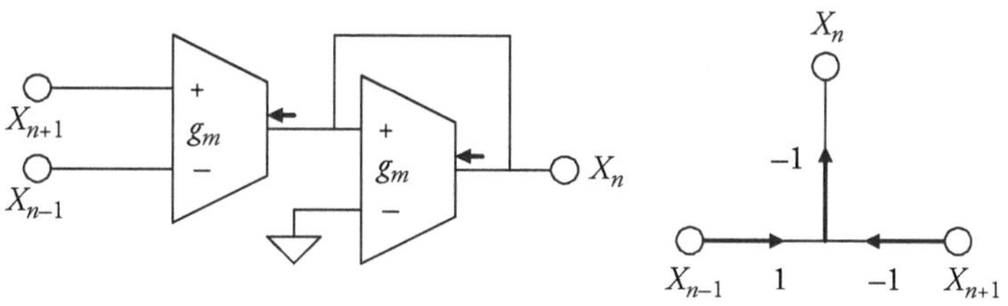

FIGURE 19.13 Gm-C circuit and SFG of addition with signed inputs.

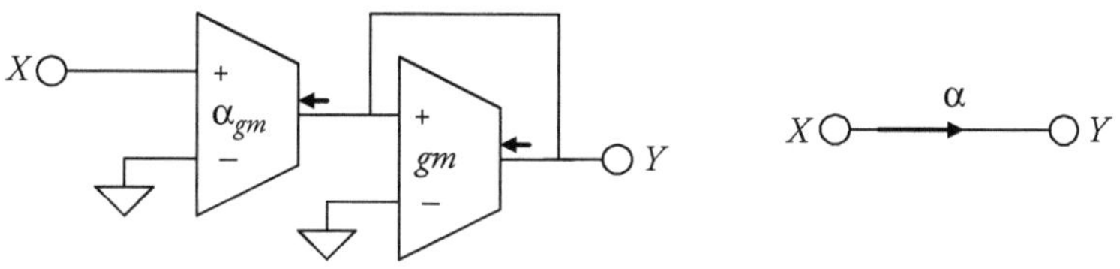

FIGURE 19.14 Gm-C circuit realization of gain stage.

19.3.3 CTI Circuit Simulation

The CTI circuit implementation of integration with signed inputs is shown in Fig. 19.15.

The CTI circuit realization of addition with signed inputs is shown in Fig. 19.16.

The CTI circuit realization of gain stage is shown in Fig. 19.17.

19.3.4 SC Circuit Simulation

The SC circuit implementation of integration with signed inputs is shown in Fig. 19.18.

The SC circuit realization of addition with signed inputs is shown in Fig. 19.19.

The SC circuit realization of gain stage is shown in Fig. 19.20.

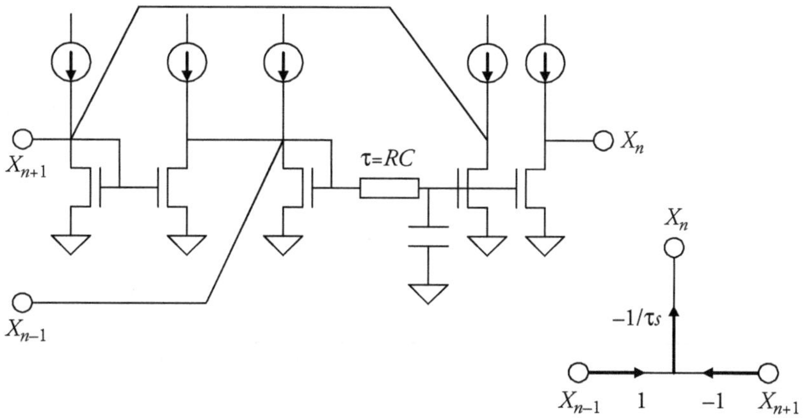

FIGURE 19.15 CTI circuit realization of integration with signed inputs.

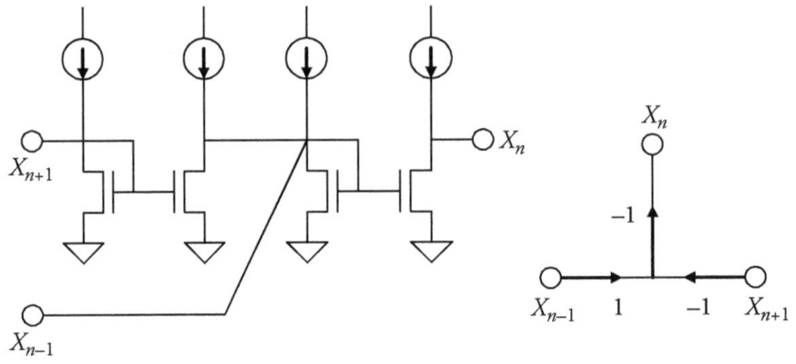

FIGURE 19.16 CTI circuit and SFG of addition with signed inputs.

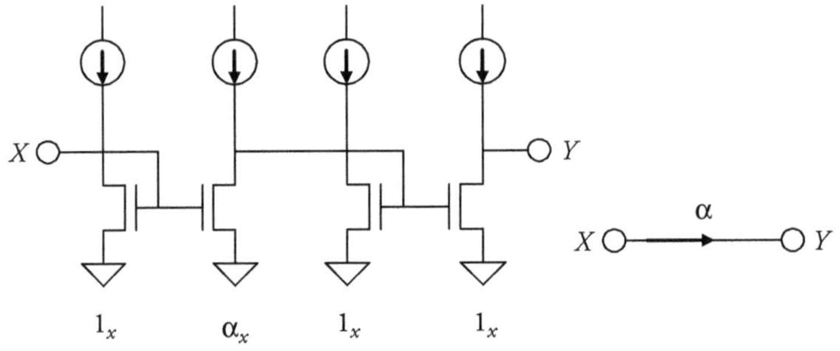

FIGURE 19.17 CTI circuit realization of gain stage.

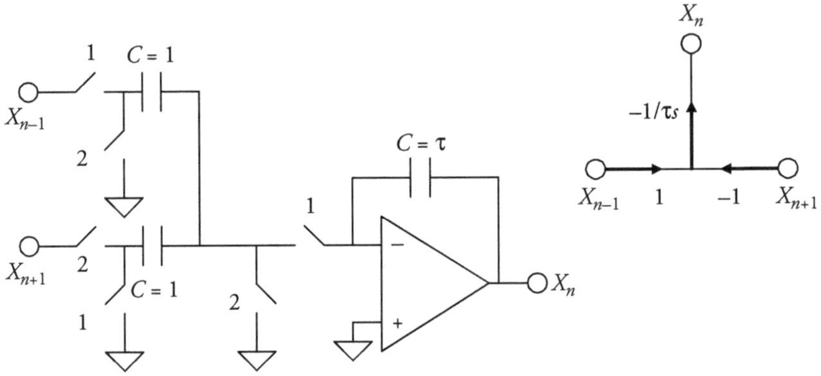

FIGURE 19.18 SC circuit realization of integration with signed inputs.

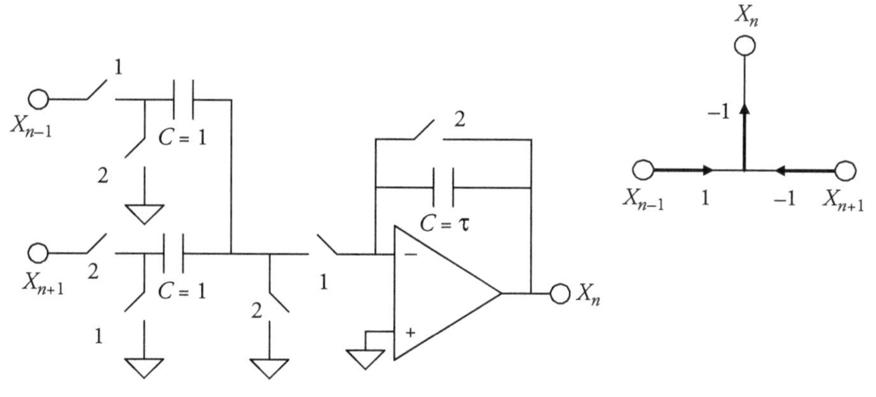

FIGURE 19.19 SC circuit and SFG of addition with signed inputs.

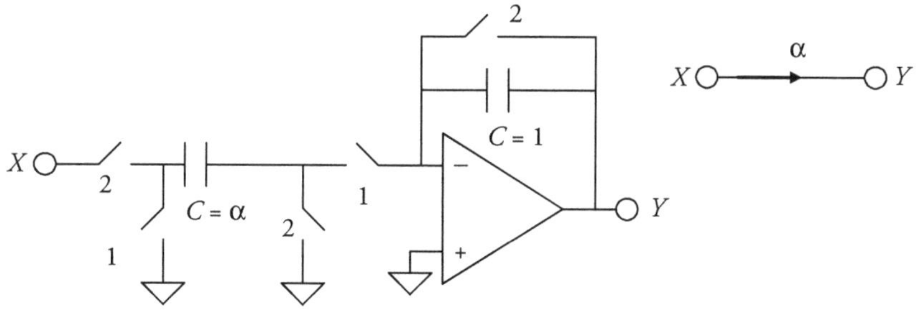

FIGURE 19.20 SC circuit realization of gain stage.

19.3.5 SI Circuit Simulation

The SI circuit implementation of integration with signed inputs is shown in Fig. 19.21.

The SI circuit realization of addition with signed inputs is shown in Fig. 19.22.

The SI circuit realization of gain stage is shown in Fig. 19.23.

19.4 VLSI Active Circuit Simulation of Band-Pass Ladder Filters

In principle, a prototype band-pass ladder can be constructed by using the frequency-transformation process directly on the inductors and capacitors from a prototype low-pass ladder filter as shown in Fig. 19.24.

$$s \rightarrow Q\left(s + \frac{1}{s}\right) \tag{19.9}$$

Figure 19.25 shows a band-pass ladder filter generated from a low-pass ladder using frequency transformation.

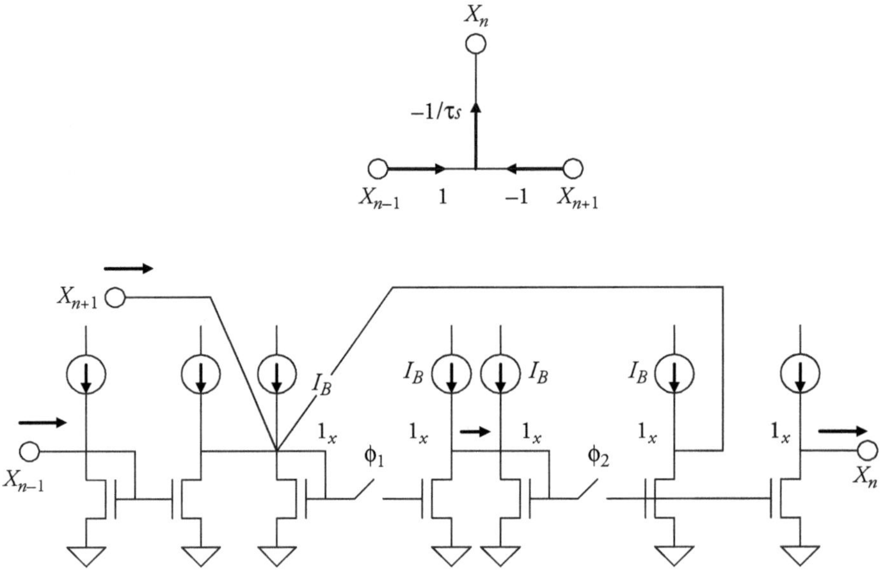

Figure 19.21 SI circuit realization of integration with signed inputs.

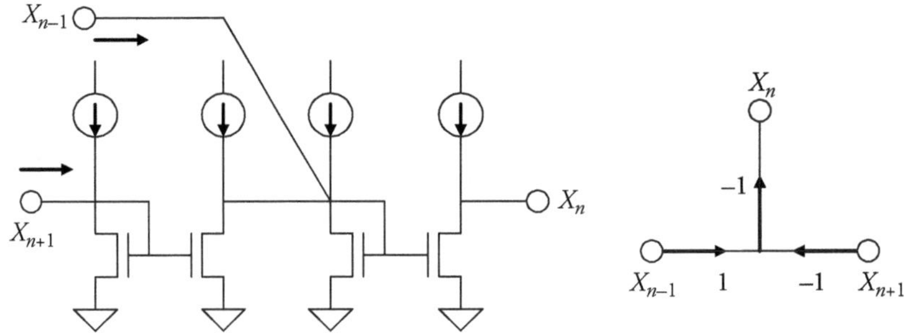

Figure 19.22 SI circuit realization of addition with signed inputs.

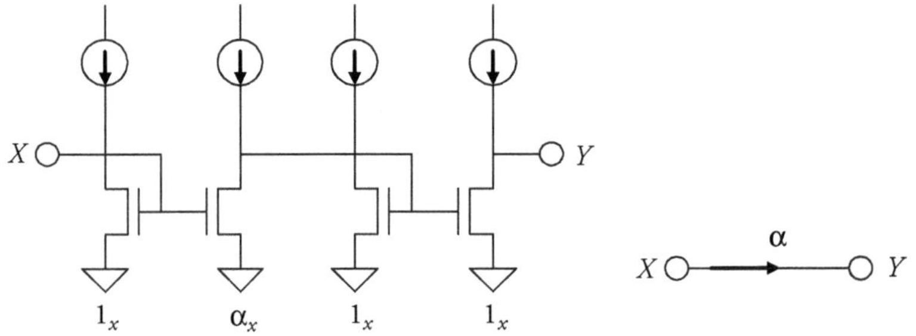

Figure 19.23 CTI circuit realization of gain stage.

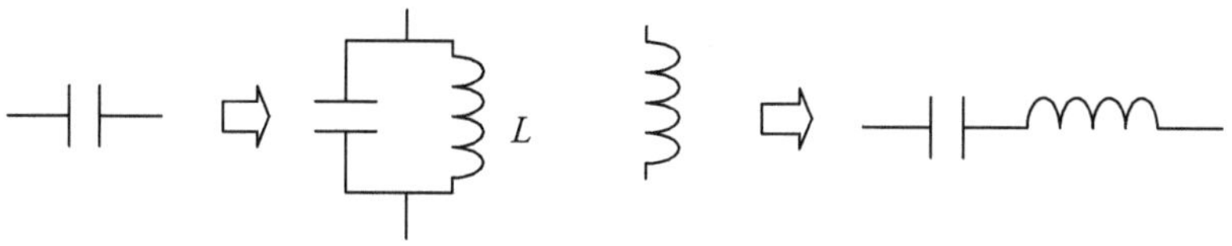

FIGURE 19.24 Low-pass-to-band-pass transformations.

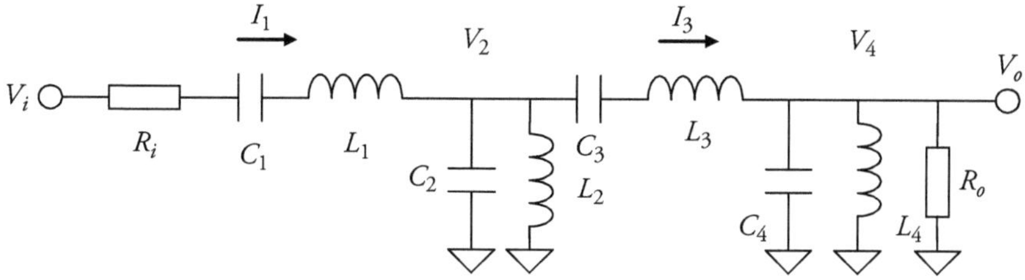

FIGURE 19.25 Band-pass ladder filter generated using the frequency transformation.

Such a band-pass ladder filter can be expressed by the following equations:

$$\begin{cases} X_1 = \dfrac{C_1 s}{L_1 C_1 s^2 + R_i C_1 s + 1}(X_0 - X_2) \\ X_2 = \dfrac{L_2 s}{(L_2 C_2 s^2 + 1)}(X_1 - X_3) \\ X_3 = \dfrac{C_3 s}{(L_3 C_3 s^2 + 1)}(X_2 - X_4) \\ X_4 = \dfrac{L_4 s}{L_4 C_4 s^2 + (L_4/R_o)s + 1} X_3 \end{cases} \tag{19.10}$$

It can be seen that the key to simulating the band-pass ladder filter is to construct the following basic building circuit blocks as follows:

$$\begin{cases} X_1 = \dfrac{AS}{S^2 + \dfrac{1}{Q}S + 1}(X_0 - X_1) \\ X_n = \dfrac{AS}{S^2 + 1}(X_{n-1} - X_{n+1}) \\ X_n = \dfrac{AS}{S^2 + \dfrac{1}{Q}S + 1}(X_{n-1}) \end{cases} \tag{19.11}$$

19.4.1 Active-RC Circuit Simulation

One of the design approaches to the active-RC simulation of band-pass ladder building blocks is based on the biquad circuit shown in Fig. 19.26.

Such a circuit can be used to realize the second-order band-pass transfer function as follows:

$$Y = \frac{AS}{S^2 + \frac{1}{Q}S + 1}(X_1 + X_2) \tag{19.12}$$

Alternatively, the Friend circuit shown in Fig. 19.27 can also be used to implement this band-pass transfer function.

The Friend circuit for infinite-Q implementation is shown in Fig. 19.28, which is based on the Q-enhancement technique.

The transfer function of the above circuit is given as follows:

$$Y = \frac{-A\left(\frac{1+2Q_o^2}{Q_o}\right)S}{S^2 + 1}(X_1 + X_2) \tag{19.13}$$

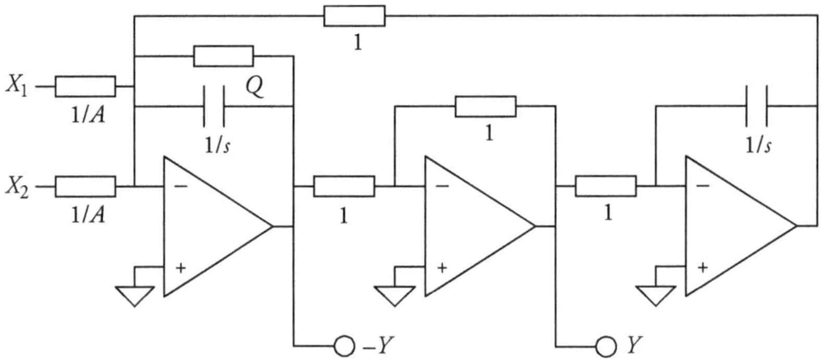

FIGURE 19.26 Active-RC band-pass biquad building block.

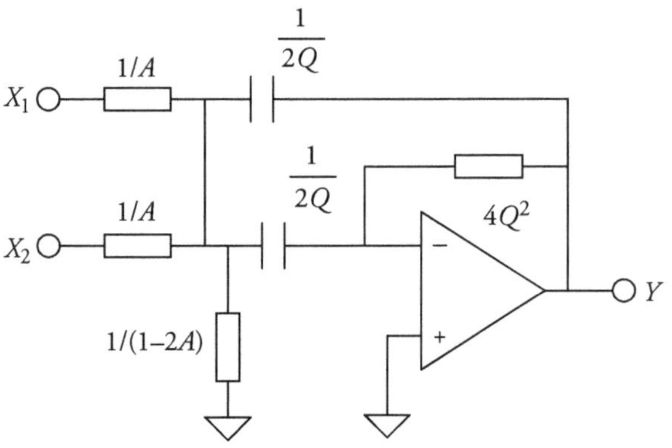

FIGURE 19.27 Delyiannis-Friend band-pass circuit structure.

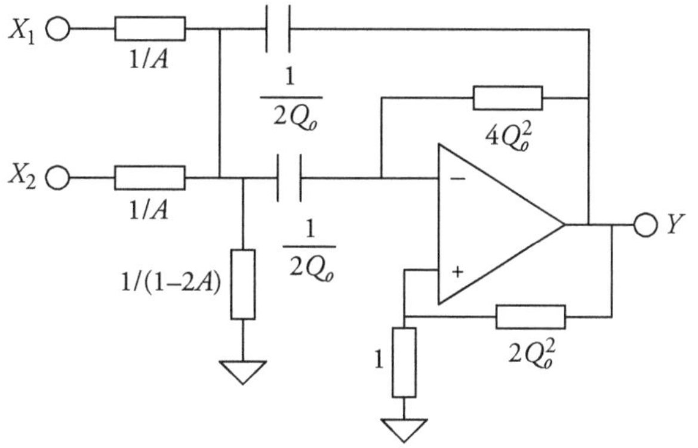

FIGURE 19.28 Friend band-pass circuit structure with enhanced Q.

19.4.2 Gm-C Circuit Simulation

The Gm-C simulation of the band-pass ladder circuit building block is shown in Fig. 9.29.

Such a circuit can be used to realize the band-pass transfer function as follows:

$$Y = \frac{AS}{S^2 + \frac{1}{Q}S + 1}(X_1 - X_2) \tag{19.14}$$

19.5 Design Examples

In the following design examples, we will demonstrate various VLSI circuit simulations for the double-terminated low-pass lossless LC ladder filter shown in Fig. 19.30.

The ladder equations of the filter can be derived as

$$\Rightarrow \begin{cases} X_1 = \beta(X_i - X_2) \\ X_2 = \dfrac{1}{\alpha_2 s}(X_1 - X_3) \\ X_3 = \dfrac{1}{\alpha_3 s}(X_2 - X_4) \\ X_4 = \dfrac{1}{\alpha_4 s}(X_3 - X_5) \\ X_5 = \dfrac{1}{\alpha_5 s}(X_4 - X_o) \\ X_o = X_5 \end{cases} \text{where} \begin{cases} X_i \equiv V_i \\ X_o \equiv V_o \\ \alpha_2 \equiv C_2 R_o \\ \alpha_3 \equiv L_3 / R_o \\ \alpha_4 \equiv C_4 R_o \\ \alpha_5 \equiv L_5 / R_o \\ \beta \equiv R_i / R_o \end{cases} \tag{19.15}$$

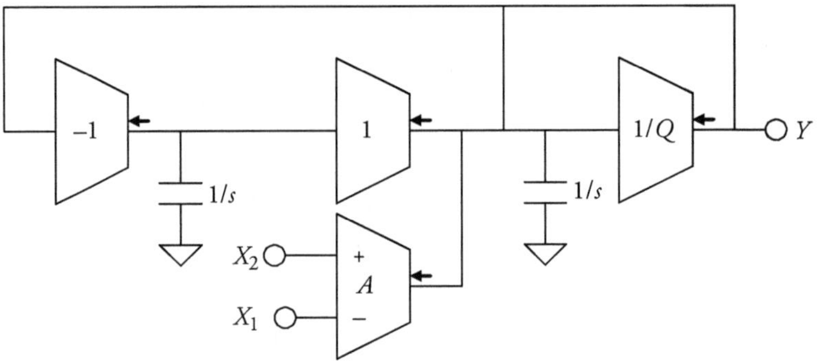

FIGURE 19.29 Gm-C circuit simulation of band-pass circuit building block.

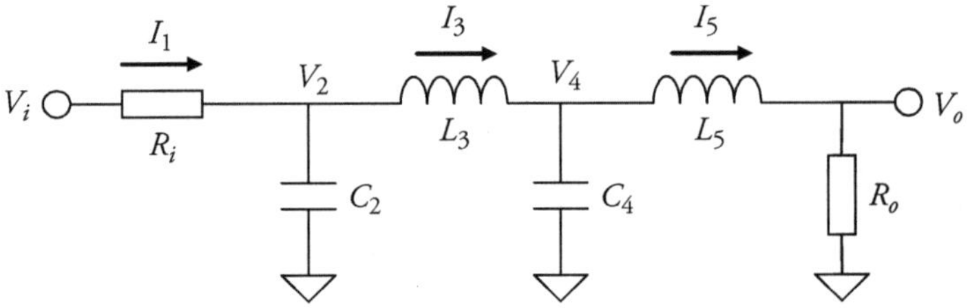

FIGURE 19.30 LC low-pass ladder filter circuit structure.

The addition (subtraction) operation in the first equation of the above equations can be eliminated using a lossy integrator to simplify the circuit implementation, and the yielded equations for the filter are given as

$$\Rightarrow \begin{cases} X_2 = \dfrac{1}{\alpha_2 s}(\beta X_i - \beta X_2 - X_3) \\ X_3 = \dfrac{1}{\alpha_3 s}(X_2 - X_4) \\ X_4 = \dfrac{1}{\alpha_4 s}(X_3 - X_5) \\ X_5 = \dfrac{1}{\alpha_5 s}(X_4 - X_5) \\ X_o = X_5 \end{cases} \tag{19.16}$$

19.5.1 Active-RC Circuit Simulation

Equation (19.16) can be directly realized using active-RC circuit building blocks as shown in Fig. 19.31.

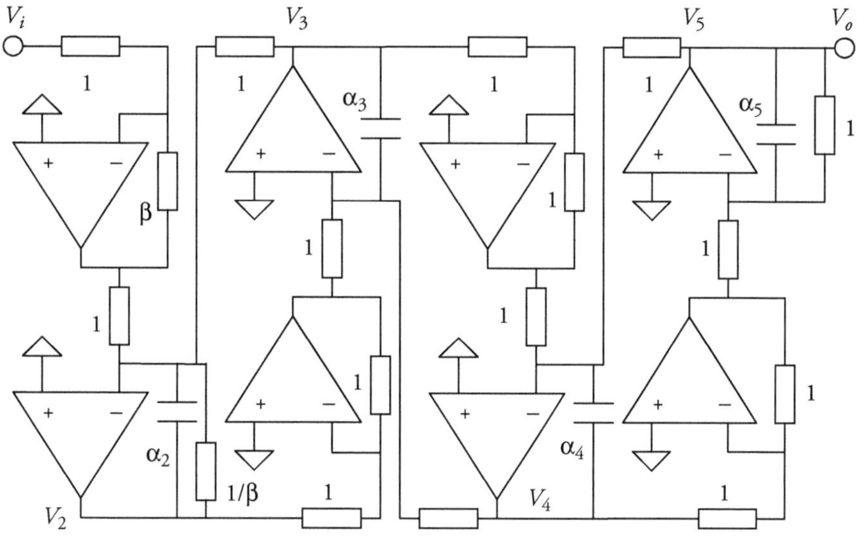

FIGURE 19.31 VLSI active-RC simulation of ladder filter.

19.5.2 Gm-C Circuit Simulation

Similarly, Eq. (19.16) can also be simulated using the Gm-C circuit as shown in Fig. 19.32.

19.5.3 SC Circuit Simulation

To realize Eq. (19.16) using the VLSI SC circuits, an active-RC circuit is first constructed as shown in Fig. 19.33; then this active-RC circuit is mapped into an SC circuit structure by replacing the resistors using switched capacitors as shown in Fig. 19.34.

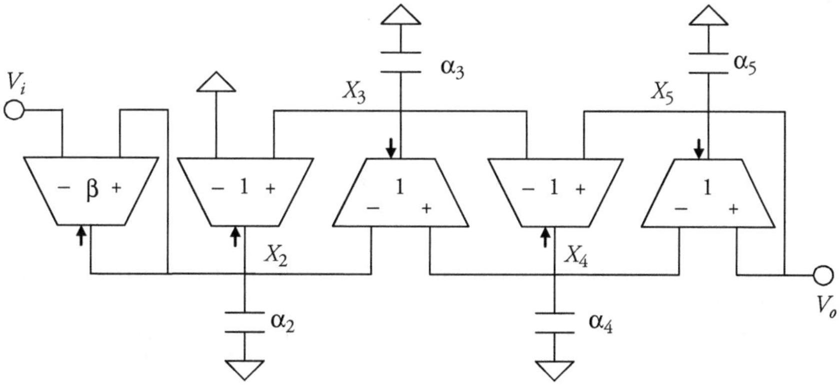

FIGURE 19.32 VLSI Gm-C circuit implementation of the ladder filter.

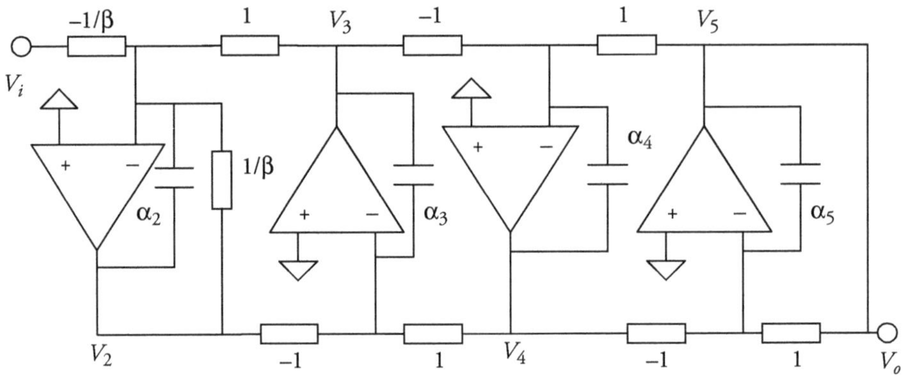

FIGURE 19.33 Active-RC circuit simulation of the ladder filter using negative resistance.

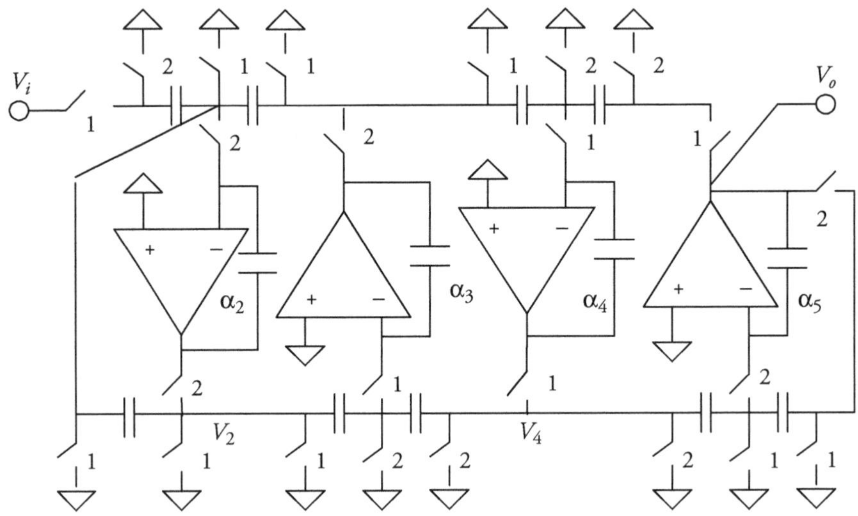

FIGURE 19.34 VLSI SC circuit simulation of the ladder filter.

References

1. R. Schaumann, "Continuous-Time Integrated Filters—A Tutorial," *IEE Proceedings*, vol. 136, pt. G, pp. 184–190, August 1989.
2. M. Banu, Y. Tsividis, "An Elliptic Continuous-Time CMOS Filter with On-Chip Automatic Tuning," *IEEE J. Solid-State Circuits*, vol. SC-20, no. 1, pp. 1114–1121, December 1985.
3. H. Khorramabadi, P. R. Gray, "High-Frequency CMOS Continuous-Time Filters," *IEEE J. Solid-State Circuits*, vol. SC-19, no. 6, pp. 939–948, December 1984.
4. F. Krummenacher, G. Van Ruymbeke, "Integrated Selectivity for Narrow-Band FM IF Systems," *IEEE J. Solid-State Circuits*, vol. SC-25, no. 3, pp. 757–760, June 1990.

5. J. M. Khoury, "Design of a 15-MHz CMOS Continuous-Time Filter with On-Chip Tuning," *IEEE J. Solid-State Circuits*, vol. SC-26, no. 12, pp. 1988–1997, December 1991.
6. G. M. Jacobs, D. J. Allstot, R. W. Brodersen, P. Gary, "Design Techniques for MOS Switched Capacitor Ladder Filters," *IEEE Trans. on Circuits and Systems*, vol. CAS-25, no. 12, pp. 1014–1021, December 1978.
7. G. M. Jacobs, D. J. Allstot, R. W. Brodersen, P. Gary, "Design Techniques for MOS Switched Capacitor Ladder Filters," *IEEE Trans. on Circuits and Systems*, vol. CAS-25, no. 12, pp. 1014–1021, December 1978.
8. R. Gregorian, K. W. Martin, G. C. Tempes, "Switched-Capacitor Circuit Design," *Proc. of the IEEE*, vol. 71, no. 8, pp. 941–966, August 1983.
9. R. B. Datar, A. S. Sedra, "Exact Design of Stray-Insensitive Switched-Capacitor Ladder Filters," *IEEE Trans. on Circuits and Systems*, vol. CAS-30, no. 12, pp. 888–898, December 1983.
10. Y. Ishibashi, K. Tsuduki, "Synthesis and RC-Active Simulation of an LR Terminated LC Filter," *IEE Proceedings, Circuits, Devices and Systems*, vol. 142, no. 1, pp. 49–52, February 1995.
11. E. Hayahara, "Design of Active RC Filters Simulating Node Equations of LC Filters," *IEEE Trans. on Circuits and Systems*, vol. 31, no. 4, pp. 394–398, April 1984.
12. J. T. Taylor, D. G. Haigh, "Generalised Switched Capacitor Simulation of LCR Ladder Filters employing state space methods," *European Conference on Circuit Theory and Design*, pp. 21–26, September 1989.
13. M. S. Tawfik, P. Senn, "A 3.6-MHz Cutoff Frequency CMOS Elliptic Low-Pass Switched-Capacitor Ladder Filter for Video Communication," *IEEE J. Solid-State Circuits*, vol. 22, no. 3, pp. 378–384, June 1987.
14. D. G. Haigh, J. T. Talflon, B. Singh, "Low-Sensitivity Switched Capacitor Simulation of Elliptic Lowpass LCR Ladder Filters," *Electronics Letters*, vol. 24, no. 1, 7, pp. 52–54, January 1988.
15. C. Ikeda, Y. Horio, S. Mori, "SC Bilinear Band-Elimination Ladder Filters with Extended Biquad," *Electronics Letters*, vol. 24, no. 7, pp. 432–434, March 1988.

CHAPTER 20

Tuning Circuit Techniques for VLSI Analog Filters

It has been a well-known fact that it is difficult to achieve the absolute accuracy for most VLSI circuit parameters such as the resistance, the capacitance, and the time constants. The variations of these circuit parameters are mainly contributed by the PVT variations and the device mismatches of the VLSI process technologies. Typical variations of these absolute circuit parameters can be as high as ±20–50 percent in a nominal VLSI process technology. On the other hand, it is fairly easy to achieve a relative (or matching) accuracy based on the physically matched VLSI device structures. Typical matching accuracies of VLSI device parameters can be better than 1–0.1 percent. Among various VLSI signal processing circuit parameters, some directly depend on the absolute values of the VLSI circuit component parameters. These parameters usually suffer from the limitations of high VLSI process variations. The others that rely only on the ratio of the same type of circuit parameters can usually be designed fairly precisely in the VLSI fabrication and in the circuit operations. Typically, bandwidths of continuous-time VLSI filter circuits—such as VLSI active-RC, MOS-C, Gm-C, and CTI filters—belong to the first group of parameters. Such parameters are determined by the products or divisions of two uncorrelated VLSI device parameters such as resistance, capacitance, and Gm values; and they usually cannot be designed accurately without employing certain types of tuning circuit techniques while the quality factors and the dc gains usually belong to the second group of parameters that are determined by the ratio of resistances, capacitances, or Gm values. These parameters usually do not need tuning unless further performance enhancements are required.

20.1 Classification of VLSI Filter-Tuning Circuits

VLSI filter-tuning circuits can be classified based on the various design aspects such as the observability, the controllability, the implementation technique, and the circuit architecture. Several commonly used definitions of VLSI analog filter-tuning circuit types are as follows:

- The F-tuning and the Q-tuning
- The resistor, capacitor, and Gm tuning
- The continuous and discrete tuning
- The direct and indirect tuning
- The open-loop and closed-loop tuning

20.1.1 F-Tuning and Q-Tuning

Bandwidth (or characteristic frequency) and quality factor are the two key filter performance parameters. Filter-tuning schemes based on the two parameters are known as either (1) the F-tuning or (2) the Q-tuning.

In the F-tuning process, the filter frequency responses are tuned proportionally without impacting the shapes of the filter response curves (or Bode plots). On the other hand, for the Q-tuning process, the filter's response curves are properly tuned without impacting the bandwidth of the filter circuit. The operations of the two filter circuit tuning schemes for a second-order filter section are shown in the pole/zero plots in Fig. 20.1.

It can be seen that during the F-tuning operation, the poles (and zeros) of the filter are adjusted close to or away from the origin proportionally without changing their angles in the s-plane. While during the Q-tuning operation, the angles of poles (and zeros) are adjusted without changing their distance to the origin. The effect of F-tuning is very similar to a frequency-scaling operation in a mathematical sense. It can be seen that the F- and Q-tuning techniques are orthogonal to each other for the second-order filter circuits since the frequency and the Q-factor are two independent parameters in this case.

In most practical filter structures, the frequency parameters of the filter depend on the absolute values of the circuit parameters, which are usually very sensitive to the PVT variations, while the quality factor for VLSI filters usually depend on the value ratio of the same type of components, which can be designed fairly accurately without tuning.

20.1.2 Resistor, Capacitor, and Gm Tuning

Capacitors, resistors, and Gm circuit elements are commonly used for the tuning of analog filter circuits. Filter-tuning schemes based on these tuning elements are usually classified as (1) the resistor tuning, (2) the capacitor tuning, or (3) the Gm tuning.

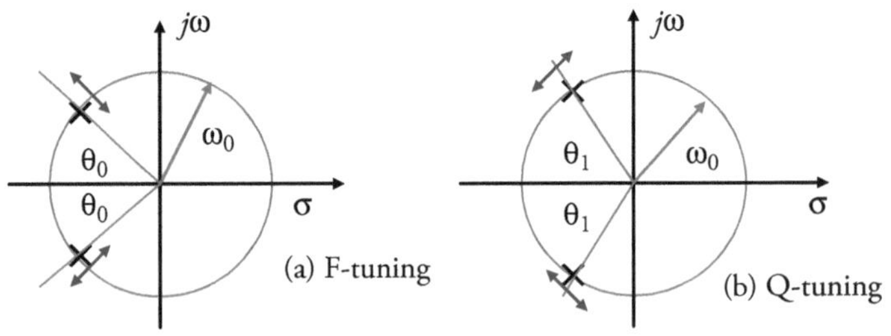

Figure 20.1 Filter pole and zero locations under the F- and Q-tuning operations.

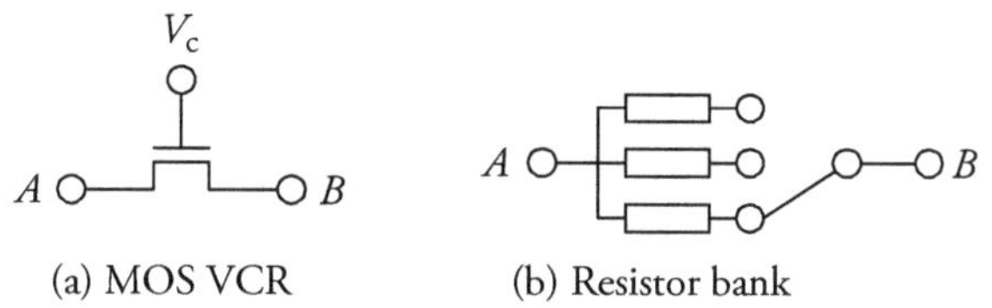

FIGURE 20.2 Resistor-based VLSI circuit tuning structures.

MOS voltage-controlled resistors (VCRs) and resistor banks are two popular methods in the resistor-tuning schemes as shown in Fig. 20.2, where the MOS VCR provides continuous tuning capability with limited linearity, and the resistor bank provides high-linearity tuning with limited tuning resolution.

MOS or PN junction diode–based varactors and linear capacitor banks are commonly used for the capacitor-tuning schemes as shown in Fig. 20.3, where a varactor provides continuous tuning capability with limited linearity, and the capacitor bank provides high-linearity tuning with limited tuning resolution.

Gm circuits can be used for the VLSI Gm-based circuit tuning where the Gm value can be controlled through the bias current or bias voltage as shown in Fig. 20.4.

20.1.3 Direct and Indirect Tuning

Based on the tuning circuit architecture, the analog filter-tuning schemes can also be classified as either (1) the direct tuning or (2) the indirect tuning (see Fig. 20.5).

When the filter circuit in the signal processing path is directly adjusted in the tuning loop, it is known as direct tuning. Alternatively, the indirect tuning circuit consists of a master-slaver circuit structure where a master filter or filter section is used within the tune loop, and the slaver filter in the signal processing path is tuned out of the loop based on device matching.

The direct-tuning architecture typically offers the advantage of higher accuracy as the signal path is directly tuned. However, during the tuning, the filter will not be available for the signal processing operation. Consequently, there is a speed penalty in implementing the direct-tuning scheme. In addition, to implement the direct-tuning architecture in multiple-filter systems, each filter will need a dedicated tuning circuit, which may result in higher area penalty in the VLSI implementation.

An advantage of the indirect-tuning architecture is that the master-tuning circuit can be shared by multiple-slaver filter circuits as long as the basic building blocks in

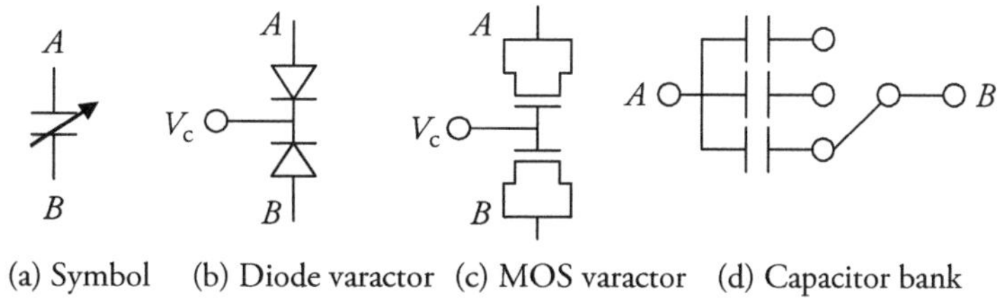

FIGURE 20.3 Capacitor-based VLSI circuit tuning structures.

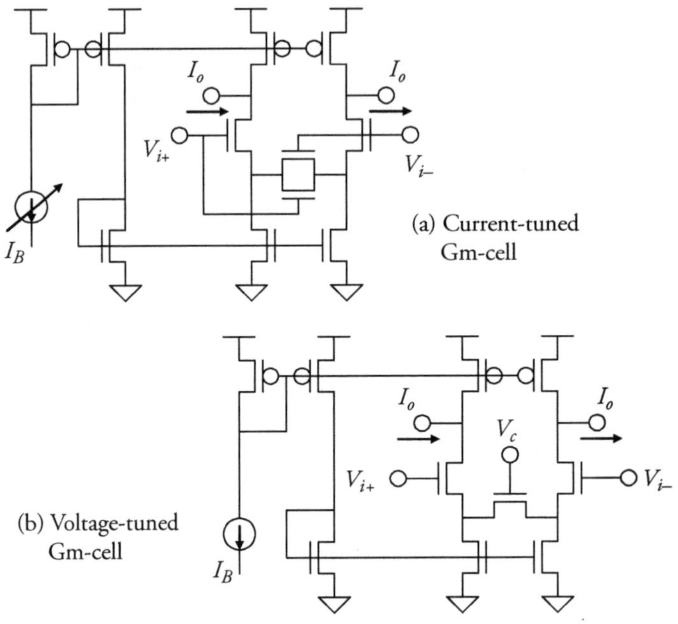

FIGURE 20.4 Gm-based VLSI tuning circuit.

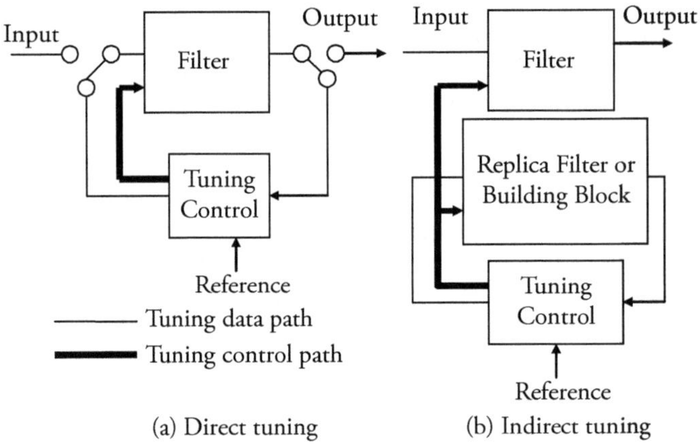

FIGURE 20.5 Direct- and indirect-tuning architectures.

these filters are identical. Such a tuning method can yield low-area penalty in VLSI implementation. Since the filter to be tuned is not part of the tuning loop, the filter operation and the tuning operation can be carried out simultaneously without interference. In most high-performance applications, dedicated time slots are usually assigned for the tuning and the normal operation for minimizing the interaction between the main filter and the tuning loop. Indirect tuning is the most commonly used tuning architecture for practical VLSI filters. A major drawback of the indirect-tuning architecture is the tuning accuracy limitation resulting from the mismatches between the master- and slaver-filter sections.

20.1.4 Closed-Loop and Open-Loop Tuning

Based on the implementation of the adjustment device, the tuning schemes can also be classified as either (1) the closed-loop tuning or (2) the open-loop tuning.

In the closed-loop tuning scheme, the elements used for tuning (such as the VCR, varactor, Gm-cell, etc.) are within the tuning loop. The changes in the elements' values during tuning are monitored and compared with the reference to determine the tuning operation.

In an open-loop tuning implementation, the tuning element is outside of the tuning loop, and the value of the tuning device is selected based on the matching. The trade-off between the closed-loop and the open-loop tuning implementations is that the closed-loop tuning provides better accuracy at the penalty of a more complicated design and that the open-loop tuning provides a simpler design at the penalty of the design's accuracy.

20.1.5 Continuous and Discrete Tuning

Based on the tuning resolution, the analog circuit tuning schemes can be classified as either (1) the continuous tuning or (2) the discrete tuning.

Examples of continuous tuning are VCRs, varactor-based tuning, and discrete tuning, which are capacitor and resistor bank–based tuning circuit implementations.

20.2 VLSI Analog Filter-Tuning Architectures

All analog filter-tuning circuits can be modeled using a transfer function under tuning (TFUT) block with the s-domain transfer function represented as $H(s, X)$, where X is the parameter under tuning, or PUT (e.g., bandwidth, time constant, Q, etc.), of the TFUT block. During the tuning operation, the response of the TFUT is measured using a given tuning circuit, and the parameter X is tuned according to a certain tuning algorithm until a given tuning target is achieved. On the other hand, this TFUT is part of the analog-filter or replica-filter sections (such as the integrator in the filter in many applications) and based on matching. The elements inside the filter are tuned in the same way as the TFUT block to achieve certain performance target.

20.2.1 Single-Slope Tuning

In the single-slope tuning architecture shown in Fig. 20.6, the TFUT block is a lossless integrator.

The PUT is the time constant ($\omega_x = 1/\tau$) of the integrator with the transfer function of the TFUT given as follows:

$$H(s,\omega_x) = \frac{1}{\left(\frac{s}{\omega_x}\right)} \tag{20.1}$$

During the tuning operation shown in Fig. 20.7, the integrator capacitor is first discharged at the reset phase ϕ_1. Then this integrator capacitor is charged with a known

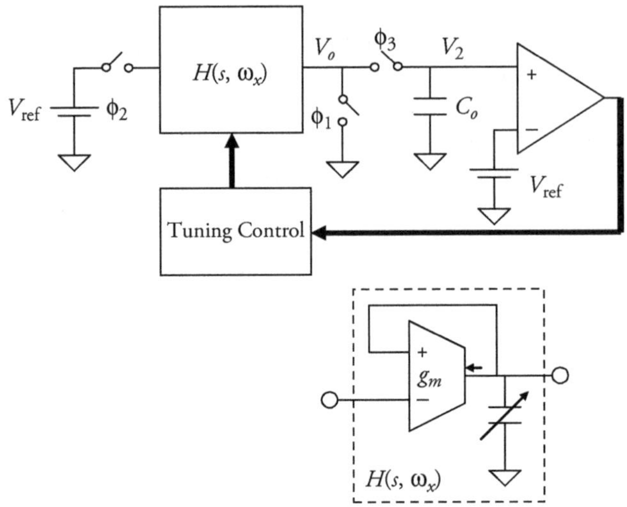

FIGURE 20.6 Single-slope tuning architecture and the Gm-C TFUT example.

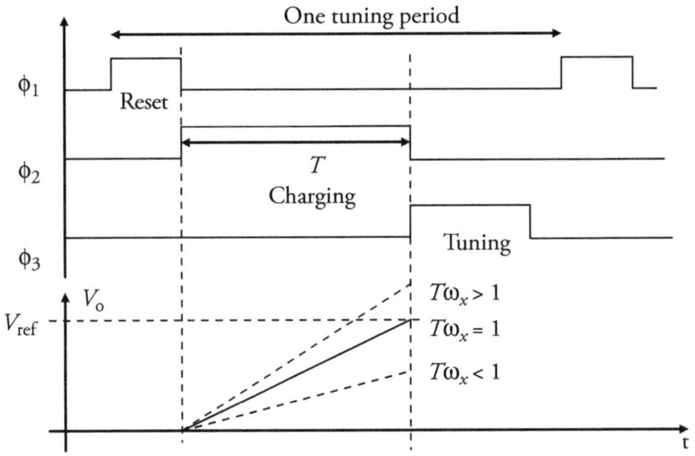

FIGURE 20.7 Timing of the single-slope tuning loop.

reference voltage V_{ref} for a given time period T controlled by phase ϕ_2. At the end of ϕ_2, the output voltage of the integrator is given as follows:

$$V_o = \omega_x T V_{ref} \tag{20.2}$$

This voltage is sampled into a hold capacitor during phase ϕ_3 and then compared with the same reference voltage V_{ref}. According to the relationship between V_o and V_{ref}, ω_x is adjusted during the tuning process. At the end of the tuning process, the output V_o is equal to V_{ref}.

Consequently, when the tuning operation is completed, the time constant of the integrator is tuned to the defined reference period T.

$$\tau \equiv \frac{1}{\omega_x} = T \tag{20.3}$$

20.2.2 Dual-Slope Tuning

The dual-slope tuning architecture consists of two identical lossless integration paths, one as the known reference and the other as the TFUT. Such a circuit operates on three nonoverlap phases as shown in Fig. 20.8.

During the tuning operation, the integrator capacitor is charged through the reference integration path in the ϕ_2 phase after being initialized to zero. Then the integrator is discharged through the second integration path in the ϕ_3 phase. The time to discharge the capacitor in the discharge phase is measured and compared with the charging period to generate the tuning control of the circuit. The tuning target is such that at the end of tuning process, the charging and discharging periods are equal as shown in Fig. 20.9.

Since the reference and TFUT transfer functions of the two integration paths are given as

$$\begin{cases} H(s,\omega_o) = \dfrac{\omega_o}{s} \\ H(s,\omega_x) = \dfrac{\omega_x}{s} \end{cases} \tag{20.4}$$

the tuning target is given as

$$\omega_x = \omega_{o1} \tag{20.5}$$

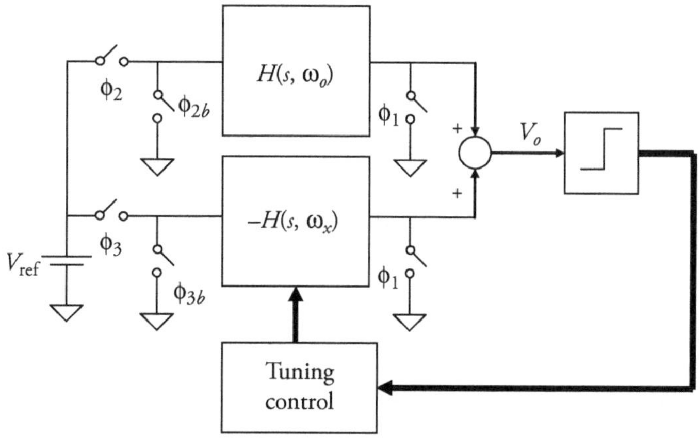

FIGURE 20.8 Dual-slope tuning architecture.

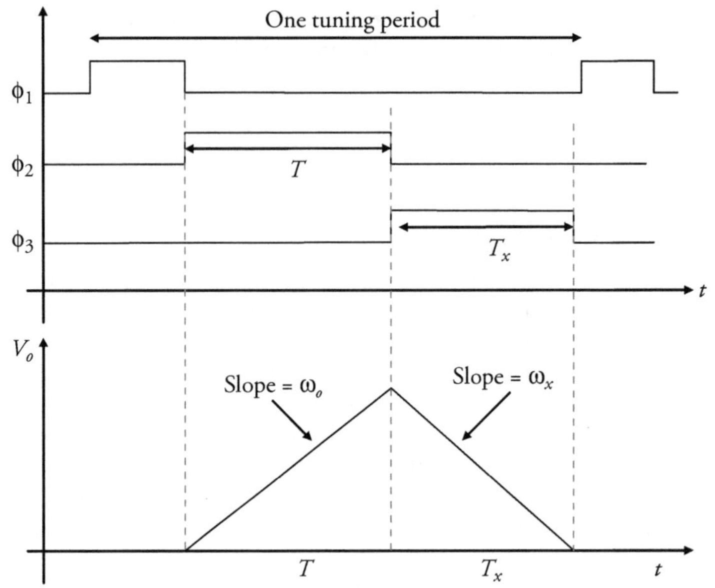

FIGURE 20.9 Timing of the dual-slope tuning circuit.

20.2.3 Phase-Lock Tuning

In the PLL-based tuning architecture shown in Fig. 20.10, the tuning block is a voltage-controlled oscillator (VCO) circuit, which generates a clock according to the analog-control voltage or digital-control code.

The TFUT in the PLL-based tuning is an oscillator with the transfer function and oscillation frequency given as follows:

$$H(s, \omega_x) = \frac{1}{\left(\frac{s}{\omega_x}\right)^2 + 1} \tag{20.6}$$

$$f = \frac{\omega_x}{2\pi} \tag{20.7}$$

During the tuning operation, the phase (and frequency) of the oscillator clock is compared with a reference clock to generate the tuning-control signal. When the tuning is completed, the oscillator clock phase is locked to the reference clock, and the time constant of the tuning block is tuned to the reference clock as follows:

$$\omega_x = 2\pi \cdot f_{\text{ref}} \tag{20.8}$$

20.2.4 Delay-Response Tuning

The delay-based tuning architecture is based on the steady-state phase delay of the TFUT. The tuning circuit is shown in Fig. 20.11.

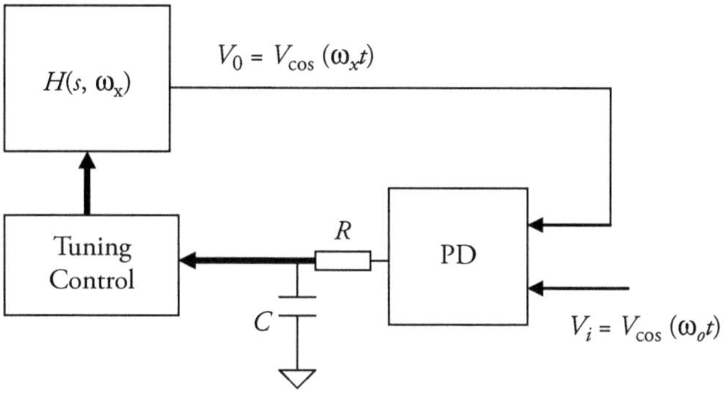

FIGURE 20.10 PLL-based filter-tuning architecture.

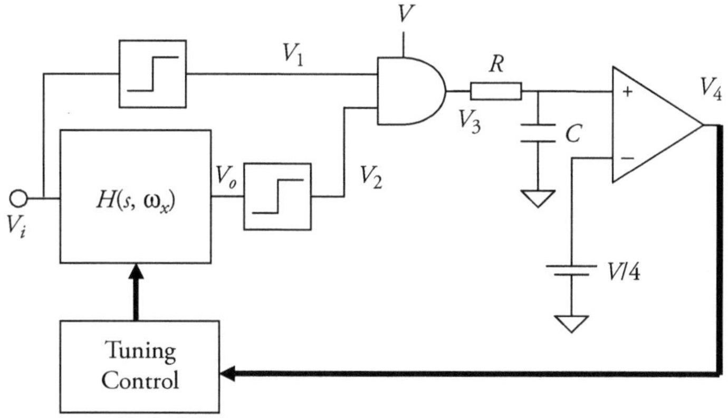

FIGURE 20.11 Delay-based tuning architecture.

The TFUT in such a circuit is a second-order low-pass filter section with the transfer function given as follows:

$$H(s, \omega_x) = \frac{1}{\left(\dfrac{s}{\omega_x}\right)^2 + \dfrac{1}{Q}\left(\dfrac{s}{\omega_x}\right) + 1} \tag{20.9}$$

The delay of the block is measured by the two-limiter amplifiers and a logic AND gate. The output is low-pass filtered and compared with the reference $V/4$ to generate the tuning control.

Such a tuning block provides $\pi/2$ delay as shown in Figs. 20.12 and 20.13 when the frequency of the reference signal is equal to the cutoff frequency ω_x of the filter. The delay through the tuning block is detected in the circuit shown in Fig. 20.11 to generate the control. When the tuning is completed, the time constant of the tuning block is tuned to the frequency of the reference as follows:

$$\omega_x = \omega_0 \tag{20.10}$$

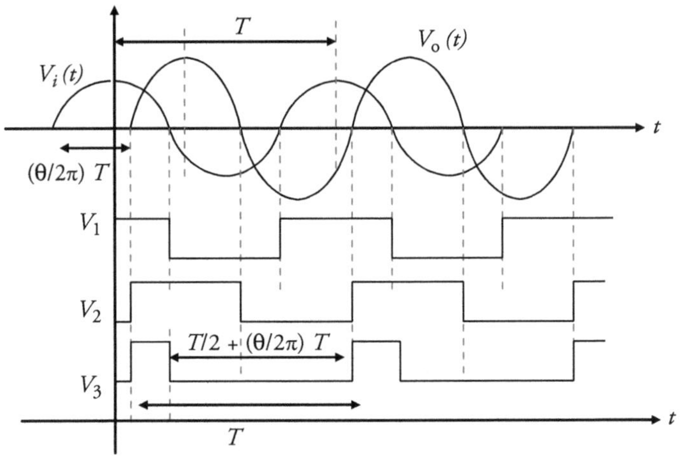

Figure 20.12 Delay-based tuning signal waveforms.

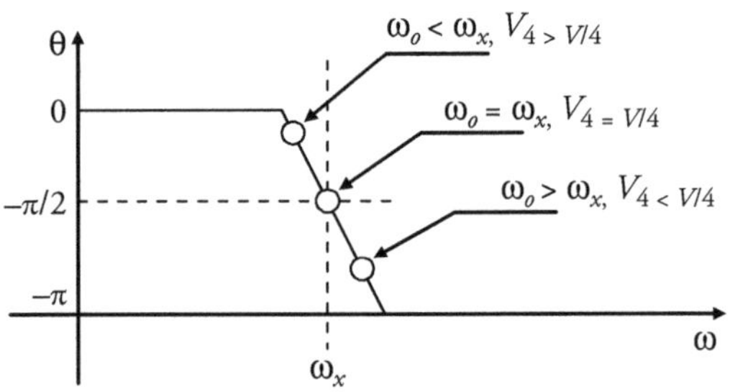

Figure 20.13 Delay-based tuning signal criteria.

20.2.5 Gain-Response Tuning

The gain-response tuning architecture is based on the gain response of the TFUT block to the steady-state input. The TFUT block in this architecture consists of an integrator as shown in Fig. 20.14.

$$H(s, \omega_x) = \frac{1}{\left(\dfrac{s}{\omega_x}\right)} \tag{20.11}$$

The magnitude of the input and output of the TFUT block is used to determine the relationship between the time constant of the tuning block and the reference as shown in Fig. 20.15. When the tuning is completed, the gain of the tuning block equals to 1, and the time constant of the tuning block is tuned to the reference frequency:

$$\omega_x = \omega_0 \tag{20.12}$$

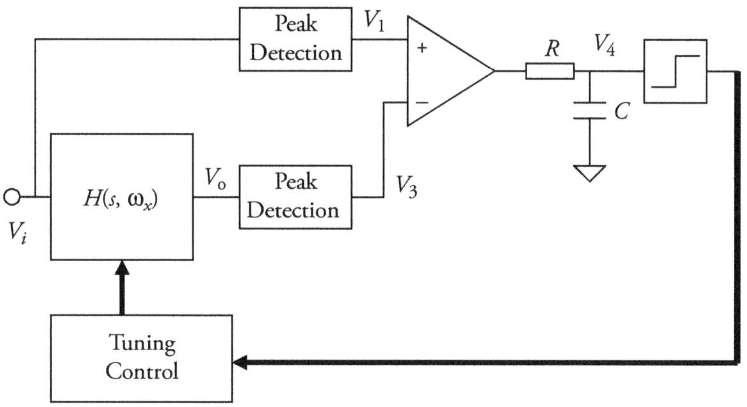

FIGURE 20.14 Gain-based tuning scheme.

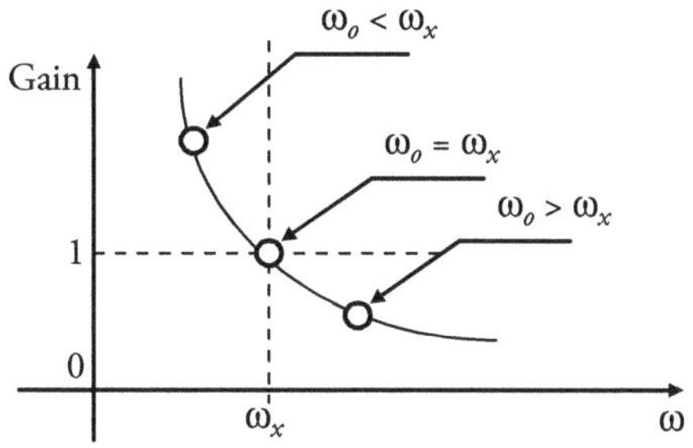

FIGURE 20.15 Detection of frequency parameter via gain response.

20.2.6 Constant-Source/Sink Tuning

The constant-source/sink tuning architecture consists of two opposite constant current paths introduced by a reference component and a tuning component as shown in Fig. 20.16. The tuning is implemented such that at the end of the tuning, the source and sink paths are balanced. Such a condition results in the tuning of the tuned component to match the reference component.

$$X_x = X_{\text{ref}} \tag{20.13}$$

20.3 VLSI Filter-Tuning Circuits Implementations

The above filter-tuning schemes can be implemented in VLSI technology for filter-tuning applications.

20.3.1 VLSI Active-RC Filter-Tuning Circuits

Figure 20.17 shows a VLSI active-RC filter-tuning circuit based on the dual-slope tuning scheme using a switched-capacitor bank used as the tuning circuit element.

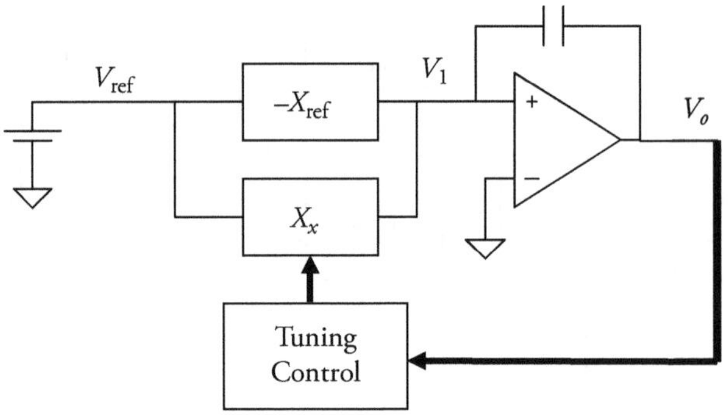

FIGURE 20.16 Constant-source/sink-based tuning architecture.

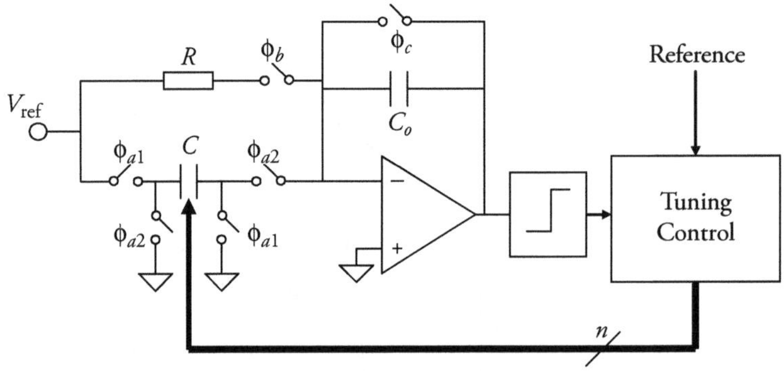

FIGURE 20.17 Discrete capacitor-tuning circuit based on dual slope with SC integrator.

The tuning circuit is controlled by a set of four nonoverlap clocks as shown in Fig. 20.18.

During the T_1 (first slope) period, the integration capacitor C_o is charged from zero through the resistor R in the following:

$$Q = -\frac{V_{ref}}{R} \cdot T_1 \tag{20.14}$$

During the T_2 period, the integration capacitor is discharged through the switched capacitor C at the rate of

$$\Delta Q = \frac{V_{ref} \cdot C}{M} \tag{20.15}$$

The number of the pulse M at which the total charge on the capacitor C_o is discharged is related to the RC product of the circuit.

$$RC = \frac{T_1}{M} \tag{20.16}$$

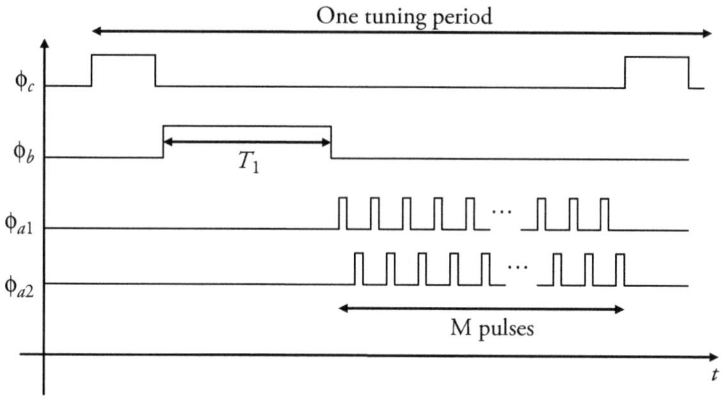

Figure 20.18 Timing diagram of dual-slope active-RC filter-tuning circuit.

Since T_1 is generated from the reference clock by a predefined method and the number of the pulse is measurable in the circuit, the circuit shown in Fig. 20.17 provides a very accurate way of measuring and controlling the RC constant of the circuit, which, in turn, can be used to tune the filter.

Alternatively, the tuning can also be implemented on the resistor as shown in Fig. 20.19 with the same tuning procedures.

20.3.2 VLSI MOS-C Filter-Tuning Circuits

A MOS-C filter continuous-tuning circuit implementation employing the MOS VCR is shown in Fig. 20.20. It operates with the control of three nonoverlap clocks as shown in Fig. 20.21.

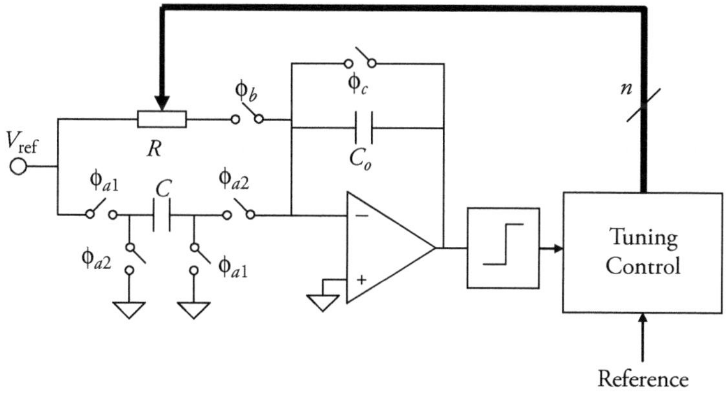

Figure 20.19 Discrete resistor-tuning circuit based on dual slope with SC integrator.

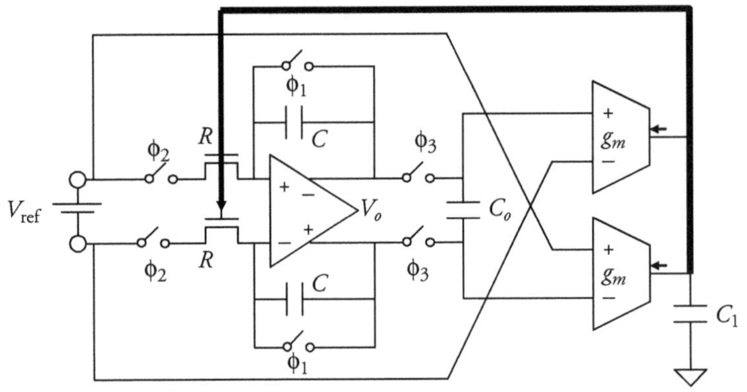

FIGURE 20.20 VLSI MOS-C filter-tuning circuit.

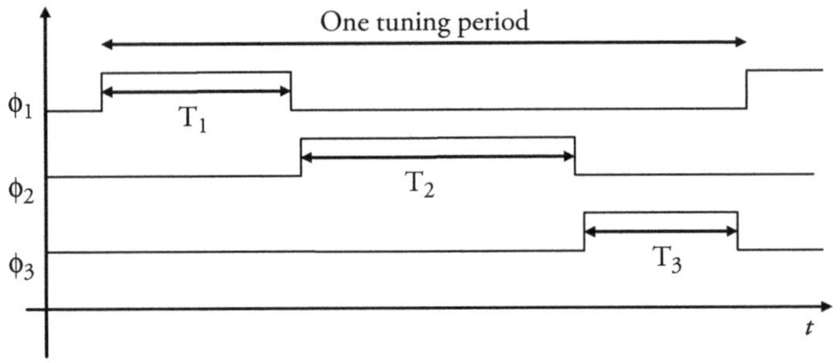

FIGURE 20.21 Timing diagram of MOS-C filter-tuning circuit.

During the tuning operation, the capacitor C is first discharged at ϕ_1. Then it is charged through the MOS VCR in ϕ_2. At the end of ϕ_2, the output voltage of the integrator is given as follows:

$$V_o = \frac{V_{\text{ref}}}{RC} \cdot T_2 \tag{20.17}$$

At ϕ_3, the above voltage is sampled to the capacitor C_o and maintained until the next sampling operation. This voltage is compared with the V_{ref}, and the difference is converted to a current to charge or discharge capacitor C_1. Due to the negative feedback of the loop, each of the tune periods will try to reduce the above voltage difference. When the tuning is completed, the voltage output of the integrator equals to the following reference voltage:

$$V_{\text{ref}} = V_o = \frac{V_{\text{ref}}}{RC} \cdot T_2 \tag{20.18}$$

that is,

$$RC = T_2 \tag{20.19}$$

20.3.3 VLSI Gm-C Filter-Tuning Circuits

In the Gm-tuning circuit implementation shown in Fig. 20.22, an external reference resistor R_{ref} is used to tune the Gm-cell. The circuit operates as follows: Two currents of opposite polarity are introduced to the input of the tuning integrator. The difference of the current is integrated to generate the Gm-tuning control. When the tuning is completed, the two currents are equal; and the node voltage V_x is equal to 0, which yields the following tuning equation:

$$g_m = \frac{1}{R_{ref}} \tag{20.20}$$

The above Gm tuning can also be realized using an alternative circuit as shown in Fig. 20.23, where the node voltage V_x at the end of the tuning sequence will be equal to the reference voltage V_{ref}. And the Gm value will be tuned to the reference resistor.

The drawback of the above two circuits is that they require an accurate resistance that for practical VLSI filter design is usually not trivial to obtain on chip. In the following Gm-tuning circuit implementation, a switched-capacitor resistor is used to implement the accurate on-chip resistor as shown in Fig. 20.24.

In this implementation, the reference resistance is given as

$$R_{ref} = \frac{T}{C} \tag{20.21}$$

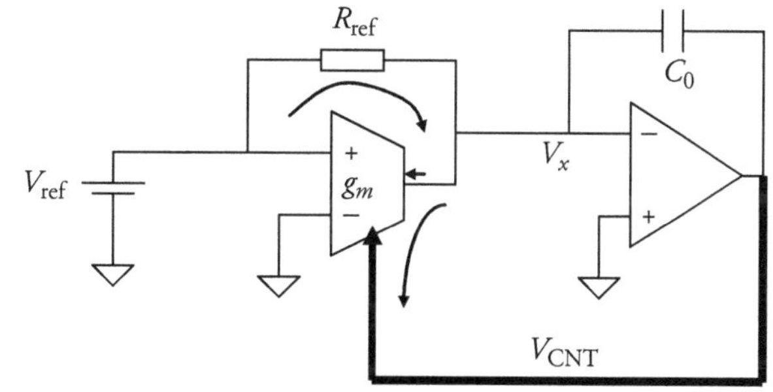

Figure 20.22 VLSI Gm-C filter-tuning circuit (I).

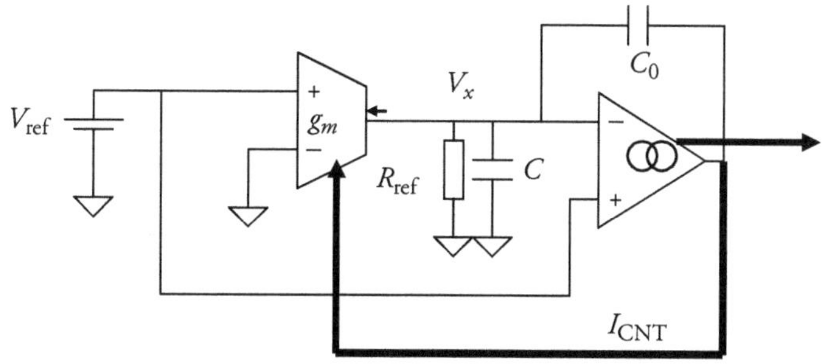

Figure 20.23 VLSI Gm-C filter-tuning circuit (II).

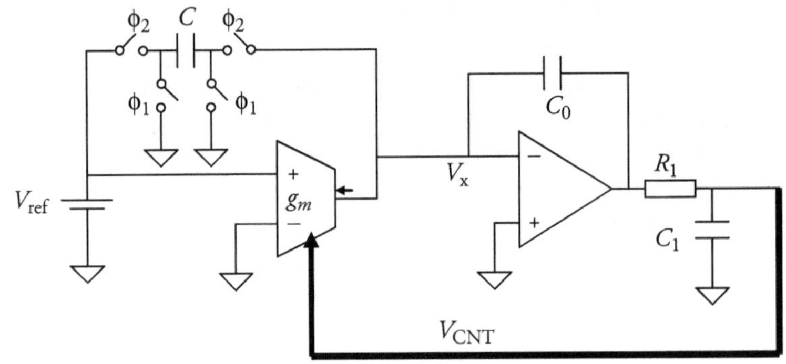

FIGURE 20.24 VLSI Gm-C filter-tuning circuit (III).

and the Gm value after tuning is

$$g_m = \frac{1}{R_{ref}} = \frac{C}{T} \tag{20.22}$$

Such a tuning circuit can be used to achieve a very accurate filter time constant, considering that the period of the on-chip clock can be tightly controlled.

20.3.4 VLSI CTI Filter-Tuning Circuits

The time constant of a VLSI CTI filter circuit usually depends on the Gm or VCR resistance of the CTI building blocks, which provides tuning capability of CTI filters through the equivalent resistance of the VCR in the circuit. Figure 20.25 shows an implementation of the open-loop CTI filter-tuning circuit structure.

Assuming all NMOS devices are identical and ignoring the body effects in this circuit, we can obtain the equivalent conductance of the VCR device as follows:

$$\frac{1}{R} = \beta\left[V_C - V_T - \left(V_T + \sqrt{\frac{2I_B}{\beta}}\right)\right] \tag{20.23}$$

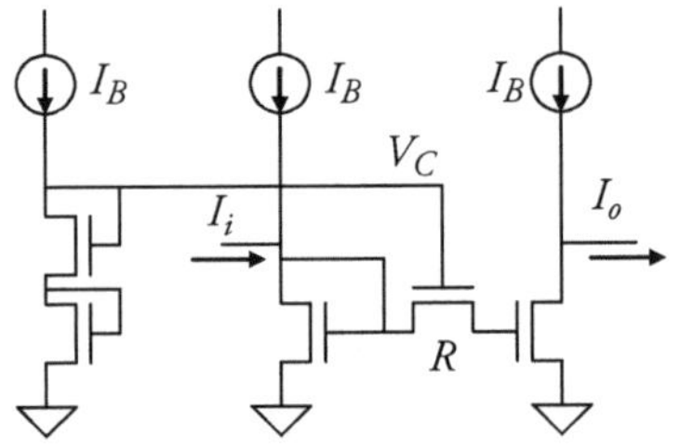

FIGURE 20.25 VLSI CTI filter-tuning circuit (I).

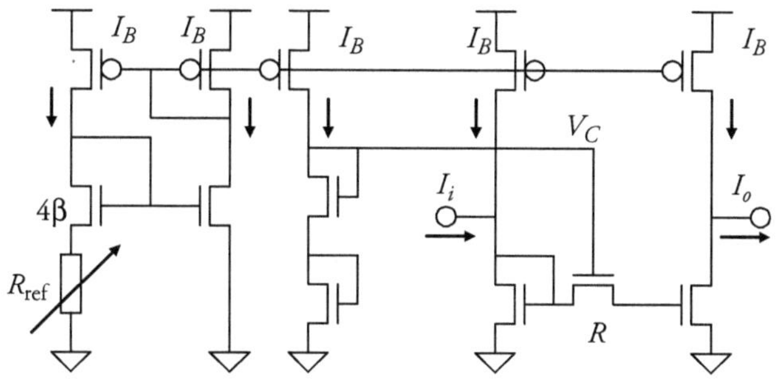

FIGURE 20.26 VLSI CTI filter-tuning circuit (II).

In the tuning circuit implementation, the control voltage V_c can be generated from the biasing circuit as follows:

$$V_C = 2V_T + 2\sqrt{\frac{2I_B}{\beta}} \tag{20.24}$$

Substituting Eq. (20.24) into Eq. (20.23), we have the following:

$$\frac{1}{R} = \sqrt{2\beta I_B} \tag{20.25}$$

It can be seen that the resistance can be controlled through the bias current of the circuit.

Figure 20.26 shows a complete open-loop resistor-tuning circuit for the CTI filters where the bias current is generated from a constant Gm-biasing circuit.

The bias current in the circuit can be derived as follows:

$$I_B = \frac{1}{2\beta R_{\text{ref}}^2} \tag{20.26}$$

Substituting Eq. (20.26) into Eq. (20.25), we have the following:

$$\frac{1}{R} = \sqrt{2\beta I_B} = \frac{1}{R_{\text{ref}}}$$

$$R = R_{\text{ref}} \tag{20.27}$$

Note that the equivalent resistance of the VRC is tuned to the reference resistance independent of the PVT variations.

20.4 S-Domain Tuning Loops

Analog filter-tuning circuits belong to a control system typically consisting of a device under tuning (DUT), a reference, a comparator, and a tuning-control unit as shown in Fig. 20.27.

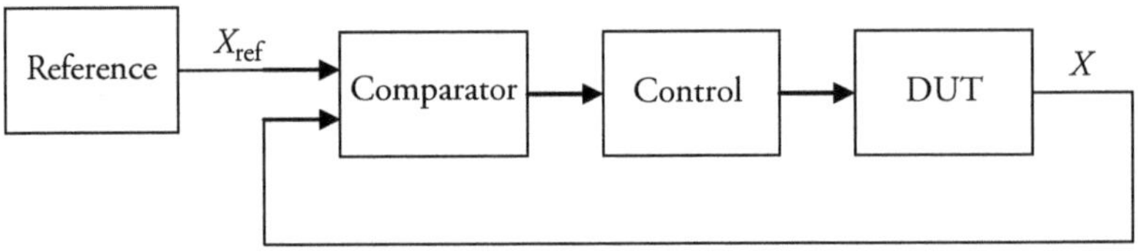

Figure 20.27 Analog filter circuit tuning loop structure.

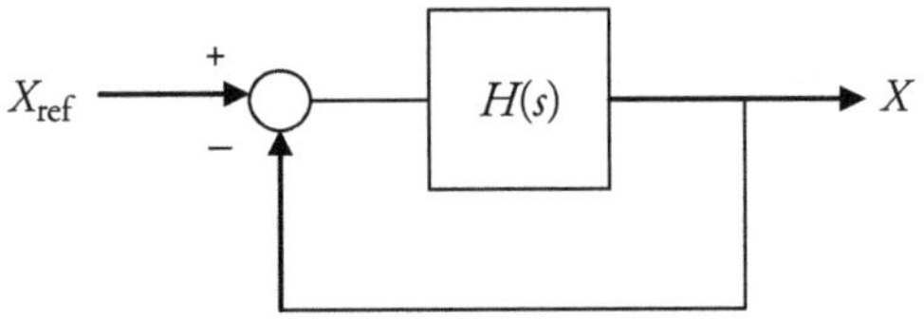

Figure 20.28 S-domain model of continuous-time tuning loop.

The tuning reference X_{ref} might be a standard resistance, a capacitance, a frequency, or a time constant, depending on the application. X is the parameter to be tuned. During the tuning process, the parameter to be tuned is compared with the reference, and the output of the comparator is processed in the control unit to generate the tuning control of the DUT unit. The tuning loop is designed to minimize the difference between the PUT and the reference parameter.

When the loop is realized in continuous time, this system represents the s-domain tuning loop as shown in Fig. 20.28, where the reference unit is represented by its reference parameter X_{ref}.

The DUT is represented by the PUT X, the comparator is represented by a subtraction unit, and the continuous-time control unit is represented by the s-domain transfer function $H(s)$. The closed-loop transfer function of the tuning loop can be expressed as follows:

$$F(s) \equiv \frac{X(s)}{X_{ref}(s)} = \frac{H(s)}{1+H(s)} \tag{20.28}$$

Alternatively, the closed-loop tuning system can also be expressed by the following tuning-error transfer function:

$$\frac{E(s)}{X_{ref}(s)} \equiv \frac{X_{ref}(s) - X(s)}{X_{ref}(s)} = \frac{1}{1+H(s)} \tag{20.29}$$

The goal of the tuning circuit, therefore, is to minimize the error function for a given reference input.

For a step reference input, such a system can be modeled in both the time domain and the s-domain as

$$X_{ref}(t) = \begin{cases} 0 & t < 0 \\ x_{ref}(0) & t \geq 0 \end{cases} \tag{20.30}$$

or

$$X_{\text{ref}}(s) = \frac{x_{\text{ref}}(0)}{s} \tag{20.31}$$

and

$$E(s) = \frac{1}{1+H(s)} \frac{x_{\text{ref}}(0)}{s} \tag{20.32}$$

It can be seen that for a stable tuning loop, it is required that all poles of the closed-loop transfer need to be on the left side of the s-plane.

Assuming p_i are the poles of a stable closed-loop transfer function

$$\frac{E(s)}{X_{\text{ref}}(s)} = \frac{1}{1+H(s)} \equiv \frac{N(s)}{(s-p_1)(s-p_2)\cdots(s-p_n)} \tag{20.33}$$

we have

$$E(s) = \frac{1}{1+H(s)} \frac{x_{\text{ref}}(0)}{s} = \sum_{i=1}^{n} \frac{C_i}{s-p_i} + \frac{1}{1+H(s)|_{s=0}} \frac{x_{\text{ref}}(0)}{s} \tag{20.34}$$

and

$$e(t) = \sum_{i=1}^{n} C_i e^{p_i t} + \frac{1}{1+H(s)|_{s=0}} x_{\text{ref}}(0) \qquad (t > 0) \tag{20.35}$$

It can be seen that the final value of the error function when time goes to infinite is approximately proportional to one over the dc loop gain of the tuning loop:

$$e(t)|_{t\to\infty} = \frac{1}{1+H(s)|_{s=0}} x_{\text{ref}}(0) \tag{20.36}$$

The acquisition time, on the other hand, is approximately proportional to one over the closest distance of the poles of the tuning loop closed-loop transfer function to the $j\omega$ axis in the s-plane:

$$\tau \sim \frac{1}{\min\{|\text{real}(p_i)|\}} \tag{20.37}$$

20.4.1 Zero-Pole Tuning Loop

Figure 20.29 shows a continuous-time tuning loop that employs a wideband high-gain amplifier circuit.

The closed-loop transfer function of the tuning circuit is given as

$$E(s) = \frac{1}{1+A} \frac{x_{\text{ref}}(0)}{s} \tag{20.38}$$

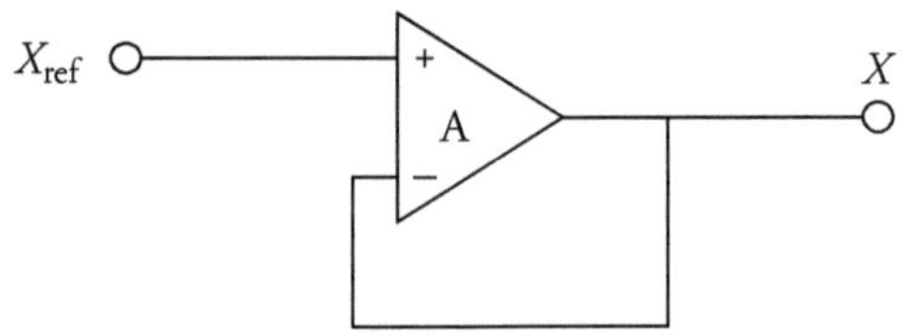

Figure 20.29 Tuning loop employing an ideal OPAMP circuit.

and

$$e(t) = \frac{1}{1+A} x_{ref}(0) \tag{20.39}$$

Note that in most practical designs, high-gain OPAMPs can be used to provide tuning accuracy better than 1–0.1 percent.

20.4.2 Single-Pole Tuning Loop

The control transfer function of a typical first-order tuning loop is given as

$$H(s) = \frac{A}{1+s/p} \tag{20.40}$$

where A is the dc gain of the control unit, and p is the pole of the closed-loop transfer function.

The error transfer function of the tuning loop is given as follows:

$$E(s) = \frac{1}{1+\dfrac{A}{1+s/p}} \frac{x_{ref}(0)}{s} = \frac{x_{ref}(0)}{1+A}\left(\frac{A}{s+p(1+A)} + \frac{1}{s}\right) \tag{20.41}$$

The time domain response of the tuning error can be derived as

$$e(t) = \frac{1}{1+A} x_{ref}(0)(1+Ae^{-p(1+A)t}) \qquad t \geq 0 \tag{20.42}$$

where we have that

$$e(t) = \begin{cases} x_{ref}(0) & t = 0 \\ \dfrac{x_{ref}(0)}{1+A} & t \to \infty \end{cases} \tag{20.43}$$

The acquisition time is approximately given as

$$\tau \sim \frac{1}{p(1+A)} \tag{20.44}$$

For the special case employing a lossless integrator as the control unit, we have

$$H(s) = \frac{\omega_o}{s} \tag{20.45}$$

The error transfer function of the tuning loop is given as follows:

$$E(s) = \frac{1}{1+\frac{\omega_o}{s}} \frac{x_{\text{ref}}(0)}{s} = \frac{x_{\text{ref}}(0)}{s+\omega_o} \tag{20.46}$$

The time domain response of the tuning error is given as

$$e(t) = x_{\text{ref}}(0)e^{-\omega_o t} \qquad t \geq 0 \tag{20.47}$$

where we have that

$$e(t) = \begin{cases} x_{\text{ref}}(0) & t = 0 \\ 0 & t \to \infty \end{cases} \tag{20.48}$$

The acquisition time is approximately given as

$$\tau \sim \frac{1}{\omega_o} \tag{20.49}$$

Note that the acquisition time is inversely proportional to the time constant of the integrator.

20.4.3 Two-Pole Tuning Loop

For a control unit with two poles in the open-loop transfer function given as

$$H(s) = \frac{\omega_o^2}{s\left(s+\frac{\omega_o}{Q}\right)} \qquad Q > 0 \tag{20.50}$$

we have the frequency-domain error response for the step function as follows:

$$E(s) = \frac{1}{1+H(s)} \frac{x_{\text{ref}}(0)}{s} = \frac{x_{\text{ref}}(0)\left(s+\frac{\omega_o}{Q}\right)}{s^2+\frac{\omega_o}{Q}s+\omega_o^2} \tag{20.51}$$

The poles of the closed-loop transfer function are given as follows:

$$\frac{p_{\pm}}{\omega_o} = -\frac{1}{2Q} \pm j\sqrt{1-\left(\frac{1}{2Q}\right)^2} \equiv -\cos\alpha \pm j\sin\alpha \tag{20.52}$$

The error function can be rewritten as follows:

$$E(s)=\frac{x_{\text{ref}}(0)}{2\omega_o j\sin\alpha}\left(\frac{e^{-j\alpha}}{s+\cos\alpha+j\sin\alpha}-\frac{e^{j\alpha}}{s+\cos\alpha-j\sin\alpha}\right) \tag{20.53}$$

The time domain response of the error function is given as

$$E(s)=\frac{x_{\text{ref}}(0)e^{-t\cdot\cos\alpha}}{\omega_o\sin\alpha}\sin((\sin\alpha)t+\alpha) \tag{20.54}$$

20.5 Z-Domain Tuning Loops

When the tuning loop is realized in discrete time, this system represents the z-domain tuning loop as shown in Fig. 20.30.

The control circuit in the z-domain tuning loop can be characterized by the z-domain open-loop transfer function $H(z)$. The z-domain closed-loop transfer function of the loop can be expressed as

$$F(z)\equiv\frac{X(z)}{X_{\text{ref}}(z)}=\frac{H(z)}{1+H(z)} \tag{20.55}$$

The tuning-error transfer function is given as

$$\frac{E(z)}{X_{\text{ref}}(z)}\equiv\frac{X_{\text{ref}}(z)-X(z)}{X_{\text{ref}}(z)}=\frac{1}{1+H(z)} \tag{20.56}$$

For the following step reference input

$$X_{\text{ref}}(nT)=\begin{cases}0 & n<0\\ x_{\text{ref}}(0) & n\geq 0\end{cases} \tag{20.57}$$

we have that

$$X_{\text{ref}}(z)=\frac{x_{\text{ref}}(0)}{1-z^{-1}} \tag{20.58}$$

The z-domain error is given as follows:

$$E(z)=\frac{1}{1+H(z)}\frac{x_{\text{ref}}(0)}{1-z^{-1}} \tag{20.59}$$

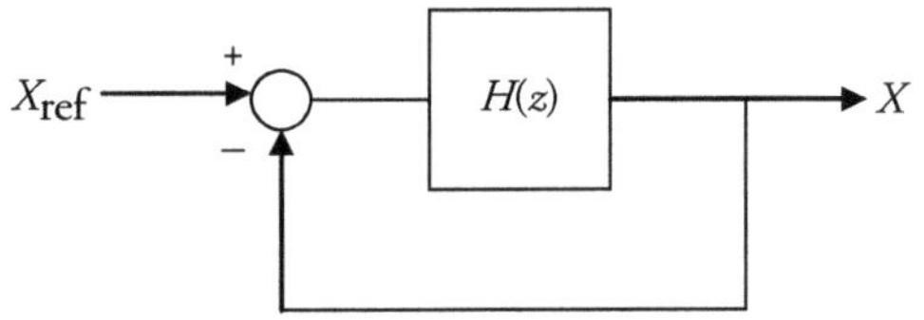

FIGURE 20.30 Z-domain model of discrete-time tuning loop.

Note that for a stable z-domain tuning loop, the poles of the closed-loop transfer function must be within the unity circle in the z-plane.

Assuming p_i are poles of a stable closed-loop transfer function

$$\frac{E(z)}{X_{\text{ref}}(z)} = \frac{1}{1+H(z)} \equiv \frac{N(z)}{(z-p_1)(z-p_2)\cdots(z-p_n)} \tag{20.60}$$

we have that

$$E(z) = \frac{1}{1+H(z)}\frac{x_{\text{ref}}(0)}{1-z^{-1}} = \sum_{i=1}^{n}\frac{C_i}{1-p_i z^{-1}} + \frac{1}{1+H(z)|_{z=1}}\frac{x_{\text{ref}}(0)}{1-z^{-1}} \tag{20.61}$$

and

$$e(k) = \sum_{i=1}^{n} C_i p_i^{\,k} + \frac{1}{1+H(z)|_{z=1}} x_{\text{ref}}(0) \qquad (k>0) \tag{20.62}$$

It can be seen that the final value of the error function as time goes to infinite is approximately proportional to one over the dc loop gain of the tuning loop:

$$e(k)\,|_{k\to\infty} = \frac{1}{1+H(z)|_{z=1}} x_{\text{ref}}(0) \tag{20.63}$$

The acquisition time of the tuning loop is approximately determined by the transfer function's closest pole to the unity circle in the z-plane as follows:

$$\tau \sim \frac{1}{\min\{|\ln|p_i||\}} \tag{20.64}$$

20.5.1 Single-Integrator Tuning Loop

One commonly used discrete-time tuning loop is shown in Fig. 20.31, where a single integrator is used in the loop. In addition, a gain A and a delay n are also included in the loop.

The closed-loop transfer function of such a tuning circuit is given as follows:

$$F(z) \equiv \frac{X(z)}{X_{\text{ref}}(z)} = \frac{Az^{-n}}{1-z^{-1}+Az^{-n}} \tag{20.65}$$

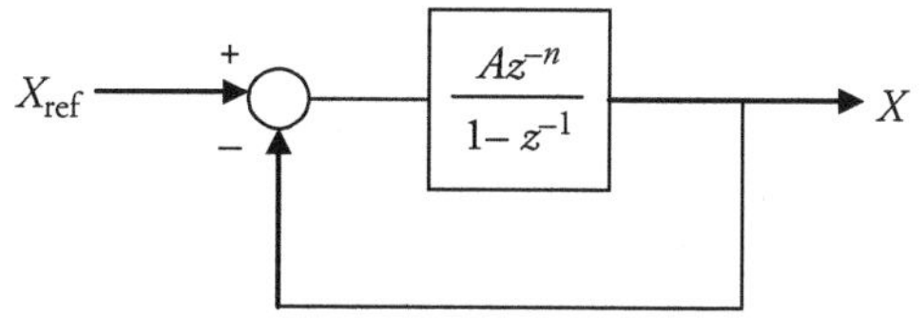

FIGURE 20.31 Z-domain model of discrete-time tuning loop with single integrator.

The tuning-error transfer function is given as

$$\frac{E(z)}{X_{\text{ref}}(z)} = \frac{1 - z^{-1}}{1 - z^{-1} + Az^{-n}} \tag{20.66}$$

For the step input, we have the following:

$$E(z) = \frac{x_{\text{ref}}(0)}{1 - z^{-1} + Az^{-n}} \tag{20.67}$$

For the simplest case of integration without delay ($n = 1$), we have the following:

$$E(z) = \frac{x_{\text{ref}}(0)}{1 - (1 - A)z^{-1}} \tag{20.68}$$

For loop stability, it is required that

$$0 < A < 2 \tag{20.69}$$

The time domain response of the tuning error is given as follows:

$$e(k) = x_{\text{ref}}(0)(1 - A)^k \qquad k \geq 0 \tag{20.70}$$

On the other hand, for the integrator with one stage delay ($n = 2$), we have the following:

$$E(z) = \frac{x_{\text{ref}}(0)}{1 - z^{-1} + Az^{-2}} \tag{20.71}$$

The gain of loop to be stable is reduced compared with the loop without delay:

$$0 < A < 1 \tag{20.72}$$

It can be seen that the region of stability is narrower, implying the effect of loop phase margin degradation due to the delay time within the control loop.

20.6 Mixed-Domain Tuning Loops

In most designs, a tuning loop may include both s- and z-domain function circuits as shown in Fig. 20.32.

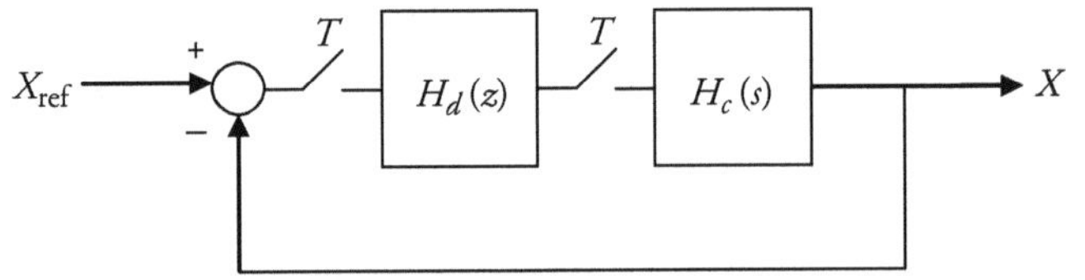

FIGURE 20.32 Mixed-domain model of discrete-time tuning loop.

For most applications, the discrete-time block is oversampled at a frequency much higher than the frequency of interest of the loop. Such a loop can be analyzed in the s-domain using the s-to-z transformation discussed in the previous chapter as

$$F(s) \equiv \frac{X(s)}{X_{ref}(s)} = \frac{H(s)}{1+H(s)} \tag{20.73}$$

where

$$H(s) \approx H_c(s) \cdot H(z)\Big|_{z=\frac{1+\frac{sT}{2}}{1-\frac{sT}{2}}} \tag{20.74}$$

References

1. R. Schaumann, "Continuous-Time Integrated Filters—A Tutorial," *IEE Proceedings*, vol. 136, pt. G, pp. 184–190, August 1989.
2. Y. Tsividis, M. Banu, J. Khoury, "Continuous-time MOSFET-C Filters in VLSI," *IEEE J. Solid-State Circuits*, vol. SC-21, no. 1, pp. 15–30, February 1986.
3. M. Banu, Y. Tsividis, "An Elliptic Continuous-Time CMOS Filter with On-Chip Automatic Tuning," *IEEE J. Solid-State Circuits*, vol. SC-20, no. 1, pp. 1114–1121, December 1985.
4. C. S. Park, R. Schaumann, "Design of a 4-MHz Analog Integrated CMOS Transconductance-C Bandpass Filter," *IEEE J. Solid-State Circuits*, vol. SC-23, no. 4, pp. 987–996, August 1988.
5. H. Khorramabadi, P. R. Gray, "High-Frequency CMOS Continuous-Time Filters," *IEEE J. Solid-State Circuits*, vol. SC-19, no. 6, pp. 939–948, December 1984.
6. A. Kaiser, "A Micropower CMOS Continuous-Time Low-Pass Filter," *IEEE J. Solid-State Circuits*, vol. SC-24, no. 3, pp. 736–743, June 1989.
7. F. Krummenacher, "Design Considerations in High-Frequency CMOS Transconductance Amplifier Capacitor (TAC) Filters," *Proc. IEEE ISCAS*, pp. 100–105, 1989.
8. Y.-T. Wang, A. A. Abidi, "CMOS Active Filter Design at Very High Frequencies," *IEEE J. Solid-State Circuits*, vol. SC-25, no. 6, pp. 1562–1574, December 1990.
9. F. Krummenacher, N. Joehl, "A 4-MHz CMOS Continuous-Time Filter with On-Chip Automatic Tuning," *IEEE J. Solid-State Circuits*, vol. SC-23, no. 3, pp. 750–758, June 1988.
10. F. Krummenacher, "Design Considerations in High-Frequency CMOS Transconductance Amplifier Capacitor (TAC) Filters," *IEEE Proc. ISCAS*, pp. 100–105, 1989.
11. K.-S. Tan, P. R. Gary, "Fully Integrated Analog Filters Using Bipolar-JFET Technology," *IEEE J. Solid-State Circuits*, vol. SC-2, no. 3, pp. 814–821, December 1978.
12. A. M. Durham, W. Redman-White, J. B. Hughes, "Low Distortion VLSI Compatible Self-Tuned Continuous-Time Monolithic Filters," *IEEE ISCAS*, pp. 1448–1451, 1991.
13. R. Schaumann, M. A. Tan, "The Problem of On-Chip Automatic Tuning in Continuous-Time Integrated Filters," *IEEE Proc. ISCAS*, pp. 106–109, 1989.

14. P. M. Van Peteghem, R. Song, "Tuning Strategies in High-Frequency Integrated Continuous-Time Filters," *IEEE J. Solid-State Circuits*, vol. CAS-36, no. 1, pp. 136–139, January 1989.
15. J. Silva-Martinez, M. Steyaert, W. Sansen, "A Novel, Approach for the Automatic Tuning of Continuous-Time Filters," *IEEE proc. ISCAS*, pp. 1452–1455, 1991.
16. C. Plett, M. A. Copeland, R. A. Hadaway, "Continuous-Time Filters using Open Loop Tunable Transconductance Amplifier," *IEEE proc. ISCAS*, pp. 1173–1176, 1986.
17. J. B. Hughes, N. C. Bird, R. S. Soin, "Self-Tuned RC-Active Filters for VLSI," *El. Letters*, vol. 22, no. 19, pp. 993–994, September 1986.
18. T. Kwan, K. Martin, "An Adaptive Analog Continuous-Time CMOS Biquadratic Filter," *IEEE J. Solid-State Circuits*, vol. SC-24, no. 6, pp. 859–867, June 1991.
19. P. M. Van Peteghem, T. L. Brooks, W. J. Adams, H. M. Fossati, K. H. Loh, S. Narayan, G. R. Spalding, R. Yin, "Fully-Integrated Correlated Tuning Processor for Continuous-Time Filters," *IEEE CICC*, pp. 25.7.1–25.7.4, 1990.
20. G. J. Smolka, U. Riedle, U. Grehl, B. Jahn, F. Parzefall, W. Veit, H. Werker, "A Low-Noise Trunk Interface Circuit with Continuous-Time Filters and On-Chip Tuning," G. J. Smolka, U. Grehl, B. Jahn, U. Riedle, W. Veit, H. Werker, " A Low-Noise Trunk Interface Circuit," ESSCIRC'91 Proc. vol. 1, pp. 257–260, 1991.
21. J. M. Khoury, "Design of a 15-Mhz CMOS Continuous-Time Filter with On-Chip Tuning," *IEEE J. Solid-State Circuits*, vol. SC-26, no. 12, pp. 1988–1997, December 1991.
22. H. Khorramabadi, M. J. Tarsia, N. S. Woo, "Baseband Filters for IS-95 CDMA Receiver Applications Featuring Digital Automatic Frequency Tuning," *Digest of Technical Papers, ISSCC*, pp. 172–173, February 1996.
23. P. M. Van Peteghem, R. Song, "Tuning Strategies in High-Frequency, Integrated Continuous-Time Filters," *IEEE Trans. on Circuits and Systems*, vol. 36, no. 1, pp. 136–139, January 1989.
24. Y. Sun, B. T. Chan, H. W. Su, "Tuning of Continuous-Time Filters using Genetic Algorithms," *IEEE APCCAS 2000*, pp. 345–347, December 2000.
25. T. Sumesaglam, A. I. Karsilayan, "A Digital Approach for Automatic Tuning of Continuous-Time High-Q Filters," *IEEE Trans. on Circuits and Systems II: Analog and Digital Signal Processing*, vol. 50, no. 10, pp. 755–761, October 2003.
26. M. Chen, J. Silva-Martinez, S. Rokhsaz, M. Robinson, "A 2-V 80-200-MHz Fourth-Order Continuous-Time Linear Phase Filter with Automatic Frequency Tuning," *IEEE J. Solid-State Circuits*, vol. 38, no. 10, pp. 1745–1749, October 2003.
27. C. Yoo, "Frequency Tuning Technique for Gm-C Filters," *Electronics Letters*, vol. 38, no. 12, pp. 544-545, June 2002.
28. H. Voorman, H. Veenstra, "Tunable High-Frequency Gm-C Filters," *IEEE J. Solid-State Circuits*, vol. 35, no. 8, pp. 1097–1108, August 2000.
29. T. H. Teo, E.-S. Khoo, D. Uday, "Fifth Order Low-Pass Transitional Gm-C Filter with Relaxation Oscillator Frequency Tuning Circuit," *2003 IEEE Conference on Electron Devices and Solid-State Circuits*, pp. 229–232, December 2003.
30. S. D'Amico, A. Baschirotto, "0.18um CMOS Gm-C Digitally Tuned Filter for Telecom Receivers," *ISCAS'03*, vol. 1, pp. I-493–I-496, May 2003.

CHAPTER 21

Practical Design Aspects of VLSI Analog Filters

Practical VLSI analog signal processing circuits suffer from various nonideal circuit effects, resulting mainly from the VLSI circuit PVT (process, supply voltage, and temperature) variations, device mismatches, device inherent and switching noises, and device nonlinearity. PVT variations may cause filter responses to deviate from the desired frequency response targets or even malfunctions. Noise and on-linearity of devices, on the other hand, may degrade the signal-to-noise (S/N) ratio of the signal processing circuit's performance. Sensitivity analysis can be used to analyze the tolerance of the signal processing circuit's performance parameters with respect to the variations of VLSI circuit element parameters under the VLSI fabrication process and the application environment. Such a technique can be used for predicating the manufacture yield of the filter that provides an effective way for defining the filter circuit specification in selecting the optimal circuit architecture, the VLSI circuit-implementation technique, and the device parameters. PVT corner simulation and statistic simulation are the two common ways for computer-aided sensitivity analysis of the filter circuits. Noise and linearity analysis are commonly used to specify the dynamic range (DR) of the VLSI signal processing circuits. The signal dynamic range of the circuit is determined by the lowest usable signal level set by noise floor and by the highest usable signal level set by harmonic distortion of the VLSI signal processing circuits.

21.1 Design for Robustness

VLSI filters suffer from performance limitations resulting from filter components variations under the PVT corners and application environment conditions. For VLSI filter circuit implementations, it is desirable to have a robust design such that high fabrication yield can be achieved in the presence of various nonidealities. The tolerance of filter performance parameters to variations in circuit components can be mathematically modeled by using the sensitivity measure of the filter.

21.1.1 Sensitivity Analysis

Mathematically, if a filter parameter y is a function of a set of independent circuit variables x_i

$$y = f(x_1, x_2, \ldots, x_i, \ldots, x_n) \tag{21.1}$$

the sensitivity of y with respect to a specific parameter x_i is defined as follows:

$$S_{x_i}^{y} \equiv \frac{\partial \ln(y)}{\partial \ln(x_i)} = \frac{x_i}{y}\frac{\partial y}{\partial x_i} \tag{21.2}$$

It can be seen that

$$S_{x_i}^{y} = \frac{x_i}{y}\frac{\partial y}{\partial x_i} \approx \frac{\dfrac{\Delta y}{y}}{\dfrac{\Delta x_i}{x_i}} \tag{21.3}$$

Note that the sensitivity provides a measure of the relative variation of specific filter parameters with respect to the relative variation of a certain variable. In practical cases, all variables may vary at the same time, and the total variation of the parameter y can be modeled as

$$\frac{\Delta y}{y} \approx S_{x_1}^{y}\frac{\Delta x_1}{x_1} + S_{x_2}^{y}\frac{\Delta x_2}{x_2} + \cdots + S_{x_i}^{y}\frac{\Delta x_i}{x_i} + \cdots + S_{x_n}^{y}\frac{\Delta x_n}{x_n} \tag{21.4}$$

In the filter design, the sensitivity of the specific node output voltage or current of the filter with respect to each and every circuit and device parameter can be simulated using the sensitivity-analysis function of the SPICE-based simulator or the parametric analysis method in the Spectre simulator.

21.1.2 Worst-Case (Corner) Simulations

The second commonly used circuit design robustness analysis method is the corner simulation technique, where the filter circuit performances are simulated against the PVT corners. Corner simulation is the simplest and most commonly used technique in performing statistic circuit analysis. To determine the worst-case circuit performance, circuit simulations are performed at points in the parameter space called corners. The process corners are typically defined as the ±3s (3 standard deviations above and below the mean) values of the representative device parameters.

However, corner simulation is limited by the device characteristic for which they were derived. And typically, circuit performance is based on many performance parameters that cannot be modeled by variation in a single-device parameter.

21.1.3 Monte Carlo Simulations

Monte Carlo simulation is a form of statistical analysis where the values of components and parameters of the circuit are varied within a specified statistical distribution. When each iteration and all varied parameter values are defined, the simulator performs one or more standard analyses in steady-state, time, or frequency domains. After each analysis, one of more circuit performance measures is calculated. At the end of all N iterations, the mean, variance, and other statistics for circuit performance measures are calculated and displayed. The result is a circuit output distribution representing the variation of circuit performance for that process. The Monte Carlo simulation can be performed on all parameters, including model parameters that have user-defined tolerances.

A major limitation in Monte Carlo simulation is the accuracy of the statistical model and the simulation time required to perform many circuit simulations. The accuracy of Monte Carlo simulation is wholly dependent on the ability of the statistical model to represent the process variation.

21.2 Designs for Low Noise

Noise effects are shown in all types of VLSI circuits. When the input terminals of practical filters are shorted to ground ($V_i = 0$), the output voltage is not zero. If a wide-bandwidth measurement equipment is used, the filter output shows fast voltage fluctuations. If a low-bandwidth or dc measurement equipment is used, the filter output shows the voltage drift. Noise is an internally generated signal inherent to any filter and arises from different sources. Random noise of the filter is a random signal. Its precise value at any future time cannot be predicted no matter how long we have been observing it in the past. We can only predict average values of the noise.

21.2.1 Thermal Noise

The most common random noise sources in electronic circuits and in any medium that dissipates energy are the random fluctuations at the microscopic level because of the thermal energy in the medium. Random charge movements yield a difference in voltage between any two points in the conductor, regardless of material, even in the absence of any connection or coupling from the conductor to any power supply.

Thermal noise is also known as Johnson noise or Nyquist noise. Thermal noise has white noise distribution, and the available noise power from a conductor at temperature T is given as

$$P_t = kTB \tag{21.5}$$

where k is Boltzmann's constant, T is the absolute temperature, and B is the noise bandwidth.

The available noise power from a resistive source is the power that can be delivered to a resistive load equal in value to the source resistance as (see Fig. 21.1)

$$P_t = \frac{V_N^2}{R} = \frac{\left(\frac{V_t}{2}\right)^2}{R} = kTB \tag{21.6}$$

or

$$V_t^2 = 4kTRB \tag{21.7}$$

$$I_t^2 = \frac{4kTB}{R} \tag{21.8}$$

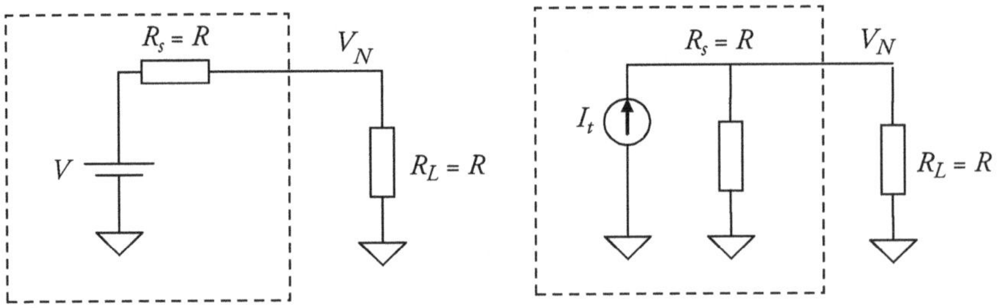

FIGURE 21.1 Circuit for determining noise voltage or noise current.

21.2.2 Shot Noise

Electric current is made of many individual charge carriers, and when there is a current across a potential barrier, the number of charge carriers crossing the barrier during each unit time is random because each carrier is independent. These random current fluctuations constitute the shot noise current with the rms value given as

$$I_t^2 = 2qI_{dc}B \tag{21.9}$$

Shot noise is white although at very high frequency, charge movements are correlated because of collisions and the PSD (Power Spectral Density) decrease.

21.2.3 Low-Frequency (1/f) Noise

At low frequency, there is noise content with PSD inversely proportional to the frequency

$$S_f(f) = \frac{V_f^2}{f^\alpha} \tag{21.10}$$

where $\alpha = 0.8\text{–}1.3$.

21.2.4 Noise Bandwidth

The noise bandwidth of a system is the frequency span of a rectangular power gain curve yielding the same output power as the actual system.

21.2.5 Modeling of Filter Noise

A practical VLSI filter can be modeled as a noiseless filter with noise voltage sources in series with the ports (see Fig. 21.2). Each voltage source represents the noise voltage measured at the respective port with the other port left as an open circuit.

21.3 Designs for High Linearity

For an ideal filter, there is a linear relationship between the steady-state input and output signals as given by the transfer function of the filter as follows:

$$\begin{cases} V_{\text{in}}(t) = V\cos(\omega t) = \text{Re}(Ve^{j\omega t}) \\ V_{\text{out}}(t) = \text{Re}(H(s)|_{s=j\omega} \cdot Ve^{j\omega t}) \end{cases} \tag{21.11}$$

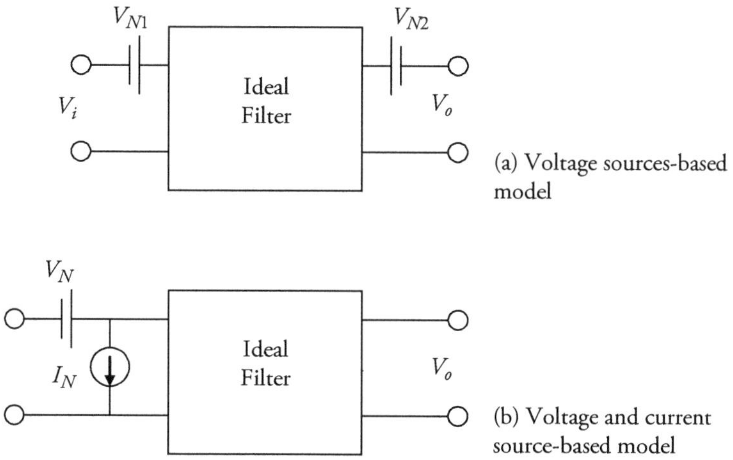

FIGURE 21.2 VLSI filter circuit noise model using equivalent noise sources.

For a practical filter, the output contains not only the linear term of the input but also the higher order terms:

$$\begin{cases} V_{\text{in}}(t) = V_i \cos(\omega t) = \text{Re}(V_i e^{j\omega t}) \\ V_{\text{out}}(t) = a_o + a_1 V_{in} + a_2 V_{in}^2 + a_3 V_{in}^3 + \cdots \end{cases} \tag{21.12}$$

These high-order terms will generate various harmonics of the input signal. The outcomes of such a phenomenon is twofold: power loss in the desired signal frequency band and unwanted power injection to the signal band from adjacent signal/noise bands. Both of them result in degradations in the signal-to-noise ratio of the filter. Mathematically, the coefficient a_i in the above equation provides the measure of the nonlinearity of a given filter. In practical cases, other equivalent linearity measures of the filter are also commonly used, such as the total harmonic distortion (THD), the second- (HD_2) and third-order (HD_3) parameters commonly used in baseband systems, the second- (IP_2) and third-order intercept points (IP_3), and −1-dB compression point ($CP-_{1\,dB}$) in the RF systems.

21.3.1 Harmonic Distortion

For a given input signal of single frequency w, the filter output will contain the following dc value and the harmonics:

$$\begin{cases} V_{\text{in}}(t) = V_i \cos(\omega t) \\ V_{\text{out}}(t) = V_o + V_1 \cos(\omega t + \theta_1) \\ + V_2 \cos(2\omega t + \theta_2) + V_3 \cos(3\omega t + \theta_3) + \cdots \end{cases} \tag{21.13}$$

The HD_2 and HD_3 and the THD of a signal are defined using the following ratios:

$$\begin{cases} HD_2 = \sqrt{\dfrac{V_2^2}{V_1^2}} \times 100\% \\ HD_3 = \sqrt{\dfrac{V_3^2}{V_1^2}} \times 100\% \\ THD = \left(\dfrac{\sqrt{V_2^2 + V_3^2 + V_4^2 + \cdots}}{\sqrt{V_1^2}}\right) \times 100\% \end{cases} \tag{21.14}$$

The THD in dB format is given as follows:

$$THD = 10 \cdot \log\left(\frac{V_2^2 + V_3^2 + V_4^2 + \cdots}{V_1^2}\right) \quad \text{(dB)} \tag{21.15}$$

For nonlinear circuits, we have that

$$\begin{aligned} V_{out}(t) &= a_o + a_1 V_{in}(t) + a_2 \cdot V_{in}(t)^2 + a_3 \cdot V_{in}(t)^3 + \cdots \\ &= \left(a_o + \frac{a_2}{2} V_i^2 + \cdots\right) + \left(1 + \frac{3}{4}\frac{a_3 V_i^2}{a_1}\right) a_1 V_i \cos(\omega t) \\ &\quad + \left(\frac{a_2}{2}\right) V_i^2 \cos(2\omega t) + \left(\frac{a_3}{4}\right) V_i^3 \cos(3\omega t) + \cdots \end{aligned} \tag{21.16}$$

We have the relationship between a_i and the harmonic distortions as follows:

$$HD_2 = \left(\frac{a_2}{2a_1}\right) V_i \tag{21.17}$$

$$HD_3 = \left(\frac{a_3}{4a_1}\right) V_i^2 \tag{21.18}$$

We can also see that due to nonlinear effects, the signal gain at the fundamental frequency will be compressed versus the ideal filter. The –1-dB compress point, when the output signal is 1 dB different from the ideal circuit, is given as

$$20 \log\left(1 + \frac{3}{4}\frac{a_3 V_i^2}{a_1}\right) = -1 \text{ dB} \tag{21.19}$$

or

$$V_i = \sqrt{\frac{\left(1 - 10^{\frac{1}{20}}\right)}{\frac{3}{4}\left|\frac{a_3}{a_1}\right|}} \tag{21.20}$$

21.3.2 Intermodulation

For ideal filters, signals of different frequencies are isolated from one another in the frequency domain. When nonlinear effects are presented in the filter circuit, signals of different frequencies will be mixed to create all possible intermodulation products. As a result, noise from the adjacent band will be injected into the signal frequency band of interest.

Such a phenomenon can be modeled by applying the following two-tone input signal:

$$V_{\text{in}}(t) = V_i(\cos(\omega_1 t) + \cos(\omega_2 t)) \tag{21.21}$$

The output of the nonlinear circuit with the two-tone input is given as follows:

$$\begin{aligned} V_{\text{out}}(t) &= a_o + a_1 V_{in}(t) + a_2 \cdot V_{in}(t)^2 + a_3 \cdot V_{in}(t)^3 + \cdots \\ &= (a_o + a_2 V_i^2 + \cdots) \\ &\quad + \left(a_1 V_i + \frac{3}{2} a_3 V_i^3\right)(\cos(\omega_1 t) + \cos(\omega_2 t)) \\ &\quad + \frac{3}{4} a_3 V_i^3 (\cos((2\omega_1 - \omega_2)t) + \cos((2\omega_2 - \omega_1)t)) \\ &\quad + \frac{1}{2} a_2 V_i^2 (\cos(2\omega_1 t) + \cos(2\omega_2 t)) + (\cos((\omega_1 + \omega_2)t) + \cos((\omega_1 - \omega_2)t)) \\ &\quad + \frac{1}{4} a_3 V_i^3 (\cos(3\omega_1 t) + \cos(3\omega_2 t)) + \frac{3}{4} a_3 V_i^3 (\cos((2\omega_1 + \omega_2)t) \\ &\quad + \cos((2\omega_2 + \omega_2)t)) \end{aligned} \tag{21.22}$$

It can be seen that for a signal with frequency ω, any two signals or noises, which have frequencies meeting the following equation, will inject noise into the signal band.

$$\pm(N\omega_1 \pm M\omega_2) = \omega \tag{21.23}$$

N and M are two integers. In most narrow-band applications, the intermodulation terms created by the following combinations are commonly used to characterize the linearity of the system as follows:

$$\begin{cases} 2\omega_1 - \omega_2 = \omega \\ 2\omega_2 - \omega_1 = \omega \end{cases} \tag{21.24}$$

$$\frac{3}{4} a_3 V_i^3 (\cos((2\omega_1 - \omega_2)t) + \cos((2\omega_2 - \omega_1)t)) \tag{21.25}$$

The magnitude of this term to the fundamental is usually defined as the third-order intermodulation (IM_3) of the system:

$$\text{IM}_3 = \frac{\frac{3}{4} a_3 V_i^3}{a_1 V_i} = \frac{3}{4} \frac{a_3 V_i^2}{a_1} = 3\text{HD}_3 \tag{21.26}$$

For most RF communication systems, a commonly used performance metric of system linearity is the third-order intercept point, which is defined as the point where the third-order intermodulation term equals to the fundamental as

$$\frac{\frac{3}{4}a_3V_{\text{IIP3}}^3}{a_1V_{\text{IIP3}}}=1 \tag{21.27}$$

or

$$V_{\text{IIP3}}=\sqrt{\frac{4}{3}\left|\frac{a_1}{a_3}\right|} \tag{21.28}$$

or in power as

$$\text{IIP3}=10\log\left(\frac{(V_{\text{IIP3}})^2}{2R}/(\text{mW})\right)=10\log\left(\frac{2}{3R}\left|\frac{a_1}{a_3}\right|/(\text{mW})\right)(\text{dBm}) \tag{21.29}$$

Similarly, we can also define IIP2 as the second harmonic equal to the fundamental as

$$a_2V_{\text{IIP2}}^2=a_1V_{\text{IIP2}} \tag{21.30}$$

or

$$V_{\text{IIP2}}=\left|\frac{a_1}{a_2}\right| \tag{21.31}$$

21.3.3 Filter Nonlinearity Characterization Methods

From the discussion above, several filter nonlinearity characterization methods can be developed depending on the transfer function of the filter (see Figs. 21.3 and 21.4).

DC Transfer-Function Method

This method is suitable in filtering with low-pass responses. In this method, the dc transfer characteristic is first obtained. The coefficients a_i of the series is extracted using the curve-fitting method.

From these coefficients, the nonlinearity parameters can be calculated.

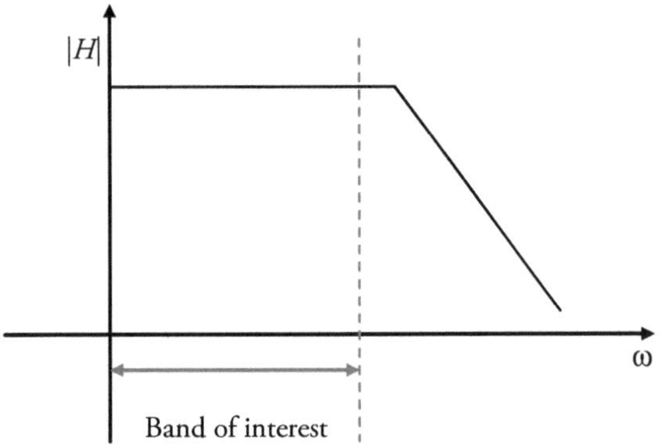

FIGURE 21.3 Filter response for dc I/O transfer-function method.

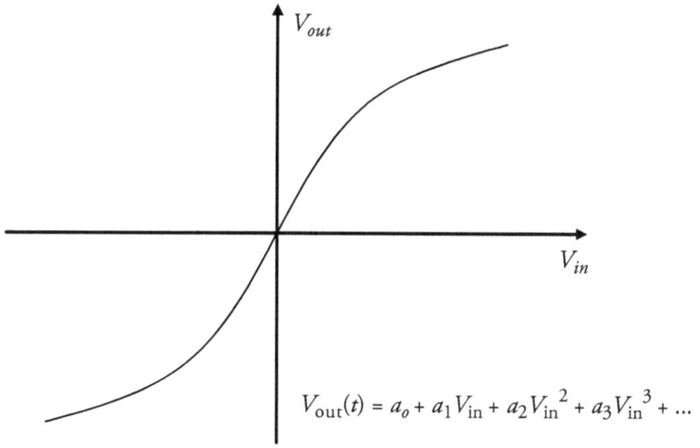

FIGURE 21.4. DC transfer-function-based nonlinearity characterization method.

Single-Tone Method

The DC I/O transfer-function method provides an easy way to characterize the nonlinearity of a filter. However, for the high-pass or band-pass filter, the dc I/O method does not work well since the dc I/O transfer function does not exist due to the dc-stopping characteristics of the filter circuit. Alternatively, an ac signal of single frequency may be used to characterize the filter linearity. This method is suitable for filters that have responses with a flat passband covering the signal and its harmonics of interests as shown in Fig. 21.5.

During the characterization, the power of the signal output and its harmonics are first obtained. The harmonic distortions are calculated. Other nonlinearity parameters can also be calculated with the equations developed in the previous sections.

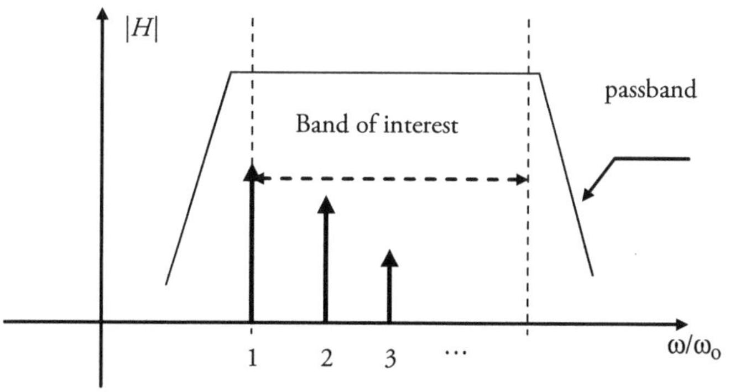

FIGURE 21.5 Band-pass response single-tone method.

Two-Tone Method

For narrow-passband filters, the above two methods are less useful since neither dc nor harmonics of interest are located in the flat passband region of the filter. In such a case, a two-tone technique can usually be used in the filter linearity characterization:

$$V_i(t) = V\cos((\omega_o + \Delta\omega/2)t) + V\cos((\omega_o - \Delta\omega/2)t) \tag{21.32}$$

Assuming ω_o is the signal frequency of interest, we have that

$$\begin{aligned} V_{\text{out}}(t) &= \left(1 + \frac{9a_3V^2}{4a_1}\right)a_1V[\cos((\omega_o + \Delta\omega/2)t) + \cos((\omega_o - \Delta\omega/2)t)] \\ &+ \frac{a_3V^3}{4}[\cos(3(\omega_o + \Delta\omega/2)t) + \cos(3(\omega_o - \Delta\omega/2)t)] \\ &+ \frac{3a_3V^3}{4}[\cos((2\omega_o + \Delta\omega/2)t) + \cos((2\omega_o - \Delta\omega/2)t)] \\ &+ \frac{3a_3V^3}{4}[\cos((\omega_o + \Delta\omega)t) + \cos((\omega_o - \Delta\omega)t)] \end{aligned} \tag{21.33}$$

If $\Delta\omega$ is selected to be much smaller than the bandwidth of the filter, then $\omega_o + \Delta\omega$ and $\omega_o + \Delta\omega$ will be located in the flat region of the passband as shown in Fig. 21.6.

It can be seen that

$$\begin{cases} A_i = V \\ A_0 = \left(1 + \dfrac{9a_3V^2}{4a_1}\right)a_1V \approx a_1V \\ A_1 = \dfrac{3a_3V^3}{4} \end{cases} \tag{21.34}$$

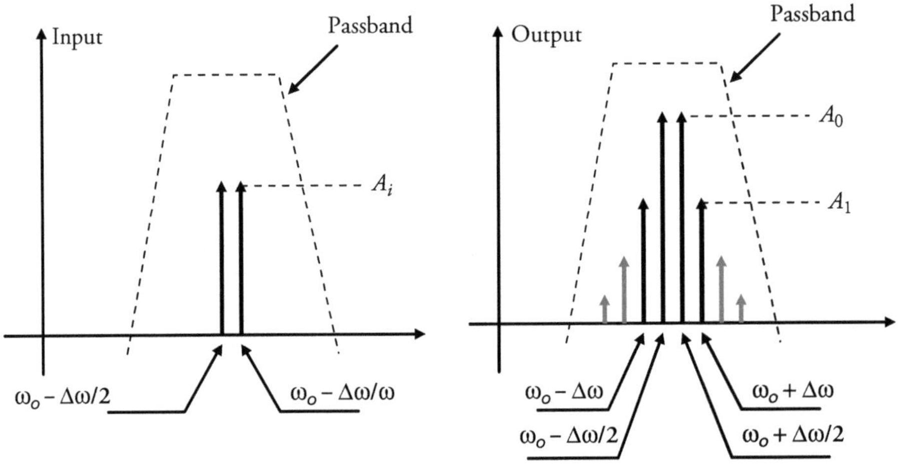

FIGURE 21.6 The input and output spectra of the filter for two-tone input signal.

By measuring the A_i, the A_0, and the A_1 at relatively low input levels, we can characterize the nonlinearity of the filter circuit as follows:

$$\begin{cases} a_1 = A_o / A_i \\ a_3 = \dfrac{4}{3}\dfrac{A_1}{A_i^3} \end{cases} \tag{21.35}$$

In some applications such as in the RF applications, we usually use the input and output powers, rather than the voltages, to characterize the nonlinearity of the system. It can be seen that such an approach is equivalent to the above method, and either way can be used to characterize the nonlinearity of the system. For a system (or filter) with input and output impedances R_i and R_o respectively, the input and output powers are given as follows:

$$\begin{cases} P_i = 10\log\left(\dfrac{1}{2}\dfrac{V^2}{R_i}\right) \\ P_1 \approx 10\log\left(\dfrac{1}{2R_o}(a_1 V)^2\right) = 10\log\left(\dfrac{R_i}{R_o}\right) + 10\log(a_1^2) + P_i \\ P_3 = 10\log\left(\dfrac{1}{2R_o}\left(\dfrac{3a_3 V^3}{4}\right)^2\right) = 10\log\left(\dfrac{R_i}{R_o}\right) + 10\log\left(\left(\dfrac{3a_3}{4}\right)^2\right) + 3P_i \end{cases} \tag{21.36}$$

Therefore, one way of getting the coefficient from the power level is by this equation:

$$\begin{cases} a_1 = 10^{\left(P_1 - P_i - 10\log\left(\frac{R_i}{R_o}\right)\right)/20} \\ a_3 = \dfrac{4}{3}\cdot 10^{\left(\frac{P_3 - 3P_i}{2} - 10\log\left(\frac{R_i}{R_o}\right)\right)/20} \end{cases} \tag{21.37}$$

Also, we may get the nonlinearity characteristics by plotting the output power versus the input power. From the plot, we can get one of the performance parameters of the system: the third-order intercept point (IP_3). The fundamental and the third-order power intercept is shown in Fig. 21.7.

We have that the IIP3 value is related to the nonlinear coefficients by

$$\frac{3a_3}{4a_1} = 10^{-\mathrm{IIP3}/10} \tag{21.38}$$

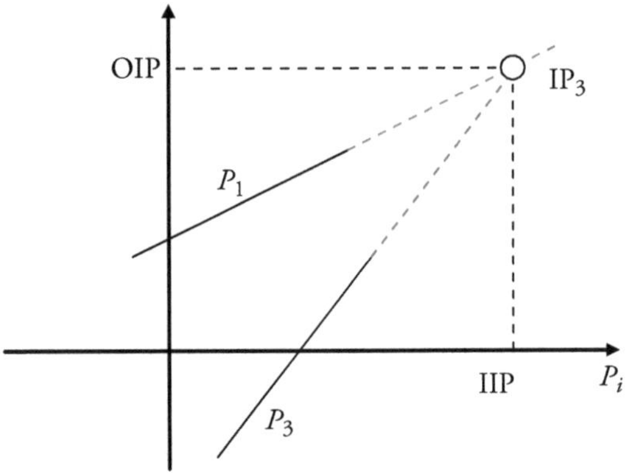

Figure 21.7 The third-order intercept point.

References

1. M. Qu, M. A. Styblinski, "Statistical Characterization and Modeling of Analog Functional Blocks," *ISCAS'94*, vol. 1, pp. 121–124, June 1994.
2. G. Espinosa, F. Montecchi, E. Sanchez-Sinencio, F. Maloberti, "Noise Performances of OTA-C Filters," *ISCAS'88*, vol. 3, pp. 2173–2176, June 1988.
3. A. N. Mohieldin, E. Sanchez-Sinencio, J. Silva-Martinez, "Nonlinear Effects in Pseudo Differential OTAs with CMFB," *IEEE Trans. on Circuits and Systems II: Analog and Digital Signal Processing*, vol. 50, no. 10, pp. 762–770, October 2003.
4. G. J. Gomez, S. H. K. Embabi, E. Sanchez-Sinencio, M. Lefebvre, "A nonlinear Macro-model for CMOS OTAs," *ISCAS'95*, vol. 2, pp. 920–923, May 1995.
5. P. Bowron, H. M. Dahir, "Modelling of Nonideal Active Devices in Continuous-Time OTA-C Filters," *European Conference on Circuit Theory and Design*, pp. 128–131, September 1989.
6. J. Silva-Martinez, G. Espinosa, "Dynamic Range Limitations in Linear OTA-Based Low-Voltage Continuous-Time Filters," *Proceedings of the 37th Midwest Symposium on Circuits and Systems*, vol. 1, pp. 89–92, August 1994.
7. W. B. Kuhn, F. W. Stephenson, A. Elshabini-Riad, "Dynamic Range of High-Q OTA-C and Enhanced-Q LC RF Bandpass Filters," *Proceedings of the 37th Midwest Symposium on Circuits and Systems*, vol. 2, pp. 767–771, August 1994.
8. A. Brambilla, G. Espinosa, F. Montecchi, E. Sanchez-Sinecio, "Noise Optimization in Operational Transconductance Amplifier Filters," *ISCAS'89*, vol. 1, pp. 118–121, May 1989.
9. G. Efthivoulidis, L. Toth, Y. P. Tsividis, "Noise in Gm-C Filters," *IEEE Trans. on Circuits and Systems II: Analog and Digital Signal Processing*, vol. 45, no. 3, pp. 295–302, March 1998.
10. J. Mahattanakul, C. Bunyakate, "Harmonic Injection Method: a Novel Method for Harmonic Distortion Analysis," *ISCAS'01*, vol. 2, pp. 85–88, May 2001.
11. R. Ganti, L. R. Carley, B. A. Myers, "A Low Distortion High Frequency Transconductor Structure," *ISCAS'99*, vol. 2, pp. 216–219, June 1999.

12. T. Curran, M. Collier, "Sensitivity Properties of SC Filters Derived from LC Ladder Prototypes," *IEEE Trans. on Circuits and Systems*, vol. 37, no. 12, pp. 1544–1546, December 1990.
13. M. Lutovac, D. Novakovic, "Efficient Low-Sensitive Selective SC-filters," *ICECS'96*, vol. 1, pp. 211–214, October 1996.
14. A. Kaelin, J. Goette, W. Guggenbuhl, G. S. Moschytz, "A Novel Capacitance Assignment Procedure for the Design of Sensitivity- and Noise-Optimized SC-Filters," *IEEE Tran. on Circuits and Systems*, vol. 38, no. 11, pp. 1255–1268, November 1991.
15. Z. Ma, K. Nakayama, G. Yamamoto, "Optimum Pairing and Ordering Conditions for Simultaneous Reduction in Total Capacitance, Sensitivity and Output Noise of Cascade SC Filters," *ISCAS'92*, vol. 3, pp. 1203–1206, May 1992.
16. Z. Q. Shang, J. I. Sewell, "Efficient Analysis of Some Nonlinearities in SC and SI Filter Networks," *IEE Colloquium on Digital and Analogue Filters and Filtering Systems*, pp. 10/1–10/5, November 1994.
17. K.-L. Lee, R. G. Mayer, "Low-Distortion Switched-Capacitor Filter Design Techniques," *IEEE J. Solid-State Circuits*, vol. 20, no. 6, pp. 1103–1113, December 1985.
18. A. S. Korotkov, "Sensitivity of Singly and Doubly Terminated Prototype Circuits," *Electronics Letters*, vol. 34, no. 2, pp. 164–165, January 1998.
19. J. C. M. Bermudez, M. C. Schneider, C. G. Montoro, "Linearity of Switched Capacitor Filters Employing Nonlinear Capacitors," *ISCAS'92*, vol. 3, pp. 1211–1214, May 1992.
20. T. Inoue, F. Ueno, S. Sonobe, T. Matsumoto, "Design of Very Low Sensitivity Fully Differential Band-Pass Switched-Capacitor Filters," *ISCAS'88*, vol. 2, pp. 1241–1244, June 1988.
21. J. Goette, A. Kaelin, W. Guggenbuhl, G. S. Moschytz, "An Approximate Noise Computation for General Integrator-Based SC Filters," *IEEE Trans. on Circuits and Systems*, vol. 38, no. 11, pp. 1249–1254, November 1991.

CHAPTER 22

Design of VLSI Polyphase Filters

VLSI analog polyphase filters belong to a class of analog signal processing circuits that offer nonsymmetric gain responses with respect to the positive and negative frequencies. Such an important property provides the selectively pass/stop behavior for signals of the opposite polarity and the same frequency. VLSI analog polyphase filters have gained much interest as a result of the recent efforts of integrating the RF circuits and systems in single silicon. Two key applications of the VLSI analog polyphase filter circuits are the quadrant phase clock generations and the baseband-based image rejection operations in low-IF RF receivers. The quadrant phase clock generations are widely used for I/Q modulation/demodulations in communication systems and the baseband-based image rejection operations are commonly adopted for the VLSI-integrated RF circuits. In addition, there has been interest in applying the analog polyphase filter circuits for the phase correction in the multiphase clocking circuits in the high-speed I/O applications.

22.1 VLSI Polyphase Filters

A VLSI continuous-time filter circuit is defined as a real filter if and only if all coefficients a_i and b_i of the filter's s-domain transfer function are real numbers:

$$\frac{Y}{X} = \frac{b_m s^{-m} + b_{m-1} s^{-(m-1)} + \cdots + b_1 s^{-1} + b_0}{a_m s^{-m} + a_{m-1} s^{-(m-1)} + \cdots + a_1 s^{-1} + 1} \tag{22.1}$$

For a real filter, the values of the poles and zeros of the transfer function are either real numbers or complex conjugates as shown in Fig. 22.1.

The gain and phase responses of the real filter are symmetrical and antisymmetrical, respectively, in the frequency domain as shown in Fig. 22.2.

$$\begin{cases} |H(j\omega)| = |H(-j\omega)| \\ \angle H(j\omega) = -\angle H(-j\omega) \end{cases} \tag{22.2}$$

On the other hand, a filter is defined as a complex filter if some of the coefficients a_i and b_i of the s-domain transfer function in Eq. (22.1) are complex numbers. The values of the pole and zero of the transfer function of a complex filter, in general, are complex numbers.

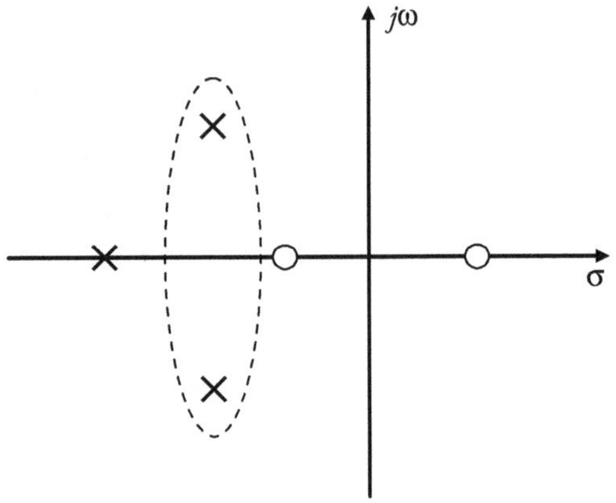

Figure 22.1 Poles and zeros of real filter transfer function.

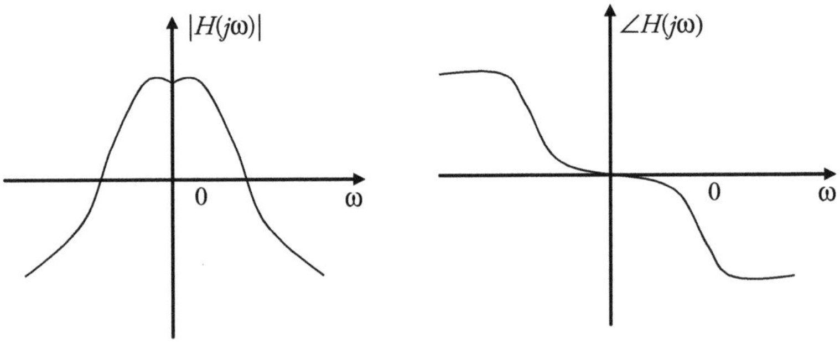

Figure 22.2 Typical gain and phase responses of the real filter.

Consequently, the gain and phase frequency responses of the complex filter are not symmetrical to the zero frequency.

$$\begin{cases} |H(j\omega)| \neq |H(-j\omega)| \\ \angle H(j\omega) \neq -\angle H(-j\omega) \end{cases} \tag{22.3}$$

22.1.1 Hilbert Filters and Hilbert Transform

There are various ways to construct a complex filter. A simple way to construct a complex filter is through the Hilbert transform. The Hilbert transform maps the poles and zeros of a real filter to the Hilbert filter in the way shown in Fig. 22.3.

$$\begin{cases} p_i \rightarrow p_i + j\omega_o \\ z_i \rightarrow z_i + j\omega_o \end{cases} \tag{22.4}$$

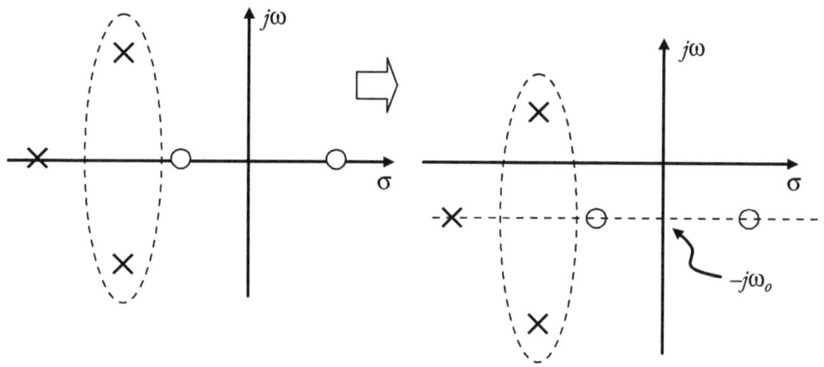

FIGURE 22.3 Poles and zeros of the Hilbert filter.

It can be seen that such a frequency transformation converts a real transfer function into a complex transfer function as follows:

$$H_{\text{Hilbert}}(s) = H_{\text{real}}(s)|_{s=s-j\omega_o} \tag{22.5}$$

It can also be seen that the Hilbert transform linearly shifts the gain and phase frequency responses of the real filter along the frequency axis by ω_o without changing the shape of the Bode plots as shown in Fig. 22.4. The filter constructed from the Hilbert transform of a real filter is known as the Hilbert filter of that real filter.

22.1.2 Polyphase Signals

Mathematically speaking, it is impossible to directly generate the complex filter response based on linear circuit components using only a single input signal phase (or path). We have to rely on the multiphase (or polyphase) signals for realizing the complex coefficients of the transfer function in the linear circuit. For a linear network with

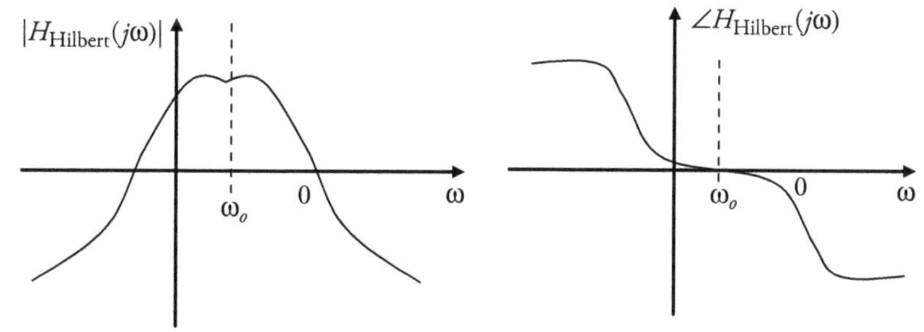

FIGURE 22.4 Gain and phase response of a Hilbert filter.

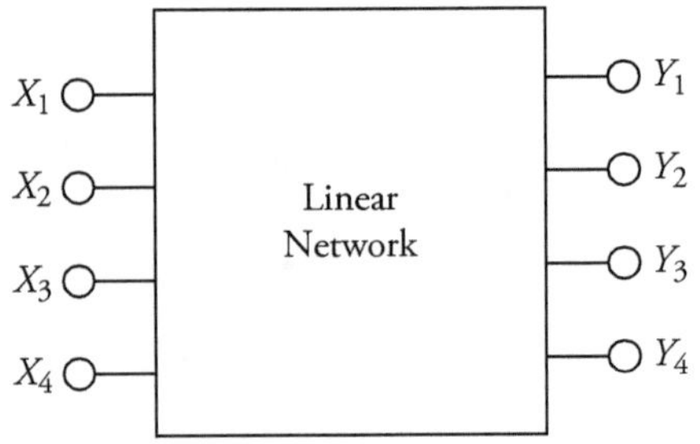

FIGURE 22.5 Four-input/four-output linear network.

four input and four output ports as shown in Fig. 22.5, the polyphase representation of signal $X(s)$ in the vector form can be defined in the s-domain as follows:

$$X_{\text{polyphase}}(s) \equiv \begin{pmatrix} X_1 \\ X_2 \\ X_3 \\ X_4 \end{pmatrix} \equiv \begin{pmatrix} X_{I+} \\ X_{Q+} \\ X_{I-} \\ X_{Q-} \end{pmatrix} = \begin{pmatrix} 1 \\ j \\ -1 \\ -j \end{pmatrix} X(s) \tag{22.6}$$

Such a signal represents a group of four signals, which are generated from the same signal with their phases equally spaced by $\pi/2$. The polyphase signal can also be represented in a two-signal (differential I/Q) form as shown in Fig. 22.6.

$$X_{\text{polyphase}}(s) \equiv \begin{pmatrix} X_1 - X_3 \\ X_2 - X_4 \end{pmatrix} \equiv \begin{pmatrix} X_{I+} - X_{I-} \\ X_{Q+} - X_{Q-} \end{pmatrix} \equiv \begin{pmatrix} X_I \\ X_Q \end{pmatrix} = \begin{pmatrix} 1 \\ j \end{pmatrix} X(s) \tag{22.7}$$

Figure 22.7 shows the phase information of the two polyphase signal representations.

22.1.3 Polyphase Filter Transfer Functions

Figure 22.8 shows the VLSI classic (passive-RC) polyphase filter circuit structure.

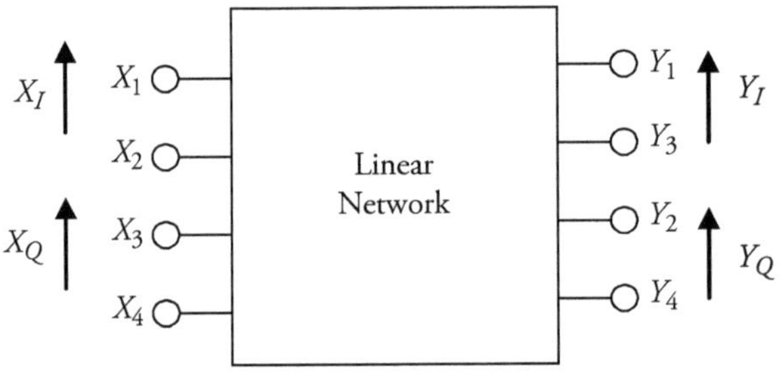

FIGURE 22.6 Two-signal linear network representation.

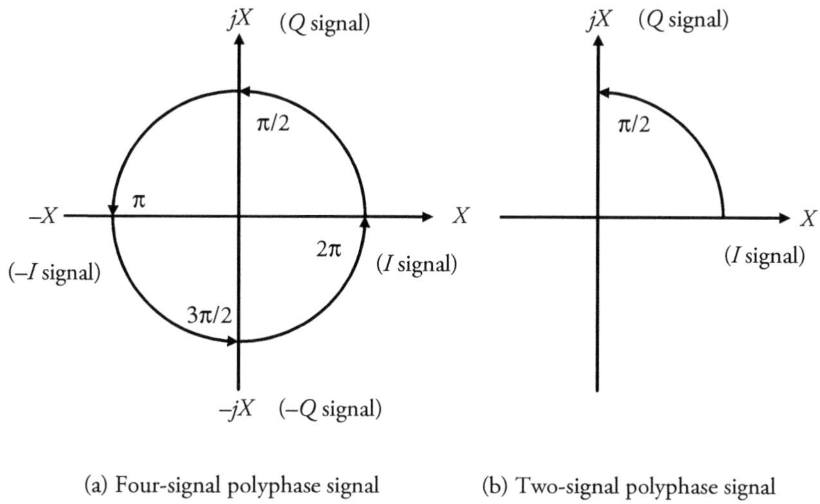

FIGURE 22.7 Phase information of polyphase signals.

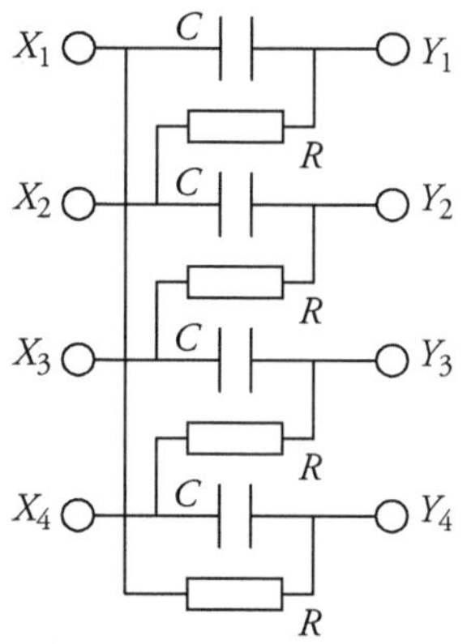

FIGURE 22.8 Classic polyphase filter structure.

Such a filter circuit structure has an important rotational symmetry property that is behaviorally identical to itself (or invariant) with respect to the rotation conversion of input ports in the following way (see Fig. 22.9):

$$\begin{cases} X_1 \to X_2, & Y_1 \to Y_2 \\ X_2 \to X_3, & Y_2 \to Y_3 \\ X_3 \to X_4, & Y_3 \to Y_4 \\ X_4 \to X_1, & Y_4 \to Y_1 \end{cases} \tag{22.8}$$

The above rotational symmetry can further be extended to the general four-input/four-output linear network as shown in Fig. 22.10, where the network is identical to itself

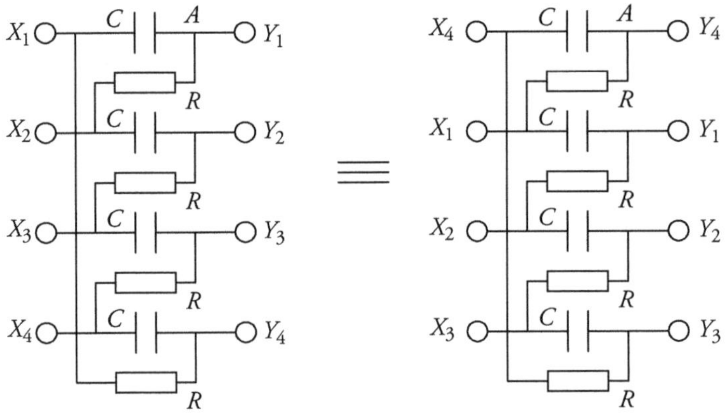

FIGURE 22.9 The rotational symmetry of classic polyphase filter circuit structure.

with respect to the rotational transformations. We can define such a type of linear circuit network as the linear rotational symmetry (LRS) network.

The rotational symmetry property of an LRS network can be extended from the VLSI passive polyphase circuits to the VLSI active polyphase filter circuits. Figure 22.11 shows a VLSI Gm-C polyphase filter circuit that also offers rotational symmetry.

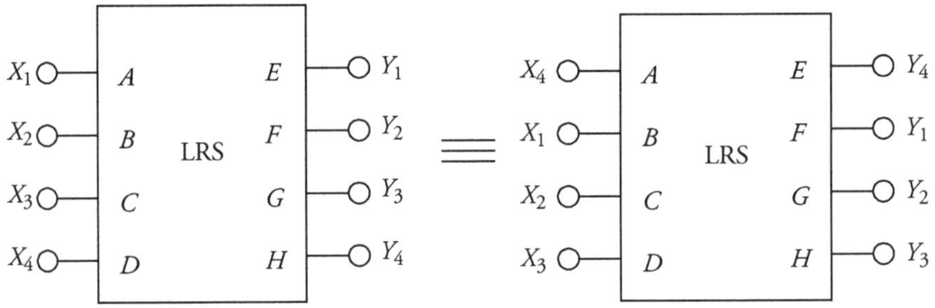

FIGURE 22.10 LRS network.

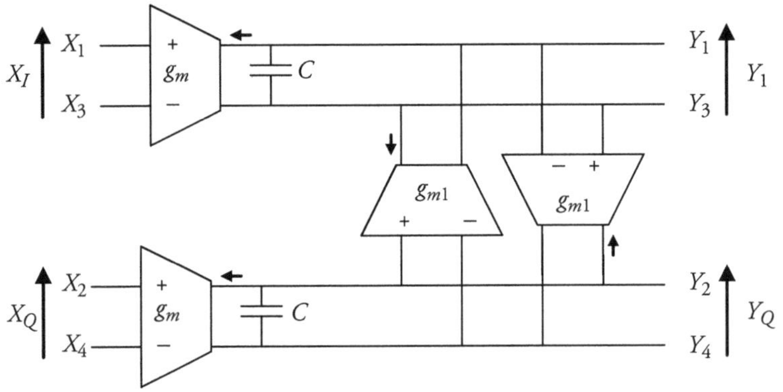

FIGURE 22.11 Complex Gm-C integrator circuit implementation.

For a general LRS network, we have

$$\begin{pmatrix} X_1 \\ X_2 \\ X_3 \\ X_4 \end{pmatrix} = \begin{pmatrix} 1 \\ j \\ -1 \\ -j \end{pmatrix} X(s) \Rightarrow \begin{pmatrix} Y_1 \\ Y_2 \\ Y_3 \\ Y_4 \end{pmatrix} = \begin{pmatrix} 1 \\ j \\ -1 \\ -j \end{pmatrix} Y(s) \tag{22.9}$$

and that

$$\begin{pmatrix} d_1 & d_2 & d_3 & d_4 \\ d_3 & d_4 & d_1 & d_2 \\ d_3 & d_4 & d_1 & d_2 \\ d_3 & d_4 & d_1 & d_2 \end{pmatrix} \begin{pmatrix} Y_1 \\ Y_2 \\ Y_3 \\ Y_4 \end{pmatrix} = \begin{pmatrix} h_1 & h_2 & h_3 & h_4 \\ h_3 & h_4 & h_1 & h_2 \\ h_1 & h_2 & h_3 & h_4 \\ h_3 & h_4 & h_1 & h_2 \end{pmatrix} \begin{pmatrix} X_1 \\ X_2 \\ X_3 \\ X_4 \end{pmatrix} \tag{22.10}$$

Substituting Eq. (22.9) into Eq. (22.10), we have that

$$\Rightarrow (d_1 + jd_2 - d_3 - jd_4) \begin{pmatrix} 1 \\ j \\ -1 \\ -j \end{pmatrix} Y(s) = (h_1 + jh_2 - h_3 - jh_4) \begin{pmatrix} 1 \\ j \\ -1 \\ -j \end{pmatrix} X(s) \tag{22.11}$$

The transfer function of a general LRS network with respect to the polyphase signal can be expressed as

$$H(s) \equiv \frac{Y(s)}{X(s)} = \frac{(h_1 + jh_2 - h_3 - jh_4)}{(d_1 + jd_2 - d_3 - jd_4)} \tag{22.12}$$

Note that such a network, in general, is a complex filter since some of the coefficients in the above transfer function are complex numbers. The above results can be used to find the time domain steady-state responses for the following sinusoidal inputs:

$$\begin{pmatrix} x_1 \\ x_2 \\ x_3 \\ x_4 \end{pmatrix} = \begin{pmatrix} 1 \\ j \\ -1 \\ -j \end{pmatrix} e^{j\omega t} \Rightarrow \begin{pmatrix} y_1 \\ y_2 \\ y_3 \\ y_4 \end{pmatrix} = H(s)_{s=j\omega} \begin{pmatrix} 1 \\ j \\ -1 \\ -j \end{pmatrix} e^{j\omega t} \tag{22.13}$$

$$\begin{pmatrix} x_1 \\ x_2 \\ x_3 \\ x_4 \end{pmatrix} = \begin{pmatrix} 1 \\ j \\ -1 \\ -j \end{pmatrix} e^{-j\omega t} \Rightarrow \begin{pmatrix} y_1 \\ y_2 \\ y_3 \\ y_4 \end{pmatrix} = H(s)_{s=-j\omega} \begin{pmatrix} 1 \\ j \\ -1 \\ -j \end{pmatrix} e^{-j\omega t} \tag{22.14}$$

The responses can also be expressed in the complex conjugate form as follows:

$$\begin{pmatrix} x_1 \\ x_2 \\ x_3 \\ x_4 \end{pmatrix} = \begin{pmatrix} 1 \\ -j \\ -1 \\ j \end{pmatrix} e^{j\omega t} \Rightarrow \begin{pmatrix} y_1 \\ y_2 \\ y_3 \\ y_4 \end{pmatrix} = (H(s)_{s=-j\omega})^* \begin{pmatrix} 1 \\ -j \\ -1 \\ j \end{pmatrix} e^{j\omega t} \tag{22.15}$$

$$\begin{pmatrix} x_1 \\ x_2 \\ x_3 \\ x_4 \end{pmatrix} = \begin{pmatrix} 1 \\ -j \\ -1 \\ j \end{pmatrix} e^{-j\omega t} \Rightarrow \begin{pmatrix} y_1 \\ y_2 \\ y_3 \\ y_4 \end{pmatrix} = (H(s)_{s=j\omega})^* \begin{pmatrix} 1 \\ -j \\ -1 \\ j \end{pmatrix} e^{-j\omega t} \tag{22.16}$$

22.2 Classic Polyphase Filters

Based on the rotational symmetry property of LRS networks, the classic polyphase filter circuit can be analyzed using the polyphase input signal mode as shown in Fig. 22.12.

We may select any of the four output ports for our analysis as

$$Y(1+s) = Xs + jX \tag{22.17}$$

The transfer function of the classic polyphase filter to the polyphase signal can be expressed as

$$H(s) \equiv \frac{Y}{X} = \frac{s+j}{1+s} \tag{22.18}$$

The pole/zero plots and the gain frequency response of the normalized classic polyphase filter are shown in Fig. 22.13.

Since at $\omega = -1$, the transfer function has zero value, the negative unity frequency is

$$H(s)|_{s=-j} = \frac{s+j}{1+s}|_{s=-j} \tag{22.19}$$

Note that a normalized classic polyphase filter will reject the following input polyphase signals:

$$x(t) = \begin{pmatrix} 1 \\ j \\ -1 \\ -j \end{pmatrix} \cdot e^{-jt} \quad \text{and} \quad x(t) = \begin{pmatrix} 1 \\ -j \\ -1 \\ j \end{pmatrix} \cdot e^{jt} \tag{22.20}$$

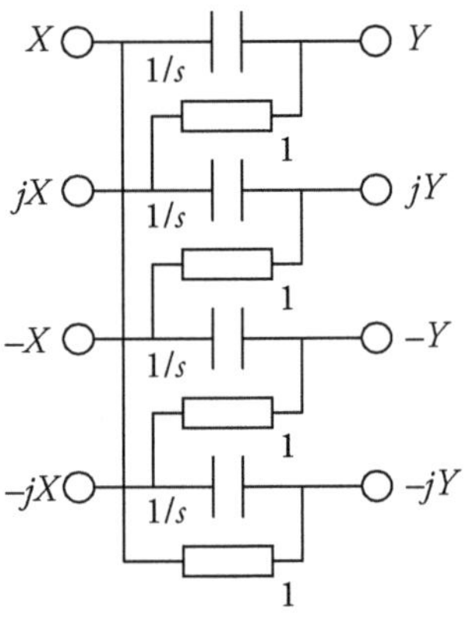

FIGURE 22.12 Classic polyphase filter response.

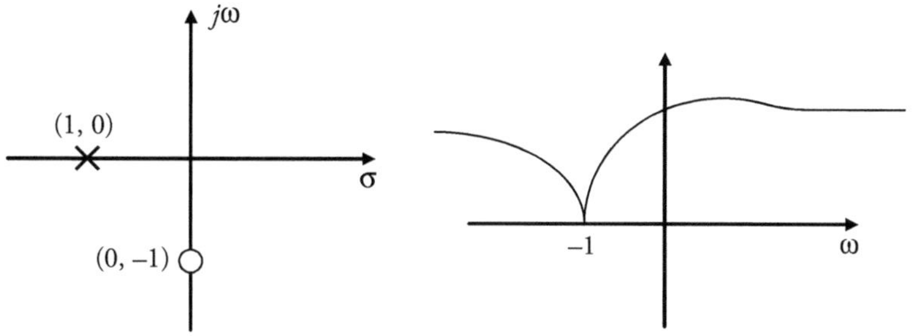

FIGURE 22.13 Pole/Zero and gain plots of the classic polyphase filter.

22.2.1 Image Rejection Based on Polyphase Filters

The results obtained in the last section can be used to explain the baseband-based image rejection operation of the VLIF RF down-conversion mixer structure employing the analog polyphase filter as shown in Fig. 22.14, where the input signal contains both the desired RF signal and its image signal.

The polyphase filter has strong attenuation for signal at frequency $\omega = -\omega_{IF}$.

$$H(s)|_{s=-j\omega IF} \approx 0 \tag{22.21}$$

The I- and Q-channel input signals at the polyphase filter input are given as

$$\Rightarrow x_I(t) = (A_{\text{signal}}\cos((\omega_{\text{LO}}+\omega_{\text{IF}})t) + A_{\text{image}}\cos((\omega_{\text{LO}}-\omega_{\text{IF}})t))\cdot\cos(\omega_{\text{LO}}t)$$

$$= \frac{A_{\text{signal}}}{4}(e^{j(\omega_{\text{LO}}+\omega_{\text{IF}})t} + e^{-j(\omega_{\text{LO}}+\omega_{\text{IF}})t})\cdot(e^{j\omega_{\text{LO}}t} + e^{-j\omega_{\text{LO}}t}) \tag{22.22}$$

$$+\frac{A_{\text{image}}}{4}(e^{j(\omega_{\text{LO}}-\omega_{\text{IF}})t} + e^{-j(\omega_{\text{LO}}-\omega_{\text{IF}})t})\cdot(e^{j\omega_{\text{LO}}t} + e^{-j\omega_{\text{LO}}t})$$

$$\Rightarrow x_I(t) = \frac{A_{\text{signal}}}{4}(e^{j\omega_{\text{IF}}t} + e^{-j\omega_{\text{IF}}t}) + \frac{A_{\text{image}}}{4}(e^{j\omega_{\text{IF}}t} + e^{-j\omega_{\text{IF}}t}) + x_I(2\omega_{\text{LO}}) \tag{22.23}$$

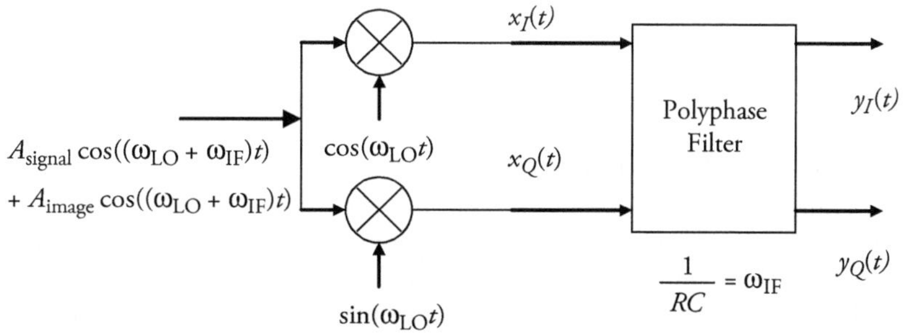

FIGURE 22.14 Image rejection down-conversion mixer.

$$x_Q(t) = (A_{\text{signal}}\cos((\omega_{\text{LO}}+\omega_{\text{IF}})t) + A_{\text{image}}\cos((\omega_{\text{LO}}-\omega_{\text{IF}})t))\cdot\sin(\omega_{\text{LO}}t)$$

$$= \frac{A_{\text{signal}}}{4j}(e^{j(\omega_{\text{LO}}+\omega_{\text{IF}})t} + e^{-j(\omega_{\text{LO}}+\omega_{\text{IF}})t})\cdot(e^{j\omega_{\text{LO}}t} - e^{-j\omega_{\text{LO}}t}) \quad (22.24)$$

$$+\frac{A_{\text{image}}}{4i}(e^{j(\omega_{\text{LO}}-\omega_{\text{IF}})t} + e^{-j(\omega_{\text{LO}}-\omega_{\text{IF}})t})\cdot(e^{j\omega_{\text{LO}}t} - e^{-j\omega_{\text{LO}}t})$$

$$\Rightarrow x_Q(t) = \frac{A_{\text{signal}}}{4j}(-e^{j\omega_{\text{IF}}t} + e^{-j\omega_{\text{IF}}t}) + \frac{A_{\text{image}}}{4j}(e^{j\omega_{\text{IF}}t} - e^{-j\omega_{\text{IF}}t}) + x_Q(2\omega_{\text{LO}}) \quad (22.25)$$

where $x_I(2\omega_{\text{LO}})$ and $x_Q(2\omega_{\text{LO}})$ are the high-frequency components in the signal resulting from up-conversion that will usually be filtered by the low-pass filter in the channel, and therefore, they can be ignored in the following discussions.

The combined polyphase filter input signal can be expressed as

$$\begin{pmatrix} x_I(t) \\ x_Q(t) \end{pmatrix} = \frac{A_{\text{signal}}}{2}\text{Re}\left[\begin{pmatrix} 1 \\ j \end{pmatrix} e^{j\omega_{\text{IF}}t}\right] + \frac{A_{\text{image}}}{2}\text{Re}\left[\begin{pmatrix} 1 \\ j \end{pmatrix} e^{-j\omega_{\text{IF}}t}\right] \quad (22.26)$$

At the output of the polyphase filter, we have the following:

$$\begin{pmatrix} y_I(t) \\ y_Q(t) \end{pmatrix} = \frac{A_{\text{signal}}}{2}\text{Re}\left[H(s)_{s=j\omega_{\text{IF}}}\begin{pmatrix} 1 \\ j \end{pmatrix} e^{j\omega_{\text{IF}}t}\right] + \frac{A_{\text{image}}}{2}\text{Re}\left[H(s)_{s=-j\omega_{\text{IF}}}\begin{pmatrix} 1 \\ j \end{pmatrix} e^{-j\omega_{\text{IF}}t}\right] \quad (22.27)$$

As a result of the selective filtering effect of the polyphase network, the image signal will be rejected:

$$\begin{pmatrix} y_I(t) \\ y_Q(t) \end{pmatrix} = \frac{A_{\text{signal}}}{2}\text{Re}\left[H(s)_{s=j\omega_{\text{IF}}}\begin{pmatrix} 1 \\ j \end{pmatrix} e^{j\omega_{\text{IF}}t}\right] \quad (22.28)$$

22.2.2 High-Order Classic Polyphase Filters

When multiple classic polyphase filter stages are combined as shown in Fig. 22.15, multiple-notch filter response can be realized with each stage providing a notch at the frequency determined by the RC constant of that stage. Such a filter structure provides a band-rejection operation for the image band as shown in Fig. 22.16.

22.3 Polyphase SFG Representations

VLSI complex filters can be implemented by employing the linear passive and active components. It can be mathematically proven that for a linear filter with single input and output ports, the transfer function can only be real. This means that a complex filter based on linear circuit components can only be realized using multiple signal path inputs. These input signals are the same input signals of different phases as shown in Fig. 22.17. In addition, in the filter implementation, it is desirable that the output signals

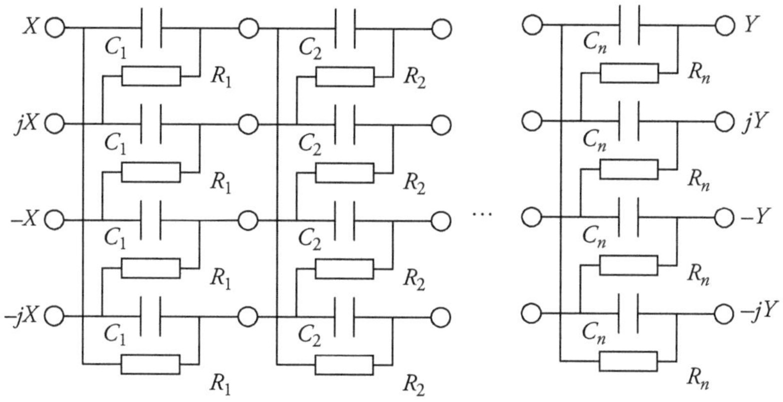

FIGURE 22.15 Cascade of classic polyphase filters.

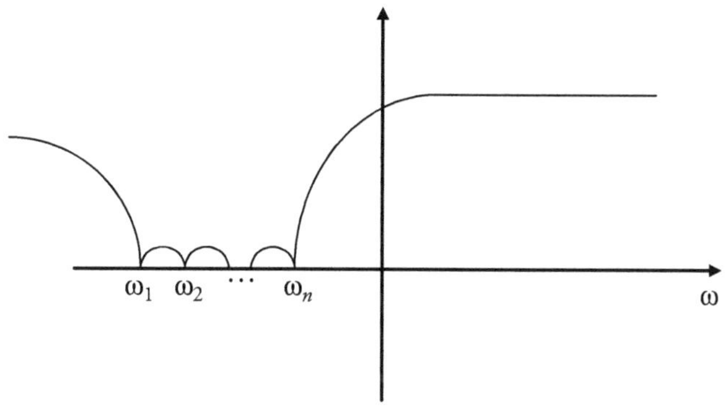

FIGURE 22.16 Frequency response of high-order classic polyphase filters.

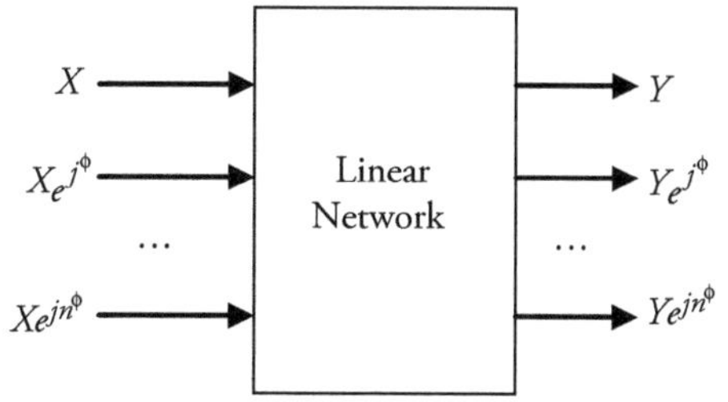

FIGURE 22.17 General complex filter response of linear network.

with the same related phases to the input are generated, serving as the inputs for the next polyphase filter stage.

There are several equivalent ways to mathematically represent an LRS network with respect to the polyphase signal inputs:

$$\begin{pmatrix} d_1 & d_2 & d_3 & d_4 \\ d_4 & d_1 & d_2 & d_3 \\ d_3 & d_4 & d_1 & d_2 \\ d_2 & d_3 & d_4 & d_1 \end{pmatrix}\begin{pmatrix} 1 \\ j \\ -1 \\ -j \end{pmatrix} Y = \begin{pmatrix} h_1 & h_2 & h_3 & h_4 \\ h_4 & h_1 & h_2 & h_3 \\ h_3 & h_4 & h_1 & h_2 \\ h_2 & h_3 & h_4 & h_1 \end{pmatrix}\begin{pmatrix} 1 \\ j \\ -1 \\ -j \end{pmatrix} X \tag{22.29}$$

or

$$\begin{pmatrix} d_1 - d_3 & d_2 - d_4 \\ d_4 - d_2 & d_1 - d_3 \end{pmatrix}\begin{pmatrix} 1 \\ j \end{pmatrix} Y = \begin{pmatrix} h_1 - h_3 & h_2 - h_4 \\ h_4 - h_2 & h_1 - h_3 \end{pmatrix}\begin{pmatrix} 1 \\ j \end{pmatrix} X \tag{22.30}$$

or

$$Y(s) = \frac{(h_1 + jh_2 - h_3 - jh_4)}{(d_1 + jd_2 - d_3 - jd_4)} X(s) \tag{22.31}$$

The equations can be expressed in SFG forms using the input and output relationships.

22.3.1 Polyphase SFG Expression

The polyphase filter expressed in Eq. (22.29) can be expressed in the SFG as shown in Fig. 22.18. Based on the rotational symmetry of the polyphase network, the SFG can be partitioned into four parallel identical SFG sections that can be easily generated from one of the sections based on the rotation symmetry of the circuit structure.

$$\begin{cases} X \rightarrow jX, \quad Y \rightarrow jY \\ jX \rightarrow -X, \quad jY \rightarrow -Y \\ -X \rightarrow -jX, \quad -Y \rightarrow -jY \\ -jX \rightarrow X, \quad -jY \rightarrow Y \end{cases} \tag{22.32}$$

Such a polyphase SFG representation provides a direct one-to-one correlation between the polyphase filter SFG expression and VLSI circuit implementation.

22.3.2 Single-Path Polyphase SFG Expression

With respect to the polyphase signal inputs, the polyphase SFG expressions contain the redundancy of the filter structures that can be further simplified by using the nonredundant forms. The single-path polyphase SFGs consisting of the basic structures of the SFG can be introduced as shown in Fig. 22.19.

One of the important applications of the single-path SFG is that it can be used to find the transfer function of the polyphase filter with respect to the polyphase signal in filter analysis.

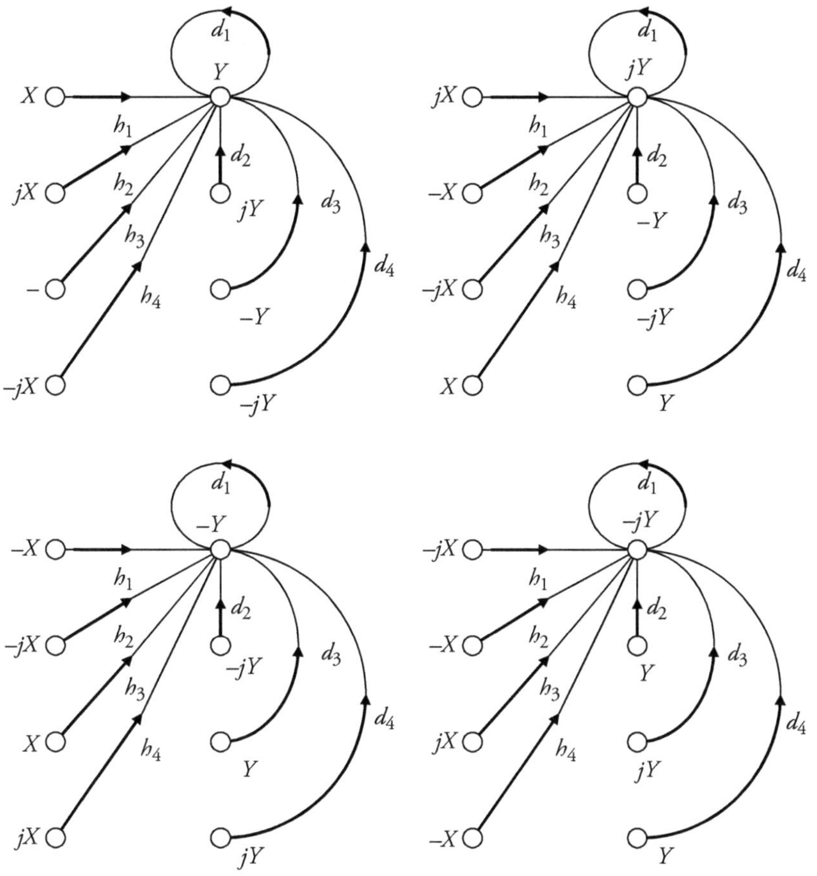

FIGURE 22.18 Full polyphase SFG representation.

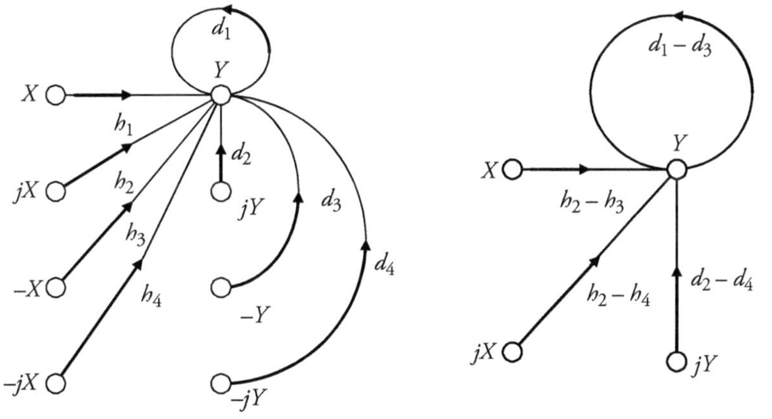

(a) Single-path SFG

(b) Differential single-path SFG

FIGURE 22.19 Single-path polyphase SFG representations.

22.4 Design of VLSI Active Polyphase Filters

With the above polyphase SFG expressions, we are now ready to introduce a systematic method for designing the VLSI polyphase filter circuits.

Similar to the SFG-based real filter circuit design, an SFG-based polyphase design procedure may be started from the construction of SFG expression that may be used to map the design specifications into the VLSI circuit structure. The only difference between the SFG-based polyphase filter design and the real filter design is that the former has to realize the image parts of the coefficients in the transfer function. Consequently, the crossover of phase signals will be seen in these SFG structures.

The design of a VLSI polyphase filter can be realized in the following steps:

- Selection of a transfer function that meets the design specification
- Construction of SFGs based on selected basic VLSI building circuit blocks (such as integrators, scalers, and adders), ignoring the complex numbers in the scaling coefficients
- Realization of complex coefficients using single-path polyphase SFGs
- Construction of full or differential polyphase SFGs using the rotational symmetry
- Mapping of polyphase SFGs to VLSI circuits

The above general polyphase filter design procedures can be demonstrated using the following filter design example. For the polyphase filter specified by the s-domain, we have the following transfer function:

$$\frac{Y}{X} = -\frac{1}{1+(s-\alpha j)} \tag{22.33}$$

The above transfer function can be modified to the form using basic VLSI building circuit blocks as

$$\Rightarrow Y = \frac{-1}{s}(X+(1-j\alpha)Y) \tag{22.34}$$

The SFG can be constructed from the above equation based on the complex number coefficients. The complex coefficient of the SFG can be further converted into a real coefficient using the single-path polyphase SFG as shown in Fig. 22.20.

Based on rotational symmetry of the polyphase filter, the single-path polyphase SFG can be converted into the polyphase or the differential polyphase SFG expressions as shown in Fig. 22.21.

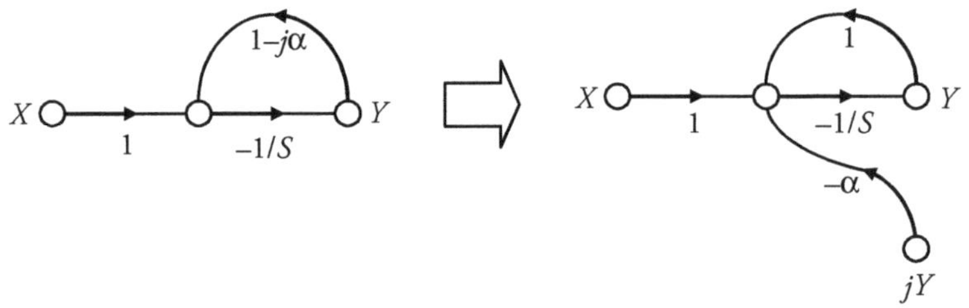

Figure 22.20 Creation of single-phase real coefficient SFG.

Finally, the SFG can be mapped into the VLSI circuit form. A Gm-C circuit implementation of the given transfer function is shown in Fig. 22.22.

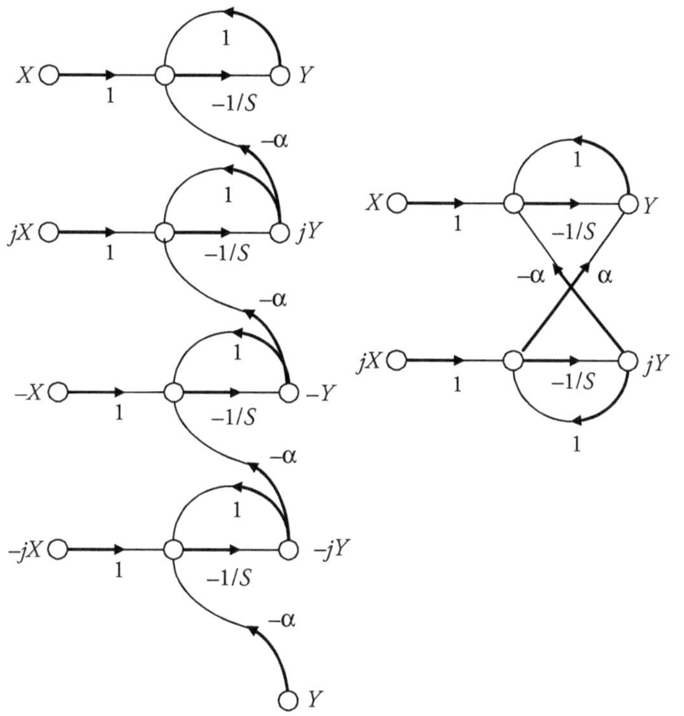

(a) Polyphase SFG (b) Differential polyphase SFG

FIGURE 22.21 Polyphase SFG expressions.

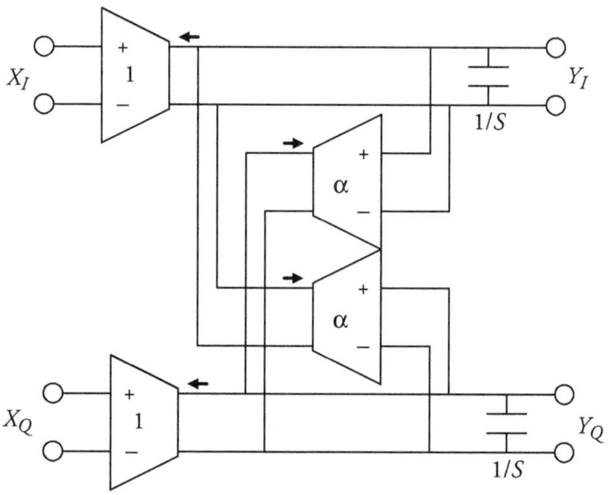

FIGURE 22.22 VLSI Gm-C circuit realization of polyphase filter.

References

1. K. Linggajaya, M. A. Do, J. G. Ma, K. S. Yeo, "A New Active Polyphase Filter for Wideband Image Reject Down Converter," *Proc. ICSE 2002. IEEE International Conference on Semiconductor Electronics*, December 2002.
2. J. Crols, M. Steyaert, "An Analog Integrated Polyphase Filter for a High Performance Low-IF Receiver," *Digest of Technical Papers, Symposium on VLSI Circuits*, June 1995.
3. D. I. Sanderson, R. M. Svitek, S. Raman, "A 5-6-GHz Polyphase Filter with Tunable I/Q Phase Balance," *IEEE Letters, Microwave and Wireless Components*, vol. 14, no. 7, pp. 364–366, July 2004.
4. E. Stikvoort, "Polyphase Filter Section with OPAMPs," *IEEE Trans. on Circuits and Systems II: Analog and Digital Signal Processing*, vol. 50, no. 7, pp. 376–378, July 2003.
5. K. Wada, Y. Tadokoro, "RC Polyphase Filter with Flat Gain Characteristic," *Proceedings, ISCAS*, vol. 1, pp. I-537–I-540, May 2003.
6. C.-Y. Chou, C.-Y. Wu, "The Design of a New Wideband and Low-Power CMOS Active Polyphase Filter for Low-IF Receiver Applications," *Circuits and Systems, 2002. APCCAS'02*, vol. 1, pp. 241–244, October 28–31, 2002.
7. H. Kobayashi, J. Kang, T. Kitahara, S. Takigami, H. Sadamura, "Explicit Transfer Function of RC Polyphase Filter for Wireless Transceiver Analog Front-end," *ASIC, 2002*, 6–8 pp. 137–140, 2002.
8. F. Behbahani, Y. Kishigami, J. Leete, A. A. Abidi, "CMOS Mixers and Polyphase Filters for Large Image Rejection," *IEEE J. Solid-State Circuits*, vol. 36, no. 6, pp. 873–887, June 2001.
9. G. Blakiewicz, J. Jakusz, "A Low-Voltage Fully-Differential BiCMOS OPAMP for Polyphase Filter," *Proceedings. ICCSC*, June 2002.
10. V. Vidojkovic, van der J. Tang, A. Leeuwenburgh, A. van Roermund, "Analysis of a 1.8–2.5 GHz Multi-standard High Image-Reject Front-End," *9th International Conference on Electronics, Circuits and Systems*, 2002, vol. 1, pp. 73–76, September 15–18, 2002.
11. M. A. F. Borremans, C. R. C. De Ranter, M. S. J. Steyaert, "A CMOS Dual-Channel, 100-MHz to 1.1-GHz Transmitter for Cable Applications," *IEEE J. Solid-State Circuits*, vol. 34, no. 12, pp. 1904–1913, December 1999.
12. F. Azevedo, M. J. Rosario, J. C. Freire, "CMOS Monolithic Wideband Image Rejection Mixer with Polyphase Filters," *Proceedings of the 2003 SBMO/IEEE MTT-S International Microwave and Optoelectronics Conference, 2003. IMOC 2003*, vol. 2, pp. 815–818, September 20–23, 2003.
13. B. Shi, W. Shan, P. Andreani, "A 57-dB Image Band Rejection CMOS Gm-C Polyphase Filter with Automatic Frequency Tuning for Bluetooth," *IEEE International Symposium on Circuits and Systems, 2002. ISCAS 2002*, vol. 5, May 26–29, 2002.
14. F. Behbahani, J. C. Leete, Y. Kishigami, A. Roithmeier, K. Hoshino, A. A. Abidi, "A 2.4-GHz Low-IF Receiver for Wideband WLAN in 6-μm CMOS-Architecture and Front-End," *IEEE J Solid-State Circuits*, vol. 35, no. 12, pp. 1908–1916, December 2000.
15. E. Tiiliharju, K. Halonen, "A Quadrature-Modulator for 0.6-2.6 GHz with Frequency Doubler," *ISCAS'02*, vol. 5, pp. V-429–V-432, May 26–29, 2002.
16. K. Stadius, P. Jarvio, P. Paatsila, K. Halonen, "Image-Reject Receivers with Image-Selection Functionality," *The 2001 IEEE International Symposium on Circuits and Systems, 2001. ISCAS 2001*, vol. 4, pp. 124–127, May 6–9, 2001.
17. D. I. Sanderson, S. Raman, "5-6 GHz SiGe VCO with Tunable Polyphase Output for Analog Image Rejection and I/Q Mismatch Compensation," *2003 IEEE Radio Frequency Integrated Circuits (RFIC) Symposium*, pp. 683–686, June 8–10, 2003.

18. C. Cuypers, N. Y. Voo, M. Teplechuk, J. I. Sewell, "The General Synthesis of Complex Analogue Filters," *9th International Conference on Electronics, Circuits and Systems, 2002*, vol. 1, pp. 153–156, September 15–18, 2002.
19. F. Behbahani, Y. Kishigami, J. Leete, A. A. Abidi, "CMOS 10 MHz-IF Downconverter with On-Chip Broadband Circuit for Large Image-Suppression," *1999 Symposium on VLSI Circuits. Digest of Technical Papers*, pp. 83–86, June 17–19, 1999.
20. C. Cuypers, N. Y. Voo, M. Teplechuk, J. I. Sewell, "The General Synthesis of Complex Analogue Filters," *9th International Conference on Electronics, Circuits and Systems*, vol. 1. pp. 153–156, September 2002.
21. S. H. Galal, H. F. Ragaie, M. S. Tawfik, "RC Sequence Asymmetric Polyphase Networks for RF Integrated Transceivers," *IEEE Trans. on Circuits and Systems II: Analog and Digital Signal Processing*, vol. 47, no. 1, pp. 18–27, January 2000.
22. E. Stikvoort, "Polyphase Filter Section with OPAMPs," *IEEE Trans. on Circuits and Systems II: Analog and Digital Signal Processing*, vol. 50, no. 7, pp. 376–378, July 2003.
23. A. A. Emira, E. Sanchez-Sinencio, "A Pseudo Differential Complex Filter for Bluetooth with Frequency Tuning," *IEEE Trans. on Circuits and Systems II: Analog and Digital Signal Processing*, vol. 50, no. 10, pp. 742–754, October 2003.
24. C. Cuypers, N. Y. Voo, M. Teplechuk, J. I. Sewell, "The General Synthesis of Complex Analogue Filters," *9th International Conference on Electronics, Circuits and Systems*, vol. 1, pp. 153–156, September 2002.
25. H. J. Song, "A General Method to VLSI Polyphase Filter Analysis and Design for Integrated RF Applications," *IEEE International SOCC Conference*, pp. 31–34, September 2006.

CHAPTER 23

Introduction to VLSI RF Signal Processing Circuits

VLSI radio frequency (RF) signal processing circuits usually operate at high frequencies, typically up to 1–12 GHz. In addition to the high operating frequency range, the signal processing circuits for RF applications have other special properties such as very wide dynamic range (as high as 10–15 orders of magnitude at the transmitter and receiver), higher sensitivity to the parasitic circuit parameters, and involves frequency translations (nonlinear circuits) that unify the signal processing circuit operations. RF circuit design and analysis traditionally adopted slightly different methods from the analog designs such as the terminologies, the design specifications, etc., mainly due to different development paths between the RF and the analog circuits in history. However, these differences are becoming smaller with the scaling of the VLSI device features and the increasing of operation speed of the VLSI circuits. Following this trend, the VLSI RF signal processing circuits are expected to become more similar to the VLSI high-speed analog signal processing circuits.

23.1 VLSI RF Circuit Component Structures

VLSI resistors, capacitors, and MOS transistors have been widely used in various (low-frequency) analog signal processing applications. VLSI inductors and varactors may be used effectively as design components for high-frequency and RF signal processing circuit applications.

23.1.1 VLSI RF Resistor Structures

The VLSI resistor structures for RF signal processing applications need to include additional device parameters for proper modeling of the high-frequency behavior. Figure 23.1 shows a typical VLSI resistor circuit model targeted at the RF applications.

23.1.2 VLSI RF Capacitor Structures

Metal-metal capacitor structures with their high-quality and high-linearity factors are widely used for RF applications. Figure 23.2 shows the metal-insulator-metal (MIM) capacitor structure, where C is the capacitance of the capacitor, R_b is the series resistance

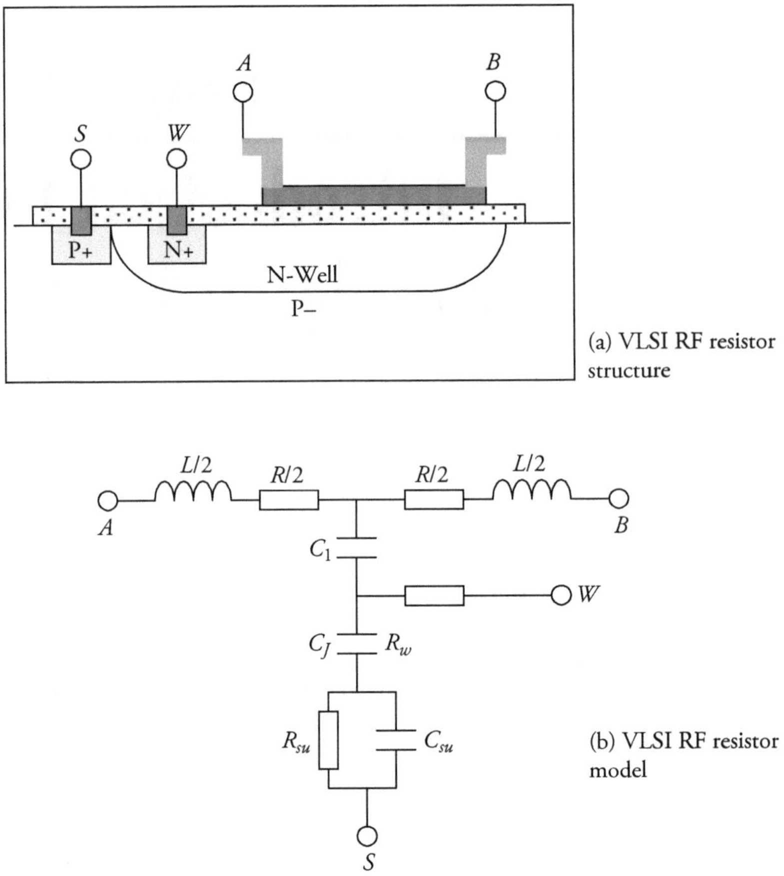

FIGURE 23.1 VLSI RF resistor structure and the circuit model.

of the bottom plate, C_B is the parasitic capacitance to substrate, and R_{su} and C_{su} are the equivalent substrate resistance and capacitance.

For standard VLSI CMOS processes without MIM capacitors, metal-metal capacitors can also be constructed from interconnect metal layers and can be used for RF applications.

23.1.3 VLSI RF Inductor and Transform Structures

VLSI spiral inductors can be constructed using metal interconnect layers in all VLSI process technologies. The commonly used on-chip inductor and transform structures and circuit models are shown in Figs. 23.3 and 23.4.

In this model, L_1 is the main inductance of the inductor, R_1 is the series resistance of the inductor, C_3 is the self-parasitic capacitance, C_1 and C_2 are the parasitic capacitances to substrate, and R_2 and R_3 and C_4 and C_5 are the resistances and capacitances of the substrate, respectively. In practical design, multiple interconnect layers provide the flexibility of the inductor construction. Generally speaking, parallel-metal-layer structures provide the highest Q and lowest inductance density, single-metal-layer structures

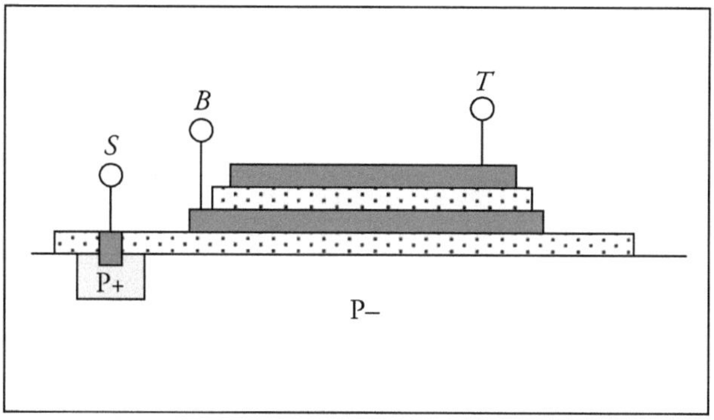

(a) VLSI RF capacitor structure

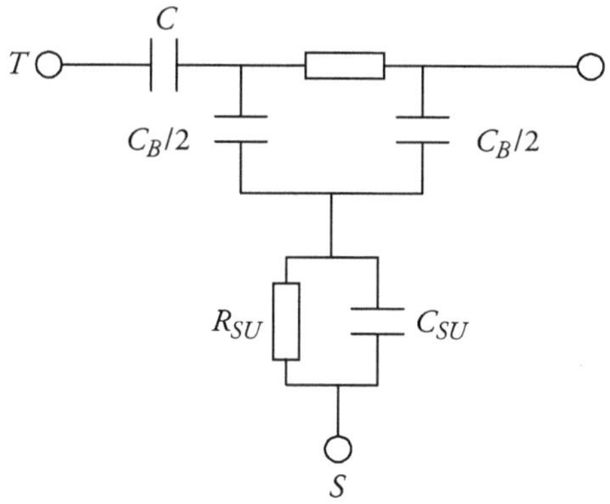

(b) VLSI RF capacitor model

FIGURE 23.2 VLSI RF capacitor structure and the circuit model.

provide medium Q and medium inductance density, and series-metal-layer structures provide the lowest Q and highest inductance density. In addition, a thin dielectric layer enhances the coupling to the substrate and substrate losses. Patterned conductive shield, meanwhile, minimizes electric field–induced power loss in the substrate.

The circuit parameters of the VLSI inductors depend on both the geometry and material parameters in constructing the inductors. A typical set of parameters used to define a VLSI inductor includes the radius of the inductor (R), the number of turns (N), the width of the metal line (W), the metal spacing (S), the metal sheet resistance (R_M), and the substrate loss (R_S). Typical value of R is between 100–200 μm, N is 4–6, S is ~1 μm, W is 4–20 μm (which provides inductance values of 3–20 nH and capacitance of 1–10 pF, resistance of 4–20 Ω), and Q of 4–8 for a typical 0.5-μm VLSI process technology circuit.

23.1.4 VLSI RF Varactor Structures

Varactors or voltage-controlled capacitors can be constructed in VLSI based on the PN diode, the accumulation mode MOS devices, or the inversion mode MOS devices. The diode-based VLSI varactor structure and its circuit model are shown in Fig. 23.5. The MOS-based VLSI varactor structure and its circuit model are shown in Fig. 23.6.

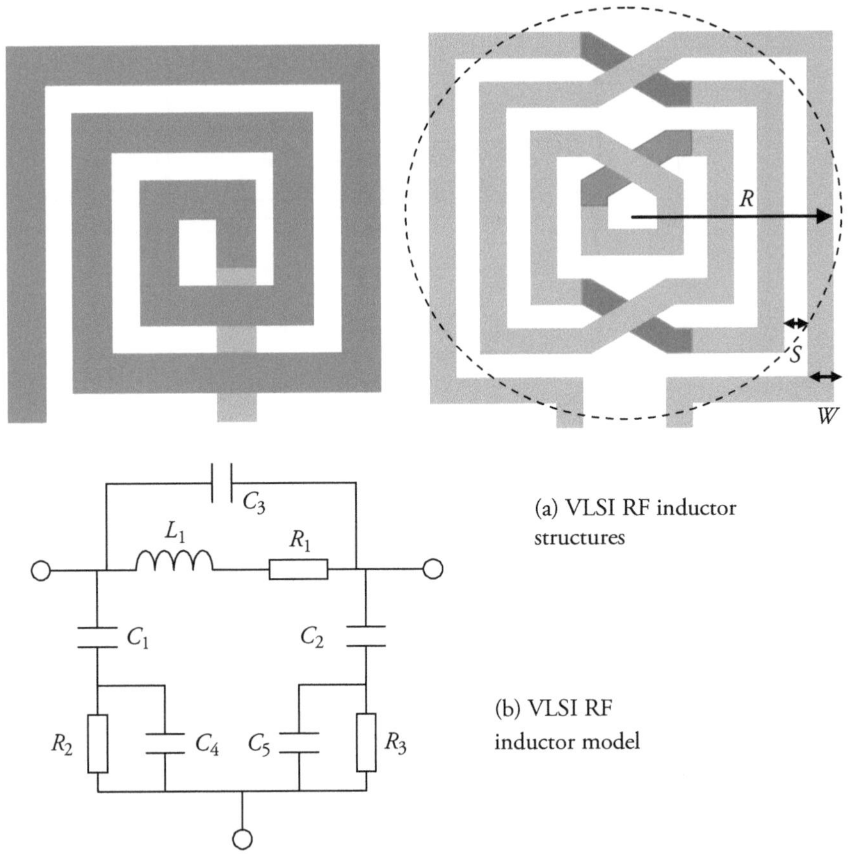

FIGURE 23.3 VLSI spiral inductor structure and the circuit model.

23.2 Impedance Transformation and Matching Techniques

For RF circuits, it is important to match the impedances of the source components with the load components since the RF signal power transfer is maximized when the source is conjugately matched to the load in the frequency domain, as shown in Fig. 23.7. In case of resistive terminations, the source resistance must be equal to the load resistance for maximum power transfer. Practical termination generally involves complex impedances, and their real parts may not be equal. In such a case, an impedance-matching circuit is required to eliminate the impedance mismatch.

23.2.1 Impedance Conversion and Matching Network

For different sources and load resistors, it is possible to introduce a match network to match the source and the load around a given frequency. The matching network is also called impedance transformer, and is shown in Fig. 23.8.

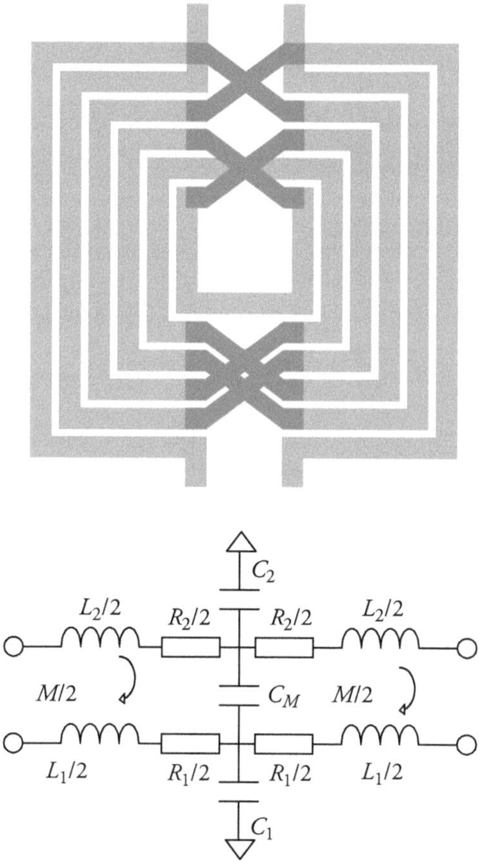

(a) VLSI RF transformer structures

(b) VLSI RF transformer model

FIGURE 23.4 VLSI transformer structure and the circuit model.

Several typical topologies of matching two unequal resistors at single frequency are shown in Fig. 23.9.

For the matching networks in Fig. 23.9*a* and *b*, we have the equivalent input impedance as follows:

$$R_L' = \frac{1}{\dfrac{1}{R_L} + \dfrac{1}{X_P}} + X_S \tag{23.1}$$

The X_p and X_s are the impedances of the parallel and serial matching components (capacitor or inductor), respectively. When the impedances are matched, we have

$$R_s = \frac{1}{\dfrac{1}{R_L} + \dfrac{1}{X_P}} + X_S = R_L \frac{1 - \dfrac{R_L}{X_P}}{1 + \left|\dfrac{R_L}{X_P}\right|^2} + X_S \tag{23.2}$$

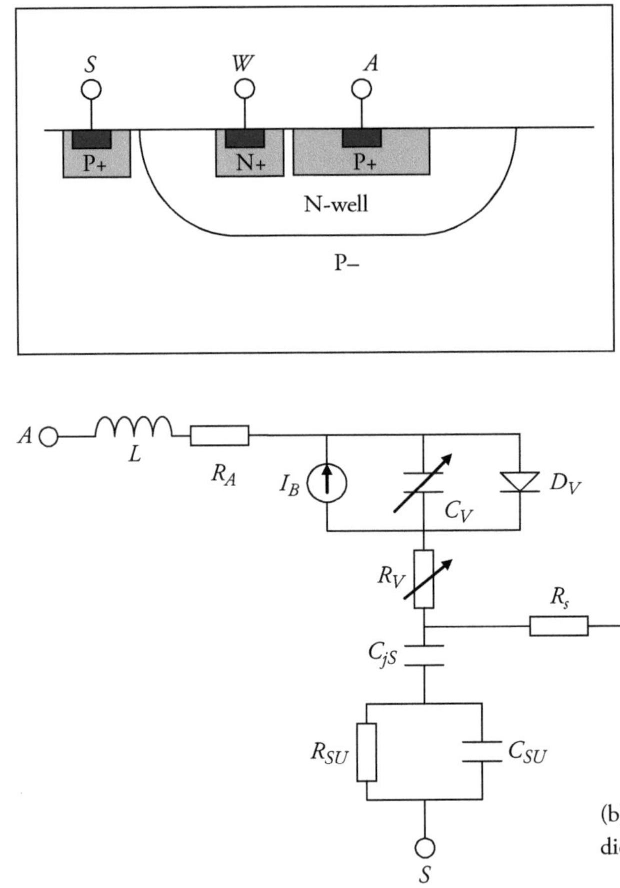

(a) VLSI PN junction diode structures

(b) VLSI PN junction diode circuit model

FIGURE 23.5 VLSI PN junction diode varactor structure and the circuit model.

Separating the real part and imaginary part of the equation yields

$$\begin{cases} R_s = R_L \dfrac{1}{1+\left|\dfrac{R_L}{X_P}\right|^2} \\ X_S = \dfrac{R_L}{X_P} \dfrac{1}{1+\left|\dfrac{R_L}{X_P}\right|^2} \end{cases} \tag{23.3}$$

Let

$$Q_P \equiv \sqrt{\frac{R_L}{R_S} - 1} \tag{23.4}$$

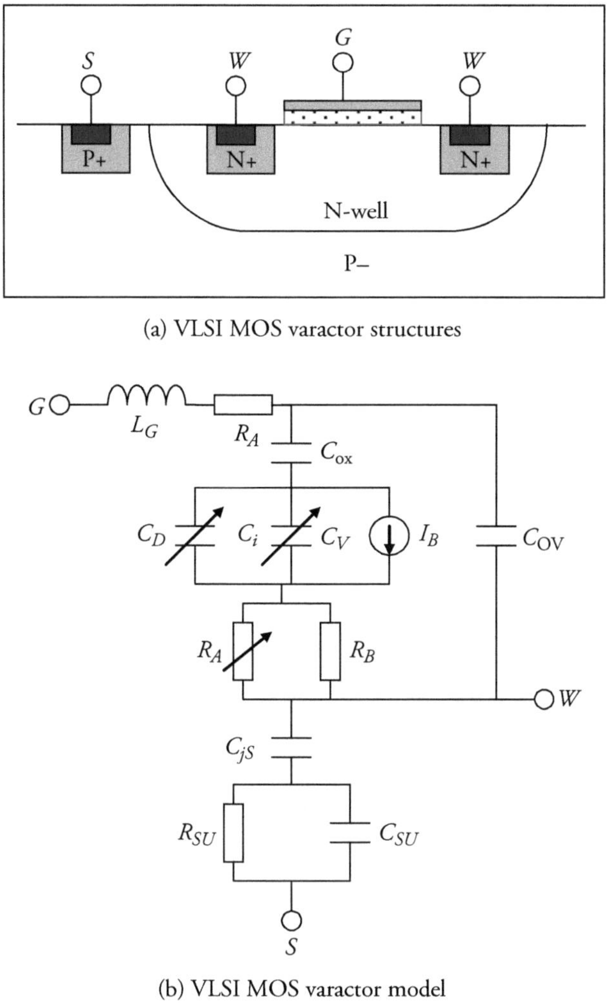

FIGURE 23.6 VLSI MOS varactor structure and SPICE model.

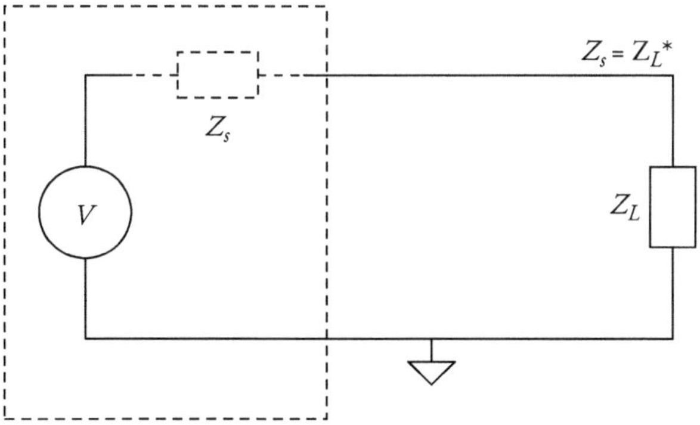

FIGURE 23.7 Impedance-matching circuits.

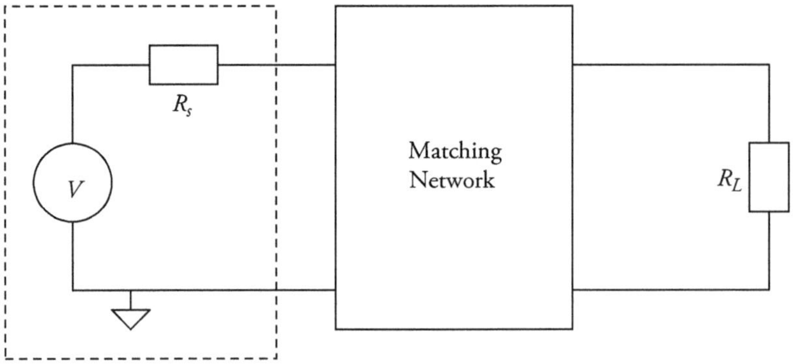

FIGURE 23.8 Impedance-matching with matching network.

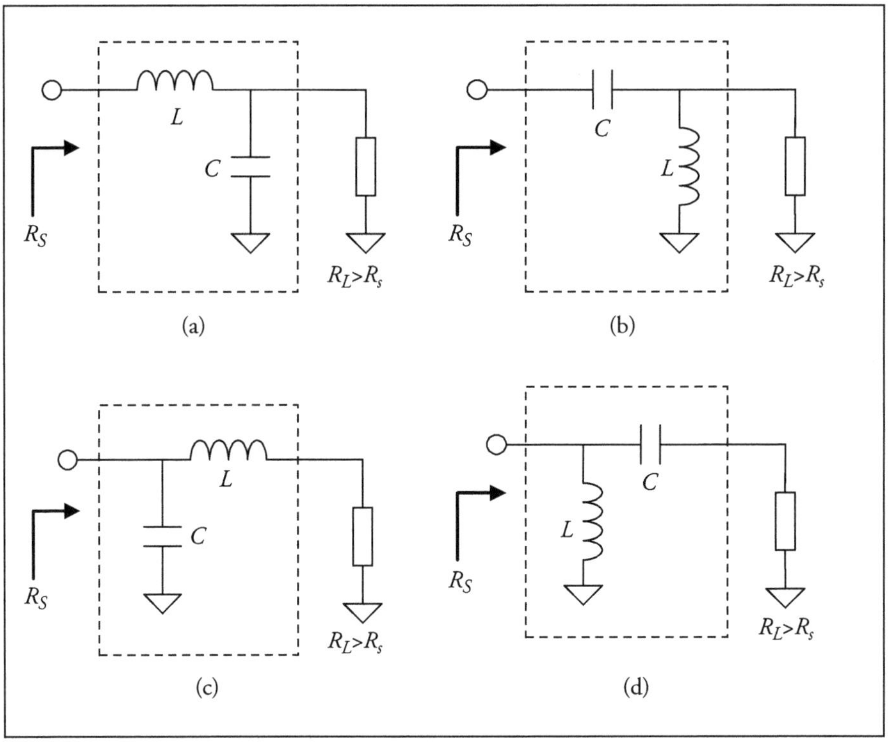

FIGURE 23.9 Lumped LC impedance-matching networks.

We have

$$\begin{cases} |X_P| = \dfrac{R_L}{Q_P} \\ X_S X_P = R_S R_L \end{cases} \tag{23.5}$$

Considering the frequency-dependent nature of X_p and X_s, the above matching network can only achieve the perfect matching at a desired frequency. Similarly, for the serial matching network that translates the smaller load resistance to larger source resistance, we have

$$Q_S \equiv \sqrt{\frac{R_S}{R_L} - 1} \tag{23.6}$$

and

$$\begin{cases} |X_P| = R_L Q_S \\ X_S X_P = R_S R_L \end{cases} \tag{23.7}$$

23.2.2 LNA Input Impedance-Matching Circuit

For the typical VLSI low-noise amplifier (LNA) circuit structure shown in Fig. 23.10, two impedance-matching inductors are typically used at both the source and the gate of the input MOS device.

For the above equivalent circuit, we have

$$V_i = V_s + \left(R_g + sL_g + \frac{1}{sC_g}\right) I_g \tag{23.8}$$

and

$$\begin{cases} V_s = sL_s \cdot (I_g + g_m V_x) \\ V_x = I_g \dfrac{1}{sC_g} \end{cases} \tag{23.9}$$

We have that

$$Z_i \equiv \frac{V_i}{I_g} = R_g + sL_g + \frac{1}{sC_g} + sL_s\left(1 + g_m \frac{1}{sC_g}\right) \tag{23.10}$$

or

$$Z_i = \left(R_g + g_m \frac{L_s}{C_g}\right) + s(L_g + L_s) + \frac{1}{sC_g} \tag{23.11}$$

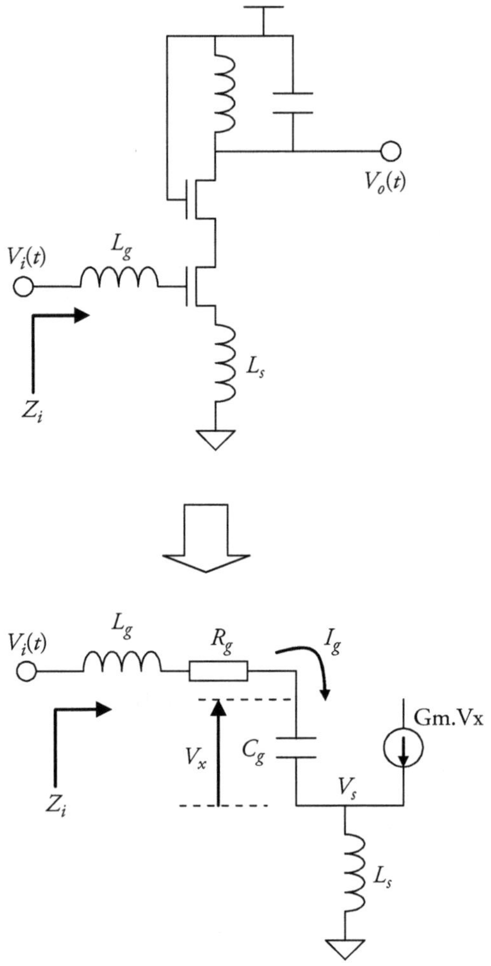

FIGURE 23.10 VLSI LNA circuit structure and equivalent circuit.

Note that for a perfect matching to the source impedance R_s, it is required that

$$\begin{cases} R_s = \left(R_g + g_m \dfrac{L_s}{C_g}\right) \\ \omega^2 (L_g + L_s) C_g = 1 \end{cases} \tag{23.12}$$

23.3 Frequency Translation Techniques

A significant difference between the RF signal processing circuits and the typical (baseband) analog circuits is that frequency translation operations as shown in Fig. 23.11 are usually employed in RF signal processing applications. Frequency

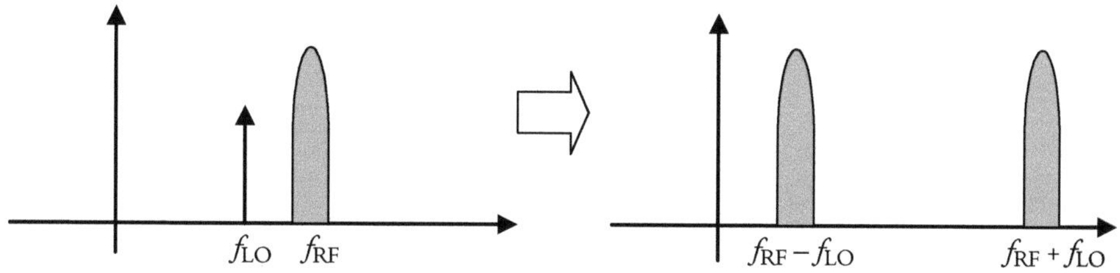

FIGURE 23.11 Frequency translation operation.

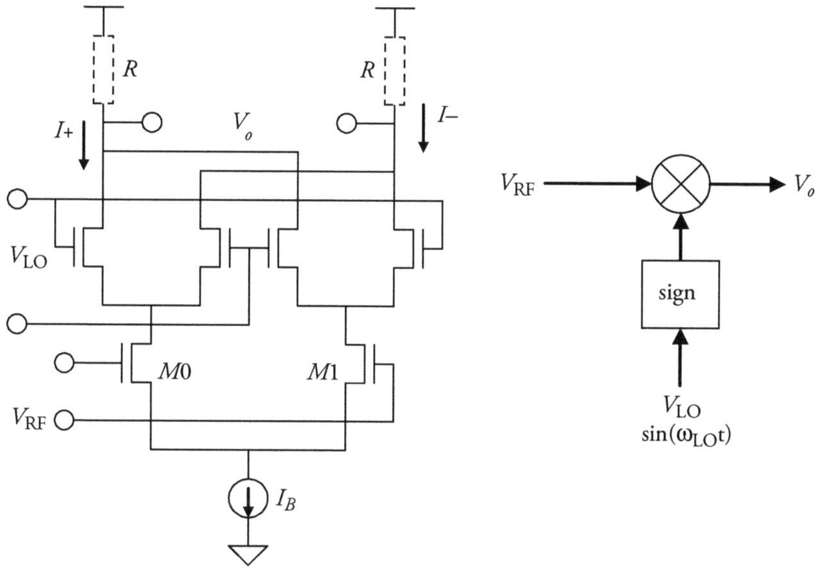

FIGURE 23.12 VLSI Gilbert mixer structure and the equivalent circuit model.

translations are realized using the mixer circuits by taking advantage of the nonlinear effects of the signal processing circuits. Figure 23.12 shows the commonly used Gilbert mixer circuit structure where the RF input signal path is the same as an input differential pair, and the local oscillator (LO) path provides swapping operation for the two outputs from the input differential pair of the RF path.

Since the sign function can be expanded using the Fourier transformation as

$$\text{sign}(\sin(\omega_{LO}t)) = \frac{4}{\pi}\left[\sin(\omega_{LO}t) - \frac{1}{3}\sin(3\omega_{LO}t) + \frac{1}{5}\sin(5\omega_{LO}t) + \cdots\right] \quad (23.13)$$

and

$$V_o(t) = V_{RF}(t) \cdot \text{Sign}(V_{LO}(t)) \quad (23.14)$$

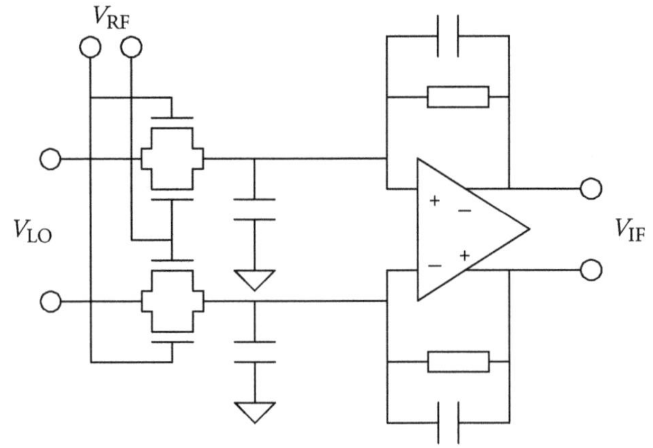

FIGURE 23.13 VLSI mixer circuit structure.

we have

$$V_o(t) = \frac{2}{\pi}\Big[\cos(\omega_{RF} + \omega_{LO})t + \cos(\omega_{RF} - \omega_{LO})t - \frac{1}{3}\cos(\omega_{RF} + 3\omega_{LO})t - \frac{1}{3}\cos(\omega_{RF} - 3\omega_{LO})t + \frac{1}{5}\cos(\omega_{RF} + 5\omega_{LO})t + \frac{1}{5}\cos(\omega_{RF} - 5\omega_{LO})t + \cdots\Big] \tag{23.15}$$

We can see that the output signal contains frequency contents of $\omega_{RF} \pm n\omega_{LO}$. For the up- or down-frequency conversion applications, either the cos $(\omega_{RF} + \omega_{LO})t$ or the cos $(\omega_{RF} - \omega_{LO})t$ component will pass through the circuits with other frequency contents being selectively filtered using a filter. It can be seen that after the selection filter, the mixer block realizes a signal frequency translation as shown below, depending on the up- or down-conversion operations:

$$\cos(\omega_{RF})t \Rightarrow \cos(\omega_{RF} + \omega_{LO})t \tag{23.16}$$

or

$$\cos(\omega_{RF})t \Rightarrow \cos(\omega_{RF} - \omega_{LO})t \tag{23.17}$$

Alternatively, the VLSI mixer circuit can also be realized using the four-transistor VCR circuit structure employing a low-speed VLSI OPAMP circuit for down-conversion as shown in Fig. 23.13.

23.4 Voltage-Controlled Oscillator (VCO) Circuits

The VLSI oscillator circuits can, in general, be viewed as special VLSI analog signal processing circuits that have poles of the transfer functions located at the $j\omega$ axis in the continuous-time circuit configurations or at the unit circle in the discrete-time circuit configurations, respectively, as shown in Fig. 23.14.

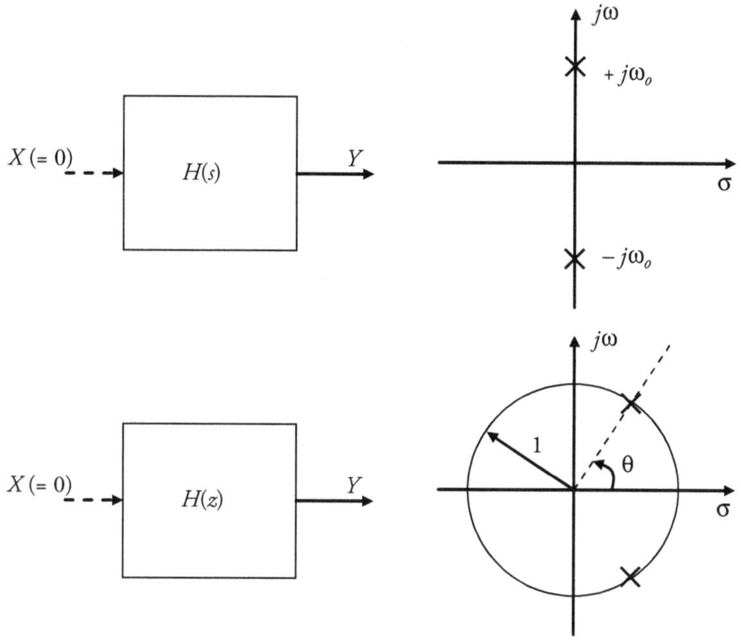

Figure 23.14 Poles of transfer function of VLSI oscillator circuits.

Based on the s-domain transfer function, we have the following:

$$Y(s) = \left(\frac{1}{s - j\omega_o} + \frac{1}{s + j\omega_o}\right) X(s) \tag{23.18}$$

$$\begin{aligned} &\text{For} \quad X(s) = 0 \quad \text{and} \quad Y(s) \neq 0 \\ &\text{Only} \quad \text{if} \quad \omega = \pm\omega_o \end{aligned} \tag{23.19}$$

Note that such output signals can exist even if there is no input signal, implying the stable oscillation condition. Also from the inverse Laplace transform, we know that the frequency of the signal that can exist is ω_o, which is the output signal frequency of the oscillator. Similarly, for a discrete-time system, we can have the following:

$$Y(z) = \left(\frac{1}{z - e^{j\theta}} + \frac{1}{z + e^{-j\theta}}\right) X(z) \tag{23.20}$$

$$\begin{aligned} &\text{For} \quad X(z) = 0 \quad \text{and} \quad Y(z) \neq 0 \\ &\text{Only} \quad \text{if} \quad \omega = \pm\theta/T \end{aligned} \tag{23.21}$$

The commonly used VLSI oscillator circuit structures include the crystal oscillator, relaxation oscillator, ring oscillator, and LC oscillator.

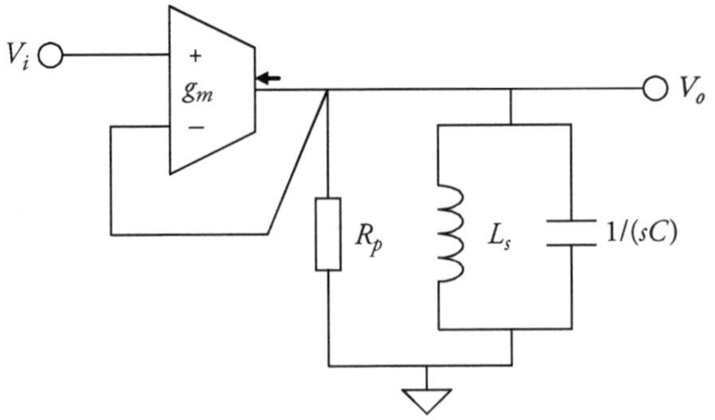

FIGURE 23.15 Conceptual LC oscillator circuit structure.

The crystal oscillator is the best known and also the most stable structure, which is based on the resonant characteristic of a crystal. However, only the driving (biasing) circuit can be integrated with it based on current VLSI process technologies, and the crystal has to be placed off-chip. On the other hand, both the VLSI relaxation oscillators and ring oscillators can be integrated on-chip. However, these oscillator circuit structures suffer from higher phase noises; therefore, they have limited applications in most high-performance RF circuit and system applications. Alternatively, the LC oscillator circuits, which can be integrated in the VLSI chip, offer high performance with very careful design.

Figure 23.15 shows a conceptual LC oscillator circuit structure that can be used to explain the operation of the LC oscillator circuits.

We can derive the s-domain transfer function of the circuit as follows:

$$\frac{V_o}{V_i} = \frac{-sL \cdot g_m}{LCs^2 + (1/R_p - g_m)Ls + 1} \tag{23.22}$$

The locations of the poles of the oscillator transfer function are given in the following equation:

$$p_{\pm} = -\frac{(1/R_p - g_m)}{2C} \pm j\sqrt{\frac{1}{LC} - \left(\frac{(1/R_p - g_m)}{2C}\right)^2} \tag{23.23}$$

Under the oscillation condition, we have

$$\begin{cases} g_m = 1/R_p \\ p_{\pm} = \pm j\sqrt{\dfrac{1}{LC}} \end{cases} \tag{23.24}$$

Note that the LC circuit will oscillate at a frequency equal to the resonant frequency of the LC tank, independent of active component parameters such as the Gm of the biasing circuit. This property makes this type of oscillator a good candidate for very low-phase noise high-frequency clock generation. In practical design, frequency adjustment is realized using tuning techniques based on the inductor or the capacitor banks, the varactors,

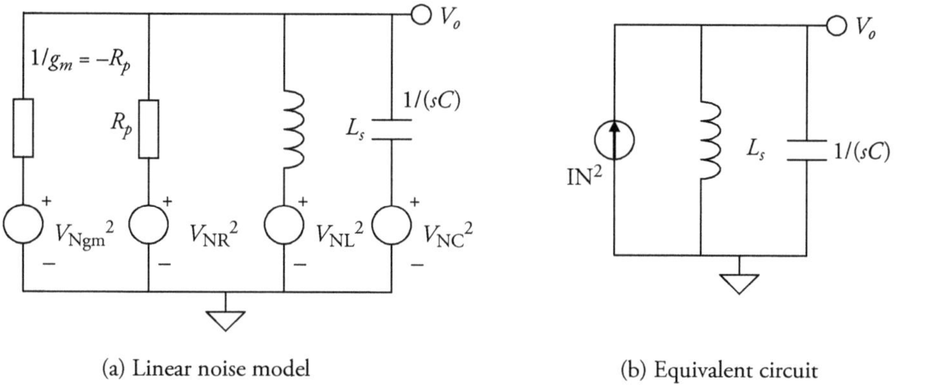

FIGURE 23.16 VCO phase noise model.

or the frequency divisions. Phase noise of the VCO is one of the key performance parameters in the VCO design, and it can be derived using the equivalent schematic shown in Fig. 23.16.

The equivalent noise current is given as

$$I_N^2 \equiv \frac{\overline{di_N}^2}{df} = 4kT\left(F_{\mathrm{Gm}}\left|G_m\right| + \frac{1}{R_P} + \frac{R_L}{(L\omega)^2} + R_C(C\omega)^2\right) \tag{23.25}$$

F_{Gm} is the access noise factor of the Gm-cell. The noise transfer function with respect to the noise current is given as

$$H_N^2(\omega) \equiv \frac{\overline{dV_{oN}}^2}{\overline{di_N}^2} = \left(\frac{1}{C\omega + \frac{1}{L\omega}}\right)^2 \tag{23.26}$$

At the $\Delta\omega$ offset from the resonant frequency ω_o, we have that

$$H_N^2(\omega_o + \Delta\omega) \equiv \frac{\overline{dV_{oN}}^2}{\overline{di_N}^2} = \left|\frac{1}{C(\omega_o + \Delta\omega) + \frac{1}{L(\omega_o + \Delta\omega)}}\right|^2 \approx \frac{\omega_o L}{4\left(\frac{\Delta\omega}{\omega_o}\right)^2} \tag{23.27}$$

The total output noise voltage is given as

$$\frac{\overline{dV_N}^2}{df} \approx 4kT\left(F_{\mathrm{Gm}}\left|G_m\right| + \frac{1}{R_P} + \frac{R_L}{(L\omega_o)^2} + R_C(C\omega_o)^2\right)\frac{\omega_o L}{4\left(\frac{\Delta\omega}{\omega_o}\right)^2} \tag{23.28}$$

By introducing the effective resistance and the quality factor of the LC tank as

$$\begin{cases} R_{\text{eff}} \equiv R_C + R_L + \dfrac{1}{R_P(\omega_o C)^2} \\ Q \equiv \dfrac{1}{R_{\text{eff}}}\sqrt{\dfrac{L}{C}} \end{cases} \tag{23.29}$$

we have the voltage noise density as

$$\frac{\overline{dV_N}^2}{df} \approx kT(F_{\text{Gm}}+1)R_{\text{eff}}\left(\frac{\omega_o}{\Delta\omega}\right)^2 \tag{23.30}$$

The single-sideband phase noise density is as follows:

$$PN \equiv \frac{\int_{\Delta\omega-1/2}^{\Delta\omega+1/2} \overline{dV_N}^2}{\text{carrier\_power}} \approx \frac{kT(F_{\text{Gm}}+1)R_{\text{eff}}\left(\dfrac{\omega_o}{\Delta\omega}\right)^2}{V_A^2/2} \tag{23.31}$$

where V_A is the signal magnitude. In practical cases, due to the nonlinearity of the VCO circuit, a typical phase noise curve will include three areas as shown in Eq. (23.31).

$$PN = 10\log\left[\frac{2FkT}{Ps}\left(1+\left(\frac{\omega_o}{2Q\Delta\omega}\right)^2\right)\left(1+\frac{\Delta\omega_{1/f^3}}{\Delta\omega}\right)\right] \tag{23.32}$$

Figure 23.17 shows some typical VLSI circuit implementations of the LC-based VCO with the diode varactors for the frequency-tuning operations.

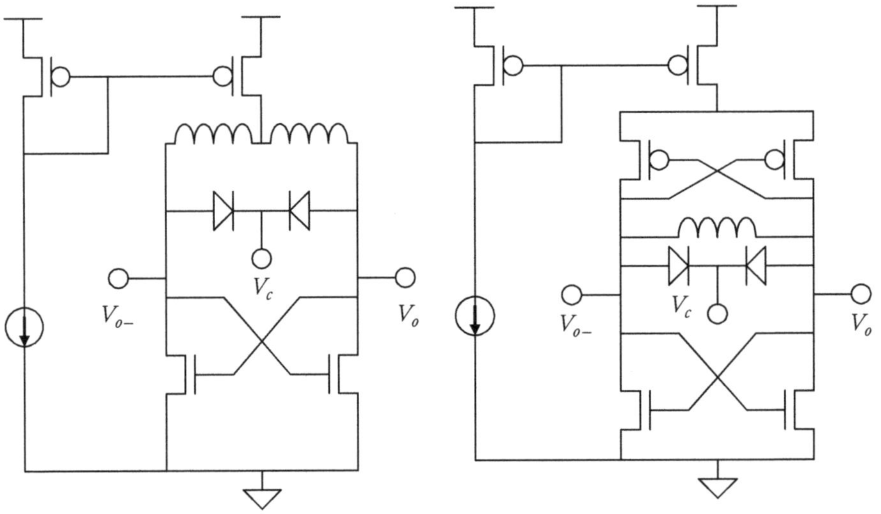

FIGURE 23.17 VLSI voltage-controlled LC oscillator structures.

23.5 Polyphase Clock Generation Techniques

Balanced polyphase clocks can be generated from single differential clocks using an RC polyphase filter circuit as shown in Fig. 23.18.

23.6 Distributed Amplifier Circuits

The VLSI distributed amplifier can be used to realize very wide-bandwidth constant-gain amplifiers. The primarily capacitive input impedance of an MOS device gate makes CMOS technology well suitable for distributed amplification. Recent interests in MOS distributed amplifiers have been fueled by the fact that a standard submicron CMOS process can reach an operational speed well into the microwave range. However, a major obstacle in the realization of useful CMOS distributed amplifiers has been due to the difficulty in realizing high-quality factor inductors and transmission lines in the standard CMOS technology. Figure 23.19 is a three-stage CMOS distributed amplifier circuit.

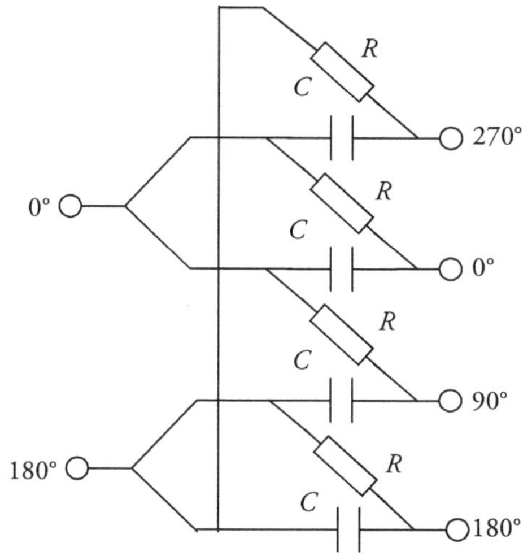

Figure 23.18 Polyphase clock generation from differential clock.

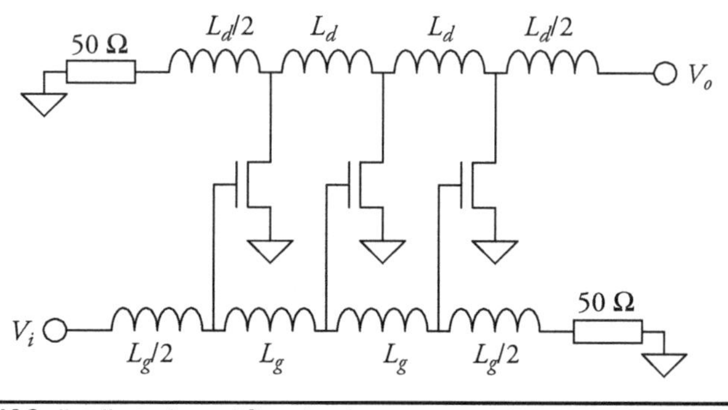

Figure 23.19 CMOS distributed amplifier circuit.

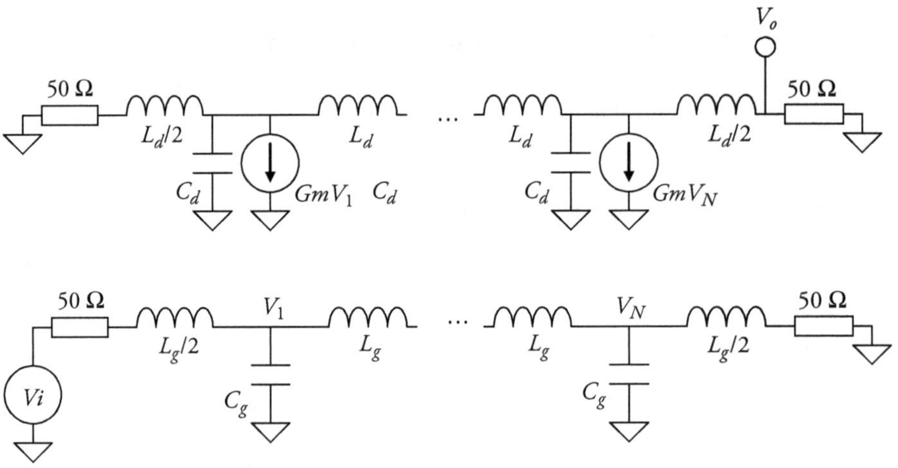

FIGURE 23.20 *N*-stage lossless distributed amplifier model.

A lossless distributed amplifier can be modeled using two networks as shown in Fig. 23.20.

In special cases, if the gate-line and the drain-line phase shifts are equal and the gate- and drain-line impedances are the same and are equal to Z_o, the gain and the bandwidth of the distributed amplifier are approximately given as

$$\begin{cases} G = \dfrac{g_m^2 N^2 Z_o^2}{4} \\ \omega_C = \dfrac{2}{\sqrt{L_g C_g}} \end{cases} \tag{23.33}$$

References

1. F. Gatta, E. Sacchi, F. Svelto, P. Vilmercati, R. Castello, "A 2-dB Noise Figure 900-MHz Differential CMOS LNA," *IEEE J. Solid-State Circuits*, vol. 36, no. 10, pp. 1444–1452, October 2001.
2. D. Feng, B. Shi, "A 2.5-V 2.45/5.25 GHz Dual-Band CMOS LNA," *International Conference on ASIC*, 2003 vol. 2, no. 21–24, pp. 1110–1113, October 2003.
3. S. Yang, R. Mason, C. Plett, "CMOS LNA in Wireless Applications," *1999 IEEE 49th Vehicular Technology Conference*, vol. 3, no. 16–20, pp. 1920–1924, May 1999.
4. R. A. Rafla, M. N. El-Gamal, "2.4-5.8 GHz CMOS LNA's Using Integrated Inductors," *The 43rd IEEE Midwest Symposium on Circuits and Systems*, vol. 1, pp. 302–304, August 2000.
5. H.-Y. Tsui, J. Lau, "A 5 GHz 56 dB Voltage Gain 0.18um CMOS LNA with Built-in Tunable Channel Filter for Direct Conversion 802.11a Wireless LAN Receiver," *2003 IEEE Radio Frequency Integrated Circuits (RFIC) Symposium*, pp. 225–228, June 2003.

6. W. Zhang, K. Chen, "Impact of Intrinsic Gate-Drain Capacitance on Noise Performance of CMOS LNA," *ICMMT 2002*, pp. 915–918, August 2002.
7. V. Krizhanovskii, N. T. Kien, S.-G. Lee, "0.18um CMOS LNA and Mixer for Wireless LAN Applications," *CriMiCo 2002*, pp. 137–138, September 2002.
8. P. Litmanen, P. Ikalainen, K. Halonen, "A 2.0-GHz Submicron CMOS LNA and a Downconversion Mixer," *ISCAS'98*, vol. 4, pp. 357–359, June 1998.
9. C.-C. Tang, C.-H. Wu, S.-I. Liu, "Miniature 3-D Inductors in Standard CMOS Process," *IEEE J. Solid-State Circuits*, vol. 37, no. 4, pp. 471–480, April 2002.
10. S. Park, W. Kim, "Design of a 1.8 GHz Low-Noise Amplifier for RF Front-End in a 0.8 μm CMOS Technology," *IEEE Trans. on Consumer Electronics*, vol. 47, no. 1, pp. 10–15, February 2001.
11. C.-C. Tang, S.-I. Liu, "Low-Voltage CMOS Low-Noise Amplifier Using Planar-Interleaved Transformer," *Electronics Letters*, vol. 37, no. 8, pp. 497–498, April 2001.
12. A. Bevilacqua, A. M. Niknejad, "An Ultra-Wideband CMOS LNA for 3.1–10.6 GHz Wireless Receivers," *ISSCC'04*, pp. 382–391, February 2004.
13. B. Mohammadi, C. A. T. Salama, "A 5.8 GHz CMOS LNA for WLAN Applications," *Radio Frequency Integrated Circuits (RFIC) Symposium*, pp. 113–116, June 2004.
14. A. N. L. Chan, C. B. Guo, H. C. Luong, "A 1-V 2.4-GHz CMOS LNA with Source Degeneration as Image-Rejection Notch Filter," *ISCAS'01*, vol. 4, pp. 890–893, May 2001.
15. Y. Ding, R. Harjani, "A +18 dBm IIP3 LNA in 0.35 um CMOS," *ISSCC*, pp. 162–163, 443, February 2001.
16. H. Kanaya, Y. Koga, G. Urakawa, K. Yoshida, "Design of HTS Coplanar Waveguide Matching Circuit for Low Noise CMOS-HTS receiver," *IEEE Trans. on Applied Superconductivity*, vol. 13, no. 2, pp. 1031–1034, June 2003.
17. R. Fujimoto, K. Kojima, S. Otaka, "A 7-GHz 1.8-dB NF CMOS Low-Noise Amplifier," *IEEE J. Solid-State Circuits*, vol. 37, no. 7, pp. 852–856, July 2002.
18. K. Cao, H. Yang, H. Wang, "The Design of CMOS RF Low Noise Amplifiers," *5th International Conference on ASIC*, vol. 2, pp. 1106–1109, October 2003.
19. K. Chung, G. Han, S. Kang, "A 0.35um CMOS Low Noise VGA," *AP-ASIC'00*, pp. 5–8, August 2000.
20. H. Fouad, K. Sharaf, E. El-Diwany, H. El-Hennaway, "An RF CMOS Modified-Cascode LNA with Inductive Source Degeneration," *NRSC 2002*, pp. 450–457, March 2002.
21. J. Long, N. Badr, R. Weber, "A 2.4GHz sub-1 dB CMOS Low Noise Amplifier with On-Chip Interstage Inductor and Parallel Intrinsic Capacitor," *RAWCON 2002*, pp. 165–168, August 2002.
22. F. Bruccoleri, E. A. M. Klumperink, B. Nauta, "Noise Cancelling in Wideband CMOS LNAs," *ISSCC'02*, vol. 1, pp. 406–407, February 2002.
23. A. Pascht, J. Fischer, M. Berroth, "A CMOS Low Noise Amplifier at 2.4 GHz with Active Inductor Load," *Silicon Monolithic Integrated Circuits in RF Systems*, pp. 1–5, September 2001.
24. C.-Y. Wu, Z.-Y. Zhou, "The Design of 2 V 2.4 GHz CMOS Low-Noise Low-Power Bandpass Amplifier with Parallel Spiral Inductors," *Proceedings of ICECS '99*, vol. 2, pp. 769–772, September 1999.
25. J. J. Zhou, D. J. Allstot, "A Fully Integrated CMOS 900 MHz LNA Utilizing Monolithic Transformers," *45th ISSCC*, pp. 132–133, February 1998.

26. B. M. Ballweber, R. Gupta, D. J. Allstot, "A Fully Integrated 0.5-5.5 GHz CMOS Distributed Amplifier," *IEEE J. Solid-State Circuits*, vol. 35, no. 2, pp. 231–239, February 2000.
27. P. J. Sullivan, B. A. Xavier, W. H. Ku, "An Integrated CMOS Distributed Amplifier Utilizing Packaging Inductance," *IEEE Trans. on Microwave Theory and Techniques*, vol. 45, no. 10, pp. 1969–1976, October 1997.
28. K. K. Moez, M. I. Elmasry, "A Novel Matrix-Based Lumped-Element Analysis Method for CMOS Distributed Amplifiers," *ISCAS'04*, vol. 1, pp. I-1048–I-1051, May 2004.
29. Z. El-Khatib, L. MacEachern, S. A. Mahmoud, "A 0.1-12 GHz Fully Differential CMOS Distributed Amplifier Employing a Feedforward Distortion Cancellation Technique," *ISCAS'04*, vol. 1, pp. I-617–I-620, May 2004.
30. R.-C. Liu, K.-L. Deng, H. Wang, "A 0.6-22-GHz Broadband CMOS Distributed Amplifier," *IEEE Radio Frequency Integrated Circuits (RFIC) Symposium*, pp. 103–106, June 2003.
31. H.-T. Ahn, D. J. Allstot, "A 0.5-8.5 GHz Fully Differential CMOS Distributed Amplifier," *IEEE J. Solid-State Circuits*, vol. 37, no. 8, pp. 985–993, August 2002.
32. R. Amaya, C. Plett, "Design of High Gain Fully-Integrated Distributed Amplifiers in 0.35um CMOS," *ESSCIRC '03*, pp. 145–148, September 2003.
33. B. Kleveland, C. H. Diaz, D. Vock, L. Madden, T. H. Lee, S. S. Wong, "Monolithic CMOS Distributed Amplifier and Oscillator," *ISSCC*, pp. 70–71, February 1999.
34. R. E. Amaya, N. G. Tarr, C. Plett, "A 27 GHz Fully Integrated CMOS Distributed Amplifier Using Coplanar Waveguides," *2004 Radio Frequency Integrated Circuits (RFIC) Symposium*, pp. 193–196, June 2004.
35. S. S. Wong, P. Yue, R. Chang, S.-Y. Kim, B. Kleveland, F. O'Mahony, "On-chip Interconnect Inductance – Friend or Foe," *Proceedings Fourth International Symposium on Quality Electronic Design*, pp. 389–394, March 2003.
36. A. Worapishet, M. Chongcheawchamnan, S. Srisathit, "Broadband Amplification in CMOS Technology Using Cascaded Single-Stage Distributed Amplifier," *Electronics Letters*, vol. 38, no. 14, pp. 675–676, July 2002.

CHAPTER 24

Introduction to Digital-Based Analog Signal Processing Circuits

Most recent developments in VLSI circuit technologies are closely related to the system-on-a-chip (SOC) solutions driven by the fast growth of worldwide markets for mobile computing, communication, and consumer electronics targeted at higher performances and lower costs. Such SOC solutions are featured by ever-increasing levels of integration. Complete systems that previously occupied one or more boards are now being integrated into a few chips or even a single chip that includes analog, digital, and even RF circuits. One of the significant changes in the signal process circuit implementations has been the increasing contents of traditional analog signal processing (ASP)–based circuit functions being implemented in the digital signal processing (DSP) circuit as a benefit from the VLSI manufacturing technology advancements and fueled by the SOC applications. On the other hand, it has been observed that in some traditional digital circuit application areas such as the high-speed I/O circuits, the analog effects are becoming increasingly important. Furthermore, the analog natures of most front-end signals in SOC signal processing (SOC-SP) circuit applications remain unchanged. As a result, we expect to see the continuous shifting of the traditional boundary between the ASP and DSP circuits; and such a trend will lead to innovative VLSI SOC-SP solutions that are based on the optimized combinations of both VLSI ASP and DSP circuit techniques for better system performance, robustness, and efficiency.

24.1 Signal Processing Circuits for SOC Applications

The VLSI SOC-SP solutions offer increased flexibilities for signal processing operation implementations based on all existing VLSI technologies for system-level design optimizations. VLSI CMOS process technologies by far are the only viable process technologies for the SOC-SP circuit implementations since the majority of devices on the SOC are digital gates. The fundamental benefits for using the VLSI digital CMOS process technologies for SOC-SP circuit implementations can be summarized as follows:

- Massive production capability of low-cost and reliable small-geometry devices that can be used to realize various signal processing functions with continuous process scaling as stated by Moore's law

- Continuous improvement of device operation speed benefited from process scaling offering better timing accuracy for signal compensating in the degradation of the signal dynamic range in voltage and current
- Continuous scaling down in active power dissipation benefited from the scaling of the device parasitic capacitance and supply voltage
- Continuous improvement in device geometry control benefited from the processing technology developments that offer better device parameter matching accuracy necessary for more robust signal processing operation

However, the VLSI digital CMOS process technologies for the SOC-SP circuit implementations suffer from several fundamental constraints in the on-chip analog circuits:

- Digital CMOS processes usually restrict the process to elements necessary for logic design. The nondigital components—such as linear resistor, capacitor, and BJT structures—are usually limited to CMOS process layers used in MOS transistors and metal connections. The absolute accuracies of these VLSI devices are usually poor compared with their discrete counterparts. The parameter adjustments, such as laser trimming, are basically unavailable.
- Signal dynamic range degradation of on-chip analog circuits results from the signal voltage headroom reduction associated with the continuous scaling down of the core power supply voltage required by device reliability and power dissipation of digital circuits along with increasing integration levels.
- Signal dynamic degradation of on-chip sensitive circuits are associated with increasing coupling of the on-chip switching noise floor of digital circuits from the substrate and connections. Such an issue is expected to be even worse with the trends of increasing devices integration and operation frequency as well as scaling down of the supply voltage.
- Design productivity degradations are associated with poor on-chip analog circuit portability across process technologies. As previously discussed, the high productivity of the VLSI circuits has greatly benefited from the excellent scalability of the digital circuits as described by Moore's law.
- Design productivity degradation associated with more complicated design verification requirements of on-chip analog circuits results from much higher PVT sensitivities of analog circuits. The combining of ASP and DSP circuits in SOC solution may result in the conflict of simulation accuracy versus simulation time.
- Higher development costs are associated with much more complicated on-chip precision analog circuit validation. This is significantly different from the testing of majority on-chip digital, which is usually based on speed or timing.

24.2 Digital-Based Analog (DBA) Signal Processing Solutions

The DBA circuit solutions are developed for the VLSI SOC-SP applications that are targeted at high design efficiency, robustness, and scalability in standard VLSI CMOS process technologies. The DBA circuit design solutions can be summarized in the following design strategies:

- *The algorithm-based signal processing strategy at the architecture level.* The algorithm-based approaches of VLSI ASP functions are employed to minimize PVT sensitivity of the VLSI circuit implementation. The basis of such implementations are the s-to-z transformation and data format (A/D, D/A, etc.) conversions that support high-signal dynamic range VLSI signal processing operations in mixed-signal forms (digital, analog, continuous time, discrete time, voltage, current, charge, etc.) and signal processing functions. The algorithm-based signal processing solutions set the base for enabling the high-signal processing efficiency, low-PVT sensitivity, high robustness, and low-cost circuit and physical implementation for VLSI SOC-SP circuit solutions.
- *The ratio-based approach at the VLSI circuit-implementation level.* The ratio-based circuit design solutions rely on the symmetry-based design constrains in the VLSI signal processing circuit implementations. This method is based on the facts that the VLSI process technology offers much better relative (matching) device parameter accuracy than the absolute accuracy and that symmetry circuits offer high tolerances to various design and operation nonidealities such as noise and device parameter mismatches. The ratio-based VLSI circuit design constraint effectively minimizes the dependency of the VLSI circuit implementation on the absolute device parameters. VLSI circuit implementations based on such a method ensure very low sensitivity to PVT sensitivity at the circuit-implementation level for high design robustness and are used to enable the symmetry-based physical implementation of the SOC-SP circuits.
- *The symmetry (matching)–based VLSI circuit physical implementation based on identical (bit-slice) structures.* The constraint in layout symmetry relies on the fact that symmetrical device layouts provide high design robustness and reliability that effectively eliminates various nonideal effects in the fabrication process and the operation of the VLSI circuit devices. The symmetry constraint in the physical implementation can also be effectively used to enhance the design productivity of the VLSI signal processing circuit implementation. The layout symmetry that is represented by the regular cell–based layout approach also provides high design reusability that can significantly reduce the VLSI layout design and design verification efforts.
- *The opportunistic signal processing function partitioning among various VLSI ASP and DSP circuit configurations and the signal format conversion circuit techniques to achieve the highest computation efficiency and design reliability.* Different from the traditional ASP and DSP circuit techniques, the SOC-SP circuit implementations offer flexible partitioning that can better fit the design applications. Such techniques can potentially provide significantly higher signal processing efficiency employing the optimized signal processing circuit partition among various available VLSI circuit implementations using the low-device-count (LDC) circuits and signal representations. The optimized mixed-signal processing operation partitioning and signal form selections—such as the wired current addition, wire voltage branching operation, as well as the pulse-width modulation (PWM) and pulse-code modulation (PCM)—are used to achieve

significantly high computation efficiency, design reliability, and robustness in VLSI implementations.

24.2.1 Basic VLSI ASP Functional Elements

The VLSI SOC-SP operations can be realized based on basic VLSI circuit elements, including the scaling, addition, delay (integration), and signal format conversion operations in the ASP, the DSP, or the mixed-signal forms. These basic signal processing circuit elements were introduced in previous chapters and are summarized in Figs. 24.18 to 24.20 in the appendix at the end of this chapter. Note that the addition and scaling operation of the VLSI analog circuits by default are ratio-based signal processing operations. Therefore, the scaling and addition operation can readily be used in low-PVT sensitivity SOC-SP circuit operation for design reliability and design robustness. However, the analog delay or integration operations usually consist of the absolute device parameters; therefore, they are not ratio-based circuit implementations.

24.2.2 Basic VLSI Signal Format Conversion Functions

To support the SOC-SP operation, the basic signal processing operations should include the signal format conversion circuits as shown in Figs. 24.1 to 24.4, including analog and digital signal format conversion circuit elements that can be used for SOC-SP operations.

24.2.3 Basic VLSI DSP Operations

The DSP operation can also be realized using basic signal processing elements as shown in Fig. 24.5.

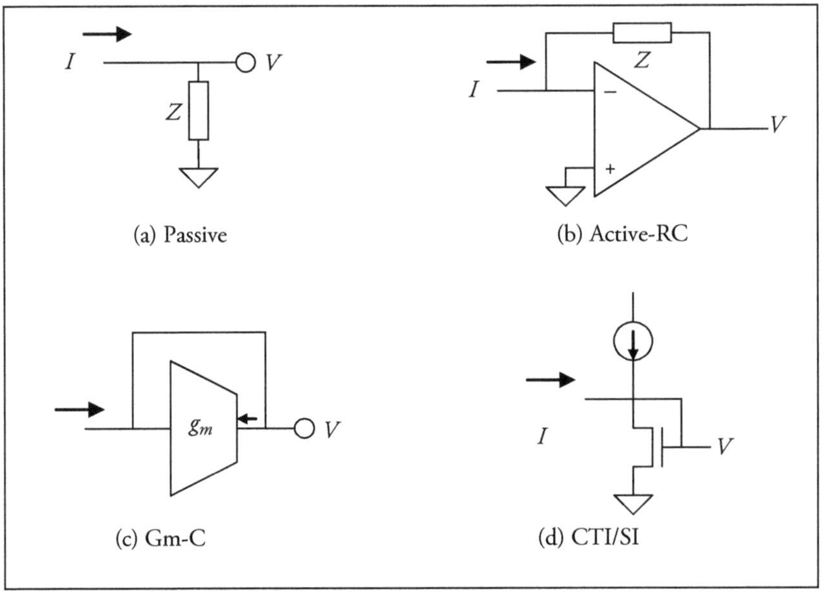

FIGURE 24.1 Basic VLSI I/V converter circuit structures.

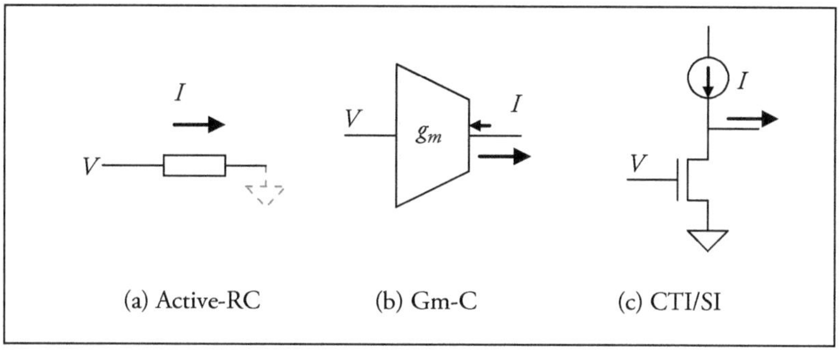

FIGURE 24.2 Basic VLSI V/I converter circuit structures.

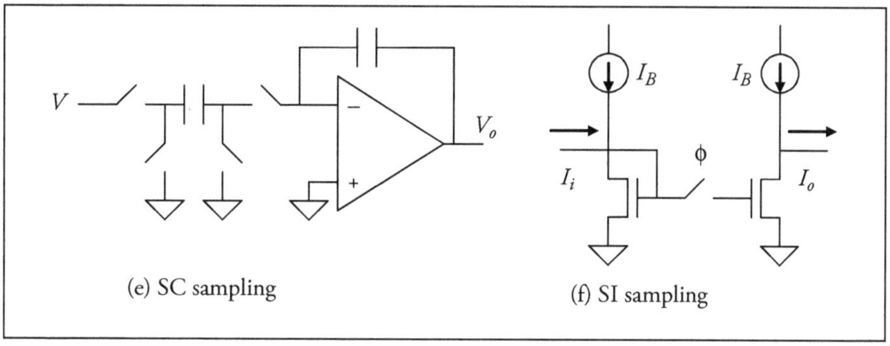

FIGURE 24.3 VLSI CT-to-DT signal conversions.

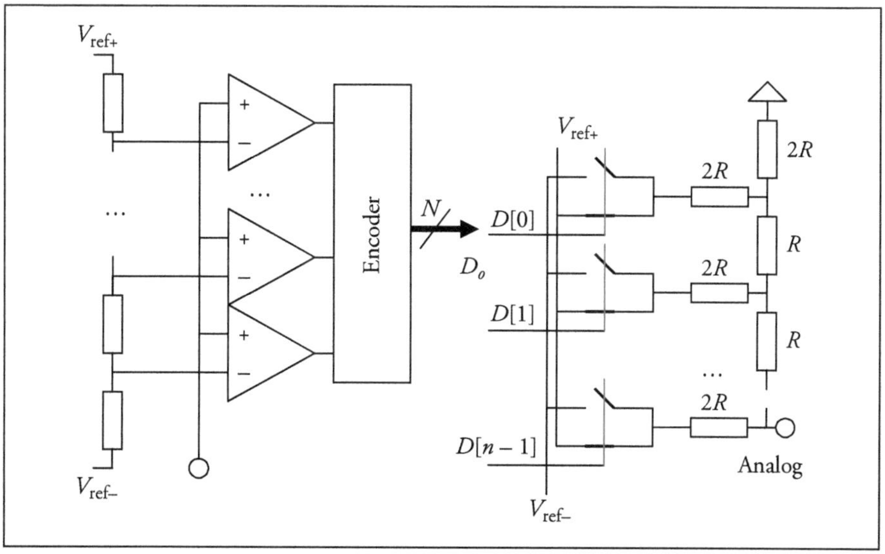

FIGURE 24.4 Basic VLSI A/D and D/A converter circuit structures.

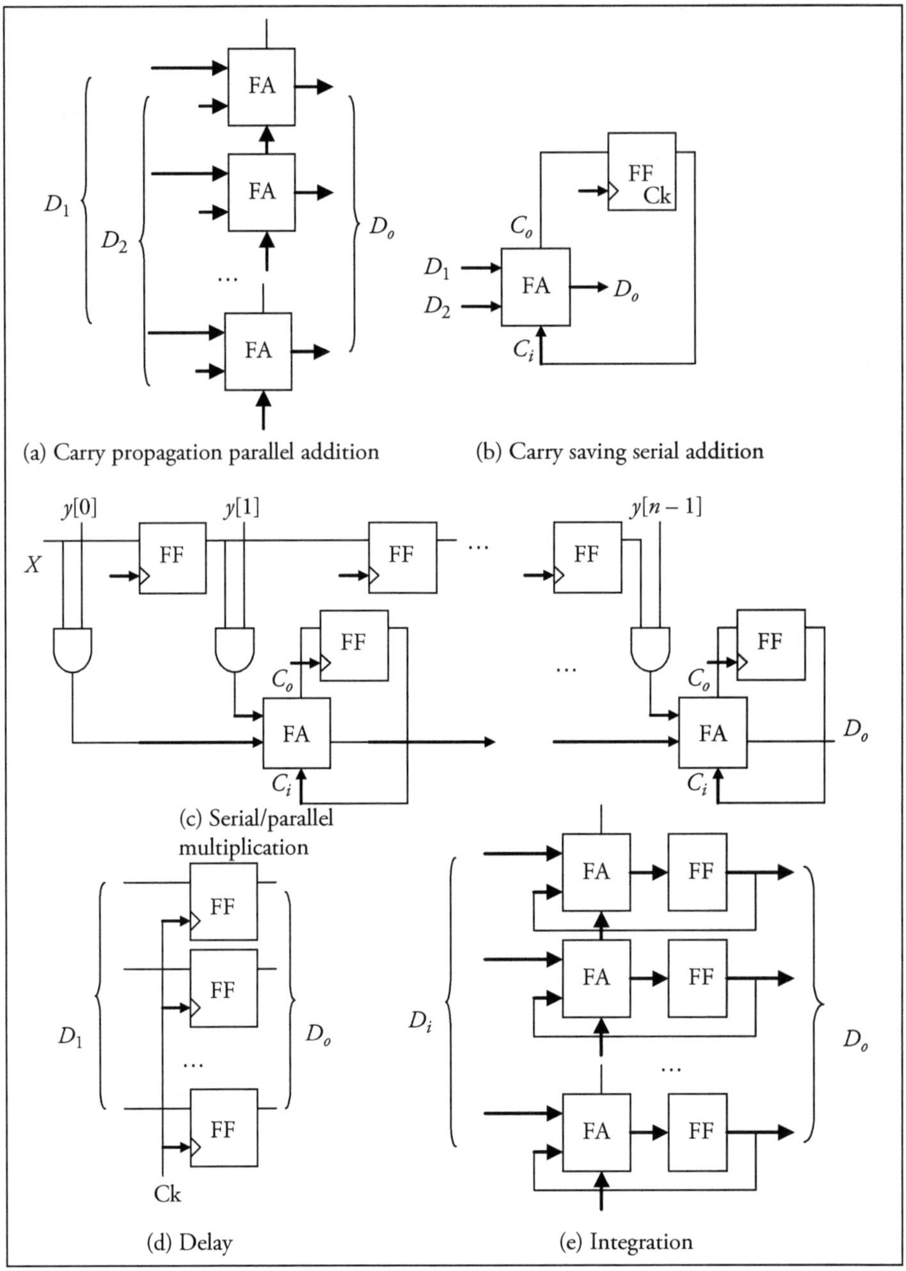

FIGURE 24.5 Basic DSP operations.

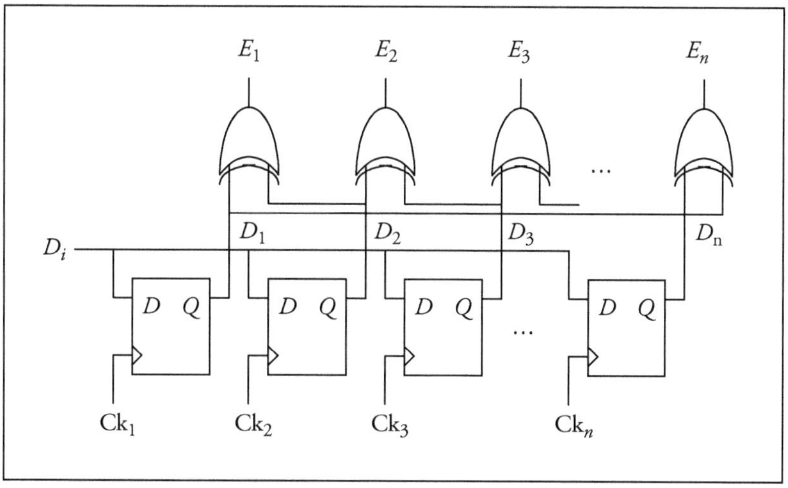

FIGURE 24.6 Basic VLSI phase to digital converter circuit structure.

Figure 24.6 shows a digital format conversion operation that converts the phase of the digital data stream into the digital code.

24.2.4 Low-Device-Count Signal Processing Circuits

In the SOC-SP circuit implementations, signal and VLSI signal processing circuits can be partitioned flexibly beyond the traditional signal processing circuit domain, including both ASP and DSP operations for enhanced signal processing efficiency. Such a property allows the realization of the given signal processing function using highly minimized device count penalty through effective partitioning of the signal processing operation among various VLSI signal processing circuit forms. Figure 24.7 shows some high-computation efficiency VLSI circuit structures that offer the LDC signal processing circuit capabilities:

- Wired branching of voltage model signals: Voltage signals can be sent to multiple locations without device penalty.
- Wired addition of current signals: Multiple current signals can be added together without device penalty.
- Scaling down of voltage signal: Voltage signals can be scaled down using a linear resistor divider.
- Single-bit A/D and D/A conversion: Analog and digital signal format conversion operations based on single-bit conversions offer very low-PVT sensitivity. In practical VLSI signal processing circuit implementation, the single-bit data conversion circuit can be combined with the oversampling circuit techniques such as the sigma-delta circuits so that high-accuracy data conversion can be realized using high-data-rate low-resolution strategies.
- Unity digital delay operation: Based on the DFF (D-Flip-Flop) using the external clock reference, the timing accuracy limiting effect of the signal processing circuit can be highly minimized.
- Temporal addition: Two oversampled digital data streams can be added through multiplexing in time.

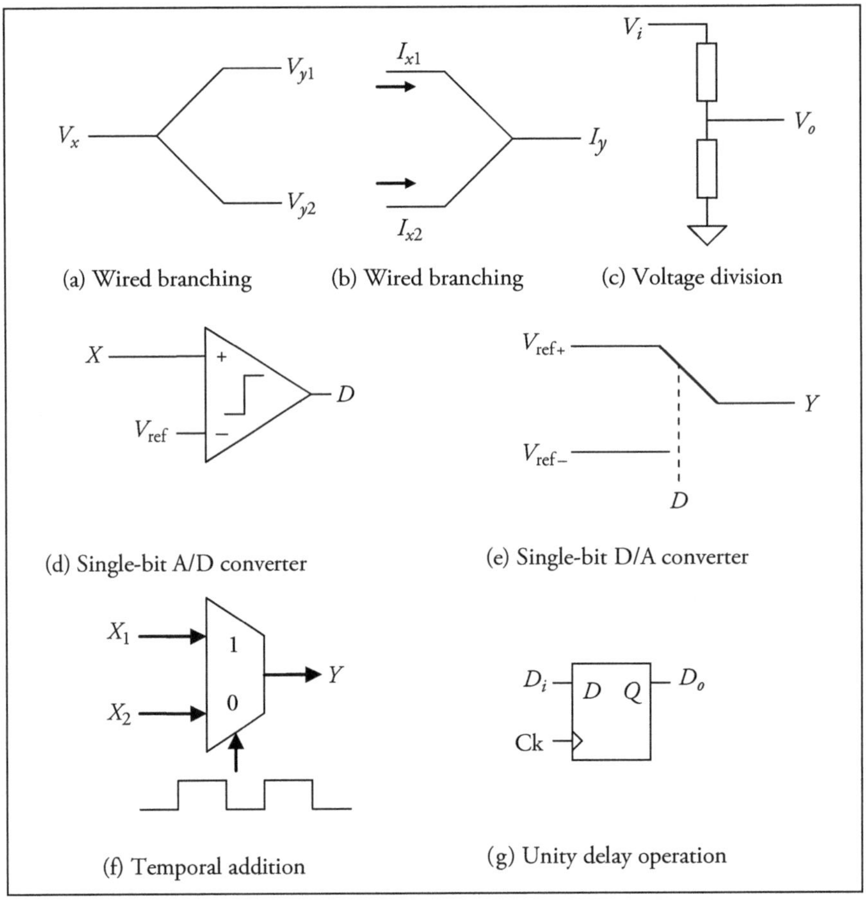

FIGURE 24.7 Example of LDC signal processing circuit structures.

Note that the signal processing operations offered by most LDC circuit configurations readily offer the low-PVT sensitivities in the VLSI circuit implementation. As a result, highly robust VLSI circuits can be implemented based on these circuit structures.

In practical VLSI circuit implementations, these LDC circuit structures can also be used as subcircuit structures to form more complicated VLSI signal processing circuit blocks. Such a method may provide an effective way for quantifying the performance of various existing or new VLSI circuit solutions.

24.3 Design Examples

Several practical VLSI signal processing circuits were primarily implemented based on the proposed DBA circuit design techniques targeted at VLSI high-speed I/O circuit applications. In addition, the DBA circuit design methodology is used for analyzing the operations of several previously developed, well-known VLSI mixed-signal circuit structures to demonstrate the DBA circuit analysis techniques.

24.3.1 High-Speed I/O Spectral Shaping Tx Circuits

A simple yet "perfect" design example to demonstrate the DBA circuit design techniques is the patented spectral-shaped high-speed I/O transmitter circuit architecture shown in Fig. 24.8 that was originally developed based on the DBA circuit design concept.

Such a transmitter circuit can be used for circuit applications such as the adaptive transmitter slew rate control, the transmitter equalization, and the noise shaping. The slew rate control circuit based on this circuit structure is significantly different from the conventional slew rate control circuits that are based on the purely analog circuit operation relying on the parasitic or explicit RC constant of the transmitter. Pre-driver or output driver circuit stages have the drawback of high-PVT sensitivity for the transmitter circuit operations. The proposed DBA transmitter circuit is based on an FIR (Finite-Impulse-Response) signal processing algorithm shown in Fig. 24.9, where the circuit realizes a discrete-time signal processing function with the z-domain transfer function given as

$$\frac{Y}{X} = c_0 + c_1 z^{-1} + \cdots + c_{n-1} z^{-(n-1)} + c_n z^{-n} \tag{24.1}$$

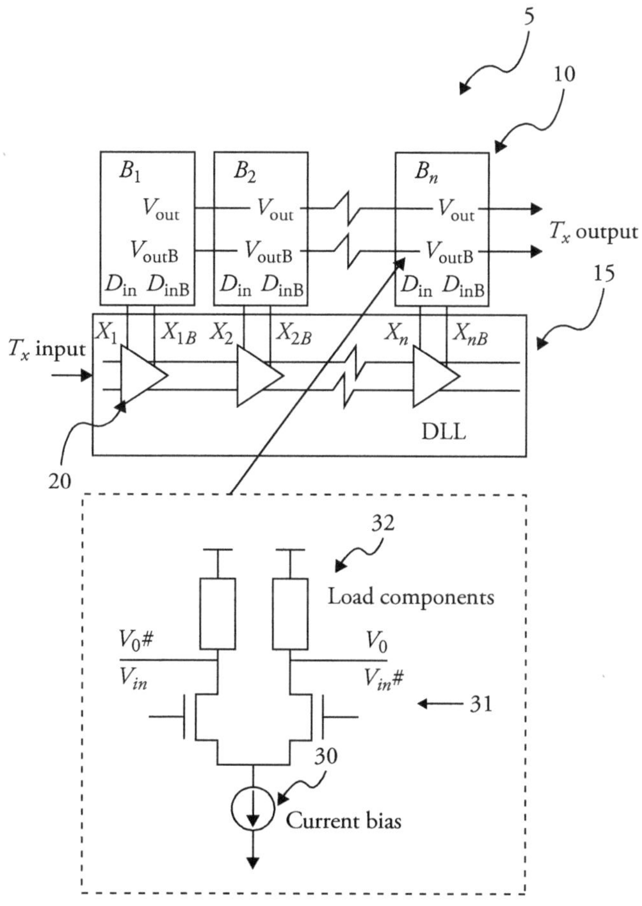

FIGURE 24.8 DBA high-speed I/O transmitter circuits.

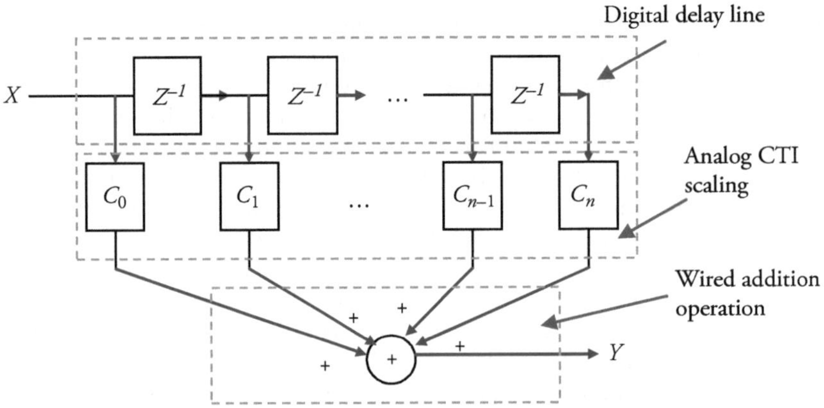

FIGURE 24.9 Signal processing algorithm of the transmitter slew rate control circuit.

Such a circuit consists of three basic signal processing operations, including the unit delay operation, the scaling operation, and the addition operation. Based on the DBA circuit approach, these signal processing operations are partitioned and implemented in VLSI circuits as follows:

- The delay operation is implemented in the digital domain since unity delay in the digital domain offers very high computational efficiency. The delay time is adaptively controlled based on the data rate. Therefore, the time constant of the circuit adaptively tracks the data rate insensitive to PVT variation that makes the slew rate control fully scalable to the data rate of the I/O circuit.
- The scaling operation is implemented in the CTI ASP circuit that offers high computational efficiency in the scaling operation. The current-mirror–based CTI circuit is fully ratio based and independent of the analog parasitic effects of the circuit. Therefore, such a circuit implementation offers very high design scalability and PVT insensitivity.
- Finally, the addition operation is realized in the current form through the wire addition operation that offers the highest signal process computational efficiency with full PVT insensitivity and scalability.

24.3.2 TMX I/O Receiver Equalization Circuit

The second design example is a DBA circuit design implementation of the receiver equalization operation for the high-speed I/O circuit applications. This circuit is based on the patented threshold multiplexing algorithm as shown in Fig. 24.10.

Such a circuit realizes as an IIR (Infinite-Impulse-Response) signal processing circuit operation with an equivalent signal processing block diagram and signal transfer function given in the form as

$$\frac{Y}{X} = \frac{1}{1-[C_1 z^{-1} + \cdots + C_{n-1} z^{-(n-1)} + C_n z^{-n}]} \tag{24.2}$$

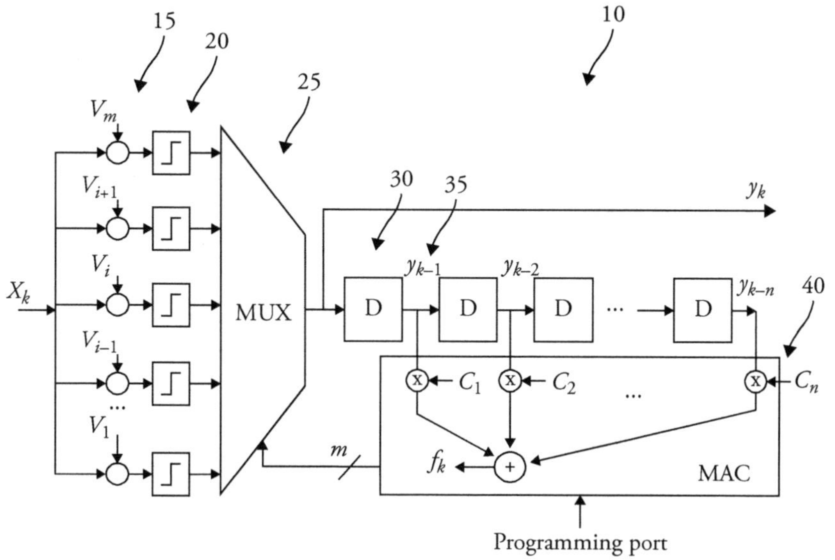

FIGURE 24.10 TMX receiver equalization algorithm.

The signal processing circuit consists of four basic signal processing operations, including the delay operation, the scaling operation, the addition operation, and the A/D (sampling and quantization) operation (see Fig. 24.11). These signal processing operations are partitioned among various ASP and DSP circuits based on the DBA design concept for achieving high computational efficiency, low-PVT sensitivity, and high scalability:

- The delay operation is implemented in the digital domain, which offers the highest computational efficiency, design robustness, and scalability. The delay time is adaptively controlled by the reference clock of the high-speed I/O circuit that makes the time constant of the signal processing circuit insensitive to PVT variation.

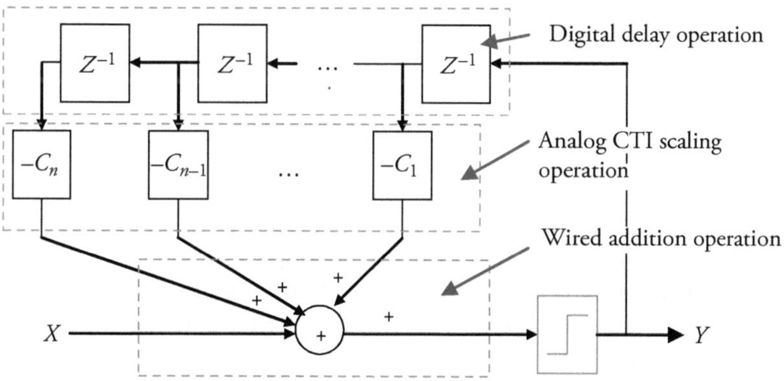

FIGURE 24.11 Signal processing algorithm of TMX receiver equalization circuit.

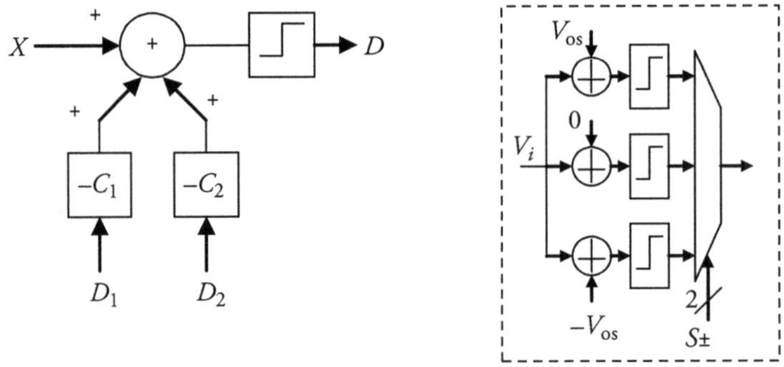

Figure 24.12 TMX sampler circuit operation.

- The scaling operation is implemented in the CTI ASP circuit based on the pure size ratio of the MOS device that offers very high computational efficiency, design robustness, and scalability in the scaling operation.
- The addition operation is realized in the current form through the wire addition.
- The A/D conversion is implemented through a single-bit differential A/D conversion circuit together with the scaling, CTI scaling, and current-mode wire addition operations as shown in Fig. 24.12 offering extremely high computation efficiency.

24.3.3 Sigma-Delta A/D and D/A Data Conversions

In this example, the proposed DBA signal processing circuit techniques are used to analyze the implementation and performance of the A/D and D/A data conversion circuits that are based on the sigma-delta modulation (see Fig. 24.13).

Sigma-delta A/D and D/A converters are introduced to solve the drawbacks of the quantization error and the linearity in the conversional VLSI A/D and D/A data converter implementation. Figure 24.14 shows a sigma-delta–based A/D converter (or PCM) circuit that converts the voltage-domain analog signal into a synchronous bit-serial digital stream based on a first-order CT ΣΔ modulator circuit.

The signal- and noise-transfer functions of the modulation can be expressed as follows:

$$Y(s) = \frac{1}{\left(\frac{s}{\omega_o}\right)+1} X(s) + \frac{\left(\frac{s}{\omega_o}\right)}{\left(\frac{s}{\omega_o}\right)+1} N(s) \tag{24.3}$$

Note that the signal- and noise-transfer functions represent the low-pass and high-pass frequency responses, respectively (see Fig. 24.15). As a result, the signal within the passband of the modulator loop can pass without attenuation, and the added noise is mainly located at a high-frequency range that can be effectively filtered when the signals are converted back to the analog form. The single-bit quantization of the circuit makes the circuit highly linear.

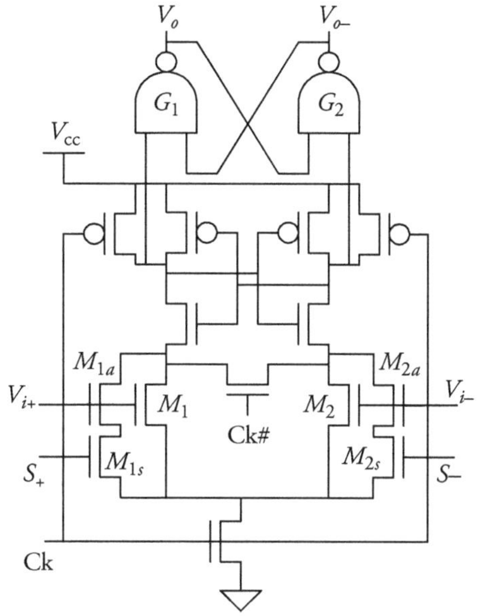

FIGURE 24.13 DBA circuit implementation of the TMX sampler circuit.

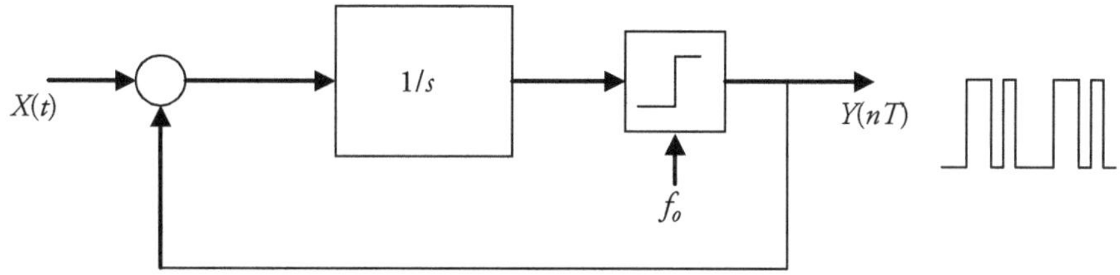

FIGURE 24.14 The PCM modulation based on the ΣΔ modulator.

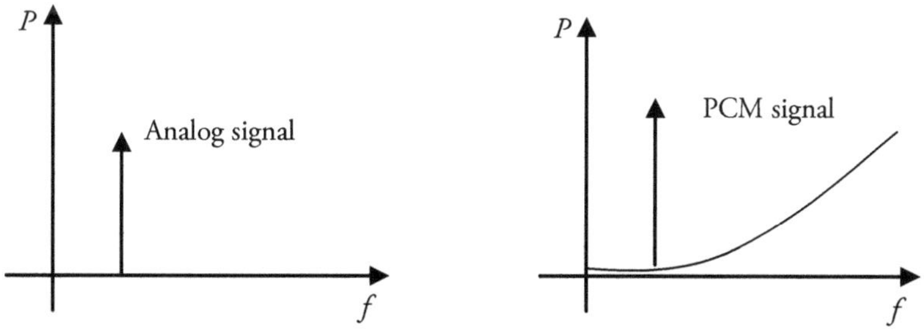

FIGURE 24.15 Power spectral densities of the sinusoidal signal and its modulated signal.

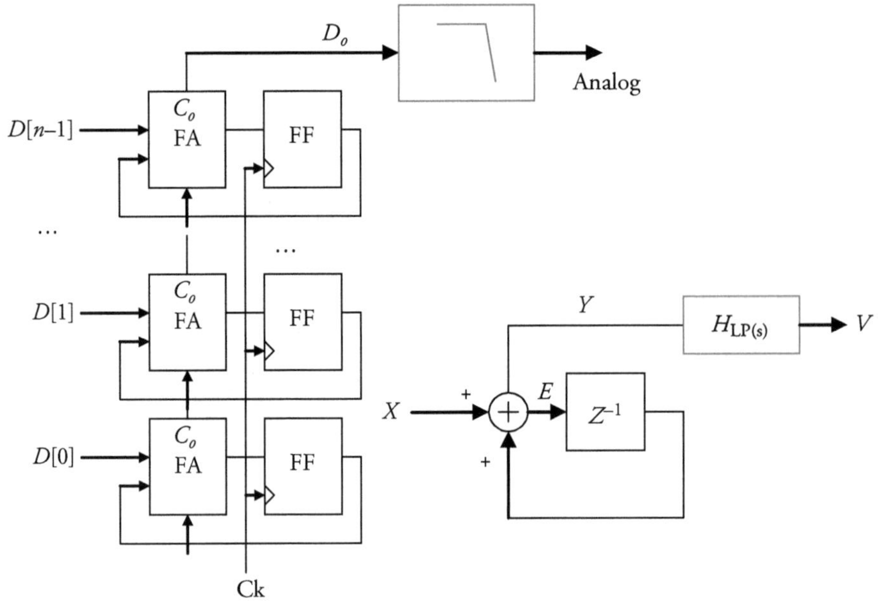

FIGURE 24.16 First-order sigma-delta D/A converter.

Although the time constant of such a circuit highly depends on the PVT corner of the process technology, the oversampling nature of the circuit highly relaxes the accuracy of the time constant of the circuit. The same concept can also be employed in the D/A converter circuit implementation as shown in Fig. 24.16, where a first-order digital sigma-delta is used.

24.3.4 Class-D Amplification

Data conversions between the analog and digital signal forms are among the important signal processing operations. One of such data format conversions is done using the PWM technique, employing a saw-waveform generator and a comparator as shown in Fig. 24.17. The PWM converts the voltage-domain analog signal into a typically asynchronous digital serial stream.

The PWM can also be approximately expressed using the signal- and noise-transfer functions:

$$Y(s) = H_{\text{signal}}(s)X(s) + H_{\text{noise}}(s)N(s) \tag{24.4}$$

where the signal transfer function also has a low-pass response, and the noise-transfer function has a high-pass (band-pass) response with high-noise content near the frequency of the saw-waveform signal.

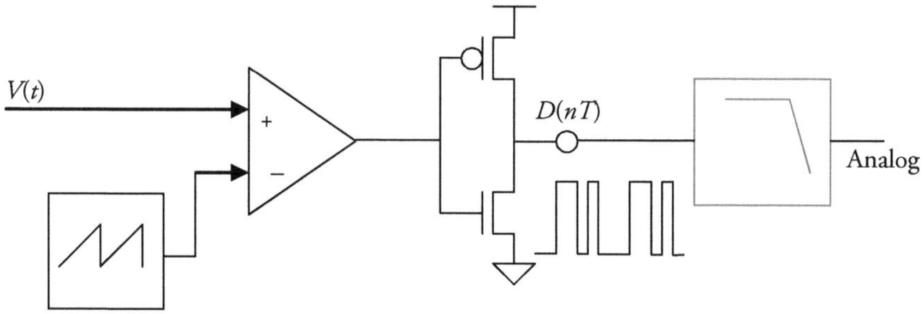

Figure 24.17 PWM modulation.

Appendix

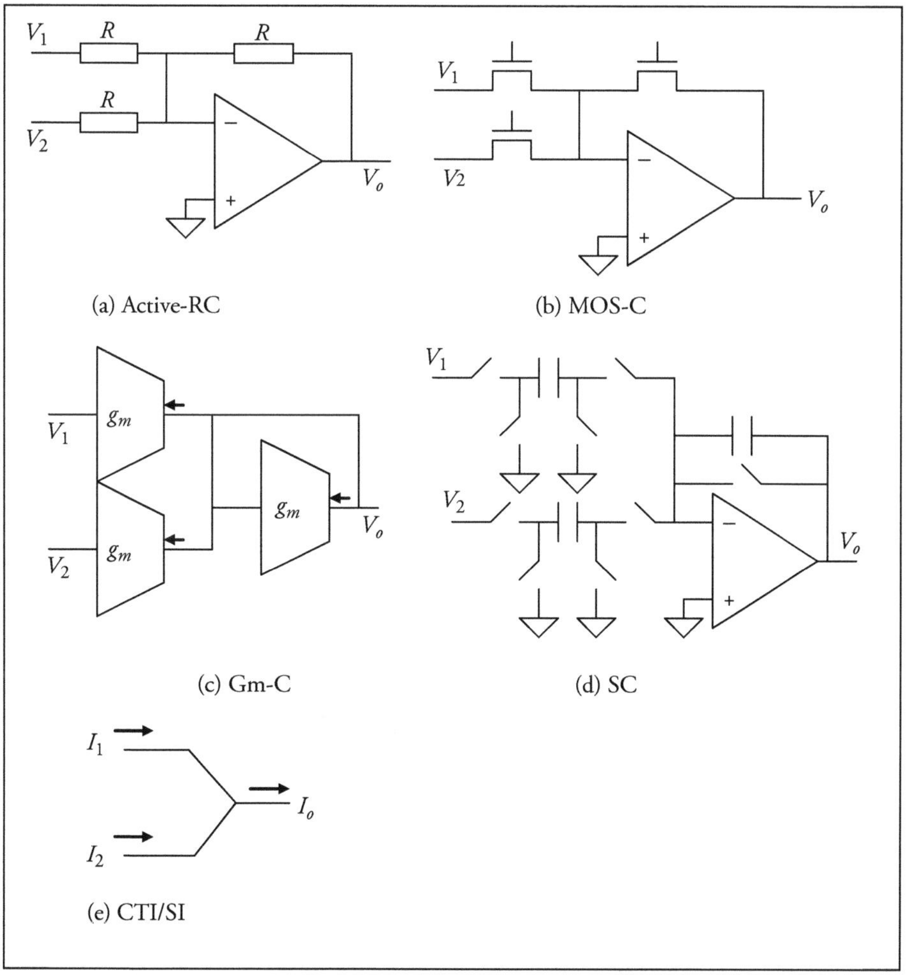

Figure 24.18 Basic VLSI analog addition circuits.

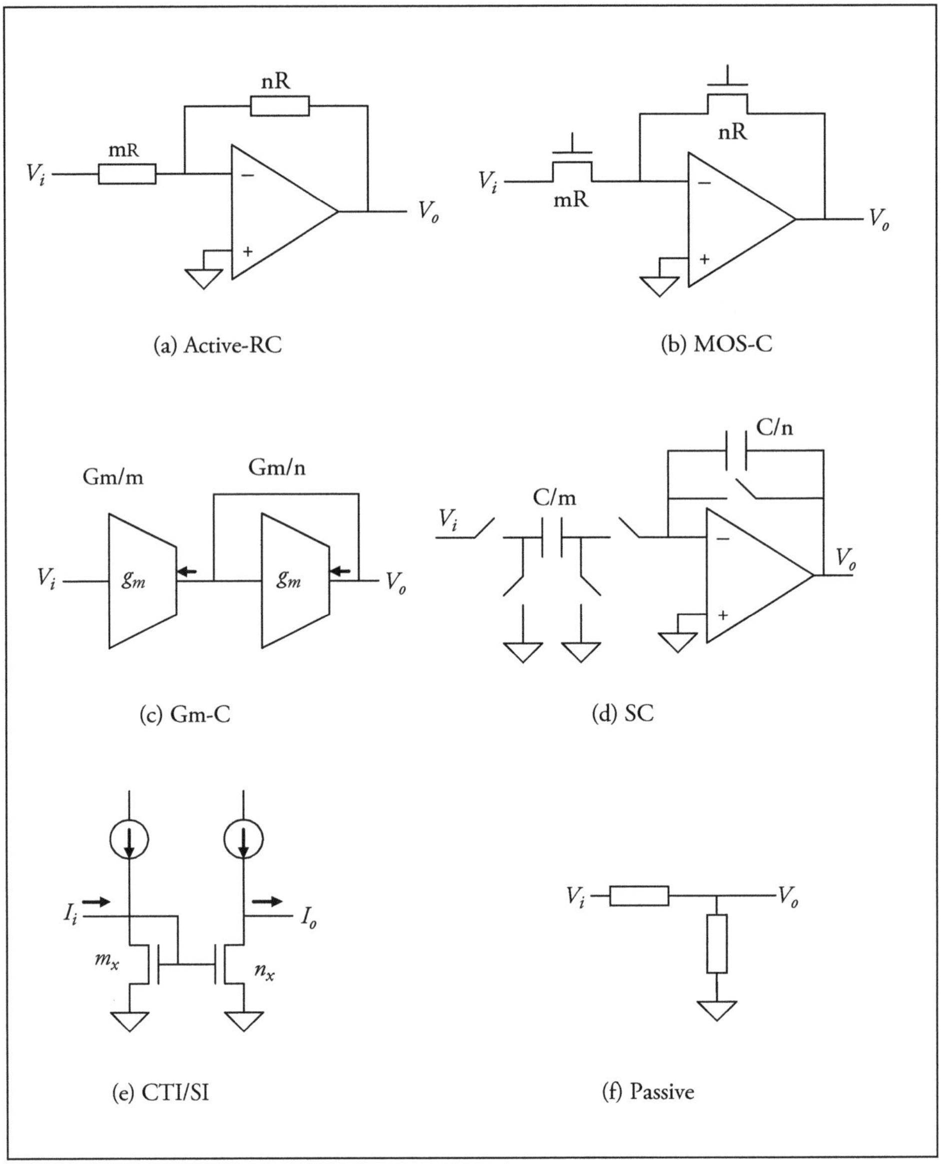

FIGURE 24.19 Basic VLSI analog scaling circuits.

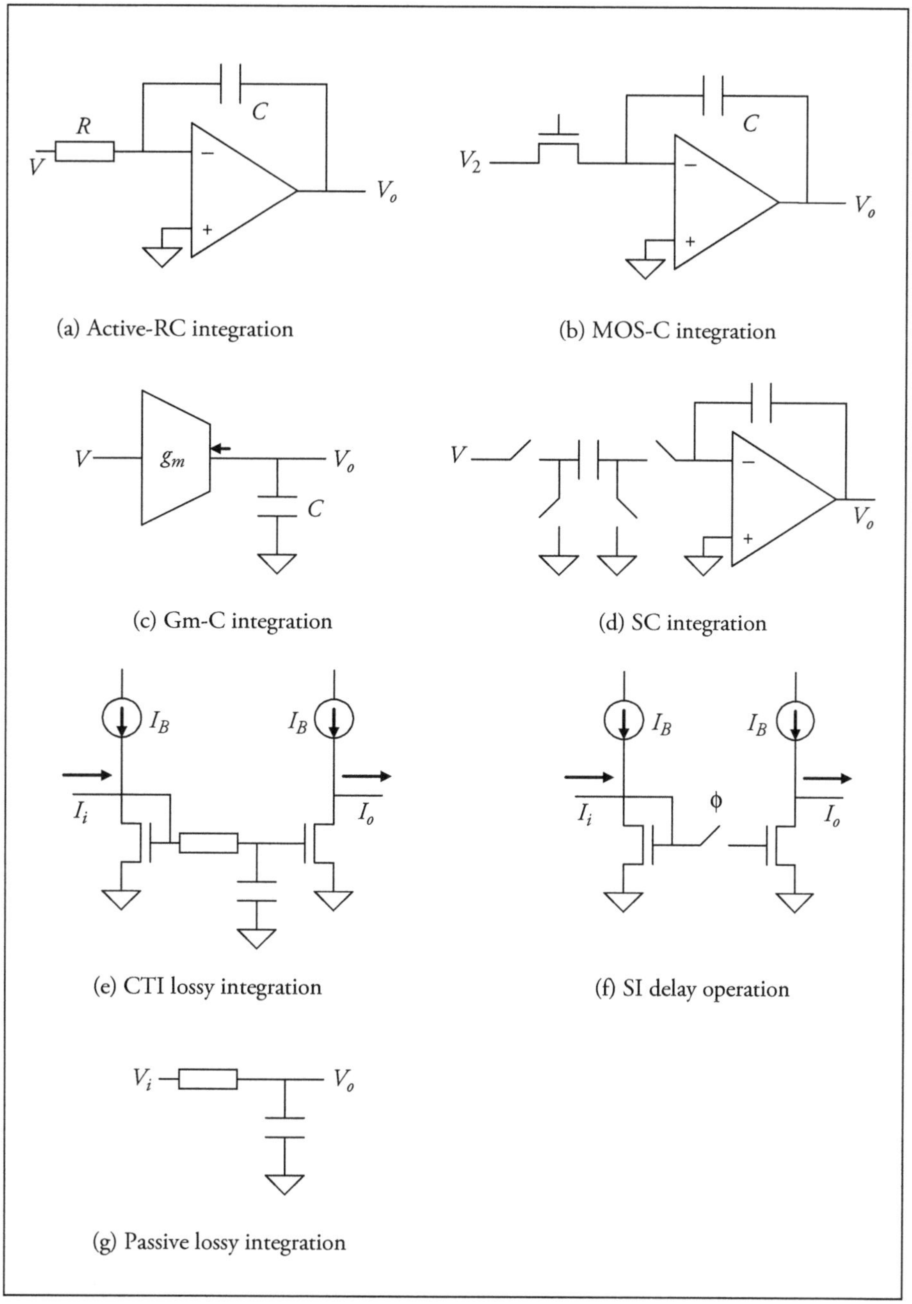

FIGURE 24.20 VLSI analog integration and delay circuits.

References

1. S. Brigati, F. Francesconi, P. Malcovati, M. Poletti, F. Maloberti, "Digital Synthesis of Analog Signals for Audiometric Applications," *ISCAS'97*, vol. 4, pp. 2625–2627, June 1997.
2. H. J. Song, "DBA – A Digital Based Analog Circuit Design Methodology for SOC Applications," Unpublished work.
3. H. J. Song, "Dual Mode Transmitter with Adaptively Controlled Slew Rate and Impedance Supporting Wide Operation Range," *IEEE SOCC2001*, pp. 321–324, 2001.
4. H. J. Song, "Architecture and Implementation of Power and Area Efficient Receiver Equalization Circuit for High-Speed Serial Data Communication," *IEEE SOCC2006*, pp. 43–46, 2006.
5. H. J. Song, "Algorithm, Architecture, and Implementation of Algorithmic Delay-Locked Loop Based Data Recovery Circuit for High-Speed Serial Data Communication," *IEEE SOCC2001*, pp. 83–87, 2001.

Index

D

E

F

9 781259 644931

Printed by Libri Plureos GmbH in Hamburg,
Germany